T0393370

Springer Texts in Business and Economics

Springer Texts in Business and Economics (STBE) delivers high-quality instructional content for undergraduates and graduates in all areas of Business/Management Science and Economics. The series is comprised of self-contained books with a broad and comprehensive coverage that are suitable for class as well as for individual self-study. All texts are authored by established experts in their fields and offer a solid methodological background, often accompanied by problems and exercises.

Patrice Poncet • Roland Portait

Capital Market Finance

An Introduction to Primitive Assets, Derivatives, Portfolio Management and Risk

Volume 2

With Contributions by Igor Toder

Patrice Poncet
ESSEC Business School
Cergy Pontoise, France

Roland Portait
ESSEC Business School
Cergy Pontoise, France

ISSN 2192-4333 ISSN 2192-4341 (electronic)
Springer Texts in Business and Economics
ISBN 978-3-030-84598-8 ISBN 978-3-030-84600-8 (eBook)
https://doi.org/10.1007/978-3-030-84600-8

Mathematics Subject Classification: 91G10; 91G15; 91G20; 91G30; 91G40; 91G60; 91G70

Translation from the French language edition: Finance de marché by Patrice Poncet, et al., © Éditions Dalloz 2008. Published by Éditions Dalloz. All Rights Reserved.

© Springer Nature Switzerland AG 2022
This work is subject to copyright. All rights are solely and exclusively licensed by the Publisher, whether the whole or part of the material is concerned, specifically the rights of reprinting, reuse of illustrations, recitation, broadcasting, reproduction on microfilms or in any other physical way, and transmission or information storage and retrieval, electronic adaptation, computer software, or by similar or dissimilar methodology now known or hereafter developed.
The use of general descriptive names, registered names, trademarks, service marks, etc. in this publication does not imply, even in the absence of a specific statement, that such names are exempt from the relevant protective laws and regulations and therefore free for general use.
The publisher, the authors, and the editors are safe to assume that the advice and information in this book are believed to be true and accurate at the date of publication. Neither the publisher nor the authors or the editors give a warranty, expressed or implied, with respect to the material contained herein or for any errors or omissions that may have been made. The publisher remains neutral with regard to jurisdictional claims in published maps and institutional affiliations.

This Springer imprint is published by the registered company Springer Nature Switzerland AG
The registered company address is: Gewerbestrasse 11, 6330 Cham, Switzerland

This book is dedicated to the dear memory of Roland Portait (1943–2021), my long-time friend, wonderful colleague and inspiring co-author, whose untimely death leaves a deep and painful void among his family and friends.

Preface

This textbook is primarily aimed at graduate students of market finance (MBA, MiM, and Ms in business schools and engineering schools at universities, first-year PhD programs in finance or economics). It is also aimed at market finance practitioners: trading rooms, asset management firms (using quantitative tools), actuaries in banks and insurance companies, risk management and control outlets, and research teams in market finance.

The objective is to make up for the relative inadequacy of most traditional finance books and mathematical finance books. The former do not offer in general the advanced mathematical techniques currently used by expert professionals. And the latter often focus on mathematical refinements at the expense of the economics behind financial contracts and products, the practical use of instruments, and the financial logics of markets; in addition, some of them are of a level inaccessible to non-mathematicians, even engineers.

This book is the result of many years of teaching market finance in the specialized programs of ESSEC (market finance track, master's in finance), CNAM (master's in market finance), University of Paris 1-Panthéon-Sorbonne (master's and PhD in finance) and within the trading rooms of Société Générale, a major French bank, and other banking institutions.

It offers a comprehensive and consistent presentation (from the point of view of analysis and notation) of the whole of market finance. In particular, it covers all primitive assets (equities, interest and exchange rates, indices, bank loans) except real estate, most vanilla and exotic derivatives (swaps, futures, options, hybrids, and credit derivatives), portfolio theory and management, and risk appreciation and hedging of individual positions as well as portfolios and firms' balance sheets. It emphasizes the methodological aspects of the analysis of financial instruments and of risk assessment and management. In particular, it devotes an important space to the probabilistic foundations of asset valuation and to credit and default risks, the poor understanding of which aggravated the 2007–2008 financial crisis originated in the subprime credit market.

The introductory chapter (Chap. 1) is devoted to the economic role played by the financial and banking markets and their organization and functioning. It is followed by four parts.

The first part, consisting of Chaps. 2 to 8, deals with primitive assets (debt securities, bonds, equities), the term structure of interest rates, floating rate instruments, and vanilla swaps. It does not use complicated mathematical tools and is accessible to undergraduate students. It is aimed at readers who are new to market finance, the most advanced of which can access the second part almost directly.

The second part, consisting of Chaps. 9 to 20, is devoted to derivatives (options and futures) and presents the main models, stochastic calculus tools, and probabilistic theories on which modern methods of valuing contingent assets or claims and financial risks are based. This part, much more technical than the previous one, is aimed at graduate students and practitioners operating in financial markets. The mathematical background required to read part of the material is that acquired during the first two years of scientific studies and quantitative economics or management studies (analysis, differential and integral calculus, linear algebra, and probability theory). These developments are signaled by one star (*) or, rarely, two stars (**). This is also the case for books referenced at the end of each chapter. Readers equipped with this mathematical background will find in the book all the complements concerning stochastic calculus and probabilistic theories, necessary and sufficient for an in-depth understanding of modern market finance.

The third part, including Chaps. 21 to 25, is devoted to portfolio theory and management. After a presentation of the standard portfolio theory (Markowitz, capital asset pricing model, and arbitrage pricing theory), various techniques of strategic and tactical asset allocation (benchmarking, portfolio insurance, alternative investment, etc.) are discussed.

Finally, the fourth part, consisting of Chaps. 26 to 30, deals with risk management with particular attention to analytical methods (simulations, value-at-risk, expected shortfall, value adjustments), credit risk (theoretical and empirical analysis of counterparty and default risks, credit VaR, credit derivatives), and new regulation regarding financial institutions and banks.

Cergy Pontoise, France

Patrice Poncet
Roland Portait

Main Abbreviations and Notations

Despite our efforts, because of the variety of themes covered in this book and its length and technicality, we could not prevent certain symbols from having distinct meanings in different chapters. Although we always define our notations before using them and the context in principle removes any ambiguity, the following list of key notations and abbreviations may prove useful.

General conventions: an underlined variable ($\underline{x}$) denotes a vector (or sequence), matrices are written in bold ($\mathbf{X}$), a ' denotes the derivative of a function or the transpose of a vector or matrix, and $[X]^+$ means max$[X, 0]$. The reading of more technical paragraphs equipped with one or two stars can be omitted or postponed.

AAO	Absence of arbitrage opportunity, no arbitrage
ABS	Asset-backed security
α	"abnormal" return rate (in asset pricing and portfolio management), or recovery rate of a debt instrument in the event of default (in credit risk analysis)
BS	Black–Scholes (formula of, or model of)
BSM	Black–Scholes–Merton (model of). Alternative name: Gaussian evaluation model
$B_T(t)$	Price on date t of a zero-coupon bond delivering \$1 or €1 on date T (duration: T-t)
$b_\theta(t)$	Discount factor or price on date t of a zero-coupon bond delivering \$1 or €1 on date $t + \theta$ (duration: θ). We have: $B_T(t) = b_{T-t}(t)$.
β	Beta of a stock (sensitivity to a stock market index) or, more generally, the slope of a linear regression
C	Usually the price of a call, sometimes a coupon, sometimes a Cap
CDO	Collateralized debt obligation
CDS	Credit default swap
CIR	Cox–Ingersoll–Ross (model of)
CRR	Cox–Ross–Rubinstein (model of). Alternative name: binomial model
D, d	In general, time to maturity (duration), or Macaulay duration
δ	The delta (sensitivity) of an option
DD	distance to default

E	mathematical expectation (E^Q, E^{QT}, E^*, E^{RN}, if the probability measure is specified)
ES	Expected shortfall (alternatively: conditional VaR or tail VaR)
F	A cash flow; the price of a floor; a forward or futures price; a distribution function
$\underline{F}$	A cash flow sequence (vector): $F_\theta \mid \theta = t1, ..., tn$
$F_T(t)$	Forward or futures price on date t for maturity T (less precise notation: $F(t)$)
$\Phi_T(t)$	Forward price when it is distinguished from the futures price denoted by $F_T(t)$
$f_{T,D}(t)$	Forward rate prevailing at t relative to the period $(T, T+D)$
$f_T(t)$	Instantaneous forward rate (limit of the previous one when D tends to 0)
$\gamma(t)$	Probability of survival at t (in case of risk of default)
$\Gamma, \theta, \rho, \nu$	The various other "Greek" parameters of an option (gamma, theta, rho, vega, respectively)
$H(t)$	Value at t of the optimal growth (logarithmic) portfolio
k	Often a fixed interest rate (nominal rate of a fixed income instrument, fixed rate of a swap,...)
K	Exercise price (strike) of an option
λ	Market price of risk, Lagrange multiplier, intensity of a jump process, hazard rate, scalar
L	Loss in default risk analysis, Lagrangian
μ	Often an expectation or a mean
$N(\mu, \sigma^2)$	Normal (Gaussian) distribution of mean μ and variance σ^2
$N(x)$	Distribution function of a reduced centered (standard) normal variable
P	Historical probability
P	Usually the price of a put
p_d	Probability of default
PDE	Partial differential equation
Q	Risk-neutral probability measure (denoted also by RN and sometimes by *)
Q_T	Forward-neutral probability measure relative to maturity T (denoted also by FN-T)
r	Generically means a rate. In general, our notation does not distinguish between proportional, compounded, discrete, and continuous rates. From Chap. 6 on, it is a zero-coupon rate, with yields to maturity of bullet bonds being denoted by y. Also, r represents a risk-free rate and $r(t)$ the instantaneous risk-free rate on date t.
r_θ	In general a zero-coupon rate of duration θ
R, R_i	Often a random return (chapters on portfolio theory)
ρ	Often a correlation coefficient; also the sensitivity of an option to interest rate variations; exceptionally an interest rate
S	Sometimes the sensitivity of an interest rate instrument to interest rate variations
s, s_θ	Often a spread (difference between rates, margin,...)

Main Abbreviations and Notations

$S(t)$, S_t	The price of a spot asset on date t
S_{k-r}	Swap fixed-rate k receiver (floating rate r payer) or lender swap
S_{r-k}	Swap fixed-rate k payer (floating rate r receiver) or borrower swap
$\boldsymbol{\Sigma}$	Return rate diffusion matrix (dynamics of a portfolio in continuous time)
σ	Volatility, or standard deviation
σ_{XY}	Covariance (X, Y)
SDE	Stochastic differential equation
T	Maturity date (distant from current date t by duration $\theta = T\text{-}t$)
τ	A date, a duration, a first-default date
U	Often a reduced centered (standard) normal variable
UA	Underlying asset (of a derivative instrument)
V	In general a market value or price
$\mathbf{V}$	A variance–covariance matrix
VaR	Value at risk
var	Variance (alternatively σ^2)
$W(t)$, W_t	Standard Brownian motion or Wiener process (value at t)
y, y_θ	Often a yield to maturity, to be distinguished from a zero-coupon rate r_θ
Z	Standardized profitability (in credit risk analysis)
ZC	Zero-coupon bond
$\underline{1}$	Unit vector
$\mathbf{1}_E$	Indicator of event E ($= 1$ if E is true, and $= 0$ if not)

Acknowledgments

The authors acknowledge the participation, assistance, and support of Igor Toder who contributed to five chapters and is the main author of half of the last one.

The authors thank Sami Attaoui, Riadh Belhaj, Frédéric Bompaire, Thierry Charpentier, Minh Chau, Guillaume Coqueret, Andras Fulop, Vincent Lacoste, Didier Maillard, Jacques Olivier, and Bertrand Tavin for their many valuable comments over the long years spent to write this book.

They also thank the students of ESSEC Business School majoring in finance and students of the master's in finance at ESSEC, the CNAM, and the Sorbonne, as well as the participants in the Société Générale (a major French global bank) training seminars, for their support and many relevant questions and remarks over the same long years. They acknowledge special support from Chloé Baraille, Fei Wang, and, particularly, Alexis Marty.

They also wish to express their gratitude to Springer's Catriona Byrne, Joerg Sixt, and Rémi Lodh for their unfailing trust, as well as for editing the book.

Of course, they are responsible for any errors and omissions.

Contents

Part III   Portfolio Theory and Portfolio Management

21   Choice Under Uncertainty and Portfolio Optimization in a Static Framework: The Markowitz Model 873
- 21.1   Rational Choices Under Uncertainty: The Criteria of the Expected Utility and Mean-Variance 874
  - 21.1.1   The Expected Utility Criterion 874
  - 21.1.2   Some Features of Utility Functions 876
  - 21.1.3   Risk Aversion and Concavity of the Utility Function . 877
  - 21.1.4   Some Standard Utility Functions 879
  - 21.1.5   The Mean-Variance Criterion 881
- 21.2   Intuitive and Graphic Presentation of the Main Concepts of Portfolio Theory 882
  - 21.2.1   Assumptions, General Framework and Efficient Portfolios 883
  - 21.2.2   Two-Asset Portfolios 884
  - 21.2.3   Portfolios with N Securities 887
  - 21.2.4   Portfolio Diversification 890
- 21.3   Mathematical Analysis of Efficient Portfolio Choices 894
  - 21.3.1   General Framework and Notations 894
  - 21.3.2   Efficient Portfolios and Portfolio Choice in the Absence of a Risk-Free Asset and of Portfolio Constraints 898
  - 21.3.3   Efficient Portfolios in the Presence of a Risk-Free Asset, with Allowed Short Positions; Tobin's Two-Fund Separation 904
- 21.4   Some Extensions of the Standard Model and Alternatives 906
  - 21.4.1   Problems Implementing the Markowitz Model; The Black-Litterman Procedure 906
  - 21.4.2   Ban on Short Positions 908
  - 21.4.3   Separation Results When Investors Maximize Expected Utility But Do Not Follow the Mean-Variance Criterion (Cass and Stiglitz) 910

xv

| | | 21.4.4 | Loss Aversion and Introduction to Behavioral Finance | 911 |

| | 21.5 | Summary | 915 |

Appendix 1: The Axiomatic of Von Neuman and Morgenstern and Expected Utility ... 916
  A1.1 The Objects of Choice ... 917
  A1.2 The Axioms Concerning Preferences ... 918
  A1.3 The Expected Utility Criterion ... 919
  A1.4 Notes and Complements ... 921
Appendix 2: A Reminder of Quadratic Forms and the Calculation of Gradients ... 922
Appendix 3: Expectations, Variances and Covariances—Definitions and Calculation Rules ... 923
  A3.1 Definitions and Reminder ... 923
  A3.2 Calculation Rules ... 924
Appendix 4: Reminder on Optimization Methods Under Constraints ... 925
  A4.1 Optimization When the Constraints Take the Form of Equalities ... 925
  A4.2 Optimization Under Inequality Constraints ... 926
Suggestions for Further Reading ... 927
  Books ... 927
  Articles ... 927

22 The Capital Asset Pricing Model ... 929
  22.1 Derivation of the CAPM ... 929
    22.1.1 Hypotheses ... 930
    22.1.2 Intermediate Results in the Presence of a Risk-Free Asset ... 931
    22.1.3 The CAPM ... 933
  22.2 Applications of the CAPM ... 943
    22.2.1 Use of the CAPM for Financial Investment Purposes ... 943
    22.2.2 Physical Investments by Firms ... 946
    22.2.3 Standard Performance Measures ... 947
  22.3 Extensions of the CAPM ... 953
    22.3.1 Merton's Intertemporal CAPM ... 953
    22.3.2 International CAPM ... 954
  22.4 Limits of the CAPM ... 955
    22.4.1 Efficiency of the Market Portfolio and Roll's Criticism ... 955
    22.4.2 Stability of Betas ... 956
  22.5 Tests of the CAPM ... 957
  22.6 Summary ... 960
  Suggestions for Further Reading ... 961
    Books ... 961
    Articles ... 962

Contents

xvii

23 Arbitrage Pricing Theory and Multi-factor Models 963
- 23.1 Multi-factor Models 964
  - 23.1.1 Presentation of Models 964
  - 23.1.2 Portfolio Management Models in Practice 966
- 23.2 Arbitrage Pricing Theory 966
  - 23.2.1 Assumptions and Notations 967
  - 23.2.2 The APT 968
  - 23.2.3 Relationship with the CAPM 976
- 23.3 APT Applications and the Fama-French Model 976
  - 23.3.1 Implementation of Multi-factor Models and APT 977
  - 23.3.2 Portfolio Selection 979
  - 23.3.3 The Three-Factor Model of Fama and French 980
- 23.4 Econometric Tests and Comparison of Models 982
  - 23.4.1 Tests of the APT 982
  - 23.4.2 Empirical and Practical CAPM-APT Comparison 983
  - 23.4.3 Comparison of Factor Models 984
- 23.5 Summary 985
- Appendix 1: Orthogonalization of Common Factors 987
- Appendix 2: Compatibility of CAPM and APT 988
- Suggestions for Further Reading 989
  - Books 989
  - Articles 990

24 Strategic Portfolio Allocation 991
- 24.1 Strategic Asset Allocation Based on Common Sense Rules . . . 992
  - 24.1.1 Common Sense Rules 993
  - 24.1.2 Reactions to the Evolution of Market Conditions and
    of the Portfolio: Convex and Concave Strategies 996
- 24.2 Portfolio Insurance 997
  - 24.2.1 The *Stop Loss* Method 998
  - 24.2.2 Option-Based Portfolio Insurance 999
  - 24.2.3 CPPI Method 1005
  - 24.2.4 Variants and Extensions of the Basic Methods 1011
  - 24.2.5 Portfolio Insurance, Financial Markets Volatility and
    Stability 1013
- 24.3 Dynamic Portfolio Optimization Models 1014
  - 24.3.1 Dynamic Strategies: General Presentation and
    Optimization Models 1014
  - 24.3.2 The Case of a Logarithmic Utility Function and the
    Optimal Growth Portfolio 1017
  - 24.3.3 The Merton Model 1020
  - 24.3.4 The Model of Cox-Huang and Karatzas-Lehoczky-
    Shreve 1022

24.4	Summary		1026

Suggestions for Further Reading ... 1028
    Books ... 1028
    Articles ... 1029

25 Benchmarking and Tactical Asset Allocation ... 1031

25.1 Benchmarking ... 1031

    25.1.1 Definitions and Classification According to the Tracking Error ... 1032

    25.1.2 Pure Index Funds and Trackers ... 1033

    25.1.3 Replication Methods ... 1033

    25.1.4 Trackers or ETFs ... 1034

25.2 Active Tactical Asset Allocation ... 1035

    25.2.1 Modeling and Solution to the Problem of an Active Manager Competing with a Benchmark ... 1035

    25.2.2 Analysis of the Performance of Active Portfolio Management: Empirical Information Ratio, Market Timing, and Security Picking ... 1038

    25.2.3 Beta Coefficient Equal to 1 ... 1038

    25.2.4 Beta Coefficient Different from 1 ... 1040

    25.2.5 Information Ratios, Sharpe Ratio, and Active Portfolio Management Theory ... 1043

    25.2.6 The Construction of a Maximum IR Portfolio from a Limited Number of Securities ... 1043

    25.2.7 The Construction of a Portfolio That Dominates the Benchmark (Higher Sharpe Ratio) ... 1045

    25.2.8 Synthesis, Interpretation and Application to Portfolio Management ... 1047

25.3 Alternative Investment Management and Hedge Funds ... 1047

    25.3.1 General Description of Hedge Funds and Alternative Investment ... 1048

    25.3.2 Definition of the Main Alternative Investment Styles ... 1049

    25.3.3 The Interest of Alternative Investment ... 1051

    25.3.4 The Particular Difficulties of Measuring Performance in Alternative Investment ... 1052

25.4 Summary ... 1054

Appendix ... 1056
    Breakdown of the Tracking Error and Performance Attribution ... 1056

Suggestion for Reading ... 1059
    Books ... 1059
    Articles ... 1059

Contents   xix

Part IV Risk Management, Credit Risk, and Credit Derivatives

26 Monte Carlo Simulations 1063
    26.1   Generation of a Sample from a Given Distribution Law 1064
          26.1.1 Sample Generation from a Given Probability
                  Distribution 1064
          26.1.2 Construction of a Sample Taken from a Normal
                  Distribution 1065
    26.2   Monte Carlo Simulations for a Single Risk Factor 1065
          26.2.1 Dynamic Paths Simulation of $Y(t)$ and $V(t, Y(t))$ in
                  the Interval $(0, T)$ 1065
          26.2.2 Simulations of $Y(T)$ and $V(T, Y(T))$ at Time T (Static
                  Simulations) 1068
          26.2.3 Applications 1070
    26.3   Monte Carlo Simulations for Several Risk Factors: Choleski
          Decomposition and Copulas 1073
          26.3.1 Simulation of a Multi-variate Normal Variable:
                  Choleski Decomposition 1073
          26.3.2 Representation and Simulation of a Non-Gaussian
                  Vector with Correlated Components Through the
                  Use of a Copula 1075
          26.3.3 General Definition of a Copula, and Student
                  Copulas (*) 1080
          26.3.4 Simulation of Trajectories 1081
    26.4   Accuracy, Computation Time, and Some Variance Reduction
          Techniques 1085
          26.4.1 Antithetic Variables 1086
          26.4.2 Control Variate 1086
          26.4.3 Importance Sampling 1088
          26.4.4 Stratified Sampling 1089
    26.5   Monte Carlo and American Options 1089
          26.5.1 General Description of the Problem and
                  Methodology 1090
          26.5.2 Estimation of the Continuation Value by Regression
                  (Carrière, Longstaff and Schwartz) 1091
          26.5.3 Overview of the Carrière Approach 1093
          26.5.4 Introduction to Longstaff and Schwartz Approach 1094
    26.6   Summary 1099
    Suggestion for Further Reading 1101
          Books 1101
          Articles 1101

27 Value at Risk, Expected Shortfall, and Other Risk Measures 1103

27.1 Analytic Study of Value at Risk 1105

    27.1.1 The Problem of a Synthetic Risk Measure and Introduction to VaR 1105

    27.1.2 Definition of the VaR, Interpretations, and Calculation Rules 1108

    27.1.3 Analytic Expressions for the VaR in the Gaussian Case 1112

    27.1.4 The Influence of Horizon h on the VaR of a Portfolio in the Absence or Presence of Serial Autocorrelation . . 1116

27.2 Estimating the VaR 1120

    27.2.1 Preliminary Analysis and Modeling of a Complex Position 1121

    27.2.2 Estimating the VaR Through Simulations Based on Historical Data 1123

    27.2.3 Partial Valuation: Linear and Quadratic Approximations (the Delta-Normal and Delta-Gamma Methods) 1130

    27.2.4 Calculating the VaR Using Monte Carlo Simulations 1138

    27.2.5 Comparison Between the Different Methods 1141

27.3 Limitations and Drawbacks of the VaR, Expected Shortfall, Coherent Measures of Risk, and Portfolio Risks 1141

    27.3.1 The Drawbacks of VaR Measures 1142

    27.3.2 An Improvement on the VaR: *Expected Shortfall* (or Tail-VaR, or C-VaR) 1145

    27.3.3 Coherent Risk Measures 1148

    27.3.4 Portfolio Risk Measures: Global, Marginal, and Incremental Risk 1150

27.4 Consequences of Non-normality and Analysis of Extreme Conditions 1155

    27.4.1 Non-normal Distributions with Fat Tails and Correlation at the Extremes 1155

    27.4.2 Distributions of Extreme Values 1159

    27.4.3 Stress Tests and Scenario Analysis 1164

27.5 Summary 1166

Suggestions for Further Reading 1168

    Books 1168

    Articles 1168

28 Modeling Credit Risk (1): Credit Risk Assessment and Empirical Analysis 1171

28.1 Empirical Tools for Credit Risk Analysis 1172

    28.1.1 Reminder of Basic Concepts, Empirical Observations, and Notations 1172

	28.1.2	Historical (Empirical) Default Probabilities and Transition Matrix	1176
	28.1.3	Risk-Neutral Default Probabilities Implicit in the Spread Curve and Discounting Methods in the Presence of Credit Risk	1179
28.2	Modeling Default Events and Valuation of Securities	1189
	28.2.1	Reduced-Form Approach (Intensity Models)	1189
	28.2.2	Structural Approach: Merton's Model and Barrier Models	1197
	28.2.3	A Practical Application: the Valuation of Convertible Bonds	1207
28.3	Summary	1216
Appendix		1219
Suggestions for Further Reading		1219
Books		1219
Articles		1220
Website		1220

29 Modeling Credit Risk (2): Credit-VaR and Operational Methods for Credit Risk Management 1221

29.1	Determining the Credit-VaR of an Asset: Overview and General Principles	1223
29.2	Empirical Credit-VaR of an Asset Based on the Migration Matrix	1224
	29.2.1	Computation of the Credit-VaR of an Individual Asset	1224
	29.2.2	Limitations of the Empirical Approach	1227
29.3	Credit-VaR of an Individual Asset: Analytical Approaches Based on Asset Price Dynamics (MKMV...) and on Structural Models	1228
	29.3.1	Asset Dynamics, Standardized Return, Default Probabilities, and Distance to Default	1229
	29.3.2	Derivation of the Rating Migration Quantiles Associated with the Standardized Return	1230
	29.3.3	Computation of the Distance to Default and Expected Default Frequency (MKMV-Moody's Analytics Method)	1233
	29.3.4	Comparing the Two Approaches	1236
	29.3.5	Estimation of the Credit-VaR of an Asset Using EDF and a Valuation Model Based on RN-FN Probabilities	1236
	29.3.6	Relationship between Historical and RN Default Probabilities	1238
29.4	Credit-VaR of an Entire Portfolio (Step 3) and Factor Models	1239

29.4.1	Marked-to-Market (MTM) Models Involving Simulations	1240
29.4.2	A Single-Factor DM Model of the Credit Risk of a Perfectly Diversified Portfolio (The Asymptotic Granular Vasicek-Gordy One-Factor Model)	1242
29.4.3	Extensions of the Asymptotic Single-Factor Granular Model	1247
29.4.4	Alternative Approach: Modeling the Default Dependence Structure with a Copula	1250
29.4.5	Probability Distribution of the Default Dates Affecting a Portfolio	1251
29.4.6	Portfolio Comprising Several Positions on the Same Obligor: Netting	1253

29.5 Credit-VaR, Unexpected Loss and Economic Capital 1254
- 29.5.1 Definition of Unexpected Loss (UL) 1254
- 29.5.2 Probability Threshold and Rating 1257

29.6 Control and Regulation of Banking Risks 1257
- 29.6.1 Regulators and the Basel Committee: General Presentation 1258
- 29.6.2 Capital and liquidity Rules under Basel 3 1260
- 29.6.3 Pillar 1 Capital Requirements under Basel 3 1261
- 29.6.4 Details on Pillar 1 Liquidity Requirements 1265
- 29.6.5 Additional Basel 3 Reflections and Reforms 1266

29.7 Summary 1269

Appendix 1. Correlation of Defaults in a Portfolio of Debt Assets 1271
Appendix 2. Regulatory Capital, Market VaR, and Backtesting 1273
Appendix 3. Calculation of Regulatory Capital under the IRB Approach: Adjustment to the Infinitely Grained One-Factor Model ... 1274
Suggestion for Further Reading 1276
- Books 1276
- Articles and Documentation 1276
- Websites 1277

30 Credit Derivatives, Securitization, and Introduction to xVA 1279
30.1 Credit Derivatives 1280
- 30.1.1 General Principles and Description of Credit Default Swaps 1281
- 30.1.2 Single-Name CDS Valuation Techniques 1285

30.2 Securitization 1306
- 30.2.1 Introduction to Securitization and ABS 1308
- 30.2.2 ABS Tranching Structuration 1311

30.3 The "xVA" Framework 1313
- 30.3.1 Counterparty Risk Exposure Measurement and Risk Mitigation Techniques 1314
- 30.3.2 Counterparty Risk Exposure Modeling Techniques ... 1319

30.3.3	Collateralized vs Non-collateralized Trades: Some Statistics	1324
30.3.4	Introduction to CVA	1327
30.3.5	Introduction to DVA	1332
30.3.6	The FVA Puzzle	1334

30.4 Summary ... 1339
Appendix 1 ... 1342
Asset Swap Analysis ... 1342
Suggestion for Further Reading ... 1345
Books ... 1345
Articles ... 1346
Website: defaultrisk.com ... 1346

Index ... 1347

Contents for Volume 1

1 Introduction: Economics and Organization of Financial Markets . . 1
  1.1 The Role of Financial Markets 1
    1.1.1 The Allocation of Cash Resources Over Time 1
    1.1.2 Risk Allocation 3
    1.1.3 The Market as a Supplier of Information 5
  1.2 Securities as Sequences of Cash Flows 5
    1.2.1 Definition of a Security (or Financial Asset) 5
    1.2.2 Characterizing the Cash Flow Sequence 6
  1.3 Equilibrium, Absence of Arbitrage Opportunity, Market
       Efficiency and Liquidity 8
    1.3.1 Equilibrium and Price Setting 8
    1.3.2 Absence of Arbitrage Opportunity (AAO) and the
           Notion of Redundant Assets 9
    1.3.3 Efficiency 11
    1.3.4 Liquidity 16
    1.3.5 Perfect Markets 18
  1.4 Organization, a Typology of Markets, and Listing 19
    1.4.1 The Banking System and Financial Markets 19
    1.4.2 A Simple Typology of Financial Markets 19
    1.4.3 Market Organization 24
  1.5 Summary 32
  Appendix: The World's Principal Financial Markets 33
       Stock markets, market indexes and interest rate instruments . . . 33
       Organized Derivative Markets (Futures and Options, Unless
       Otherwise Indicated) 34
  Suggestion for Further Reading 35
       Books 35
       Articles 35

Part I Basic Financial Instruments

2 Basic Finance: Interest Rates, Discounting, Investments, Loans . . 39
  2.1 Cash Flow Sequences 39

xxv

2.2	Transactions Involving Two Cash Flows		40
	2.2.1	Transactions of Lending and Borrowing Giving Rise to Two Cash Flows over One Period	41
	2.2.2	Transactions with Two Cash Flows over Several Periods	42
	2.2.3	Comparison of Simple and Compound Interest	47
	2.2.4	Two "Complications" in Practice	49
	2.2.5	Continuous Rates	53
	2.2.6	General Equivalence Formulas for Rates Differing in Convention and the Length of the Reference Period	53
2.3	Transactions Involving an Arbitrary Number of Cash Flows: Discounting and the Analysis of Investments		55
	2.3.1	Discounting	57
	2.3.2	Yield to Maturity (YTM), Discount Rate and Internal Rate of Return (IRR)	63
	2.3.3	Application to Investment Selection: The Criteria of the NPV and the IRR	66
	2.3.4	Interaction Between Investing and Financing, and Financial Leverage	69
	2.3.5	Some Guidelines for the Choice of an Appropriate Discount Rate	72
	2.3.6	Inflation, Real and Nominal Cash Flows and Rates	74
2.4	Analysis of Long-Term Loans		76
	2.4.1	General Considerations and Definitions: YTM and Interest Rates	76
	2.4.2	Amortization Schedule for a Loan	79
2.5	Summary		84
Appendix 1: Geometric Series and Discounting			85
Appendix 2: Using Financial Tables and Spreadsheets for Discount Computations			87
	1. Financial Tables		87
Suggested Reading			92

3 The Money Market and Its Interbank Segment 93

3.1	Interest Rate Practices and the Valuation of Securities		94
	3.1.1	Interest Rate Practices on the Euro-Zone's Money Market	95
	3.1.2	Alternative Practices and Conventions	99
3.2	Money Market Instruments and Operations		100
	3.2.1	The Short-Term Securities of the Money Markets	100
	3.2.2	Repos, Carry Trades, and Temporary Transfers of Claims	101
	3.2.3	Other Trades	103
3.3	Participants and Orders of Magnitude of Trades		105
	3.3.1	The Participants	105

	3.3.2	Orders of Magnitude	106
3.4	Role of the Interbank Market and Central Bank Intervention		107
	3.4.1	Central Bank Money and the Interbank Market	107
	3.4.2	Central Bank Interventions and Their Influence on Interest Rates	110
3.5	The Main Monetary Indices		113
	3.5.1	Indices Reflecting the Value of a Money-Market Rate on a Given Date	113
	3.5.2	Indices Reflecting the *Average* Value of a Money-Market Rate During a Given Period	115
3.6	Summary		117
	Suggestions for Further Reading		118

4 The Bond Markets 119

4.1	Fixed-Rate Bonds		121
	4.1.1	Financial Characteristics and Yield to Maturity at the Date of Issue	121
	4.1.2	The Market Bond Value at an Arbitrary Date; the Influence of Market Rates and of the Issuer's Rating	126
	4.1.3	The Quotation of Bonds	131
	4.1.4	Bond Yield References and Bond Indices	135
4.2	Floating-Rate Bonds, Indexed Bonds, and Bonds with Covenants		135
	4.2.1	Floating-Rate Bonds and Notes	136
	4.2.2	Indexed Bonds	137
	4.2.3	Bonds with Covenants (Optional Clauses)	137
4.3	Issuing and Trading Bonds		138
	4.3.1	Primary and Secondary Markets	138
	4.3.2	Treasury Bonds and Treasury Notes Issues: Reopening and STRIPS	139
4.4	International and Institutional Aspects; the Order of Magnitude of the Volume of Transactions		141
	4.4.1	Brief Presentation of the International Bond Markets	142
	4.4.2	The Main National Markets	143
4.5	Summary		146
	Suggested Readings		148

5 Introduction to the Analysis of Interest Rate and Credit Risks ... 149

5.1	Interest Rate Risk		149
	5.1.1	Introductory Examples: The Influence of the Maturity of a Security on Its Sensitivity to Interest Rates	150
	5.1.2	Variation, Sensitivity and Duration of a Fixed-Income Security	152
	5.1.3	Alternative Expressions for the Variation, Sensitivity and Duration	156

	5.1.4	Some Properties of Sensitivity and Duration	158
	5.1.5	The Sensitivity of a Portfolio of Assets and Liabilities or of a Balance Sheet: Sensitivity and Gaps	161
	5.1.6	A More Accurate Estimate of Interest Rate Risk: Convexity	167
5.2	Introduction to Credit Risk		170
	5.2.1	Analysis of the Determinants of the Credit Spread	170
	5.2.2	Simplified Modeling of the Credit Spread; the Credit Triangle	172
5.3	Summary		174
Appendix 1			175
	Default Probability, Recovery Rate and Credit Spread		175
Suggested Reading			176

6 The Term Structure of Interest Rates 177

6.1	Spot Rates and Forward Rates		177
	6.1.1	The Yield Curve	177
	6.1.2	Yields to Maturity and Zero-Coupon Rates	179
	6.1.3	Forward Interest Rates Implicit in the Spot Rate Curve	184
6.2	Factors Determining the Shape of the Curve		187
	6.2.1	The Curve Shape	187
	6.2.2	Expectations Hypothesis with Term Premiums	188
	6.2.3	Influence of the Credit Spread on Yield Curves	191
6.3	Analysis of Interest Rate Risk: Impact of Changes in the Slope and Shape of the Yield Curve		192
	6.3.1	The Risk of a Change in the Slope of the Yield Curve	192
	6.3.2	Multifactor Variation and Sensitivity and Models of Yield Curves	194
6.4	Summary		203
Suggested Readings			204

7 Vanilla Floating Rate Instruments and Swaps 205

7.1	Floating Rate Instruments		206
	7.1.1	General Discussion and Notation	206
	7.1.2	"Replicable" Assets: Valuation and Interest Rate and Spread Risks	212
7.2	Vanilla Swaps		228
	7.2.1	Definitions and Generalities About Swaps	228
	7.2.2	Replication and Valuation of an Interest Rate Swap	232
	7.2.3	Interest Rate, Counterparty and Credit Risks for an Interest Rate Swap	241
	7.2.4	Summary of the Various Types of Swaps	247
7.3	Summary		251

Appendix 252
 Proof of the Equivalence Between Eq. (7.2') and
 Proposition 1 252
Suggested Reading 253
 Books 253
 Articles 253

8 Stocks, Stock Markets, and Stock Indices 255
 8.1 Stocks 255
  8.1.1 Basic Notions: Equity, Stock Market Capitalization,
   and Share Issuing 256
  8.1.2 Analysis of Stock Issues, Dilution, and Subscription
   Rights 261
  8.1.3 Market Performance of a Share and Adjusted Share
   Price 263
  8.1.4 Introduction to the Valuation of Firms and Shares;
   Interpretation and Use of the PER 266
 8.2 Return Probability Distributions and the Evolution of Stock
  Market Prices 273
  8.2.1 Stock Price on a Future Date, Stock Return, and Its
   Probability Distribution: Static Analysis 273
  8.2.2 Modeling a Stock Price Evolution with a Stochastic
   Process: Dynamic Analysis 276
 8.3 Placing and Executing Orders and the Functioning of Stock
  Markets 284
  8.3.1 Types of Orders 284
  8.3.2 The Clearing and Settlement System 286
  8.3.3 Investment Management 288
  8.3.4 The Main Stock Markets 293
 8.4 Stock Market Indices 294
  8.4.1 Composition and Calculation 295
  8.4.2 The Main Indices 300
 8.5 Summary 302
 Appendix 1 304
  Skewness and Kurtosis of Log-Returns 304
 Appendix 2 305
  Modeling Volatility with ARCH and GARCH 305
 Suggestions for Further Reading 308
  Book Chapters 308
  Articles 308
  For an Online Comparative Description of Investment Funds
  from Different Countries 308
  For an Online Description and Analysis of the Asset
  Management Industry 308

Part II Futures and Options

9 Futures and Forwards 311
  9.1 General Analysis of Forward and Futures Contracts 312
    9.1.1 Definition of a Forward Contract: Terminology and Notation 312
    9.1.2 Futures Contracts: Comparison of Futures and Forward Contracts 314
    9.1.3 Unwinding a Position Before Expiration 317
    9.1.4 The Value of Forward and Futures Contracts 318
  9.2 Cash-and-carry and the Relation Between Spot and Forward Prices 319
    9.2.1 Arbitrage, Cash-and-Carry, and Spot-Forward Parity .. 319
    9.2.2 Forward Prices, Expected Spot Prices, and Risk Premiums 324
  9.3 Maximum and Optimal Hedging with Forward and Futures Contracts 325
    9.3.1 Perfect or Maximum Hedging 326
    9.3.2 Optimal Hedging and Speculation 333
  9.4 The Main Forward and Futures Contracts 334
    9.4.1 Contracts on Commodities 335
    9.4.2 Contracts on Currencies (Foreign Exchanges) 338
    9.4.3 Forward and Futures Contracts on Financial Securities (Stocks, Bonds, Negotiable Debt Securities), FRA, and Contracts on Market Indices 340
  9.5 Summary 348
  Appendix 349
    The Relationship Between Forward and Futures Prices 349
  Suggestions for Further Reading 352
    Books 352
    Articles 352

10 Options (I): General Description, Parity Relations, Basic Concepts, and Valuation Using the Binomial Model 353
  10.1 Basic Concepts, Call-Put Parity, and Other Restrictions from No Arbitrage 354
    10.1.1 Definitions, Value at Maturity, Intrinsic Value, and Time Value 354
    10.1.2 The Standard Call-Put Parity 358
    10.1.3 Other Parity Relations 361
    10.1.4 Other Arbitrage Restrictions 364
  10.2 A Pricing Model for One Period and Two States of the World 367
    10.2.1 Two Markets, Two States 368
    10.2.2 Hedging Strategy and Option Value in the Absence of Arbitrage 369

	10.2.3	The "Risk-Neutral" Probability	372
	10.2.4	The Risk Premium and the Market Price of Risk	374
10.3	The Multi-period Binomial Model		377
	10.3.1	The Model Framework and the Dynamics of the Underlying's Price	377
	10.3.2	Risk-Neutral Probability and Martingale Processes	379
	10.3.3	Valuation of an Option Using the Cox-Ross-Rubinstein Binomial Model	381
10.4	Calibration of the Binomial Model and Convergence to the Black-Scholes Formula		386
	10.4.1	An Interpretation of Premiums in Terms of Probabilities of Exercise	387
	10.4.2	Calibration and Convergence	389
10.5	Summary		391
Appendix 1			393
	Calibration of the Binomial Model		393
*Appendix 2			395
Suggestions for Further Reading			397
	Books		397
	Articles		397

11 Options (II): Continuous-Time Models, Black–Scholes and Extensions 399

11.1	The Standard Black-Scholes Model		399
	11.1.1	The Analytical Framework and BS Model's Assumptions	400
	11.1.2	Self-Financing Dynamic Strategies	401
	11.1.3	Pricing Using a Partial Differential Equation and the Black–Scholes Formula	402
	11.1.4	Probabilistic Interpretation	406
11.2	Extensions of the Black–Scholes Formula		412
	11.2.1	Underlying Assets That Pay Out (Dividends, Coupons, etc.)	412
	11.2.2	Options on Commodities	421
	11.2.3	Options on Exchange Rates	422
	11.2.4	Options on Futures and Forwards	424
	11.2.5	Variable But Deterministic Volatility	427
	11.2.6	Stochastic Interest Rates: The Black–Scholes–Merton (BSM) Model	429
	11.2.7	Exchange Options (Margrabe)	434
	11.2.8	Stochastic Volatility (*)	437
11.3	Summary		443
Appendix 1			444
	Historical and Risk-Neutral Probabilities and Changes in Probability		444

Appendix 2		446
Changing the Probability Measure and the Numeraire		446
Appendix 3		449
Alternative Interpretations of the Black–Scholes Formula		449
Suggested Reading		451
Books		451
Articles		451

12 Option Portfolio Strategies: Tools and Methods 453

12.1	Basic Static Strategies	454
	12.1.1 The General P&L Profile at Maturity	454
	12.1.2 The Main Static Strategies	455
	12.1.3 Replication of an Arbitrary Payoff by a Static Option Portfolio (*)	457
12.2	Historical and Implied Volatilities, Smile, Skew and Term Structure	457
	12.2.1 Historical Volatility	458
	12.2.2 The Implied Volatility	459
	12.2.3 Smile, Skew, Term Structure, and Volatility Surface	461
12.3	Option Sensitivities (Greek Parameters)	463
	12.3.1 The Delta (δ)	464
	12.3.2 The Gamma (Γ)	467
	12.3.3 The Vega (υ)	468
	12.3.4 The Theta (θ)	470
	12.3.5 The Rho (ρ)	471
	12.3.6 Sensitivity to the Dividend Rate	472
	12.3.7 Elasticity and Risk-Expected Return Tradeoff	473
12.4	Dynamic Management of an Option Portfolio Using Greek Parameters	475
	12.4.1 Variation in the Value of a Position in the Short Term and General Considerations	475
	12.4.2 Delta-Neutral Management	476
	12.4.3 A Tool for Risk Management: The P&L Matrix	485
12.5	Summary	486
Appendix 1		488
Computing Partial Derivatives (Greeks)		488
Appendix 2		493
Option Prices and the Underlying Price Probability Distribution		493
Appendix 3		496
Replication of an Arbitrary Payoff with a Static Option Portfolio		496
Suggestions for Further Reading		498
Books		498
Articles		498

Contents for Volume 1   xxxiii

13 American Options and Numerical Methods ... 501
- 13.1 Early Exercise and Call-Put Parity for American Options ... 502
  - 13.1.1 Early Exercise of American Options ... 502
  - 13.1.2 Call-Put "Parity" for American Options ... 515
- 13.2 Pricing American Options: Analytical Approaches ... 516
  - 13.2.1 Pricing an American Call on a Spot Asset Paying a Single Discrete Dividend or Coupon ... 517
  - 13.2.2 Pricing an American Option (Call and Put) on a Spot Asset Paying a Continuous Dividend or Coupon ... 520
  - 13.2.3 Prices of American and European Options: Orders of Magnitude ... 528
- 13.3 Pricing American Options with the Binomial Model ... 529
  - 13.3.1 Binomial Dynamics of Price S: The Case of a Discrete Dividend ... 529
  - 13.3.2 Binomial Dynamics of Price S: The Continuous Dividend Case ... 531
  - 13.3.3 Pricing an American Option Using the Binomial Model ... 531
  - 13.3.4 Improving the Procedure with a Control Variate ... 533
- 13.4 Numerical Methods: Finite Differences, Trinomial and Three-Dimensional Trees ... 533
  - 13.4.1 Finite Difference Methods (*) ... 534
  - 13.4.2 Trinomial Trees ... 539
  - 13.4.3 Three-Dimensional Trees Representing Two Correlated Processes ... 541
- 13.5 Summary ... 543
- Appendix 1 ... 545
  - Proof of the Smooth Pasting (Tangency) Condition (13.5b) ... 545
- Appendix 2 ... 546
  - Orthogonalization of the Processes In S_1 and ln S_2 and Construction of a Three-Dimensional Tree ... 546
- Suggestion for Further Reading ... 547
  - Books ... 547
  - Articles ... 548

14 *Exotic Options ... 549
- 14.1 Path-Independent Options ... 550
  - 14.1.1 The Forward Start Option (with Deferred Start) ... 550
  - 14.1.2 Digital and Double Digital Options ... 551
  - 14.1.3 Multi-underlying (Rainbow) Options (*) ... 555
  - 14.1.4 Options on Options or "Compounds" ... 559
  - 14.1.5 Quantos and Compos ... 560
- 14.2 Path-Dependent Options ... 567
  - 14.2.1 Barrier Options ... 567
  - 14.2.2 Digital Barriers ... 573

	14.2.3	Lookback Options (*)		576
	14.2.4	Options on Averages (Asians)		578
	14.2.5	Chooser Options (*)		586
14.3	Summary			587

Appendix 1 589
    **Value of a Compo Call 589
Appendix 2 591
    **Lemmas on Hitting Probabilities for a Drifted Brownian
    Motion 591
Appendix 3 594
    **Proof of the "Inverses" Relation for Barrier Options 594
Appendix 4 595
    **Valuing a Call Up-and-Out with L (Barrier) $>$ K (Strike) . . . 595
Appendix 5 597
    **Valuing Rebates 597
Appendix 6 598
    **Proof of the Price of a Lookback Call 598
Appendix 7 601
    **Options on an Average Price 601
Appendix 8 602
    **Options with an Average Strike 602
Suggestions for Further Reading 604
    Books 604
    Articles 604

15 Futures Markets (2): Contracts on Interest Rates 607

15.1	Notional Contracts		608
	15.1.1	Basket of Deliverable Securities (DS) and Notional Security	608
	15.1.2	The Euro-Bund Contract	609
	15.1.3	Settlement and Conversion Factors	611
	15.1.4	Cheapest to Deliver and Quoting Futures at Expiration	613
	15.1.5	Arbitrage and Cash-Futures Relationship	618
	15.1.6	Interest Rate Sensitivity of Futures Prices	624
	15.1.7	Hedging Interest Rate Risk Using Notional Bond Contracts	629
	15.1.8	The Main Notional Contracts	637
15.2	Short-Term Interest Rate Contracts (STIR) (3-Month Forward-Looking Rates and Backward-Looking Overnight Averages)		641
	15.2.1	STIR 3-Month Contracts (LIBOR Type, Forward-Looking)	641
	15.2.2	Futures Contracts on an Average Overnight Rate	647
	15.2.3	Hedging Interest Rate Risk with STIR Contracts	656
15.3	Summary		659

Appendices 661
1 Valuation of the Delivery Option 661
2 Relationship Between Forward and Futures Prices 662
Suggestions for Further Reading 666
Books 666
Articles 666
Internet Sites 666

16   Interest Rate Instruments: Valuation with the BSM Model,
     Hybrids, and Structured Products 667
     16.1   Valuation of Interest Rate Instruments Using Standard Models . .   667
            16.1.1   Principles of Valuation and the Black-Scholes-Merton
                     Model Generalized to Stochastic Interest Rates   668
            16.1.2   Valuation of a Bond Option Using the BSM-Price
                     Model   671
            16.1.3   Valuation of the Right to a Cash Flow Expressed as
                     a Function of a Rate and the BSM-Rate Model   672
            16.1.4   Convexity Adjustments for Non-vanilla Cash
                     Flows (*)   676
     16.2   Nonstandard Swaps and Swaptions   680
            16.2.1   Review of Swaps and Notation   680
            16.2.2   Some Nonstandard Swaps   682
            16.2.3   Swap Options (or Swaptions)   686
     16.3   Caps and Floors   688
            16.3.1   Vanilla Caps   689
            16.3.2   A Vanilla Floor   692
     16.4   Static Replications and Combinations; Structured Contracts . . .   693
            16.4.1   Basic Instruments: Notation and General Remarks . . .   693
            16.4.2   Replication of a Capped or Floored Floating-Rate
                     Instrument Using a Standard Asset Associated with
                     a Cap or a Floor   696
            16.4.3   Collars   699
            16.4.4   Non-standard Caps and Floors   702
            16.4.5   Other Static Combinations; Structured Products;
                     Contracts on Interest Rates with Profit-Sharing   705
     16.5   Bonds with Optional Features and Hybrid Products   707
            16.5.1   Convertible Bonds   708
            16.5.2   Other Bonds with Optional Features   711
     16.6   Summary   713
     Appendix   715
            The Q_a-Martingale Measure   715
     Suggestions for Further Reading   716
            Books   716
            Articles   716

		Contents for Volume 1

17 Modeling Interest Rates and Options on Interest Rates 719
   17.1 Models Based on the Dynamics of Spot Rates 720
      17.1.1 One-Factor Models (Vasicek, and Cox, Ingersoll and Ross) 721
      17.1.2 Fitting the Initial Yield Curve; the Hull and White Model 726
      17.1.3 Multifactor Structures 728
   17.2 Models Grounded on the Dynamics of Forward Rates 729
      17.2.1 The Heath–Jarrow–Morton Model (1992) 730
      17.2.2 The Libor (LMM) and Swap (SMM) Market Models 737
   17.3 Summary 749
   Appendix 1 751
      *The Vasicek Model 751
   Appendix 2 754
      *The LMM and SMM Models 754
   Suggestions for Further Reading 762
      Books 762
      Articles 762

18 Elements of Stochastic Calculus 765
   18.1 Definitions, Notation, and General Considerations About Stochastic Processes 766
      18.1.1 Notation 766
      18.1.2 Stochastic Processes: Definitions, Notation, and General Framework 766
   18.2 Brownian Motion 769
      18.2.1 The One-Dimensional Brownian Motion 769
      18.2.2 Calculus Rules Relative to Brownian Motions 775
      18.2.3 Multi-dimensional Arithmetic Brownian Motions 777
   18.3 More General Processes Derived from the Brownian Motion; One-Dimensional Itô and Diffusion Processes 779
      18.3.1 One-Dimensional Itô Processes 779
      18.3.2 One-Dimensional Diffusion Processes 780
      18.3.3 Stochastic Integrals (*) 782
   18.4 Differentiation of a Function of an Itô Process: Itô's Lemma . . . 785
      18.4.1 Itô's Lemma 785
      18.4.2 Examples of Application 787
   18.5 Multi-dimensional Itô and Diffusion Processes (*) 789
      18.5.1 Multivariate Itô and Diffusion Processes 790
      18.5.2 Itô's Lemma (Differentiation of a Function of an n-Dimensional Itô Process) 791
   18.6 Jump Processes 793
      18.6.1 Description of Jump Processes 793
      18.6.2 Modeling Jump Processes 793
   18.7 Summary 794

Contents for Volume 1 xxxvii

Suggestions for Further Reading 796
Books 796

19 *The Mathematical Framework of Financial Markets Theory 797
19.1 General Framework and Basic Concepts 798
19.1.1 The Probabilistic Framework 798
19.1.2 The Market, Securities, and Portfolio Strategies 799
19.1.3 Portfolio Strategies 800
19.1.4 Contingent Claims, AAO, and Complete Markets 802
19.1.5 Price Systems 804
19.2 Price Dynamics as Itô Processes, Arbitrage Pricing Theory and
the Market Price of Risk 805
19.2.1 Price Dynamics as Itô Processes 806
19.2.2 Arbitrage Pricing Theory in Continuous Time 806
19.2.3 Redundant Securities and Characterizing the Base of
Primitive Securities 808
19.3 The Risk-Neutral Universe and Transforming Prices into
Martingales 809
19.3.1 Martingales, Driftless Processes, and Exponential
Martingales 809
19.3.2 Price and Return Dynamics in the Risk-Neutral
Universe, Transforming Prices into martingales
and Pricing Contingent Claims 813
19.3.3 Characterizing a Complete market and Market Prices
of Risk 816
19.4 Change of Probability Measure, Radon-Nikodym derivative
and Girsanov's Theorem 818
19.4.1 Changing Probabilities and the Radon-Nikodym
Derivative 818
19.4.2 Changing Probabilities and Brownian Motions:
Girsanov's Theorem 820
19.4.3 Formal Definition of RN Probabilities 822
19.4.4 Relations between Viable Price Systems, RN
Probabilities, and MPR 823
19.5 Changing the Numeraire 825
19.5.1 Numeraires 825
19.5.2 Numeraires and Probabilities that yield martingale
Prices 826
19.6 The P-Numeraire (Optimal Growth or Logarithmic Portfolio) . . . 832
19.6.1 Definition of the Portfolio $(\underline{h}, H)$ as the P-Numeraire . . 832
19.6.2 Characterization and Composition of the P-Numeraire
Portfolio $(\underline{h}, H)$ 833
19.7 ** Incomplete Markets 836
19.7.1 MPR and the Kernel of the Diffusion Matrix $\sum(t)$ 837
19.7.2 Deflators 840

xxxviii   Contents for Volume 1

19.8 Summary   842
Appendix   844
 Construction of a One-to-one Correspondence between
 $\mathbb{Q}$ and $\mathbf{\Pi}$ 844
Suggestions for further reading   845
 Books   845
 Articles   846

20 The State Variables Model and the Valuation Partial Differential
Equation   847
20.1 Analytical Framework and Notation   847
 20.1.1 Dynamics of State Variables   847
 20.1.2 The Asset Pricing Problem   848
20.2 Factor Decomposition of Returns   849
 20.2.1 Expressing the Return dR as a Function of the dX_j . . .   849
 20.2.2 Expressing the Return dR as a Function of the dW_k . . .   850
20.3 Expected Asset Returns and Arbitrage Pricing Theory (APT)
 in Continuous Time   850
 20.3.1 First Formula for Expected Returns   851
 20.3.2 Continuous Time APT in a State variables Model   851
20.4 The General valuation PDE   853
 20.4.1 Derivation of the General valuation PDE   853
 20.4.2 Market Prices of Risk and Risk Premia   854
 20.4.3 The Relation between MPR and Excess Returns
  on Primitive Securities and the Condition for Market
  Completeness   855
20.5 Applications to the Term Structure of Interest Rates   857
 20.5.1 Models with One State Variable   857
 20.5.2 Multi-Factor models and valuation of Fixed-Income
  Securities   859
20.6 Pricing in the Risk-Neutral Universe   862
 20.6.1 Dynamics of Returns, of Brownian Motions and
  of State Variables in the Risk-Neutral Universe   862
 20.6.2 The Valuation PDE   863
20.7 Discounting under Uncertainty and the Feynman–Kac
 Theorem   864
 20.7.1 The Cauchy-Dirichlet PDE and the Feynman-Kac
  Theorem   864
 20.7.2 Financial Interpretation of the Feynman–Kac Theorem
  and Discounting under Uncertainty   865
20.8 Summary   866
Appendix   868
Suggestions for Further Reading   869
 Books   869
 Articles   869

Part III

Portfolio Theory and Portfolio Management

This third part, which includes Chaps. 21–25, is devoted to portfolio theory, portfolio management, and equilibrium asset prices. Initially designed for equity portfolios, this theory is actually suitable for all types of portfolios and is therefore very general. The financial concepts and the mathematical tools used are generally of intermediate difficulty between those, simpler, of the first part, and those, more complicated, of the second, with the exception of some more arduous developments, which could be skipped at first reading. This part is accessible to undergraduate students or graduate students at the early stage, and is intended for all readers who have assimilated the content of the first part.

Chapter 21 is devoted to the exposition of the standard portfolio theory developed by Harry Markowitz (1959) and based on the theory of decision under uncertainty pioneered by Von Neuman and Morgenstern (1947). Various extensions of the model and some recent developments concerning the so-called "behavioral" finance are also exposed.

Chapter 22 presents the equilibrium model making it possible to obtain the expected return and the price of risky assets (Capital Asset Pricing Model, CAPM) from the aggregation of the optimal individual choices studied in the previous chapter. Many applications and some extensions of the model are then introduced.

Chapter 23 is devoted to the examination of Arbitrage Pricing Theory (APT), less demanding than the CAPM in that it is based solely on the requirement of absence of arbitrage opportunity in the market and not on equilibrium. The chapter then studies the multi-factor models which are derived from APT and used as alternatives to the single-factor model derived from the CAPM.

Chapter 24 presents various theories and techniques of strategic asset allocation (rules used by professional fund managers, portfolio insurance, dynamic portfolio management in continuous time). Strategic asset allocation focuses on a small number of asset *classes* (typically stocks, bonds and money market instruments), depends on the investor's risk aversion and wealth, and is designed for long term horizons.

Chapter 25, lastly, concerns tactical asset allocation which focuses on *individual* securities, constitutes a short-term bet by the investor on their ability to take advantage of temporary opportunities present in the market or to forecast the reversal of a market trend.

Choice Under Uncertainty and Portfolio Optimization in a Static Framework: The Markowitz Model

21

Previous analyses of the value and risk of an asset have in common that they consider a given asset as *isolated*. One of the key results of investment choice theory is that it is possible, by the appropriate combination of several assets in a *portfolio*, to reduce the total risk for a given expected return through diversification. Therefore, the interest of investing in financial security should not be evaluated separately but in the context of the investor's entire portfolio and a market where many savings vehicles compete.

This chapter analyses how investors optimally allocate their savings[1] between the different risky assets available on the market.[2]

We, therefore, consider a financial market on which securities (shares, bonds, etc.) can be bought or sold (or issued), i.e., long or short positions can be adopted. Portfolios are formed by combinations of these securities. The returns of the different securities, and therefore the returns on the portfolios, are generally uncertain.

Under certainty, the choice of investments is made using the simple NPV criterion. In a random world, the problem is much more complex because the investments differ from one another not only in the mathematical expectation of the flows they generate but also in their risk (defined below) and general criteria to support rational choices under uncertainty must be established beforehand.

The first section deals with the theory of choice under uncertainty; it contains a brief exposition of the expected utility criterion (Von Neuman and Morgenstern) and the Mean-Variance paradigm. The second section is devoted to an intuitive presentation of the main concepts (efficient frontier, diversification, ...) based on very

---

[1] After economic theory, individuals make two interdependent decisions: according to their tastes and preferences, and their initial wealth, they divide their income between consumption and savings. Given the latter, wealth is allocated among the various assets available. The portfolio theory presented in this chapter is silent about the first decision, which is supposed to be already taken.

[2] Portfolio theory was initially developed with reference to equities. However, the following analyses apply to all assets.

© The Author(s), under exclusive license to Springer Nature Switzerland AG 2022
P. Poncet, R. Portait, *Capital Market Finance*, Springer Texts in Business and Economics, https://doi.org/10.1007/978-3-030-84600-8_21

873

simple graphs and statistical arguments. The third section rigorously presents the mathematical analysis of the static portfolio selection strategies by an investor who respects the Mean-Variance criterion (the Markowitz model) under the assumption that short positions are allowed. The fourth section presents various extensions of the standard model: forbidden short positions, preferences described by HARA utilities (whose compliance with the Mean-Variance criterion is only a special case), loss aversion and other behaviors described by "Behavioral Finance," not always consistent with the rationality of Von Neuman and Morgenstern.

21.1 Rational Choices Under Uncertainty: The Criteria of the Expected Utility and Mean-Variance

This section presents the main aspects of choice theory under uncertainty. The problem consists in determining the optimal decision among alternatives leading to different random gains (or losses) $\widetilde{W}$. The basic concepts presented in this section are simply explained by focusing on the case of random variables $\widetilde{W}$ taking a finite number of values $(w_1 \ldots, w_N)$ with probabilities respectively equal to $(p_1, ..., p_N)$. The random variable $\widetilde{W}$ can then be interpreted as the algebraic value of the gain generated by a lottery and it is a matter of establishing a criterion that makes it possible to compare different lotteries in order to choose the "best" one.

The first four subsections succinctly describe the expected utility criterion and analyze risk aversion. The fifth subsection is devoted to the mean-variance paradigm on which the classical portfolio theory is based.

21.1.1 The Expected Utility Criterion

Before the work of Bernoulli and Cramer (at the beginning of the eighteenth century) the attractiveness of a lottery was supposed to be based on the expected value of its gain:

$$E\left(\widetilde{W}\right) = \sum_{i=1}^{N} p_i w_i$$

According to such a conception, a rational individual should be indifferent between the lottery with the uncertain outcome $\widetilde{W}$ and a certain sum equal to $E(\widetilde{W})$ and, between several lotteries, should prefer the one that gives the highest expected gain.

Different counterexamples, concerning the behavior of individuals facing risk, contradict this simplistic conception.

Counterexample 1

Consider a lottery $\widetilde{X}$ that gives:

$$\widetilde{X} = \begin{cases} 0 \ \text{€} & \text{with probability } 1/2 \\ \\ 20,000 \ \text{€} & \text{with probability } 1/2 \end{cases}$$

Most people would prefer a certain sum of 10,000 € to the uncertain sum $\widetilde{X}$ even though $E(\widetilde{X}) = 10,000$ €.

This preference for the certain outcome reflects the risk aversion that characterizes most economic agents. This risk aversion is linked to the fact that the marginal utility of the extra euro is decreasing. Indeed, the rational individual classifies her expenditure projects in decreasing order of priority: the first 10,000 € is allocated to projects that are more "useful" than the next 10,000 € and, as a result, the utility of 20,000 € is less than double the utility attributable to 10,000 €.

In fact, for most economic agents, *the certainty equivalent* of the lottery $\widetilde{X}$ in the previous example is less than 10,000 €.

Counterexample 2: *St. Petersburg paradox*

Consider the following game devised by D. Bernoulli in 1732 in St. Petersburg: a coin is tossed until it falls on "head"; the game stops as soon as head is obtained the first time. We call n the number of draws. The probability that exactly n draws is needed to get the first head is equal to $\left(\frac{1}{2}\right)^n$. In addition, the lottery gives $\widetilde{W} = 2^n$ € (i.e. 2 € if head appears in the first draw, 4 € if two draws are necessary, 8 € if the first "head" is obtained in the third draw ...).

Therefore: $E(\widetilde{W}) = \sum_{n=1}^{\infty} \frac{1}{2^n} 2^n = \infty$.

Although the expected gain is infinite, any rational individual, however rich, will accept to pay only a limited sum, possibly very small (a few euros), to acquire such a lottery.

As in the case of counterexample 1, this behavior is explained by the fact that the utility of the additional euro decreases with wealth and that, as a result, the utility of $2x$ € is less than twice the utility of x €. That is why Cramer and D. Bernoulli offered to apply to the gain 2^n (obtained in the case of n draws) an increasing but concave *utility function* U (thus with decreasing marginal utility) and to measure the attractiveness of such a lottery by the expected utility of the gain:

$$E[U(\widetilde{W})] = \sum_{n=1}^{\infty} \frac{1}{2^n} U(2^n).$$

This sum is possibly defined ($< \infty$) for U concave.

For example, for the logarithmic utility function, we have: $E\left(\ln\left(\widetilde{W}\right)\right) = \sum_{n=1}^{\infty} \times$ $\frac{1}{2^n} \ln 2^n = \ln 2 \sum_{n=1}^{\infty} \frac{n}{2^n}$; we can show[3] that $\sum_{n=1}^{\infty} \frac{n}{2^n} = 2$. Therefore, $E\left(\ln\left(\widetilde{W}\right)\right) = 2\ln 2$. The certain sum k that gives the same utility as $\widetilde{W}$ is thus such that $\ln k = 2\ln 2$, i.e. $k = 4$. The "logarithmic" agent will therefore accept to pay only 4 € to participate in this lottery although its expected gain is infinite.

These different ideas have been systematized by Von Neuman and Morgenstern in a fundamental work[4] which, based on an axiomatic of choices under uncertainty, establishes expected utility as a criterion of rational choice (see Appendix 1 for an explanation of this theory).

Indeed, based on a few principles that should govern any rational behavior, it is possible to show that any individual who obeys these principles seeks to maximize the expected value of a utility function of wealth. This utility function $U(.)$ is specific to each individual and reflects in particular their aversion to risk. Schematically, a rational individual whose preferences are represented by the function U and who must make a decision d in a set D of decisions, with d leading to a random gain $\widetilde{W}(d)$, solves the program:

$$\underset{d \in D}{Max} \mathrm{E}\left[U\left(\widetilde{W}(d)\right)\right]$$

The reader interested in this fundamental question, mostly theoretical in nature, will refer to Appendix 1 of this chapter.

21.1.2 Some Features of Utility Functions

We will first notice that two utility functions U and V linked by an affine linear relation, such that: $V(W) = a\, U(W) + b$ where a is a positive constant and b is any constant, lead to the same decisions. Indeed, d maximizes $\mathrm{E}[U(\widetilde{W}(d))]$ if and only if it maximizes $\mathrm{E}[V(\widetilde{W}(d))] = a\, \mathrm{E}[U(\widetilde{W}(d))] + b$. The behavior of an individual characterized by the utility U cannot be distinguished from the one whose utility is V: in this sense, the utility functions are defined only up to a linear affine transformation.

It will also be noted that as a general rule, $\mathrm{E}[U(\widetilde{W})] \neq U(\mathrm{E}(\widetilde{W}))$. It will be shown, moreover, that the utility function of an individual is concave and that, in this case,

---

[3]We start with the expansion $\frac{1}{1-x} = \sum_{n=0}^{\infty} x^n$; we derive each of the two terms of this relation with respect to x, we multiply both sides of the equation by x, and we express the result for $x = \frac{1}{2}$.

[4]*The Theory of Games and Economic Behavior* (1947).

21.1 Rational Choices Under Uncertainty: The Criteria of the Expected...

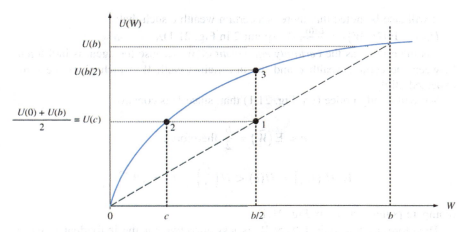

Fig. 21.1 Risk aversion and concavity of U

$E[U(\widetilde{W})] < U(E(\widetilde{W}))$. An individual allergic to risk will therefore prefer the certain sum equal to $E(\widetilde{W})$ to the lottery giving the random sum $\widetilde{W}$.

21.1.3 Risk Aversion and Concavity of the Utility Function

21.1.3.1 The Form of the Utility Function
To reflect the behavior of a "standard" economic agent, a utility function $U(.)$ must have, at least, the following two characteristics:

- It must be increasing with wealth: assuming that the function U is differentiable, the marginal utility $U'(W)$ is therefore positive.
- The marginal utility of wealth, $U'(W)$ must be decreasing, which implies that $U(.)$ is concave, as in Fig. 21.1.[5]

It has already been observed that this last characteristic of the utility function is related to risk aversion. We now examine this relationship more precisely.

Consider a lottery that generates the following gain $\widetilde{W}$:

$$\widetilde{W} = \begin{cases} 0 & \text{with proba } 1/2 \\ b & \text{with proba } 1/2 \ (b > 0) \end{cases}$$

Thus: $E(\widetilde{W}) = \frac{b}{2}$.

Consider the agent u whose utility function U is concave, as in Fig. 21.1.
First note that: $E[U(\widetilde{W})] = \frac{U(0)+U(b)}{2}$ (point 1 in Fig. 21.1).

---
[5] Some behaviors contradict this hypothesis, as it is indicated in Sect. 21.4.3.

It will also be noted that there is a certain wealth c such that:
$U(c) = \mathrm{E}\,[U\,(\widetilde{W})] = \frac{U(0)+U(b)}{2}$ (point 2 in Fig. 21.1).

c is interpreted as the *certainty equivalent* of $\widetilde{W}$ because the agent is indifferent between the certain wealth c and the uncertain wealth $\widetilde{W}$ which gives the same expected utility.

We will finally notice (*see* Fig. 21.1) that, since U is concave:

$$c < \mathrm{E}\left(\widetilde{W}\right) = \frac{b}{2}; \text{therefore}:$$

$$\mathrm{E}\,[U\,(\widetilde{W})] = U(c) < U\left(\frac{b}{2}\right) = U\,[\mathrm{E}\,(\widetilde{W})]$$

(compare points 2 and 3 in Fig. 21.1).

Therefore, an uncertain lottery $\widetilde{W}$ is less attractive for the individual u than a certain sum equal to $\mathrm{E}(\widetilde{W})$.

This result, which reveals u's risk aversion (financial interpretation), results from the concavity of U: thus the concavity of the utility function mathematically reflects risk aversion.

21.1.3.2 Local Measure of the Degree of Risk Aversion

Consider again an individual u, whose preferences are represented by a utility function U of the wealth that we will assume (at least) twice differentiable with:

$U'(\widetilde{W}) > 0$ (utility grows with wealth)

$U''(\widetilde{W}) < 0$ (the marginal utility is decreasing, so U is concave and u is risk-averse).

We will assume further that u has a wealth $\widetilde{W}$ subject to a small risk $\widetilde{\varepsilon}$ such that:

$$\widetilde{W} = W_0 + \widetilde{\varepsilon}$$

with: $W_0 = \mathrm{E}(\widetilde{W})$ therefore $\mathrm{E}(\widetilde{\varepsilon}) = 0$; and σ_ε^2 assumed small.

The risk aversion of u implies that the risk that adds to its wealth without affecting its expected value reduces its expected utility: u is, therefore, ready to pay a certain amount (risk premium) to get rid of this unwanted risk.

The maximum amount π of risk premium that u is willing to pay to eliminate $\widetilde{\varepsilon}$ results from the equation:

$$U(W_0 - \pi) = \mathrm{E}[U(W_0 + \widetilde{\varepsilon})]$$

The left term represents the utility of the (certain) wealth available after payment of the premium and the right one the expected utility of the (random) wealth in the event that u does not pay the premium and thus bears the risk $\widetilde{\varepsilon}$.

An expansion of the two sides of this last equation, around W_0, gives:

21.1 Rational Choices Under Uncertainty: The Criteria of the Expected...

$$U(W_0) - \pi U'(W_0) + o(\pi) = \mathrm{E}\left[U(W_0) + \tilde{\varepsilon}U'(W_0) + \frac{1}{2}\tilde{\varepsilon}^2\, U''(W_0) + o(\tilde{\varepsilon}^2)\right]$$

$$= U(W_0) + \frac{1}{2}\sigma_\varepsilon^2\, U''(W_0) + o(\sigma_\varepsilon^2)$$

where $o(\sigma_\varepsilon^2)$ stands for a term of second order when σ_ε^2 is of first order.

Therefore: $\lim\limits_{\sigma_\varepsilon^2 \to 0} \frac{\pi}{\sigma_\varepsilon^2} = -\frac{1}{2}\frac{U''(W_0)}{U'(W_0)}$ or, when σ_ε^2 and π are small, approximately:

$$\pi \approx -\frac{1}{2}\frac{U''(W_0)}{U'(W_0)}\, \sigma_\varepsilon^2$$

The risk premium that the agent u would agree to pay to get rid of the risk $\tilde{\varepsilon}$ is therefore proportional to σ_ε^2 and to $-\frac{U''(W_0)}{U'(W_0)}$.

σ_ε^2 represents the "intensity" of the risk ε whereas $-\frac{U''(W_0)}{U'(W_0)}$ is specific to the agent u and characterizes their degree of risk aversion.

Definition $A(W_0) \equiv -\frac{U''(W_0)}{U'(W_0)}$ *is called the absolute risk aversion coefficient.*[6]

As $U''(W_0) < 0$ for a concave utility U and as $U' > 0$ always, the aversion coefficient $A(W_0)$ is positive.

It is also useful to define:

- **The relative risk aversion:** $R(W_0) \equiv W_0\, A(W_0)$;
- **The coefficients of risk tolerance (respectively absolute and relative)** $\frac{1}{A(W_0)}$ **and** $\frac{1}{R(W_0)}$.

21.1.4 Some Standard Utility Functions

The following utility functions are frequently considered in the financial literature.

- **Utility with constant absolute risk aversion (CARA):** $U(W) = -e^{-AW}$, where A is a positive constant and $-\frac{U''(W)}{U'(W)} = A$.
- **Utility with constant relative risk aversion (CRRA):** $U(W) = \frac{1}{1-R}W^{1-R}$, where R is constant (positive and $\neq 1$) and $-\frac{U''(W)}{U'(W)} = \frac{R}{W}$, where R is the coefficient of relative risk aversion.

---

[6]Coefficient of Pratt–Arrow, named after these authors who introduced it in the early 1960s.

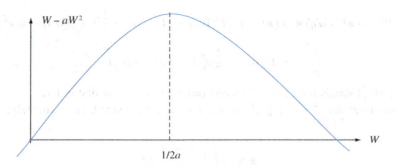

Fig. 21.2 Quadratic utility

- **Logarithmic utility:** $U(W) = \ln W$, thus $-W\frac{U''(W)}{U'(W)} = 1$. The coefficient of relative risk aversion of a logarithmic agent is therefore equal to 1 (limit case of the previous one for R = 1).
- **Quadratic utility:** $U(W) = W - aW^2$, where a is a positive constant, represented by the parabola of Fig. 21.2.

However, quadratic utility, although very often (explicitly or implicitly) assumed, has three serious defects:

- It is only growing for $W < \frac{1}{2a}$ and can therefore only be used if the different wealth to be considered and compared have a negligible probability of exceeding the saturation threshold $\frac{1}{2a}$.
- The coefficient of absolute risk aversion, $\frac{2a}{1-2aW}$ grows with wealth, yet, empirically, this coefficient decreases with wealth for most agents.
- It implies that the agent is indifferent to a possible asymmetry of the probability distribution of wealth. This point will be developed in the following paragraph.
- **"HARA" utility** *(Hyperbolic Absolute Risk Aversion)*: $U(W) = \frac{\gamma}{1-\gamma}\left(\frac{W-\theta}{\gamma}\right)^{1-\gamma}$, with some restrictions[7] on the coefficients γ and θ and on the domain of definition.

The absolute tolerance to risk is linear in wealth and is written: $-\frac{U'(W)}{U''(W)} = \frac{W-\theta}{\gamma}$.

For $\theta = 0$, HARA utility becomes a utility with constant relative aversion equal to γ and for $\theta = 0$ and $\gamma = +1$, asymptotically, it is logarithmic; for $\gamma = -1$, it is quadratic; when θ approaches minus infinity, it approaches a utility with constant absolute aversion.

The family of HARA utility functions, therefore, encompasses all (four) utility classes previously described.

---

[7] See, for example, Elton et al. (2010) for more details on these restrictions.

21.1.5 The Mean-Variance Criterion

21.1.5.1 Presentation of the Criterion

Current experience indicates that most investors are risk-averse, i.e., between different wealth distributions with the same mathematical expectation, they prefer those whose risk is lowest. The investor who follows the mean-variance criterion (M-V below) measures the risk affecting her wealth $\widetilde{W}$ using its variance $[\sigma^2 (\widetilde{W})]$; she will make her decisions according to the expectation, $E(\widetilde{W})$, that she desires the greatest possible, and to the variance, $\sigma^2(\widetilde{W})$, that she wants the lowest possible. Let us consider the case where a decision d must be taken in a set $\boldsymbol{D}$ of possible decisions. The wealth the agent will have, $\widetilde{W}(d)$, will depend on the decision d she chooses. In such a situation, the decider that follows the M-V criterion solves the following program (P):

$$\underset{d \in D}{\text{Max}}\, f(E(\widetilde{W}(d)), \sigma^2(\widetilde{W}(d))) \qquad \text{(P)}$$

where f is an increasing function of E and a decreasing function of σ^2: For a given variance $\sigma^2(W)$, the agent will make the decision that leads to the maximum expected wealth; for a given expectation $E(W)$, it will minimize $\sigma^2(W)$. Therefore, the solution d^* of (P) also solves the following two programs P_E and P_σ:

$$\text{Min } \sigma^2(W(d)), \text{under the constraint}: E(W(d)) = K_E \qquad \text{(P}_E\text{)}$$

$$\text{Max } E(W(d)), \text{under the constraint}: \quad \sigma^2(W(d)) = K_\sigma \qquad \text{(P}_\sigma\text{)}$$

where K_E and K_σ are two constants, respectively, equal to $E(W(d^*))$ and $\sigma^2(W(d^*))$.

21.1.5.2 Mean-Variance Criterion and Expected Utility

The expected utility criterion has a theoretical grounding, while the mean-variance criterion is ad hoc and open to criticism in a number of ways. In addition to the previously mentioned shortcomings, risk assessment using variance leads to considering positive deviations from the mean and negative deviations as equivalent. For example, the two probability distributions of the two wealth W_a and W_b in Fig. 21.3, which have the same mean E and the same dispersion around E, are equivalent for the investor who follows the M-V criterion.

By construction, these two distributions are asymmetrical but symmetrical to one another with respect to a vertical axis passing through E, their common mean. They, therefore, have the same mean E and the same variance but the asymmetry is positive for W_a and negative for W_b.

In fact, the agents are not usually indifferent to the third moment of the distribution of their wealth (called *skewness*) which expresses a dissymmetry of the distribution relative to its mean (see Appendix 1 of Chap. 8). In general, risk aversion is associated with a preference for wealth (or income) whose distribution has a positive asymmetry, such as W_a, with a low risk of very high losses.

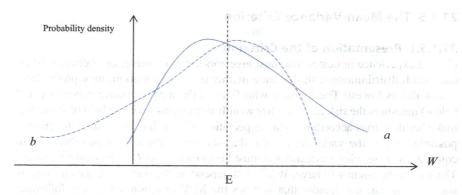

Fig. 21.3 Two equivalent wealth distributions for a M-V investor

Nevertheless, the M-V criterion is involved by the maximization of the expected utility in two cases: that of a quadratic utility function on the one hand; that of wealth distributed according to a Gaussian distribution (irrespective of the utility function), on the other hand.

In the case of an individual whose preferences can be represented using a quadratic utility function ($W - aW^2$; $a > 0$), the maximization of the expected utility $E(\widetilde{W}) - a\, E(\widetilde{W}^2)$ leads, for a given $E(\widetilde{W}) = k$, to prefer the wealth minimizing $E(\widetilde{W}^2)$, hence the one that minimizes $\sigma^2(\widetilde{W}) = E(\widetilde{W}^2) - k^2$: the quadratic preference agent, therefore, applies the M-V criterion.

In the case where the wealth is distributed according to a Gaussian distribution, the whole distribution of $\widetilde{W}$ is characterized by the two parameters $E(\widetilde{W})$ and $\sigma^2(\widetilde{W})$; we can then write, for any utility function U:

$$E\left[U(\widetilde{W})\right] = f\left(E(\widetilde{W}), \sigma^2(\widetilde{W})\right).$$

All economic agents who compare several Gaussian wealth, therefore, apply the M-V criterion. In addition, any risk-averse investor maximizes a function that positively depends on expectation and negatively on variance.

21.2 Intuitive and Graphic Presentation of the Main Concepts of Portfolio Theory

The main concepts that emerge from the theory of optimal choices by individuals who apply the mean-variance criterion as well as those relating to diversification and its limits can be easily understood from graphs and very simple mathematical and statistical tools. This section develops these technically basic analyses, the more precise mathematical study of efficient portfolios being developed in Sect. 21.3. The reader familiar with these fundamental concepts can directly approach that section.

21.2.1 Assumptions, General Framework and Efficient Portfolios

21.2.1.1 General Framework and Representation of Long and Short Positions

Hereafter, we consider an individual who combines different securities to build a portfolio. The investment involves a period of arbitrary duration, beginning at $t = 0$ and ending at $t = 1$ (we can think of a period of 1 year of duration). The investor takes positions at the beginning of the period ($t = 0$) and then does not carry out any transactions for the remainder of the period. He is interested in the value of this portfolio at the end of the period ($t = 1$). For a given initial investment, this terminal value is uniquely linked to its return between 0 and 1. The (arithmetic) rate of return over the period $(0, 1)$ of a security or portfolio whose price is P_0 at $t = 0$, P_1 at $t = 1$ and which distributes a dividend D_1 at $t = 1$ is equal to:

$R = \frac{P_1 + D_1 - P_0}{P_0}$. Thereafter, so as to simplify, P_1 represents most often the global value including the dividend; total return[8] then simply writes: $R = \frac{P_1 - P_0}{P_0}$.

Positions can be long or short. In the case of a long position on an asset with a price P_0 at time 0, P_1 (random) at time 1, the investor buys the security in 0 and has a wealth of P_1 in 1: it is an investment generating the sequence $(-P_0, P_1)$ whose return is R.

In the case of a short position, the agent sells or issues the security at $t = 0$, which results in the sequence $(+P_0, -P_1)$ (at $t = 0$ the operator receives P_0 € and at $t = 1$ the operation leads to a reduction in the value of their wealth by P_1 €): it is, therefore, analogous to borrowing (at the rate R), possibly intended to finance long positions taken on other securities. Short positions are taken in different contexts and according to different procedures such as:

- A borrowing of the security at $t = 0$ (to be returned in $t = 1$) immediately followed (at $t = 0$) by its sale for cash (see Sect. 3.4.1, on security lending and borrowing): this is a short sale;
- A forward sale (of maturity $t = 1$) of the security, plus simultaneously borrowing (between 0 and 1) an amount equal to the spot price P_0;
- An issue of the security at price P_0, P_1 being equal to the market value of the debt in $t = 1$;
- A simple sale in $t = 0$ of a security previously present in the portfolio of the operator; the sequence $+P_0$, $-P_1$ then is interpreted as a *difference* with respect to the flows or values prevailing in the absence of the operation.

In general, a position on n securities has a value nP, with $n > 0$ or $n < 0$ depending on whether the position is long or short: e.g., a long position generates the sequence $-nP_0$, nP_1.

---

[8]To simplify, we will consider the terms "rate of return" and "return" as equivalent.

884   21   Choice Under Uncertainty and Portfolio Optimization in a Static Framework:...

21.2.1.2 Efficient Portfolios

We will assume that the investor assesses the risk of their portfolio by the variance of its return and applies the M-V criterion.

Definition *Portfolios with maximum expected return for a given return variance (or with minimal return variance for a given expected return) are called efficient portfolios.*

The main concepts of efficient portfolios were introduced in the early 1950s by Harry Markowitz.

We will analyze efficient portfolios by starting with two assets only before addressing the general case of portfolios comprising any number of assets.

21.2.2 Two-Asset Portfolios

21.2.2.1 Notations and Analytic Forms of a Portfolio Return, Its Expected Value and Its Variance

Consider two assets A and B whose random returns are denoted by R_A and R_B.

The following notations are used: the random variables are in upper case and the non-random elements (scalars) in lower case; $\mu_A \equiv E(R_A)$, $\mu_B \equiv E(R_B)$, $\sigma_A \equiv \sigma(R_A)$ and $\sigma_B \equiv \sigma(R_B)$ designate the expected values and standard deviations of the returns of these two assets and $\sigma_{AB} \equiv cov(R_A, R_B)$ their covariance. Let us remember that $\sigma_{AB} = \rho_{AB}\,\sigma_A\sigma_B$ where ρ_{AB} is the correlation coefficient between R_A and R_B.

A portfolio combining A and B is characterized by the weights allocated to each asset: a fraction x of the sum invested is invested in security A and the additional fraction $(1-x)$ in B. By normalizing the investment to \$1, x and $(1-x)$ dollars are therefore initially invested in the securities A and B, respectively.

If R_P is the return of the portfolio, its terminal value (for \$1 invested) will be equal to:

$$1 + R_P = x(1 + R_A) + (1 - x)(1 + R_B) = 1 + xR_A + (1 - x)R_B.$$

The return on the portfolio is therefore equal to the weighted average of the returns on the securities that make it up: $R_P = x\,R_A + (1-x)R_B$.

Let us insist on the fact that, ex ante, the returns R are uncertain while the weights x are known (chosen) from the beginning.

The expected return of the portfolio, μ_P, and its variance σ^2_P are given by:

$$\mu_P = x\mu_A + (1 - x)\mu_B \tag{21.1}$$

$$\sigma^2_P = x^2\sigma^2_A + (1 - x)^2\,\sigma^2_B + 2x(1 - x)\sigma_{AB} \tag{21.2}$$

Thus, the expectation and variance of the portfolio are expressed in terms of the value of x, which allows investors to choose the weight x appropriately.

21.2 Intuitive and Graphic Presentation of the Main Concepts of Portfolio Theory

Example 1

Suppose that in portfolio P, the weight x of asset A is equal to 2/3 and that of B is 1/3. Suppose further that:

- The expected return on asset A is 12% and that on asset B is 18%
- The standard deviations of the two returns are equal to 40% and their correlation coefficient equal to 0.5.

Let us compute the expected return μ_P, and the standard deviation, σ_P, of this portfolio.

By application of the formula (21.1): $\mu_P = (\frac{2}{3} \times 12) + (\frac{1}{3} \times 18) = 14\%$,

$\sigma_{AB} = \rho_{AB}\, \sigma_A \sigma_B = 0.5 \times 0.4 \times 0.4 = 0.08$. The formula (21.2) then gives:

$$\sigma^2{}_P = \left(\frac{2}{3}\right)^2 \times (0.4)^2 + \left(\frac{1}{3}\right)^2 \times (0.4)^2 + 2 \times \frac{2}{3} \times \frac{1}{3} \times 0.08 = 0.12444; \text{ i.e.}$$

$$: \sigma_P = 0.353.$$

It should be noted that the standard deviation of the return on this portfolio (0.353) is lower than the average standard deviation of the returns of its components (0.4). This risk reduction is due to diversification, a fundamental concept discussed later.

21.2.2.2 Geometric Representation of the Combinations of Two Assets

Consider the two Eqs. (21.1) and (21.2): to each x corresponds a point in the space (σ, μ) and, by varying x, one generates a curve which is the locus of the representative points of the portfolios obtained by combinations of A and B. In fact, Eqs. (21.1) and (21.2) constitute the parametric equation of this curve which is the *hyperbola* represented in Fig. 21.4.

The portfolio consisting only of asset A ($x = 1$) is represented by point A. By moving on the AB curve from A to B, the portfolios contain fewer and fewer securities A and more and more securities B. Point B corresponds to a portfolio including only the security B ($x = 0$, therefore $1 - x = 1$). The points on the curve to the right of B represent portfolios for which $x < 0$ (short sale of A). The points to the left of A represent portfolios for which $1 - x < 0$ (short position on B). If the short positions are forbidden or impossible to achieve, only the points of the curvilinear segment AB correspond to reachable portfolios.

The shape of the hyperbola AB depends on the correlation coefficient ρ_{AB} between the two returns.

Four particular cases deserve to be examined.

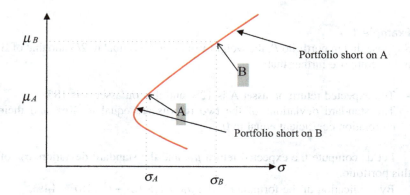

Fig. 21.4 Combination of two assets A and B, general case

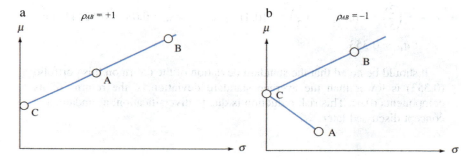

Fig. 21.5 Perfect correlation

- First case: the returns of the two assets are perfectly correlated ($\rho_{AB} = 1$). This case is shown on Fig. 21.5a.
  The hyperbola AB then degenerates into a straight line[9] that cuts the vertical axis in C: in this extreme case of perfect correlation, it is possible to build a risk-free portfolio (of zero standard deviation of return), taking a short position on B (so a weight >1 is assigned to A).
- Second particular case (Fig. 21.5b): the returns of the two securities are perfectly negatively correlated ($\rho_{AB} = -1$).[10] It is then possible to build a portfolio C long on both securities without any risk because, for a particular value ($x = x_C$), the two risks ($x_C\ R_A$ and $(1-x_C)R_B$) are neutralized perfectly. It is a perfect

---

[9] Equation (21.2) is reduced to: $\sigma^2_P = x^2\sigma^2_A + (1-x)^2 \sigma^2_B + 2x(1-x)\sigma_A\sigma_B$, a perfect square that yields: $\sigma_P = x\sigma_A + (1-x)\sigma_B = \sigma_B + x(\sigma_A - \sigma_B)$; furthermore: $\mu_P = x_A\ \mu_A + (1-x)\ \mu_B$. By eliminating x from the last two equations, we obtain a linear relationship between μ_P and σ_P.

[10] Equation (21.2) is reduced to: $\sigma^2_P = x^2\sigma^2_A + (1-x)^2\sigma^2_B - 2x(1-x)\sigma_A\sigma_B = [x\sigma_A - (1-x)\sigma_B]^2$, or: $\sigma_P = |\ x\sigma_A - (1-x)\sigma_B\ | = |-\sigma_B + x(\sigma_A + \sigma_B)\ |$ which equals 0 for $x = \sigma_B/(\sigma_A + \sigma_B)$ and that, associated with $\mu_P = x_A\mu_A + (1-x)\ \mu_B$, leads to the equation of the two segments shown in Fig. 21.5b.

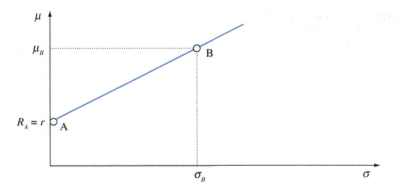

Fig. 21.6 Combinations of A and B when A is risk-free

diversification (obtained with only two securities) which is due to the fact that these two random variables are perfectly negatively correlated.
- Third particular case: $\rho_{AB} = 0$ (the returns of the two assets are not correlated). This case does not give rise to a particular geometrical representation (it could be represented by Fig. 21.4) and constitutes an intermediate case between the two extreme cases represented in Fig. 21.5a, b. As in the general case, the reachable portfolios are represented, in the space (σ, μ), by the hyperbola AB.
- Fourth particular case (Fig. 21.6): one of the two assets, for example, A, is risk-free; the standard deviation of its return is, therefore, zero and it is represented by point A on the vertical axis. It could be a fixed-rate Treasury bill or bond with the same maturity as the investor's horizon. Therefore, not only $\sigma_A = 0$ but $\rho_{AB} = 0$. The risk-free rate will be denoted by r ($R_A = r$). In this case, the representative points of the reachable portfolios are aligned on the straight half-line going through A and B.

21.2.3 Portfolios with N Securities

In general, when portfolios combine any number of risky securities ($N > 2$), the efficient frontier can take on two forms, depending on the absence or presence of a risk-free asset.

Hereafter, we will consider these two cases sequentially. We will assume that short positions are permitted (the consequences of a ban on short positions will be discussed in Sect. 21.4.1).

21.2.3.1 First Case: All Assets Are Risky

It is possible to show (see next section) that the set of representative points of *all the possible portfolios in space* (σ, μ) *is constituted by the blue surface S* represented in Fig. 21.7 and delimited by a hyperbola. However, only certain portfolios are relevant

Fig. 21.7 Efficient frontier with N risky assets in the absence of a risk-free asset

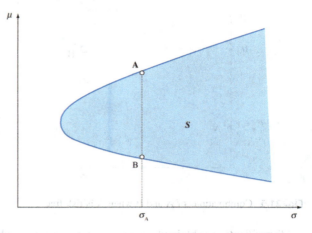

to the investor subject to the M-V criterion. For example, all portfolios on the AB segment with the same level of risk σ_A offer different expected returns; in particular, portfolio A with the highest expected return will be preferred by all investors to the other portfolios represented on the AB segment. Considering all the possible values of the standard deviation σ_A, it is clear that the locus of the representative points of the "dominant portfolios" is the upper branch of the hyperbola delimiting the surface S shaded in blue in Fig. 21.7. This curve, called *efficient frontier*, is the locus of points representing efficient portfolios.

21.2.3.2 Second Case: Existence of a Risk-Free Asset

Suppose now that it is possible over the period to lend (or invest funds) and borrow (or issue securities) over the period at the same riskless interest rate r. To simplify, we will call this risk-free asset r. It is represented in Fig. 21.8 by the point of zero abscissa and ordinate r.

In addition, there are N risky assets, such as those considered in the preceding Sect. 21.2.3.1 and all their possible combinations generate, again, the blue area S.

Consider a given portfolio k consisting of only risky assets (its representative point is therefore in S) and a portfolio P containing a proportion x of k and a proportion $(1 - x)$ of the risk-free asset with a certain return r. It is easy to show that:[11]

$$\mu_P = r + \frac{\mu_k - r}{\sigma_k}\sigma_P \qquad (21.3)$$

---

[11] Equations (21.1) and (21.2) applied to P imply: $\mu_P = x\mu_k + (1-x)r$; $\sigma_P = x\sigma_k$, thus: $\mu_P = r + \frac{\mu_k - r}{\sigma_k}\sigma_P$, or a linear relation between μ_P et σ_P, represented by the straight line going through r and k, with a slope $[\mu_k - r]/\sigma_k$.

21.2 Intuitive and Graphic Presentation of the Main Concepts of Portfolio Theory

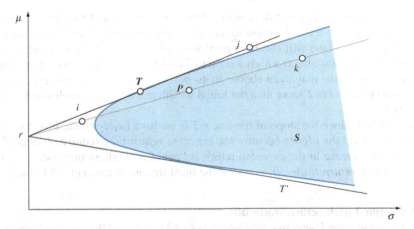

Fig. 21.8 Achievable portfolios and efficient frontier (*r–T*) in the presence of *N* risky assets and a risk-free asset

The representative point of *P* is therefore located on the half-line starting from *r* and passing through *k* (see Fig. 21.8) and, by varying *x* between 0 and +∞, this whole half-line is generated.

In general, any *P* portfolio is obtained by combining the risk-free asset and a portfolio of risky assets such as *k*. It is therefore represented by a point located on one of the straight lines such that *r-k* issued from *r* (and forming the cone *T'*-*r*-*T* shown in Fig. 21.8) constituting the set of representative points of all the realizable portfolios. The *efficient* portfolios are represented by the straight line of the *T'*-*r*-*T* beam, the slope of which is the largest, i.e. the tangent *r-T*. Indeed, each level of risk (to each value of σ) corresponds a point of this line which represents a portfolio whose expected return is maximum for this level of risk. *The efficient frontier is therefore the half-line r-T, tangent to S from the point r.*

In accordance with Eq. (21.3), the equation of the efficient frontier writes:

$$\mu_P = r + [\mu_T - r]\frac{\sigma_P}{\sigma_T}. \tag{21.3'}$$

T represents the tangent portfolio which, like all the portfolios of *S*, exclusively consists of risky assets.

A fundamental result, due to Tobin and Markowitz and called the *two-fund separation theorem*, is simply deduced from this graphical analysis: *all efficient portfolios are combinations of the risk-free asset and the tangent portfolio T*: although the market offers *N* + 1 different assets (*N* risky and one risk-free) all efficient portfolios are built from the same two funds (the risk-free asset and *T*).

It is important to understand that investors sharing the same investment horizon and the same beliefs about expectations, variances, and covariances (they have the same efficient frontier) all hold a combination of the *same* portfolio *T* of risky assets and the risk-free asset, but that the respective weights they allocate to *T* and the risk-free asset in this combination depends on their risk aversion and wealth. A cautious individual allocates a small weight to the risky asset portfolio T and high weight to the risk-free asset (she will choose a portfolio such as i in Fig. 21.8) while a more adventurous investor affects a high weight to T and a low weight to r to obtain a high expected return. She may even choose to go into debt (negative weight on the risk-free asset) to invest in T more than her initial wealth and build a portfolio such as j in Fig. 21.8.

In addition, since the slope of the line r-T is positive $(\mu_T > r)$, *the investor must accept a risk all the higher because the expected return they require is large.* The fact that an increase in the expected return is "paid for" with an increased risk, the so-called *risk-return trade-off*, is one of the most important concepts of Finance.

Example 2 (risk-return trade-off)
Suppose that the 1-year risk-free rate, r, is equal to 4%, and that, for the tangent portfolio, $\mu_T = 10\%$, $\sigma_T = 21\%$; what risk (standard deviation of return) should be assumed in order to obtain an expected return of 6%? of 8%?

To obtain $\mu_P = 6\%$ we need to accept $\sigma_P = \sigma_T (\mu_P - r)/(\mu_T - r) = 7\%$ and to obtain $\mu_P = 8\%$ we must face a standard deviation σ_P of 14%. The greater the expected return, the higher the risk that must be accepted.

21.2.4 Portfolio Diversification

21.2.4.1 General Considerations

Let P be any portfolio, defined by the weights $x_1, x_2, \ldots, x_N$ allocated to the N risky assets (the sum of the weights x_i is equal to 1) so that its return R_P is given by:

$$R_P = x_1 R_1 + x_2 R_2 + \ldots + x_N R_N \tag{21.4}$$

The expected return of the portfolio is the weighted average (by the weights x_i) of the expected return on each of the securities that compose it; the contribution of each security i to the return of P is therefore directly proportional to its return. This result demonstrated in the case of two securities in the previous Subsect. 21.2.2.1 is easily generalized in the case of a portfolio consisting of any number N of securities.

Indeed, let P be a portfolio whose value is \$1 at $t = 0$, with x_i \$ allocated to asset i. This ith component of portfolio P takes the value $x_i (1+R_i)$ \$ for $t = 1$. Therefore, for \$1 invested at $t = 0$, the total value of P thus is $\sum_{i=1}^{N} x_i(1 + R_i) = \sum_{i=1}^{N} x_i + \sum_{i=1}^{N} x_i R_i = 1 + \sum_{i=1}^{N} x_i R_i$ for $t = 1$; therefore, $R_P = \sum_{i=1}^{N} x_i R_i$.

We can measure the risk of this portfolio P by the variance (or standard deviation) of R_P. But what is true at the portfolio level is wrong at the level of individual

21.2 Intuitive and Graphic Presentation of the Main Concepts of Portfolio Theory

security: Indeed the risk induced by individual security i for the investor holding the portfolio P must be measured by the contribution of i to the overall risk affecting the performance R_P. A misconception consists in measuring the risk induced by i by the variance or the standard deviation of R_i, whereas it is in fact *its correlation with R_P that determines this risk.*

To facilitate the intuitive understanding of this last and important assertion, consider a stock i that is negatively correlated with the portfolio P: when the performance of P is low, that of i has a high probability of being good and *reciprocally.* The security i, therefore, tends to *push the overall return R_P towards its mean* and consequently to reduce the amplitude of its variations, i.e., to reduce the overall risk. On the other hand, if stock i is strongly and positively correlated with P, the fluctuations of its return generally "add" to those of the other securities and its holding accentuates the overall risk.

These intuitive considerations, therefore, lead to *associating the risk induced by a security with the covariance of its return with that of the portfolio ($cov(R_i, R_P)$).*

We now clarify these considerations by examining the particular case of securities whose returns are statistically related to each other by a single factor.

21.2.4.2 Diversification in the Context of the Market Model (also Called Diagonal Model or Sharpe Model)

The following hypothesis, which leads to relation (21.5) below, is the *market model* or the *diagonal model*, due to William Sharpe (1963); this hypothesis is not necessary for the validity of the results, which are more general, but it allows an intuitive explanation of diversification and its limits.

We will assume below that the uncertainty affecting the return R_i of each security is due:

- Partly to the overall economic situation represented by the return R_m of the stock market as a whole and which can possibly be measured from a stock market index. This risk is common to all R_i.
- Partly to a factor ε_i, specific to the security i considered, independent from the previous one and independent from the factor ε_j specific to the security $j \neq i$.

Therefore, the return on any security i is consistent with the following equation:

$$R_i = \mu_i + \beta_i(R_m - \mu_m) + \varepsilon_i \tag{21.5}$$

μ_i is non-random and represents the expected value of R_i. The term $\beta_i(R_m - \mu_m)$, expresses the reaction of the return R_i to market fluctuations: when R_m is greater (lower) than its mean μ_m, R_i is increased (decreased) by a fraction β_i of the difference $R_m - \mu_m$. The coefficient β_i which thus reflects the sensitivity of R_i to the variations of R_m is called the *beta* of the security i and plays a key role in portfolio theory. The risk represented by $\beta_i R_m$ is called the *systematic or non-diversifiable risk*, for reasons developed below.

The term ε_i is random (although represented by a lowercase letter), of zero expected value, and represents the *specific* risk of asset i. It is assumed to be independent from the market index R_m. The different ε_i are also assumed to be independent from each other. The ε_i are called *diversifiable* for reasons that will be clarified soon.

The return R_p on portfolio P is then written as the sum of three components. Indeed, Eqs. (21.4) and (21.5) imply that:

$$R_p = (x_1\mu_1 + x_2\mu_2 + \cdots + x_N\mu_N) + (x_1\beta_1 + \cdots + x_N\beta_N)(R_m - \mu_m) + (x_1\varepsilon_1 + x_2\varepsilon_2 + \cdots + x_N\varepsilon_N),$$

that is:

$$R_p = \underbrace{\mu_P}_{\text{expected return}} + \underbrace{\beta_P(R_m\mu_m)}_{\text{systematic risk}} + \underbrace{\varepsilon_P}_{\text{diversifiable risk}} \tag{21.6}$$

with:

$\mu_P = x_1\mu_1 + x_2\mu_2 + \ldots + x_N\mu_N$ represents the expected return on portfolio P, equal to the weighted average of the expected returns on the different securities;

$\beta_P(R_m - \mu_m) = (x_1\beta_1 + \ldots + x_N\beta_N)(R_m - \mu_m)$ corresponds to the reaction of the return on portfolio P to market fluctuations and $\beta_P = x_1\beta_1 + \ldots + x_N\beta_N$ denotes the beta of the portfolio P which *is therefore equal to the weighted average of the betas of the securities that compose it*;

$\varepsilon_P = x_1\varepsilon_1 + x_2\varepsilon_2 + \ldots + x_N\varepsilon_N$ is equal to the weighted sum of the independent and specific risks of the individual securities; ε_P is independent of R_m since all its components ε_i are.

Therefore, in accordance with the additivity of variances of ε_P and R_m resulting from the independence of these variables, the total variance affecting R_P, $\sigma^2(R_P)$, can be written as the sum of two components:

$$\sigma^2(R_P) = \beta_P{}^2\sigma^2(R_m) + \sigma^2(\varepsilon_P)$$

The first component $(\beta_P{}^2\sigma^2(R_m))$ constitutes the systematic variance, irreducible by diversification, while the second $(\sigma^2(\varepsilon_P))$ represents the specific or diversifiable variance that may be canceled by diversification. Indeed, the *law of large numbers that can be applied to the sum of independent risks ε_i composing ε_P allows to almost cancel this last variance as long as portfolio P is well diversified (i.e., N needs to be big enough and each x_i small enough).*[12]

---

[12]For instance, consider the special case of an equally weighted portfolio with N risky assets (each weight is thus equal to $1/N$). We can write: $\varepsilon_P = \frac{1}{N}(\varepsilon_1 + \varepsilon_2 + \ldots + \varepsilon_N)$.

And, by independence of the ε_i :

$\sigma^2(\varepsilon_P) = \frac{1}{N^2}\sum_{i=1}^{n}\sigma^2(\varepsilon_i) = \frac{1}{N}v^2$, where $v^2 = \frac{1}{N}\sum_{i=1}^{n}Z\sigma^2(\varepsilon_i)$ is the average variance of the specific risks.

21.2 Intuitive and Graphic Presentation of the Main Concepts of Portfolio Theory

One intuitively understands that, in the case of a large number of independent ε_i, some of which are positive and others are negative but whose expected value is zero, their weighted average is practically equal to zero. That is why the component ε_i of each return R_i is called "diversifiable risk" as it is practically eliminated in a well-diversified portfolio.

By contrast, the second component $\beta_P(R_m - \mu_m)$ cannot be eliminated by diversification and is the only relevant risk affecting the return of a well-diversified portfolio. For the latter, the following approximate relationships prevail:

$$R_P \approx \mu_P + \beta_P(R_m - \mu_m); \qquad \sigma^2(R_P) \approx \beta_P{}^2 \sigma^2(R_m).$$

Example 3

Consider N securities and assume that $\sigma(R_m) = 20\%$, that the market model is valid and that the N securities all have a return with a diversifiable risk ε_i whose standard deviation $\sigma(\varepsilon_i)$ is equal to 50% and with a beta equal to 0.9. The market model yields:

$$R_i = \mu_i + \beta_i(R_m - \mu_m) + \varepsilon_i, \text{ then } \sigma^2(R_i) = \beta_i{}^2 \sigma^2{}_m + \sigma^2(\varepsilon_i) = 0.2824;$$

$\sigma^2(R_i)$ breaks down into:

- A systematic variance $= \beta_i{}^2 \sigma^2(R_m) = (0.9)^2 \times 0.04 = 0.0324;$
- A specific variance $= \sigma^2(\varepsilon_i) = 0.25.$

Consider now an equally-weighted portfolio composed of these N securities (the weight of each security is set at $1/N$). The beta of this portfolio is equal to the average of its components, i.e. $\beta_P = 0.9$, and its return is: $R_P = \mu_p + 0.9 \times (R_m - \mu_m) + \varepsilon_P$ with $\varepsilon_P = \frac{1}{N}(\varepsilon_1 + \varepsilon_2 + \ldots + \varepsilon_N)$ and $var(\varepsilon_P) = 0.25/N$.

As ε_P et R_m are independent: $\sigma^2{}_P = var(0.9 \times R_m) + var(\varepsilon_P)$.

The systematic variance can be measured by $var(0.9 \times R_m) = [0.9 \times 0.2]^2 = 0.0324$ and the specific variance by $var(\varepsilon_P) = 0.25/N$.

The specific « risk » is generally overwhelming for a single security ($N = 1$); high for a poorly diversified portfolio (for $N = 10$ it represents 75% of the systematic risk in our example); negligible compared with the systematic risk for a well-diversified portfolio (for $N = 80$, it is ten times lower than the systematic risk in our example).

The risk $\sigma^2(\varepsilon_P)$ is therefore proportional to the average specific risk and *inversely proportional to the number of securities*: approximately 95% of diversifiable risk is eliminated, on average, with 20 securities, and 99% with 100 securities.

21.3 Mathematical Analysis of Efficient Portfolio Choices

This section is based on a more elaborate mathematical and statistical apparatus than the one used in the previous section. The reader must be familiar with matrix calculus, quadratic forms, and optimization under constraints using the Lagrangian. Appendices 2, 3, and 4 provide some brief reminders about these tools.

21.3.1 General Framework and Notations

We consider investments over a period (0, 1) where 0 is the investment date (today) and 1 is the future terminal date. We assume that short positions are allowed. Underlined variables are vectors; a prime denotes a transpose. Except in a case defined below, variables and random vectors are represented by capital letters while non-random parameters are in lower case. The matrices, although not random, are indicated by capital letters in bold.

21.3.1.1 Assets

N risky assets denoted by $i = 1,\ldots, N$ exist; the price of asset i is $S_i(t)$ with $t = 0$, 1 and its return is $R_i = \frac{S(1)i - S_i(0)}{S_i(o)}$ (any dividends are incorporated into $S_i(1)$);

$\underline{S}(t) \equiv (S_1(t)\ldots,S_N(t))'$ denotes the price column vector and $\underline{R} \equiv (R_1(t)\ldots,R_N(t))'$ the return column vector;

$\mu_i \equiv E(R_i)$ is the expected return of the security i, $\sigma_i^2 \equiv \text{variance}(R_i)$ and $\mu \equiv (\mu_1,\ldots, \mu_N)'$ is the column vector of the N expected return; $\sigma_{ij} \equiv \text{cov}(R_i,R_j)$ and $\mathbf{V}$ is the variance–covariance matrix ($N \times N$) of the N return whose general term is σ_{ij}.

We distinguish the case where a risk-free asset exists from the one where such an asset does not.

In its presence, the $N + 1$st asset, called asset 0, gives a certain return denoted by r (the interest rate). The purchase of this asset may be interpreted as lending at a rate r and its sale (or issue or short position) as borrowing at the same rate r.

Its price is denoted $S_0(t)$ ($t = 0$, 1) and we have: $S_0(1) = (1+r) S_0(0)$; although S_0 is certain, it is exceptionally denoted by a capital letter.

In the presence of a risk-free asset $\underline{S}(t) \equiv (S_0(t), S_1(t)\ldots, S_N(t))'$ will denote the vector of the $N+1$ prices while $\underline{R} \equiv (R_1(t)\ldots, R_N(t))'$ will represent *only the N random return*.

21.3.1.2 Portfolios

A portfolio P can be characterized either by the number of securities (of each type) it contains, or by the weight attributed to each security.

In the first case, n_i represents the number of securities i contained in portfolio P. Note that $n_i > 0$ denotes a long position and $n_i < 0$ a short position on i. Note also that $n_i(0) = n_i(1) \equiv n_i$ because no transaction occurs outside the initial time 0 (static strategy).

21.3 Mathematical Analysis of Efficient Portfolio Choices

In the case where the portfolio is characterized by weights, x_i is the *initial* weight allocated to the security i: $x_i = \frac{n_i S_i(0)}{X(0)}$ where $X(0)$ denotes the portfolio's initial value.

We start with the case where a risk-free asset does not exist.

(a) *Absence of a risk-free asset*

Portfolio P can then be characterized by the N-dimensional vector: $\underline{n} = (n_1, \ldots, n_N)'$.

The value of portfolio P, denoted by $X(t)$, is therefore:

$$X(t) = \underline{n}' \underline{S}(t) \equiv \sum_{i=1}^{N} n_i S_i(t) \quad t = 0, 1 \tag{21.7}$$

($\underline{n}'\underline{S}$ therefore is a scalar product).

The return on portfolio P is:

$$R_p = \frac{X(1) - X(0)}{X(0)} \quad \text{or} \quad 1 + R_p = \frac{X(1)}{X(0)} \tag{21.8}$$

Furthermore $\underline{x} = (x_1, \ldots, x_N)'$ is the vector (column) of the initial weights that characterizes the portfolio P, up to a scale factor.

By definition, the sum of the weights is equal to 1:

$$\underline{x}' \underline{1} \equiv \sum_{i=1}^{N} x_i = 1 \ \left(\underline{1} \text{ denotes the unit vector of } \mathbb{R}^N \right).$$

Fundamental property: the return R_p on the portfolio is equal to the weighted (by the x_i) sum of the different returns R_i on its components:

$$R_P = \underline{x}' \underline{R} \tag{21.9}$$

This equation, which is none other than Eq. (21.4) of Sect. 21.1, can be formally proved:

$$1 + R_p = \frac{X(1)}{X(0)} = \sum_{i=1}^{N} \frac{n_i S_i(1)}{X(0)} \equiv \sum_{i=1}^{N} \left(\frac{n_i S_i(0)}{X(0)} \right) \frac{S_i(1)}{S_i(0)}$$

$$= \sum_{i=1}^{N} x_i(1 + R_i) = 1 + \sum_{i=1}^{N} x_i R_i,$$

thus $R_P = \sum_{i=1}^{N} x_i R_i$, which is Eq. (21.9).

Expressions for $E(R_P) = \mu_P$ and variance$(R_P) = \sigma^2_P$ come from Eq. (21.9):

$$\mu_P = \underline{x}'\mu \tag{21.10}$$

$$\sigma^2{}_P = \sum_{i=1}^{N} \sum_{j=1}^{N} x_i x_j \sigma_{ij} = \underline{x}'\mathbf{V}\underline{x}. \tag{21.11}$$

(b) *Presence of a risk-free asset*

Here, an asset denoted 0, of certain return r, exists. Moreover, as in (a), N risky assets exist and are characterized by $\underline{R}, \mu$ (vectors of $\mathbb{R}^N$) and $\mathbf{V}$ (variance-covariance matrix $N \times N$).

A given portfolio P can then be defined in two equivalent ways:

- Either by the "number of securities" vector: $\underline{n} = (n_0, n_1 \ldots, n_N)'$, $(N+1)$-dimensional;
- Or, up to a scale factor, by the weight vector of only the risky assets:

$\underline{x} = (x_1 \ldots, x_N)'$, is an N-dimensional column vector. The weight x_0 on the risk-free asset is deduced from $\underline{x}$ since the sum of the $N+1$ weights is equal to 1:

$$x_0 = 1 - \underline{x}'\underline{1} \tag{21.12}$$

The fundamental property according to which the return on the portfolio is equal to the weighted average of the return on its components writes, in this case:

$$R_P = x_0 r + \underline{x}'\underline{R} = r + \underline{x}'(\underline{R} - r\underline{1}) \tag{21.13}$$

Therefore:

$$\mu_p = r + \underline{x}'\left(\mu - r\underline{1}\right) \tag{21.14}$$

$$\sigma^2{}_P = \underline{x}'\mathbf{V}\underline{x} \tag{21.15}$$

Note that Eqs. (21.13) and (21.14) differ from their counterparts (21.9) and (21.10) while Eq. (21.15) is identical to (21.11).

21.3.1.3 Properties of the Variance–Covariance Matrix and Concept of Asset Redundancy

The variance-covariance matrix $\mathbf{V}$, whose general term is $\sigma_{ij} = cov(R_i, R_j)$, always has the following properties:

- It is symmetrical because $cov(R_i, R_j) = cov(R_j, R_i)$;
- It is semi-definite positive (s.d.p.). Let us remember that $\mathbf{V}$ is s.d.p. if for any $\underline{x} \neq \underline{0} \in \mathbb{R}^N : \underline{x}'\mathbf{V}\underline{x} \geq 0$

21.3 Mathematical Analysis of Efficient Portfolio Choices

V is thus s.d.p because, for any $\underline{x} \in \mathbb{R}^N : \underline{x}' V \underline{x} = \sum_{i=1}^{N} \sum_{j=1}^{N} x_i x_j \, \sigma_{ij} = var(\underline{x}'\underline{R}) \geq 0$

(a variance cannot be negative).

Let us also remember that the matrix V is definite positive (d.p.) if, for any $x \neq \underline{0} \in \mathbb{R}^N : \underline{x}'V\underline{x} > 0$; it is a stronger condition than that which defines a s.d.p matrix and we can show that a s.d.p matrix is d.p. if and only if it is invertible. The character d.p. of the variance-covariance matrix V has a financial interpretation in terms of redundant assets.

Definition *We say that the risky assets* $1,\ldots, N$ *are redundant if there exist* $N + 1$ *scalars* $\lambda_0, \lambda_1, \ldots, \lambda_N$, *such that:* $\sum_{i=1}^{N} \lambda_i R_i = \lambda_0.$

By undertaking, if necessary, a change of scale of the λ_i, we can assume without loss of generality that $\sum_{i=1}^{N} \lambda_i = 1$ and interpret $\lambda = (\lambda_1, \ldots, \lambda_N)'$ as the weights of a portfolio composed of N risky assets. The return $\lambda'\underline{R}$ on such portfolio is risk-free (because $\lambda'\underline{R} = \lambda_0$). This portfolio, therefore, replicates the risk-free asset; moreover, if the latter exists and yields the interest rate r, the absence of arbitrage opportunity implies: $\lambda_0 = r$.

As already noted, the concept of redundancy has become fundamental in modern finance and is used in particular to evaluate redundant assets by arbitrage. Many redundant assets are the subject of important developments in this book. Examples are: a European call and a European put of the same strike and maturity which together with their underlying asset constitute a set of three redundant securities making it possible to replicate the riskless asset; a forward contract and its underlying; a share of a mutual fund and the securities comprising the portfolio of the latter.

The following proposition makes it possible to decide on the redundancy or not of N risky assets from the variance-covariance matrix of their returns.

Non-redundancy criterion *The necessary and sufficient condition for the N risky assets* $i = 1, \ldots, N$ *not to be redundant is that the variance-covariance matrix of their returns is positive definite.*

Proof

The N assets are not redundant $\Leftrightarrow$ variance $(\lambda'\underline{R}) > 0$ for any $\lambda \neq 0$ (because otherwise $\lambda'\underline{R} = \lambda_0$ not random for at least one $\lambda \neq 0$) $\Leftrightarrow$ V d.p.

In the following we will assume that the N risky assets are non-redundant, so that V is d.p. and therefore invertible.

21.3.1.4 Definition of Efficient Portfolios

Definition *A portfolio $\underline{x}^*$ is efficient if, for any portfolio $\underline{y}$: $\sigma_y < \sigma_{x*}$ implies $\mu_y < \mu_{x*}$ and $\sigma_y = \sigma_{x*}$ implies $\mu_y \leq \mu_{x*}$.*

In accordance with the previous notations, μ_{x*} and μ_y represent the expected return on the two portfolios $\underline{x}^*$ and $\underline{y}$ whereas σ_{x*} and σ_y denote the standard deviations of the two returns.

Similarly $\underline{x}^*$ is efficient if for any portfolio $\underline{y}$: $\mu_y > \mu_{x*}$ implies $\sigma_y > \sigma_x^*$ and $\mu_y = \mu_{x*}$ implies $\sigma_y \geq \sigma_{x*}$.

21.3.2 Efficient Portfolios and Portfolio Choice in the Absence of a Risk-Free Asset and of Portfolio Constraints

By definition, an efficient portfolio $\underline{x}^*$ solves the following optimization program:

$$\text{Max } E(R_x) \text{ under the constraint } \text{var}(R_x) = \sigma^2{}_{x^*} \equiv k$$

In the absence of a risk-free asset and constraints on the sign of the positions (short positions are allowed), this program is written in terms of weights:

$$\underset{\underline{x}\in\mathbf{R}^N}{\text{Max }} \underline{x}'\underline{\mu}, \text{ under two constraints} : \underline{x}'V\underline{x} = k; \quad \underline{x}'\underline{1} = 1. \quad \text{(P)}$$

21.3.2.1 First Order Conditions and General Form of the Solution to (P)

Let $L(\underline{x}, \theta, \lambda)$ be the Lagrangian of (P):

$$L(\underline{x},\theta,\lambda) = \underline{x}'\underline{\mu} - \frac{\theta}{2}\left(\underline{x}'V\underline{x} - k\right) - \lambda(\underline{x}'\underline{1} - 1)$$

where $\frac{\theta}{2}$ and λ denote the positive Lagrange multipliers associated with the two constraints.

The first-order conditions, therefore, imply that there are two positive scalars θ and λ such that the solution $\underline{x}^*$ satisfies:

$$\frac{\partial L}{\partial \underline{x}} = \underline{\mu} - \theta\, V\, \underline{x}^* - \lambda\, \underline{1} = \underline{0} \tag{21.16}$$

Moreover the two constraints must be satisfied:

$$\underline{x}^{*'}\, V\underline{x}^* = k; \tag{21.17a}$$

$$\underline{x}^{*'}\, \underline{1} = 1 \tag{21.17b}$$

21.3 Mathematical Analysis of Efficient Portfolio Choices

The first-order conditions (21.16) and (21.17a, 21.17b) form a set of $N + 2$ equations; given the concavity of the Lagrangian, they constitute necessary and sufficient conditions so that $\underline{x}^*$ is solution of (P) and yield $\underline{x}^*$ as well as θ and λ.

Indeed, since we have assumed that the N risky assets are non-redundant (they cannot, therefore, be used to synthesize the risk-free asset, which, besides, is presumed not to exist), the matrix $\mathbf{V}$ is invertible and Eq. (21.16) is equivalent to:

$$x^* = \frac{1}{\theta} \mathbf{V}^{-1} \left[\underline{\mu} - \lambda \underline{\mathbf{1}} \right] \tag{21.18}$$

The value of θ and λ can then be obtained from the two constraints (21.17a and 21.17b).

21.3.2.2 Efficient Portfolios and Quadratic Investors

It will be noted that, up to a constant, the Lagrangian of the program (P) is also that of the following program (P'):

$$\text{Max} \quad \underline{x}'\underline{\mu} - \frac{\theta}{2} \quad \underline{x}'\mathbf{V}\underline{x} \quad \text{with } \underline{x}'\underline{\mathbf{1}} = 1. \tag{P'}$$

Therefore (P) and (P') must yield the same solution provided that we choose in (P') a value of θ equal to that of the Lagrange multiplier resulting from (P), and which corresponds to the chosen level of risk $\sigma^2_{x^*} = k$.

The objective function of (P'), $E(R_x) - \frac{\theta}{2}\sigma^2(R_x)$, is that of a quadratic investor who seeks the optimal trade-off between the expectation and the variance of the portfolio return and $\frac{\theta}{2}$ may be interpreted as their risk aversion (because the variance weighs all the heavier, compared to the expectation, because θ is high).[13]

The first-order conditions of (P') are still (21.16) and the form of the solution is expressed by Eq. (21.18), as well as for the program (P); but θ is a *known parameter* in (P'), not an *unknown* as in (P).

In fact (P') makes it possible to characterize all the efficient portfolios: for each $\theta > 0$, (P') corresponds to a particular efficient portfolio, x_θ, which will be chosen by the investor whose risk aversion is characterized by θ. Conversely, any efficient portfolio is a solution of (P') for a particular value of $\theta > 0$.

The following proposition precisely defines the portfolio $\underline{x}_\theta$, solution to (P'):

Proposition 1
The solution $\underline{x}_\theta$ to (P') writes:

---

[13] This is true only in a static, one-period, model. In a dynamic model (in continuous or discrete time), quadratic utility and mean-variance criterion are not in general equivalent (except under the highly unrealistic assumption that the portfolio variance is constant over time).

$$\underline{x}_\theta = \underline{k}_1 + \widehat{\theta}\underline{k}_2 \qquad (21.19)$$

where $\widehat{\theta} \equiv \frac{1}{\theta}$ is the investor's risk tolerance

$$\underline{k}_1 \equiv \frac{\mathbf{V}^{-1}\mathbf{1}}{\mathbf{1}'\mathbf{V}^{-1}\mathbf{1}} \qquad (21.20)$$

$\underline{k}_1$ *is the vector of the minimum variance portfolio weights;*

$$\underline{k}_2 \equiv \mathbf{V}^{-1} \left[\underline{\mu} - \frac{\mathbf{1}'\mathbf{V}^{-1}\underline{\mu}}{\mathbf{1}'\mathbf{V}^{-1}\mathbf{1}} \, \underline{\mathbf{1}} \right] \qquad (21.21)$$

$\underline{k}_2$ *represents a "position" whose sum of weights is zero* ($\underline{\mathbf{1}}' \underline{k}_2 = 0$).
The set of efficient portfolios is therefore: { $\underline{k}_1 + \widehat{\theta}\underline{k}_2 \mid \widehat{\theta} > 0$ }.

Proof

Let us consider first the program whose solution $\underline{v}$ is the portfolio of minimal variance Min $\underline{x}'\mathbf{V}\underline{x}$ under the constraint $\underline{x}' \, \underline{\mathbf{1}} = 1$ and show that $\underline{v}$ is indeed $\underline{k}_1$ given by Eq. (21.20).

If we call 2γ the Lagrange multiplier associated with the constraint, the Lagrangian is:
$\underline{x}'\mathbf{V}\underline{x} - 2\gamma \, \underline{x}'\underline{\mathbf{1}}$ and the solution $\underline{v}$ obeys the first-order conditions:

$$\mathbf{V} \underline{v} = \gamma \, \underline{\mathbf{1}} \quad \Leftrightarrow \quad \underline{v} = \gamma \, \mathbf{V}^{-1} \, \underline{\mathbf{1}}$$

The constraint implies: $\underline{\mathbf{1}}'\underline{v} = 1$, therefore: $\gamma \, \underline{\mathbf{1}}' \, \mathbf{V}^{-1} \, \underline{\mathbf{1}} = 1$, or: $\gamma = \frac{1}{\mathbf{1}' \, \mathbf{V}^{-1} \, \mathbf{1}}$.
Thus: $\underline{v} = \frac{\mathbf{V}^{-1} \, \mathbf{1}}{\mathbf{1}' \, \mathbf{V}^{-1} \, \mathbf{1}}$, that is $\underline{k}_1$ in accordance with Eq. (21.20).

Now let us rewrite the first-order condition in the form (21.18) which implies that the solution x_θ of (P') (and of (P)) satisfies: $\underline{x}_\theta = \widehat{\theta}\mathbf{V}^{-1} \, (\underline{\mu} - \lambda\underline{\mathbf{1}})$ with $\widehat{\theta} \equiv \frac{1}{\theta}$

The multiplier λ is obtained by writing that $\underline{x}_\theta$ satisfies the constraint of (P'):

$$\underline{\mathbf{1}}'\underline{x}_\theta = \theta\underline{\mathbf{1}}' \, \mathbf{V}^{-1} \, (\underline{\mu} - \lambda \, \underline{\mathbf{1}}) = 1, \text{therefore}: \lambda = \frac{\mathbf{1}'\mathbf{V}^{-1} \, \mu - \theta}{\mathbf{1}'\mathbf{V}^{-1} \, \mathbf{1}}$$

Thus: $\underline{x}_\theta = \widehat{\theta} \, \mathbf{V}^{-1}[\underline{\mu} + \frac{\theta - \underline{\mathbf{1}}'\mathbf{V}^{-1}\mu}{\mathbf{1}'\mathbf{V}^{-1}\,\mathbf{1}} \, \underline{\mathbf{1}}]$, which is Eq. (21.19), due to Eqs. (21.20) and (21.21).

Moreover, $\underline{\mathbf{1}}'\underline{k}_2 = 0$ since $\underline{\mathbf{1}}'\underline{x}_\theta = 1$ and $\underline{\mathbf{1}}'\underline{k}_1 = 1$.

Interpretation

Proposition 1 is interpreted as follows: *all investors* who follow the M-V criterion and who share the same investment horizon (instant 1) build their portfolio *from the*

21.3 Mathematical Analysis of Efficient Portfolio Choices

same portfolios $\underline{k}_1$ *and* $\underline{k}_2$. The portfolio $\underline{k}_1$ is the minimum variance portfolio that is chosen by an investor whose risk aversion is infinite ($\widehat{\theta} = 0$) and thereby minimizes the variance without considering the expected return. A "normal" individual, characterized by a tolerance for risk $\widehat{\theta}$, "adds" the position $\widehat{\theta}\underline{k}_2$ to $\underline{k}_1$, which increases his risk but also his expected return. Since $\mathbf{1'}\underline{k}_2 = 0$, $\underline{k}_2$ is a "pseudo-portfolio" or rather a vector of positions with *a sum of weights equal to zero,* which means that it requires no initial investment, with short positions financing long positions.

Note that the relation (21.19) is equivalent to:

$$\underline{x}_\theta = (1 - \theta)\,\underline{k}_1 + \widehat{\theta}\underline{s} \tag{21.19'}$$

where $\underline{s} \equiv \underline{k}_1 + \underline{k}_2$ is an efficient portfolio (chosen by the investor whose risk tolerance $\widehat{\theta} = 1$), whose sum of weights is equal to 1 (because $\mathbf{1's} = \mathbf{1'}\underline{k}_1 + \mathbf{1'}\underline{k}_2 = 1$), that can be shown to be the one that maximizes the *Sharpe ratio* $(E(R)/\sigma(R))$.[14]

Finally, relation (21.19) also implies that the locus of the representative points of efficient portfolios, or efficient frontier, is:

- A parabola in the space (σ^2, μ)
- A hyperbola in the space (σ, μ)

Indeed, it follows directly from Eq. (21.19) that the efficient frontier, in the space (σ^2, μ), has for parametric equations those of a parabola:

$$\mathrm{E}(R_\theta) = a + \widehat{\theta}b; \quad \sigma^2(R_\theta) = c + 2\widehat{\theta}d + \widehat{\theta}^2\,e$$

with: $\widehat{\theta} \equiv \frac{1}{\theta}$; $a \equiv \mathrm{E}(R_{k1})$; $b \equiv E(R_{k2})$; $c \equiv \sigma^2(R_{k1})$; $d \equiv cov(R_{k1}, R_{k2})$; $e \equiv \sigma^2(R_{k2})$.

Replacing $\widehat{\theta}$ by $\frac{E(R_\theta)-a}{b}$ in the expression of $\sigma^2(R_\theta)$ we get a parabolic relation linking $\sigma^2(R_\theta)$ to $\mathrm{E}(R_\theta)$ (and a hyperbolic relation linking $\sigma(R_\theta)$ to $\mathrm{E}(R_\theta)$).

Now let us rewrite the vector equation (21.16) "row by row":

$$\mu_i - \lambda = \theta \sum_{j=1}^{N} x_j^* \, \mathrm{cov}\big(R_i, R_j\big) \text{ for } i = 1, \ldots, N \tag{21.16'}$$

These conditions lead, almost directly, to an efficiency criterion which, given its importance (notably for the justification of the CAPM in the next chapter), will be formulated in the form of a proposition.

Proposition 2

The necessary and sufficient condition for any given portfolio $\underline{x}^*$ to be efficient is that there exist two scalars $\theta \geq 0$ and $\lambda \geq 0$ such that, for any security $i = 1, \ldots, N$:

---

[14] In the presence of a risk-free asset, the Sharpe ratio is equal to $(E(R)-r)/\sigma(R)$; in the absence of a risk-free asset it can be defined as $E(R)/\sigma(R)$.

$$\mu_i \lambda = \theta \, \mathrm{cov}(R_i, R_{x^*}) \tag{21.22}$$

Note that both scalars θ and λ depend on $\underline{x}^*$ but, for a given portfolio $\underline{x}^*$, are the same for each of the N securities i.

Proof
Starting from Eq. (21.16'), x^* is efficient if and only if $\exists \, \theta \geq 0$ and $\lambda \geq 0$ such that:

$$\mu_i \lambda = \theta \sum_{j=1}^{N} x_j^* \, \mathrm{cov} \, (R_i, R_j) = \theta \, \mathrm{cov} \, (R_i, \sum_{j=1}^{N} x_j^* \, R_j) = \theta \, \mathrm{cov}(R_i, R_{x^*})$$

(using the bi-linearity of the covariance operator).

$\mu_i - \lambda$ is interpreted as an expected excess return (compared to λ) and $cov(R_i, R_{x^*})$ as the contribution of security i to the risk of the portfolio x^*; indeed, again using the bi-linearity of the covariance operator:

$$\sigma^2_{x^*} = \sum_{i=1}^{N} \sum_{j=1}^{N} x_i^* x_j^* \sigma_{ij} = \sum_{j=1}^{N} x_j^* \, \mathrm{cov}(R_i, \sum_{j=1}^{N} x_j^* \, R_j)$$

$$\text{i.e.}: \qquad \sigma^2_{x^*} = \sum_{j=1}^{N} x_j^* \, \mathrm{cov}(R_i, R_{x^*}).$$

The term $cov(R_i, R_{x^*})$ does appear as the contribution of the security i (per unit of weight) to the overall risk $\sigma^2_{x^*}$ that affects R_{x^*}.

Proposition 2 can be interpreted as follows: a portfolio $\underline{x}^*$ is efficient if and only if its weight is allocated so that the excess return $(\mu_i - \lambda)$ of the different securities is proportional to the risk that they induce (equal to $cov(R_i, R_{x^*})$ for asset i): hence, *an affine linear relationship links the expected return of any individual security to the covariance of its return with that of any efficient portfolio.*

21.3.2.3 The Two-Fund Separation
The first-order conditions (21.16) lead to the two-fund separation theorem of Fischer Black (1972).

Proposition 3
Black's two-fund separation theorem
    Consider any two efficient portfolios $\underline{a}$ and $\underline{b}$.

(i) *Any convex combination $u\underline{a} + (1 - u)\underline{b}$ with $u \in [0,1]$ is an efficient portfolio;*
(ii) *Any efficient portfolio $\underline{e}$ is a combination (not necessarily convex) of the two portfolios $\underline{a}$ and $\underline{b}$.*

Proof
Let $\underline{a}$ and $\underline{b}$ be *any* two efficient portfolios. According to Proposition 2, there are two positive scalars θ_a and θ_b such that: $\underline{a} = \underline{k}_1 + \theta_a \, \underline{k}_2$; $\underline{b} = \underline{k}_1 + \theta_b \, \underline{k}_2$.

21.3 Mathematical Analysis of Efficient Portfolio Choices

(i) Consider a convex combination $\underline{x}$ of $\underline{a}$ and $\underline{b}$, that is, long on $\underline{a}$ and on $\underline{b}$ such that: $\underline{x} = u\underline{a} + (1-u)\underline{b}$ with $u \in [0,1]$, i.e.:

$$\underline{x} = \underline{k}_1 + (u\theta_a + (1-u)\theta_b)\underline{k}_2.$$

We have $\theta_x = u\,\theta_a + (1-u)\,\theta_b$ and note that $\theta_x \geq 0$ because $\theta_a \geq 0$, $\theta_b \geq 0$, $u \geq 0$ and $(1-u) \geq 0$.

Therefore, according to proposition 2, $\underline{x} = \underline{k}_1 + \theta_x\,\underline{k}_2$ is efficient.

(ii) Now consider any efficient portfolio $\underline{e}$ and let us show that it is written as a combination of $\underline{a}$ and $\underline{b}$ (not necessarily convex). Since $\underline{e}$ is efficient, there exists $\theta_e \geq 0$ such that: $\underline{e} = \underline{k}_1 + \theta_e\,\underline{k}_2$.

Consider a combination of $\underline{a}$ and $\underline{b}$: $u\underline{a} + (1-u)\underline{b} = \underline{k}_1 + (u\theta_a + (1-u)\theta_b)\underline{k}_2$.
For this portfolio to replicate $\underline{e}$, we just need to choose u such that:

$$u\theta_a + (1-u)\theta_b = \theta_e, \text{i.e.} : u = \frac{\theta_e - \theta_b}{\theta_a - \theta_b}.$$

More generally, it can be shown that any portfolio corresponding to any point on the hyperbola (including those of the lower branch representative of the most inefficient portfolios) is a combination of $\underline{a}$ and $\underline{b}$.

Comments and interpretations

Proposition 3-(i) implies that any portfolio that is a combination of the two efficient funds $\underline{a}$ and $\underline{b}$ with positive weight on these two funds, such that $\underline{e}$ *inside* the curvilinear segment $(\underline{a}, \underline{b})$, is efficient, therefore located on the EF efficient frontier (see Fig. 21.9).

In contrast, a non-convex combination involving weights $u < 0$ and $(1-u) > 1$ (like $\underline{f}$), is not necessarily efficient ($\underline{f}$ is on the lower branch of the representative hyperbola of *the most inefficient portfolios*).

Proposition 3-(ii), reciprocal of (i), states that any efficient portfolio (such as $\underline{e}$ or $\underline{e}'$) is a combination of $\underline{a}$ and $\underline{b}$ but the weights allocated to these two portfolios are not necessarily positive (the efficient portfolio $\underline{e}'$, for example, is obtained by using a short position on $\underline{a}$; $u < 0$, $(1-u) > 1$).

In addition, it can be shown that all the portfolios of the hyperbola (upper and lower branches) are generated by the combinations of any two portfolios of the hyperbola.

This result of two-fund separation is very important because it means that all M-V investors sharing the same investment horizon are indifferent between a situation in which they can trade on a very large number of assets ($N =$ several thousands) and a situation where the choices are reduced to the combinations of any two efficient

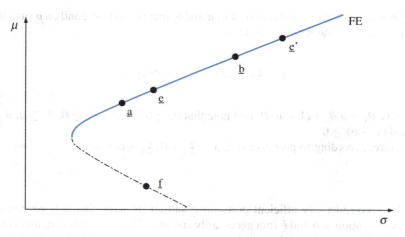

Fig. 21.9 Hyperbolic Efficient Frontier (by the way, replace, inside the figure, FE by EF)

funds only: *any two well-managed mutual funds are therefore sufficient to satisfy all investors* with a common investment horizon; they simply combine them according to weights that depend on their risk aversions.

21.3.3 Efficient Portfolios in the Presence of a Risk-Free Asset, with Allowed Short Positions; Tobin's Two-Fund Separation

We assume here that a risk-free asset (asset 0 that pays the interest rate r and is both an investment and a mean of financing) is added to the N risky assets. Remember that the portfolios are defined by the weight x^* on the risky assets only, that $x_0^* = 1 - \mathbf{1}' \underline{x}^*$, and that $\underline{\mu}$ is the vector of the expected returns on the N risky assets.

Proposition 4 sets out the main results that are obtained in this context.

Proposition 4
(i) *The efficient frontier in the space (σ, μ) is the half-line r-T starting from $(0, r)$ and tangent on T to the hyperbola representative of efficient portfolios in the absence of a risk-free asset (see Fig. 21.8).*
(ii) *Point T represents the (only) efficient portfolio $\underline{t}$ that contains only risky assets.*
(iii) *An efficient portfolio can be characterized as a combination of the tangent portfolio and the risk-free asset (Tobin's two-fund separation); more generally, any two efficient portfolios make it possible to obtain, by combination, any efficient portfolio.*
(iv) *A portfolio $\underline{x}^*$ is efficient if and only if there exists $\widehat{\theta} \geq 0$ such that:*

21.3 Mathematical Analysis of Efficient Portfolio Choices — 905

$$\underline{x}^* = \widehat{\theta}\mathbf{V}^{-1}\left[\underline{\mu} - r\,\underline{\mathbf{1}}\right] \quad \left(\text{and } \mathrm{x}_0 = 1 - \underline{\mathbf{1}}'\,\underline{x}^*\right) \tag{21.23}$$

(v) *The portfolio $\widehat{\theta}\mathbf{V}^{-1}\left[\underline{\mu} - r\,\underline{\mathbf{1}}\right]$ is chosen by the M-V investor characterized by a risk tolerance parameter equal to $\widehat{\theta}$.*

(vi) *The tangent portfolio is characterized by the weights:* $\underline{t} = \dfrac{\mathbf{V}^{-1}\left[\underline{\mu}-r\underline{\mathbf{1}}\right]}{\underline{\mathbf{1}}'\mathbf{V}^{-1}\left[\underline{\mu}-r\underline{\mathbf{1}}\right]}$ *(and $t_0 = 0$).*

Proof

Results (i), (ii), and (iii) were obtained in Sect. 21.2; we show here (iv), (v), and (vi).

Consider the investor who solves the M-V program:

$$\operatorname*{Max}_{\underline{x}}\left(1 - \underline{x}'\,\underline{\mathbf{1}}\right)r + \underline{x}'\,\underline{\mu} - \frac{\theta}{2}\underline{x}'\,\mathbf{V}\underline{x} \quad \text{with } \theta > 0, \text{ without constraint.}$$

Note that no constraint on weights is required here because $\underline{x}$ represents allocations on risky assets only and $x_0 = 1 - \underline{x}'\underline{\mathbf{1}}$ can take any positive or negative values.

At the optimum, $\underline{x}^*$ therefore satisfies: $-r\,\underline{\mathbf{1}} + \underline{\mu} - \theta\,\mathbf{V}\,\underline{x}^* = 0$, hence, with still $\widehat{\theta} = \frac{1}{\theta}$:

$$\underline{x}^* = \widehat{\theta}\mathbf{V}^{-1}\left[\underline{\mu} - r\underline{\mathbf{1}}\right], \quad \text{which proves (iv) and (v).}$$

The tangent portfolio being itself efficient, there exists $\widehat{\theta}_t > 0$ such that: $\underline{t} = \widehat{\theta}_t\mathbf{V}^{-1}$ $(\underline{\mu} - r\underline{\mathbf{1}})$.

The tangent portfolio containing only risky assets: $\underline{t}'\underline{\mathbf{1}} = 1$, we have:

$$\widehat{\theta}_t = \frac{1}{\underline{\mathbf{1}}'\mathbf{V}^{-1}\left(\underline{\mu}-r\underline{\mathbf{1}}\right)}, \quad \text{which proves (vi).}$$

Comments and interpretations

Section 21.2 is referred to for various comments and interpretations of items (i), (ii), and (iii). We note here the very strong analogy of *separation results prevailing in the absence and in the presence of a risk-free asset* (separation of Black (1972) and Tobin (1958), respectively). *In either case, any two efficient portfolios are sufficient to build any efficient portfolio.*

The result (iv) means that the weight vectors of all the efficient portfolios are homothetic to $\mathbf{V}^{-1}\left[\underline{\mu} - r\,\underline{\mathbf{1}}\right]$ and homothetic to each other; according to (v), the investor characterized by a tolerance $\widehat{\theta}$ simply chooses the weights $\widehat{\theta}\mathbf{V}^{-1}\left[\underline{\mu} - r\,\underline{\mathbf{1}}\right]$ (these are the weights on the risky assets only).

The result (vi) gives the exact composition of the tangent portfolio $\underline{t}$ which allows us to characterize any efficient portfolio in two equivalent ways:

- As a portfolio combining the risk-free asset and the tangent portfolio, in proportions that depend on the investor's risk tolerance;
- As a portfolio homothetic to the tangent portfolio $\underline{t}$.

21.4 Some Extensions of the Standard Model and Alternatives

The Markowitz model presented in the previous sections is the standard model. Its implementation as an asset allocation tool faces a major problem related to its sensitivity to the parameters. This problem, and the Black and Litterman (1992) model that attempts to answer it, are discussed in Sect. 21.4.1.

In addition, the Markowitz model is based on various assumptions that are not always true to reality:

- Short positions can be formed without constraints and on all securities.
- The investor applies the mean-variance criterion.
- The investment covers only one period; more precisely, as soon as the allocations are made at the initial time, no transaction is authorized until the end of the investment period (the strategy is called static or *buy and hold*).

We will examine the consequences of relaxing the first two assumptions (in Sects. 21.4.2 and 21.4.3), the case of dynamic strategies involving inter-temporal reallocations being covered in Chap. 24.

On the other hand, some behaviors are not correctly reflected by maximizing a concave expected utility. However, some adaptations of the postulated form for the utility function make it possible to conform to some of these behaviors. In other respects, other observations reveal preferences in contradiction with the theory of expected utility and are based on alternative approaches (behavioral finance). These issues are discussed in Sect. 21.4.4.

21.4.1 Problems Implementing the Markowitz Model; The Black-Litterman Procedure

Hereafter, we assume the existence of a risk-free asset. In its simplest form, the Markowitz model is implemented in three stages, as follows:

- The variance-covariance matrix $\mathbf{V}$ is econometrically estimated from a history of returns (Chapter 23 describes how such an estimate can be conducted, notably using factorial models).

21.4 Some Extensions of the Standard Model and Alternatives 907

– Expected returns μ are estimated from the views of financial analysts.
– *The desired portfolio $\underline{x}^*$ is homothetic to the tangent portfolio:* $\underline{x}^* = \dfrac{\mathbf{V}^{-1}[\underline{\mu}-r\underline{1}]}{\theta_x}$

The main practical difficulty of such a procedure comes from the *high sensitivity* of the structure $\mathbf{V}^{-1}(\underline{\mu} - r\underline{1})$ of this portfolio to the choice of parameters, notably to the presumed expected returns $\underline{\mu}$ of risky securities. Indeed, a slight modification of $\underline{\mu}$ can result in a very strong variation of the optimal weights. Moreover, for a broad spectrum of values of $\underline{\mu}$, the structure of optimal portfolios can be very "unrealistic": Weights way greater than 1 on some stocks and large short positions on others, for example. This drawback is all the more serious as the estimates of $\underline{\mu}$ are inevitably questionable and differ from one analyst to another.

Black and Litterman's model (1992), developed within Goldman Sachs, attempts to provide practical answers to these problems. Crudely, it is structured as follows.

– It is based on estimates that are implicit in the composition of the market portfolio. We describe in detail this market portfolio in the next chapter and limit ourselves here to characterize it by its weight $\underline{m}$ in line with the market average (the weight of a security is the weight of its market value in the total market capitalization: $m_i = V_i / \sum_j V_j$). This "average" portfolio would be chosen by an "average investor" whose expected returns are equal to $\underline{\mu}_m$ (μ_{mi} for security i). These "average" expectations can be estimated in several ways, notably with the help of the CAPM (see Chap. 22).[15]
– Moreover, the financial analysts that the manager (or investor) follows, have different views characterized by μ_{ki} for the analyst k and μ_{ai} for the average μ_{ki}. The final parameter μ_i that is used in the calculations is a weighted average of market expectations μ_{mi} and μ_{ai} of the analysts. More precisely, one can write: $\mu_i = (1-\lambda\gamma_i)\mu_{mi} + \lambda\gamma_i\mu_{ai}$. The weight coefficient $\lambda\gamma_i$ assigned to the analysts for the security i depends on two elements: an overall relative confidence coefficient λ which is granted to them, between 0 and 1 and identical for all the securities (with a coefficient $\lambda = 0$, market expectations are used; with $\lambda = 1$, the weight given to analysts in the estimate of μ_i is γ_i). The coefficient γ_i is specific to each security and reflects the agreement between the different analysts k; it can be taken inversely proportional to the empirical variance of μ_{ki}: the greater the consensus between analysts, the lower the variance, so the higher γ_i and the weight attributed to the opinion of analysts.

---

[15] We can also start from: $\underline{m} = \dfrac{\mathbf{V}^{-1}\left[\underline{\mu}_m - r\underline{1}\right]}{\theta_m}$, where θ_m represents the average market risk aversion (Eq. 21.23 applied to the "average investor" who holds the market portfolio). It follows that: $(\underline{\mu}_m - r\underline{1}) = \theta_m \mathbf{V}^{-1}\underline{m}$. The value of θ_m is also estimated to be between 1 and 2 depending on the context (using for instance a risk premium $E(R_M) - r = 6\%$ and a market volatility $= 20\%$, we obtain $\theta_m = E(R_M) - r) / \sigma^2_m = 1.5$).

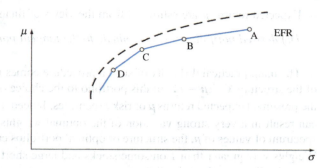

Fig. 21.10 Two efficient frontiers

21.4.2 Ban on Short Positions

In previous analyses, short positions were presumed to be allowed. When they are not, the choices are obviously limited and the previous results no longer hold. In fact, no analytical result can be obtained in this case, but we can use numerical methods. The latter give the composition of effective portfolios that can be described qualitatively, as in what follows. We successively examine the cases of absence and presence of a risk-free asset.

21.4.2.1 Absence of a Risk-Free Asset

In this case, the efficient frontier is no longer a hyperbola such as that shown in Figs. 21.7, 21.8 or 21.9. In fact, bans on short positions make the risk/return trade-off less favorable: for the same standard deviation, the expected return will be lower in the presence of constraints. Geometrically, this situation, which is less advantageous, results in an efficient frontier southeast of the one prevailing in the absence of constraints, in accordance with Fig. 21.10 where the non-constrained frontier is dashed and the frontier in their presence is in full, blue line (it is the efficient frontiers in the absence of a risk-free asset; note that the two frontiers could partially coincide).

We will qualitatively describe the composition of efficient portfolios ranging from the riskiest (which is also the one that gives the highest expected return) to the less risky (from A to D in the figure). In the absence of a risk-free asset, the portfolio whose expectation is maximum (it is therefore efficient) is A, consisting *exclusively* of one security, called 1: the one whose expected return is maximum (it is, therefore, impossible to build a position to the right of A). To obtain a return with a standard deviation slightly lower than $\sigma_A = \sigma(R_1)$, at least one additional security 2 must be introduced; the combinations of 1 and 2 are represented by the points of the branch of hyperbola AB, the shifts from A towards B being obtained by an increase of the weight allocated to 2. If one wishes to obtain a standard deviation lower than σ_B, it is optimal to introduce new securities. Risk reduction is permitted by combining these securities up to reaching σ_C, from which further risk reduction involves a new set of securities, and so on. In the absence of a risk-free asset, the efficient frontier is thus made up of a series of hyperbolic segments AB, BC, CD, ..., forming a continuous

EFR (Efficient Frontier without a Risk-free asset) curve. Portfolios A, B, C, from which a risk reduction involves the introduction of new securities, are called *corner portfolios*.

21.4.2.2 Presence of a Risk-Free Asset

In the presence of the riskless asset n° 0, allowing to lend and borrow at rate r, represented in Fig. 21.11 by the ordinate point r on the vertical axis, the efficient frontier is the tangent r-T to EFR (EFR is the succession of hyperbola branches described in Sect. 21.4.2.1). When the borrowing and lending rates are equal, points on r-T to the right of T are achievable, all efficient portfolios are combinations of the tangent portfolio and the risk-free asset and a result of two-fund separation prevails, as in the standard case where short positions are permitted: the efficient frontier then is the full half-line going from r to T and beyond T. This is not the case when short positions on asset n° 0 (i.e., borrowing intended to finance the acquisition of risky securities) are prohibited: the efficient frontier is then r-TA constituted by the segment rT extended by the succession of hyperbola branches starting from T (see Fig. 21.11).

When borrowing is possible, but at a higher rate r' than lending ($r' > r$), the efficient frontier is rT-TT'-$T'E$, i.e., the segment rT extended by one or several branches of hyperbola such that TT', then by the half-line $T'E$, where T' is the point of contact with EFR of its tangent starting from r'. If the amount of borrowing is limited (which amounts to having a constraint on the short position on the risk-free asset, yet at a non-zero level), only a segment to the right of T', such as $T'E$, is attainable.

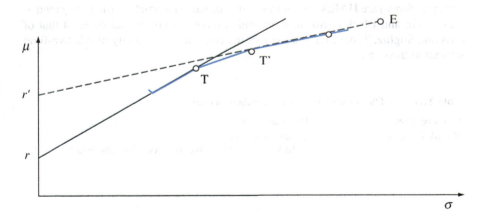

Fig. 21.11 Presence of risk-free assets

21.4.3 Separation Results When Investors Maximize Expected Utility But Do Not Follow the Mean-Variance Criterion (Cass and Stiglitz)

The results of Black and Tobin, presented above, are the most well-known separation theorems usually taught in basic finance courses. They apply when all investors are presumed:

(i) To use a mean-variance criterion in their portfolio selection (or if asset returns are normal)
(ii) To have an identical investment horizon, typically 1 year
(iii) To implement static portfolio strategies with no revision of any kind (*buy and hold*)
(iv) To have homogeneous expectations (same estimates of expectations, variances, and correlations of returns)

Assumptions (ii), (iii), and (iv) will be relaxed in Chap. 24 and more general separation results presented. In this paragraph, we simply analyze the consequences of a more general and more "rational" behavior: that of investors who do not follow the M-V criterion but do maximize an expected utility.

Cass and Stiglitz (1970) generalized Black–Tobin's separation theorem and characterized the conditions required to obtain a two-fund separation theorem that prevails when investors maximize the expectation of a HARA utility function. Let us remember that a utility function is of this type if it writes: $U(X) = \frac{\gamma}{1-\gamma} \left(\frac{X-\theta}{\gamma} \right)^{1-\gamma}$ with some restrictions on the coefficients and domains of definition.

Cass and Stiglitz have shown that all HARA investors *sharing the same parameter γ* (but differing in their initial wealth and the parameter θ) can choose their portfolio optimally by combining the same two funds.[16] The risk-free asset, if it exists, can be chosen as one of the two funds. Given that all the quadratic investors (mean-variance) are HARA investors sharing the same parameter γ (making γ tend to -1), Tobin and Black's two-fund separation theorems are special cases of that of Cass and Stiglitz. Table 21.1 summarizes the conditions of validity of this two-fund separation theorem.

Table 21.1 Conditions for the two-fund separation theorem to hold

Risk-free asset	HARA functions
No risk-free asset	Quadratic ($\gamma = -1$)
	or CRRA (Constant Relative Risk Aversion, obtained for $\theta = 0$)

---

[16]See Ingersoll (1987) for an in-depth presentation of the results of Cass and Stiglitz.

21.4.4 Loss Aversion and Introduction to Behavioral Finance

Some observations and experiments highlight the fact that the choices under uncertainty of many individuals and in many situations are not faithfully described by maximizing the expectation of a concave utility function. Some types of behavior can be described by maximizing expected utility, provided that the concavity assumption is relaxed for all values of wealth. Other behaviors, although widespread as many surveys show, violate axioms of rationality (such as the axiom of independence or of transitivity). In addition, individuals process information in a biased manner, which distorts their subjective probabilities. These behaviors led to alternative theories of attitude towards risk based not on the presumed rationality of economic agents but on their observed behavior. These theories, among which the best known are Kahneman and Tversky's prospect theory and Thaler's research, are in the realm of behavioral finance. We first present *loss aversion*, which can be obtained by simply adapting the standard form of the utility function and that allows to explain many observed behaviors. We then discuss the approach of behavioral finance.

21.4.4.1 Loss Aversion

There are two characteristics of many individuals' behavior with respect to risk that cannot be explained by the simple concavity of their utility function:

- They perceive a loss and a gain of equal amplitude asymmetrically,[17] the (negative) effect of a welfare loss being stronger than the (positive) effect of a gain of the same amount. This phenomenon has been described as loss aversion.[18]
- Some individuals look for risk when it comes to losses while they are allergic to risk when it comes to earnings. This explains the behavior of some individuals towards gambling (after registering losses, they gamble even more in an attempt to recoup these losses).

These two attitudes towards risk are reflected by the following utility function:

$$U(W) = \begin{cases} V(W) & \text{for } W \geq W^* \\ \\ V(W) + L(W) & \text{for } W < W^* \end{cases} \tag{21.24}$$

where W^* is a tolerance threshold, called the reference level, below which losses are felt specifically. $V(.)$ is a standard, increasing and concave or linear utility function. The function $L(.)$, which expresses the attitude towards losses, is increasing and *convex*. Its increasing character reflects loss aversion ($L(.)$ accentuates the "disutility" of a loss); its convexity reflects the attractiveness of the risk that manifests itself

---

[17] Beyond what can be explained by the concavity of their utility function.

[18] See Jarrow and Zhao (2006) for an analysis of loss aversion and the resulting portfolio decision.

Fig. 21.12 Loss aversion

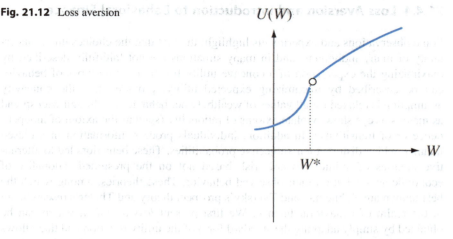

in a context of losses. If we want U to be continuous at the point W^*, we must have $L(W^*) = 0$. Moreover, $L'(W^*) > 0$ is often assumed so that utility is kinky at point W^* (the marginal utility is greater for $W < W^*$ than for $W > W^*$). Figure 21.12 reflects such a situation.

The expression (21.24) of the utility function implies that its expectation writes:

$$E(U(W)) = E(V(W)) + \int_{-\infty}^{W^*} L(w) dF_W(w)$$

F_W is the distribution function of W and the term $\int_{-\infty}^{W^*} L(w) dF_W(w)$ is called the *Downside Risk Measure* or DRM.

The solution to the program Max $E[U(W)]$ where U is given by Eq. (21.24) is also solution to the program:

$$\text{Max } E[V(W)] \text{ under the constraint}: DRM(W) = k.$$

Different approaches and models are special cases of Eq. (21.24).
- The prospect theory of Kahneman and Tversky (see next paragraph), with:

$$V(W) = W - W^* \text{ and } L(W) = b(W - W^*);$$

- The models with partial lower moments with $L(W) = -k(W^* - W)^m$ and $k > 0$; in particular:

21.4 Some Extensions of the Standard Model and Alternatives

- The semi-variance model with $V(W) = W$, $W^* = 0$ and $L(W) = -kW^2$;[19]
- The models where risk is measured by the *Expected Shortfall*, also called *Conditional VaR* (*Value at Risk*, see Chap. 27), with $L(W) = W - W^*$ and $W^* = -VaR$.[20]

21.4.4.2 Elements of Behavioral Finance[21]

Various observations on the behavior of economic agents reveal that, in addition to the previously mentioned loss aversion, they often depart from some of the principles of rationality on which the standard theory is based.

Although these "biases" have been observed for a long time,[22] behavioral finance is considered to have been initiated by Kahneman and Tversky (1973, 1979, 1992 ...), designated by KT below. Behavioral finance attempts to structure the observations collected "in the field and in the laboratory" and to construct a descriptive theory of choices under uncertainty according to the way economic agents actually behave. This approach can be described as pragmatic and psychological.

Various reasons have been put forward for explaining the differences between observed behavior and expected utility theory. Some lie in the "irrationality" of the implementation of preferences and others in the way individuals perceive or process information. Among the first ones are:

- The specific "pain," "regret," or "psychic cost" caused by a wrong decision;
- The "money illusions" that can affect preferences depending on how the alternative choices are presented.

As for the errors of appreciation of "probabilities," they find their source in the lack of information and analytical difficulties which make it difficult to assess the "objective" probabilities.[23]

Thus, behavioral finance has highlighted a number of "typical situations" in which the attitude of a large number of individuals departs from the expected utility theory in its standard form, i.e., in terms of the exercise of preferences, or on the assessment of probabilities.

---

[19] Which is increasing with W for $W < 0$ $[L(W) = -k(-W)^2]$.

[20] It will be seen in Chap. 27 that the coherent measure of ES_p risk (*Expected Shortfall* at probability level p) writes $ES_p = E(W \mid W < -VaR_p) = \frac{1}{p} \int_{-\infty}^{-VaR(p)} w dF_W(w)$; therefore, if $L(W) = W - W^*$ and $W^* = -VaR_p$, we have: $DRM = \int_{-\infty}^{-VaR(p)} (w + VaR_p) dF_W(w) = p\, ES_p + p VaR_p$.

[21] The presentation of behavioral finance that follows being sketchy only, the interested reader is invited to refer to the suggestions for further reading at the end of the chapter for supplements on this issue, in particular Thaler (1993, 2015).

[22] Maurice Allais (Nobel laureate in Economics in 1988) highlighted such "biases" as early as 1953.

[23] We thus distinguish situations of risk from situations of uncertainty: the "objective" probabilities are known in the first and unknown in the second. This important distinction is due to Frank Knight (1921).

Five of the biases, highlighted by KT, often affect the exercise of preferences.

- Choices among risky alternatives depend on how their consequences are presented; for example, if the results of the choices are presented as random gains, individuals tend to be more risk-averse than if they are presented as random losses. This observation leads KT to introduce the concept of *"frame."*
- Financial flows are treated differently according to their origin or their assignment, as if the individuals managed compartmentalized budgets, allocated to their various activities by "mental accounting." Thus, individuals who lose their cinema ticket give up going to the movies, while if they lose banknotes of a value equal to the price of the ticket they do not. This observation leads KT to introduce the notion of *mental accounting.*
- With identical financial consequences, the psychic cost of making a bad decision is higher than that of not having made a good one. This finding leads KT to highlight the notion of *regret.* In addition, a decision that goes wrong is all the more regretted that it is unconventional. Thus, a bad investment is "more painful" if it concerns a start-up or a small company than a large capitalization (possibly because it can be more difficult to be blamed for bad luck).
- This notion of regret can be associated with the reluctance of many individuals to sell assets on which they have suffered losses (*disposition effect*). They will therefore prefer to sell securities that have well performed rather than endorse losses by liquidating those that have not.
- A greater value is often attributed to an object that is possessed than to an object with the same market value that is not (*endowment effect*).

The "biases" in the formation of subjective probabilities often revealed by surveys include those related to:

- "Over-confidence" (individuals overestimate their skills and investments)[24]
- "Conservatism" (economic agents do not sufficiently revise their opinions and prior estimates, following new information)
- The tendency to overestimate the representativeness of small samples.

Several of the behavioral characteristics that have just been described can be integrated into a model of choice that KT have named *prospect theory.* Schematically, according to this theory, individuals maximize a "pseudo-expected" utility of gains or losses, calculated with:

- The "valuation" function U defined by Eq. (21.24) with $V(W) = W - W^*$; $L(W) = b(W - W^*)$; $b > 0$; it reflects some of the characteristics of the preferences mentioned above.

---

[24]Most of these field studies were conducted in the US; the conclusions could differ for other countries.

Fig. 21.13 Weighting of probabilities

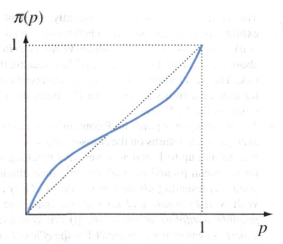

- "Probabilities" adjusted by a weighting function π that reflects some of the biases mentioned above. In particular, the function π over-weights the low probabilities, as in Fig. 21.13 below.

The objective function to maximize is thus written: $\int_{-\infty}^{+\infty} \pi(p(w))U(w)dw$.

Finally, let us note that behavioral finance is not normative but descriptive. It would obviously be absurd, for example, to choose one's portfolio with biased probabilities, but it may be useful to integrate these irrationalities to understand how markets actually behave. Thus, certain "anomalies" or "inefficiencies" found in markets could be explained by behavioral economics.

21.5 Summary

- Appropriately combining several assets in a portfolio reduces through diversification total risk for a given expected rate of return.
- Any rational economic agent may be thought as seeking to *maximize the expected utility of terminal wealth*. The utility function $U(W)$ is specific to each agent and reflects in particular their risk aversion.
- Two utility functions U and V linked by the affine linear relation $V(W) = a\,U(W) + b$, where a is a positive constant and b is any constant, lead to the same decisions.
- Utility $U(W)$ is increasing with wealth, which implies, if it is differentiable, that the marginal utility $U'(W)$ is *positive and decreasing for a risk-averse individual*.
- Decreasing $U'(W)$ implies that $U(.)$ is concave and that $U''(W)$ is negative (if the second derivative U'' exists). *Risk aversion* is *characterized* by the *concavity* of U.
- $A(W) \equiv -U''(W)/U'(W)$ is called the *absolute* risk aversion, and $R(W) \equiv W \cdot A(W)$ is the *relative* risk aversion. These coefficients are positive for risk averters, and the risk premium that individuals are ready to pay to eliminate a « small risk » can be shown to be proportional to their absolute risk aversion A.

- The utility functions more frequently encountered in the economic literature exhibit constant absolute or relative risk aversion or are quadratic: $U(W) = W - aW^2$. Investors follow the *Mean-Variance criterion* (M-V) when they only care about the mean and the variance of their wealth, the variance being the measure of risk. The M-V criterion has *several drawbacks* and is only theoretically justified for quadratic utility investors or if returns are Gaussian. The Markowitz model relies on the M-V criterion.
- The return R_P on a portfolio P containing N assets $1, \ldots, N$ is equal to the weighted average of the returns on the N assets: $R_P = x_1 R_1 + x_2 R_2 + \ldots + x_N R_N$, the weights x_i summing up to 1, and an x_i negative meaning a short position in i. Expressions for the mean μ_P and the variance σ^2_P then obtain easily. In the (σ, μ) space, the points representing *all the combinations of any two assets* form a *hyperbola*.
- With *N risky assets and no risk-free asset*, the representative points of *all the possible portfolios in space* (σ, μ) are contained in a surface delimited by a *hyperbola* whose upper branch EF is the *efficient frontier*. EF is the set of all *efficient portfolios* that yield the *higher expected return μ_P for a given risk σ_P*, hence those chosen by M-V investors. Moreover, *any* efficient portfolio can be obtained by a combination of *any given two* efficient portfolios (Black's *two-fund separation*).
- When a *risk-free asset* yielding r is added, the efficient frontier becomes the *half-line r-T*, tangent at T to the upper branch of the hyperbola EF. Tobin's *two-fund separation theorem* follows: *all efficient portfolios are combinations of the risk-free asset and the tangent portfolio T*. They differ by the weight on T which decreases with the investor's risk aversion.
- T contains risky assets only and is characterized by the weights: $t = \dfrac{\mathbf{V}^{-1}\left[\mu - r\mathbf{1}\right]}{\mathbf{1}'\mathbf{V}^{-1}\left[\mu - r\mathbf{1}\right]}$,

  with $\mathbf{V}$ the variance-covariance matrix of the risky returns and $\underline{1}$ the N-dimensional unit vector.
- In *Sharpe's market model*, the return of any security i obeys the following equation: $R_i = \mu_i + \beta_i (R_m - \mu_m) + \varepsilon_i$. The N random ε_i are assumed i.i.d. and vanish by aggregation in a diversified portfolio by the law of large numbers: they are thus diversifiable (non-systematic) risks. By contrast, the risks represented by $\beta_i R_m$ are non-diversifiable (systematic).
- Markowitz's results have been extended in several directions (such as a ban on short selling, non-M-V investors, behavioral finance). Multi-period portfolio strategies are studied in Chap. 24.

Appendix 1: The Axiomatic of Von Neuman and Morgenstern and Expected Utility

The theory of Von Neuman and Morgenstern (1947) is based on axioms concerning agents' preferences for risky alternatives. We describe the objects of choice to which these preferences relate before presenting the axioms that lead to the expected utility criterion.

Appendix 1: The Axiomatic of Von Neuman and Morgenstern and Expected Utility

A1.1 The Objects of Choice

Individuals compare different risky alternatives or lotteries, as in Sect. 21.1.1, which will be defined here more precisely.

Lotteries are built from prizes that can consist of physical or monetary assets or any "useful" object (basket of consumer goods, gifts, dollars ...). We denote by lowercase italics (e.g., x, y ...) these different prizes forming a set C from which the lotteries will be defined.

A simple lottery X gives a prize x_1, with a probability p_1, a prize x_2 with a probability p_2, ..., a prize x_N with a probability p_N (with $\sum_{i=1}^{N} p_i = 1$). We denote $X \equiv (\underline{x}, \underline{p})$ such a lottery.

A special case of a simple lottery is that of a lottery with a certain result that gives prize x with a probability equal to 1. A lottery with a certain result $(x; 1)$, will be denoted simply by x (like the prize it yields with certainty); the set of all lotteries with a certain result can therefore be identified with the set of all the prizes C. Composite lotteries will also be considered, for example, the one that gives either a simple lottery X ($\equiv (\underline{x}, \underline{p})$) with a probability q, or a simple lottery Y ($\equiv (\underline{y}, \underline{p'})$), with a probability $(1-q)$; we will denote $(X, Y; q)$ such a composite lottery, which can be represented by the following diagram:

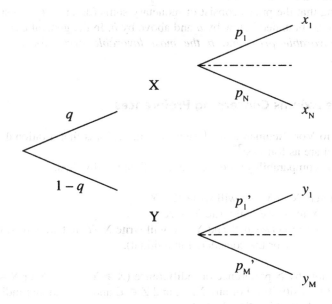

As will be seen in the sequel (see axiom 4 below), the rational individual is supposed to be indifferent between the composite lottery $(X, Y; q)$ and the simple lottery Z which gives the same final results, with the same probabilities, defined by the graph below:

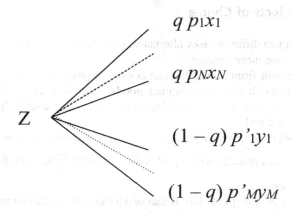

This principle of independence postulates that a rational individual is concerned only by the final result (and its probability of occurrence) and not by the procedure of attribution of this result (simple or composite lottery).

In the following, $\mathcal{L}$ represents the set of all the lotteries built on C (with certain, simple, or composite results) which must be compared and prioritized in order of preference, by the individuals considered. In particular, $\mathcal{L}$ contains all lotteries with certain results, that is $\mathcal{L} \supset C$.

Assuming that the prizes consist of monetary sums (dollars), C is a subset of $\mathbb{R}$ presumed to be bounded below by a and above by b. In the general case, a denotes *the least favorable prize* and b *the most favorable prize*, for the individual considered.

A1.2 The Axioms Concerning Preferences

According to Von Neuman and Morgenstern the axioms that a rational individual must respect are as follows:[25]

Axiom 1 (comparability): For X and Y $\in \mathcal{L}$, any individual:

- either prefers X to Y; we will write X $\succ$ Y
- or prefers Y to X; we will write Y $\succ$ X
- or is indifferent between X and Y; we will write X~Y and we will say that X is equivalent to Y (for the considered individual).

We will note by $\succcurlyeq$ preference or indifference (X $\succcurlyeq$ Y if X $\succ$ Y or X ~ Y).
Axiom 2 (transitivity): For any X, Y, and Z $\in \mathcal{L}$ and any rational individual: X $\succcurlyeq$ Y and Y $\succcurlyeq$ Z implies: X $\succcurlyeq$ Z.

---

[25] We do not present below the most parsimonious system of axioms but that which leads to a relatively simple justification of the expected utility criterion.

Appendix 1: The Axiomatic of Von Neuman and Morgenstern and Expected Utility   919

This axiom expresses simply a condition of choice coherence.

Axiom 3 (reflexivity): $X \succcurlyeq X$ for any $X \in \mathcal{L}$.

This axiom is almost tautological.

Because of these first three axioms, the relationship $\succcurlyeq$, representing the preferences of any rational individual, is a pre-order on $\mathcal{L}$.

Axiom 4 (independence): For any $X, Y \in \mathcal{L}$:

$(X, Y; q) \sim Z$ where Z is the simple lottery $(x_1 \ldots, x_N, y_1 \ldots, y_M; \underline{\pi})$, with:

$$\underline{\pi} = (qp^{X_1}, \ldots, qp^{X_N}, (1-q) \, p^{Y_1}, \ldots, (1-q) \, p^{Y_M}),$$

where p^{X_i} is the probability of occurrence of x_i (lottery X) and p^{Y_i} that of y_i (lottery Y).

The meaning of this principle of independence has been explained in Sect. A1.1 above.

Axiom 5 (continuity):

For any lottery $X, Y, Z \in \mathcal{L}$ with $X \succ Z$ and $X \succcurlyeq Y \succcurlyeq Z$, there exists a single parameter $p \in [0,1]$ such that $(X, Z; p) \sim Y$.

Of course, p depends on the preferences of the individual as well as on X, Y, and Z.

Axiom 6 (dominance 1):

Let $X = (x_1, x_2; p)$ and $Y = (x_1, x_2; p')$ with $x_1 > x_2$.

$$X \succ Y \text{ if and only if } p > p'; \quad X \sim Y \text{ if and only if } p = p'.$$

Axiom 7 (dominance 2):

Consider two composite lotteries $L = (X, Y; p)$ and $L' = (X, Z; p)$. Thus:

$$L \succ L' \text{ if and only if } Y \succ Z; \text{ and } L \sim L' \text{ if and only if } Y \sim Z.$$

This axiom implies in particular that by replacing in a lottery one of the prizes by an equivalent prize, one obtains a second lottery equivalent to the first.

A1.3 The Expected Utility Criterion

On the basis of these seven axioms, it is possible, in two steps, to establish the validity of the expected utility criterion.

Step 1. Building a utility function U reflecting preferences on the certain results C

We will build a particular function U, of C in $\mathbb{R}$ which preserves (or reflects) preferences on C, which means that for any x and $y \in C$:

$U(x) > U(y)$ if and only if $x \succ y$; $U(x) = U(y)$ if and only if $x \sim y$.

Any function U that preserves the preferences qualifies as a valid utility function.

Let us start by arbitrarily fixing the two bounds $U(a)$ and $U(b)$, recalling that a is the worst prize and b the best. As a thermometer that is graded by setting the temperature of the melting ice to 0 and the temperature of the boiling water to 100, let us set $U(a) = 0$; $U(b) = 1$. Now consider any certain result x in C. In accordance with axiom 5 (continuity), there exists a single $p(x) \in [0, 1]$, such that $(b, a; p(x)) \sim x$. This defines a mapping of C on $[0, 1]$ which will be the utility indicator *defined on C* because we will set, for any $x \in C : U(x) = p(x)$.

In accordance with axiom 6 (dominance 1), such a function $U(x)$ preserve preferences; indeed:

$x \succ y$ if and only if $p(x) = U(x) > U(y) = p(y)$; $x \sim y$ if and only if $U(x) = U(y)$.

The function U, thus defined, thus constitutes a valid utility function on C.

Note that the function U is only one possible choice among an infinity of valid utility functions on C. Thus, any function V deduced from U by any increasing monotonic transformation $(V(x) = f(U(x))$ with f increasing) would also preserve preferences on C.[26]

Step 2. Justification of the expected utility criterion as an expression of preferences on $\mathcal{L}$

We will now compare lotteries of $\mathcal{L}$ leading to uncertain results.

Consider $X = (x_1, \ldots, x_N; p_1, \ldots, p_N)$ and let $E(U(X)) \equiv \sum_{i=1}^{N} p_i \, U(x_i)$ where U designates the utility function, defined on C, built in Step 1.

Proposition

$X \to E(U(X))$ *is a mapping of $\mathcal{L}$ in $\mathbb{R}$ that preserves preferences, that is, for any X and $X' \in \mathcal{L}$:*

$X \succ X'$ *if and only if $E[U(X)] > E[U(X')]$*

$X \sim X'$ *if and only if $E[U(X)] = E[U(X')]$*

Proof

Consider the simple lottery $X = (x_1 \ldots x_N; p_1, \ldots, p_N)$ represented by graph X (below on the left).

---

[26] But it could not serve as a basis for the representation of preferences on uncertain results (using the criterion $E(V(\bullet))$, the expected value of $V(\bullet)$), unless f is an affine linear transformation (see below).

Appendix 1: The Axiomatic of Von Neuman and Morgenstern and Expected Utility

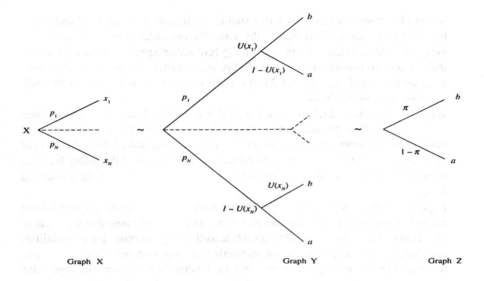

Graph X   Graph Y   Graph Z

Remember that $x_i \sim (b, a; U(x_i))$ for $i = 1,\ldots,N$.

Therefore, from axiom 7 (dominance 2), X is equivalent to the composite lottery Y represented on the graph Y (each x_i of X has been replaced by the equivalent lottery $(b, a; U(x_i))$); Moreover, in accordance with axiom 4 (independence), Y is itself equivalent to the lottery $Z = (b, a; \pi)$ of the graph Z, with:

$$\pi = \sum_{i=1}^{N} p_i \, U(x_i) = \mathrm{E}\,[U(X)].$$

Consider now another lottery X' of $\mathcal{L}$ with $X' = (x'_1,\ldots x'_M\,;\,p'_1,\ldots, p'_M)$ and $\mathrm{E}[U(X')] = \sum_{j=1}^{M} p'_j \, U(x'_j)$.

For the same reason as that used for X, we have:

$X' \sim (b, a; \pi')$ with $\pi' = \mathrm{E}\,[U(X')]$,

and, under axiom 6 (dominance 1):

$X \succ X'$ if and only if $\pi = \mathrm{E}[U(X)] > \mathrm{E}\,[U(X')] = \pi'$

$X \sim X'$ if and only if $\pi = \mathrm{E}\,[U(X)] = \mathrm{E}\,[U(X')] = \pi'$

which proves that $\mathrm{E}\,[U(\bullet)]$ preserves preferences on $\mathcal{L}$. QED.

A1.4 Notes and Complements

- The concept of "prizes" used in previous developments is very general. It thus accommodates very diverse situations such as vectors of consumer goods, or more simply aggregate consumption, considered in economics. We have already noted that in the special case where these prizes are sums of money, C is simply a subset of $\mathbb{R}$ bounded below by a (the smallest gain, possibly negative) and above by b; we have also noticed that, in this case, the pre-order $\succcurlyeq$ on C translates, for every rational individual, by the relation $\geq$ (greater or equal) because the rational individuals are presumed to desire always more wealth. In this case, the utility function U defined in Sect. 21.1.3 is *the utility of wealth* which is, in fact, the only meaning used in this book.

– We noticed, in Sect. 21.1 of this chapter, that the utility function is only defined up to a linear affine transformation ($V = k_1 U + k_2$ with $k_1 > 0$ leads to the same ranking as U in terms of expected utility); in the construction of the U function of Sect. 21.1.3, these two degrees of freedom in the definition of the utility function correspond to the flexibility conferred by setting the two extreme values $U(a)$ and $U(b)$.

Expected utility theory is presented in this work neither in its most general form nor with the most parsimonious set of axioms. Let us simply note that a version of this theory, due to Savage (1954), avoids resorting to prior objective probabilities but *deduces* the necessary use of subjective or objective probabilities for any rational decision-making under uncertainty. Savage thus answers the objections of those who question the relevance of the notion of probability (objective or even subjective) for real-world decision-making.

– The validity of Von Neuman and Morgenstern's theory is often disputed, using both theoretical and empirical convincing arguments. Let us quote on this subject the paradox of Allais, as well as the recent developments of Behavioral Finance (initiated by the works of Kahneman and Tversky and developed by Thaler) that we have briefly exposed in Sect. 21.4.4. The paradigm of expected utility remains nonetheless the cornerstone of the economic and financial theories of choice under uncertainty.

Appendix 2: A Reminder of Quadratic Forms and the Calculation of Gradients

– Consider an element $\underline{x} \equiv (x_1, x_2, \ldots, x_N)'$ of $\mathbb{R}^N$, a prime designating a transpose and the underlining a vector ($\underline{x}$ is thus a column vector) and a square matrix $N \times N$ denoted by $\mathbf{A}$ and general term a_{ij}. Recall that $\mathbf{A}$ defines a quadratic form by:

$$\underline{x}'\mathbf{A}\,\underline{x} = \sum_{i=1}^{N} x_i \left(\sum_{j=1}^{N} a_{ij} x_j \right) = \sum_{i=1}^{N} \sum_{j=1}^{N} a_{ij} x_i x_j.$$

In the case of $N = 2$ for example: $\underline{x}'\mathbf{A}\,\underline{x} = x_1^2 a_{11} + x_1 x_2 a_{12} + x_2 x_1 a_{21} + x_2^2 a_{22}$. In the special case of a symmetric matrix A ($a_{12} = a_{21}$): $\underline{x}'\mathbf{A}\,\underline{x} = x_1^2 a_{11} + 2\,x_1 x_2 a_{12} + x_2^2 a_{22}$.

Appendix 3: Expectations, Variances and Covariances—Definitions... 923

- Consider a function f of $\mathbb{R}^N$ in $\mathbb{R}$, differentiable with respect to each of its components, and let $\frac{\partial f}{\partial x}\left(\underline{x}\right)$ be the gradient of f, i.e.:

$$\frac{\partial f}{\partial x}\left(\underline{x}\right) = \left(\frac{\partial f}{\partial x_1}\left(\underline{x}\right), \frac{\partial f}{\partial x_2}\left(\underline{x}\right), \ldots, \frac{\partial f}{\partial x_N}\left(\underline{x}\right)\right)'$$

In the special case of the quadratic form $\underline{x}'\mathbf{A}\,\underline{x}$ we can write:

$$\frac{\partial}{\partial x_i}\left(\underline{x}'\mathbf{A}\underline{x}\right) = \frac{\partial}{\partial x_i}\sum_{k=1}^{N} x_k\left(\sum_{j=1}^{N} a_{kj}x_j\right) = \sum_{j=1}^{N} a_{ij}x_j + \sum_{k=1}^{N} x_k a_{ki},$$ i.e. the ith line of the vector $\mathbf{A}\underline{x}+\mathbf{A}'\underline{x}$. We can thus write:

$$\frac{\partial}{\partial x}\left(\underline{x}'\mathbf{A}\underline{x}\right) = \mathbf{A}\,\underline{x} + \mathbf{A}'\underline{x}$$

and when A is symmetric (which is the case of a variance-covariance matrix):

$$\frac{\partial}{\partial x}\left(\underline{x}'\mathbf{A}\underline{x}\right) = 2\mathbf{A}\,\underline{x}.$$

Appendix 3: Expectations, Variances and Covariances— Definitions and Calculation Rules

A3.1 Definitions and Reminder

We first consider two random variables X and Y whose expectations (or means) are denoted by $E(X)$ and $E(Y)$, the variances σ^2_X and σ^2_Y (or $var(X)$ and $var(Y)$) and the covariance $cov(X, Y) \equiv \sigma_{XY}$. Let us remember that:

$$\sigma^2_X = E\left\{[X - E(X)]^2\right\} = E[X^2] - [E\,(X)]^2;$$

$$\sigma_{XY} = E\{[X - E(X)]\,[Y - E(Y)]\} = E(X\,Y) - E\,(X)E(Y).$$

Moreover, the correlation coefficient between X and Y, denoted ρ_{XY}, is given by:

$$\rho_{XY} = cov(X, Y)/\sigma_X\sigma_Y$$

The coefficient of determination (R^2) of the linear regression of X on Y (or vice versa) is equal to $(\rho_{XY})^2$.

A3.2 Calculation Rules

The following rules result from the previous definitions.

- The operator $cov(.,.)$ is symmetric: $cov(X, Y) = cov(Y, X)$ and $cov(X, X) = var(X)$.
- Consider a real number (non-random scalar) λ, and the random variable λX.

$$E(\lambda X) = \lambda E(X) \text{ (linear operator)}$$

$$var(\lambda X) = \lambda^2 var(X)$$

- Consider two random variables R_1 and R_2, two scalar x_1 and x_2 and the random variable $x_1R_1 + x_2R_2$. We obtain:

$$E(x_1R_1 + x_2R_2) = x_1E(R_1) + x_2E(R_2)$$

$$var(x_1R_1 + x_2R_2) = x_1{}^2var(R_1) + x_2{}^2var(R_2) + 2x_1x_2cov(R_1, R_2)$$

and in the special case of uncorrelated variables:

$$var(x_1R_1 + x_2R_2) = x_1^2var(R_1) + x_2^2var(R_2)$$

The operator $cov(.,.)$ is linear with respect to each of its arguments (it is bilinear):

$$cov(x_1R_1 + x_2R_2, X) = cov(X, x_1R_1 + x_2R_2) = x_1 cov(R_1, X) + x_2 cov(R_2, X)$$

- The preceding formulas extend to the case of N random variables $R_1 \ldots, R_N$ and N scalars $x_1 \ldots, x_N$.

We denote: $\mu_i \equiv E(R_i)$; $\sigma_i^2 \equiv var(R_i)$; $\underline{\mu} \equiv (\mu_1, \ldots, \mu_N)'$ (a column vector); $\mathbf{V}$ is the variance–covariance matrix $(N \times N)$ of the R_i, whose general term is $cov(R_i, R_j)$.

$$E\left(\sum_{i=1}^{N} x_iR_i\right) = \sum_{i=1}^{N} x_i\mu_i = \underline{x}'\underline{\mu} \text{ (linearity)}$$

$$cov\left(\sum_{i=1}^{N} x_iR_i, X\right) = cov\left(X, \sum_{i=1}^{N} x_iR_i\right) = \sum_{i=1}^{N} x_icov(R_i, X) \text{ (bilinearity)}$$

$$\mathrm{var}\left(\sum_{i=1}^{N} x_i R_i\right) = \mathrm{cov}\left(\sum_{i=1}^{N} x_i R_i, \sum_{j=1}^{N} x_j R_j\right) = \sum_{i=1}^{N} x_i \mathrm{cov}\left(R_i, \sum_{j=1}^{N} x_j R_j\right)$$

$$= \sum_{i=1}^{N} \sum_{j=1}^{N} x_i x_j \mathrm{cov}(R_i, R_j) = \underline{x}' \mathbf{V} \underline{x}.$$

In the special case of variables R_i uncorrelated to each other:
$var\left(\sum_{i=1}^{N} x_i R_i\right) = \sum_{i=1}^{N} x_i^2 var(R_i)$.

Appendix 4: Reminder on Optimization Methods Under Constraints

A4.1 Optimization When the Constraints Take the Form of Equalities

Our aim is to find the element of $\mathbb{R}^N$ denoted $\underline{x}^* = (x_1{}^*, x_2{}^*, \ldots, x_N{}^*)$ which maximizes (or minimizes) a function of $\mathbb{R}^N$ in $\mathbb{R}$, whose values are denoted $f(x_1, x_2, \ldots, x_N) = f(\underline{x})$, imposing that $\underline{x}^*$ satisfies M *constraints that are written* $g_1(\underline{x}^*) = c_1$; $\ldots$; $g_M(\underline{x}^*) = c_M$ (the c_i being constants).

The two maximization and minimization programs are written compactly:

$$\underset{\underline{x}}{\mathrm{Max}}\, f(\underline{x}) \text{ s.t. : } g_1(\underline{x}) = c_1; \ldots; g_M(\underline{x}) = c_M; \text{ or : } \underset{\underline{x}}{\mathrm{Min}}\, f(\underline{x}) \text{ s.t. : } g_1(\underline{x})$$

$$= c_1; \ldots; g_M(\underline{x}) = c_M.$$

In the case of program *Max* one will assume that f is a concave or linear function and the g_i are convex or linear, and in the case of program *Min* one will assume f to be convex or linear and the g_i concave or linear; in both cases f and the g_i are presumed differentiable with respect to their N arguments.

The solution to these optimization problems can be obtained by the Lagrange multipliers method which consists in:

- Associating to each constraint c_i a parameter λ_i called Lagrange multiplier;
- Defining a function of $N+M$ variables, called the Lagrangian, by:

$$L(\underline{x}, \underline{\lambda}) = f(\underline{x}) - \lambda_1 \left[g_1(\underline{x}) - c_1\right] - \ldots - \lambda_M \left[g_M(\underline{x}) - c_M\right]$$

- Looking for the values $\underline{x}^*$ and $\underline{\lambda}^*$ ($N+M$ unknown) that maximize (or minimize) $L(\underline{x}, \underline{\lambda})$ *without constraints* (program L) and of which we can show the following properties:
  - $\underline{x}^*$, that maximizes (or minimizes) L, *maximizes (minimizes) also f under constraints*;

- $\lambda_i^* dc_i$ is equal to the increase df obtained by a relaxation dc_i of the ith constraint (which thus become $g_i(\underline{x}) = c_i + dc_i$); λ_i^* is interpreted as the cost of the ith constraint (its *shadow price*). The values $\underline{x}^*$ and $\underline{\lambda}^*$ which satisfy the following relations, said the first-order conditions of L, constituting a system of $N+M$ equations with $N+M$ unknown, are extrema (maxima or minima) of L:

$$\frac{\partial L(\underline{x}^*, \underline{\lambda}^*)}{\partial \underline{x}} = \underline{\mathbf{0}} \ (N \text{ equations}); \quad g_i(\underline{x}^*) = c_i \quad \text{for } i = 1, \ldots, M \ (M \text{ equations}).$$

- If $L(\underline{x}, \underline{\lambda})$ is concave, the extremum $\underline{x}^*$ satisfying the first-order conditions is a maximum of L and solves program *Max*. If $L(\underline{x}, \underline{\lambda})$ is convex, $\underline{x}^*$ minimizes L and solves program *Min*. Moreover, $L(\underline{x}, \underline{\lambda})$ is concave (convex) if $f(\underline{x})$ is concave (convex) or linear and the $g_i(\underline{x})$ are convex (concave) or linear.[27]
- The usual technique for determining the optimum is to express the N values x_i^* as a function of $\underline{\lambda}^*$ using the N equations $\dfrac{\partial L\left(x^*, \ \lambda^*\right)}{\partial x} = \underline{\mathbf{0}}$, then to substitute the $\underline{x}^*(\underline{\lambda}^*)$ thus obtained in the M constraints $g_i(\underline{x}^*) = c_i$ in order to determine the M multipliers λ_i^*.

A4.2 Optimization Under Inequality Constraints

The program writes:

$$\operatorname*{Max}_{\underline{x}} f(\underline{x}) \text{ s.t.} : g_1(\underline{x}) \le c_1; \ \ldots; g_m(\underline{x}) \le c_m; g_{m+1}(\underline{x}) = c_{m+1}; \ \ldots; g_M(\underline{x}) = c_M$$

$$\text{or } \operatorname*{Min}_{\underline{x}} f(\underline{x}) \text{ s.t.} : g_1(\underline{x}) \le c_1; \ \ldots; g_m(\underline{x}) \le c_m; g_{m+1}(\underline{x}) = c_{m+1}; \ \ldots; g_M(\underline{x}) = c_M$$

The first m constraints are therefore inequalities and the last $(M - m)$ are equalities. It is often clear that some of the first m constraints are binding at the optimum $(g_i(\underline{x}^*) = c_i)$ and we can then replace the inequality constraint with an equality. It may occur that all constraints are binding and the program can be treated as in Sect. A4.1 above; this is the situation of the optimization programs discussed in Sect. 21.3 of this chapter. Assuming that the binding of the constraints is not obvious a priori, the solution is obtained with the following method: The Lagrangian is

---

[27] Therefore, either f, or one of the g_i, must have at least the *correct* convexity. If this convexity prevails in a domain D containing all $\underline{x}$ respecting the M constraints, the $\underline{x}^*$ satisfying the first-order conditions is unique and is a global optimum. On the other hand, if the condition of convexity is satisfied only in a neighborhood of $\underline{x}^*$, the latter may be only a local optimum (it is this last property that the second-order conditions, not presented in this brief presentation, allow to check).

Suggestions for Further Reading 927

defined as in Sect. A4.1; the $M+N$ first-order conditions (conditions of Kuhn and Tucker) write:

$$\frac{\partial L(\underline{x}^*, \underline{\lambda}^*)}{\partial \underline{x}} = \mathbf{0}; \quad \lambda_i^*(g_i(\underline{x}^*) - c_i) = 0$$

(with either $\lambda_i^* > 0$ and $g_i(\underline{x}^*) = c_i$, or $\lambda_i^* = 0$ and $g_i(\underline{x}^*) < c_i$), for $i = 1, \ldots, m$; In addition $g_j(\underline{x}^*) = c_j$ for $j = m+1, \ldots, M$.

The condition $\lambda_i^*(g_i(\underline{x}^*) - c_i) = 0$, relative to the ith inequality constraint, is easily understood thanks to the interpretation of the multiplier λ_i^* as the cost of this ith constraint: Either it is binding, so $g_i(\underline{x}^*) - c_i = 0$ and $\lambda_i^* > 0$ (the constraint is effective and its relaxation would allow obtaining a better optimum and therefore a higher value of $f(\underline{x}^*)$[28]); Or it is not ($g_i(\underline{x}^*) < c_i$) and its relaxation would leave the optimum $\underline{x}^*$ unchanged, so it would not be possible to increase the value of f.

Suggestions for Further Reading

Books

Bodie, Z., Kane, A., & Marcus, A. (2010). *Investments* (9th ed.). Irwin.
*Dumas, B., & Luciano, E. (2017). *The economics of continuous-time finance*. The MIT Press.
Elton, E., Gruber, M., Brown, S., & Goetzmann, W. (2010). *Modern portfolio theory and investment analysis* (8th ed.). Wiley.
Ingersoll, J. (1987). *Theory of financial decision making*. Rowman and Littlefield.
Markowitz, H. (1959). *Portfolio selection: Efficient diversification of investment*. Wiley.
Markowitz, H. (1987). *Mean-variance analysis in portfolio choice and capital markets*. Basil Black-well.
Sharpe W. 2000. *Portfolio theory and capital markets*. Mc Graw Hill.
Thaler, R. 1993. *Advances in behavioral finance*. Russell Sage Foundation.
Thaler, R. (2015). *Misbehaving: The making of behavioral economics*. W. W. Norton.
Von Neuman, J., & Morgenstern, O. (1947). *The theory of games and economic behavior*. Princeton University Press.

Articles

Barberis, N., & Thaler, R. (2006). A survey of behavioral finance. In G. Constantinides, M. Harris, & R. Stulz (Eds.), *Handbook of economics and finance*.
Best, M., & Grauer, R. (1992). On the sensitivity of mean-variance portfolios to changes in asset means: Some analytical and computational results. *Review of Financial Studies, 4*, 315–342.
Black, F. (1972). Equilibrium with restricted borrowing. *Journal of Business, 45*, 444–454.
Black, F., & Litterman, R. (1992). Global portfolio optimization. *Financial Analysts Journal, 48*, 28–43.

---

[28] For a maximum; a small variation Δc_i would lead to $\Delta f(x^*) = \lambda_i^* \, \Delta c_i$.

Cass, D., & Stiglitz, J. E. (1970). The structure of investor preferences and asset returns, and separability in portfolio allocation: A contribution to the pure theory of mutual funds. *Journal of Economic Theory, 2*, 122–160.

Jarrow, R., & Zhao, F. (2006). Downside loss aversion and portfolio management. *Management Science, 52*, 558–566.

Jobson, J., & Korkie, B. (1981). Putting Markowitz theory to work. *Journal of Portfolio Management., 7*, 70–74.

Kahneman, D., & Tversky, A. (1973). Prospect theory: an analysis of decisions under risk. *Psychology Review, 80*, 237–251.

Kahneman, D., & Tversky, A. (1979). On the psychology of prediction. *Econometrica., 47*(2), 263–291.

Merton, R. C. (1979). An analytic derivation of the efficient portfolio frontier. *Journal of Financial and Quantitative Analysis, 7*, 1851–1872.

Odean, T. (1998). Are investors reluctant to realize their losses? *Journal of Finance, 53*, 1775–1798.

Pratt, J. W. (1964). Risk aversion in the small and in the large. *Econometrica, 2*, 122–136.

Schefrin, H., & Statman, M. (1985). The disposition to sell winners too early and ride losers too long: Theory and evidence. *Journal of Finance, 40*, 777–790.

Sharpe, W. (1963). A simplified model for portfolio analysis. *Management Science, 9*, 277–293.

Sharpe, W. (2000). Behavioral portfolio theory. *Journal of Financial and Quantitative Analysis, 35*(2), 127–151.

Sortino, F., & Van den Meer, R. (1991). Downside risk. *Journal of Portfolio Management*, 27–32.

Tobin, J. (1958). Liquidity preference as behavior toward risk. *Review of Economic Studies*, 65–86.

Tversky, A., & Kahneman, D. (1992). Advances in prospect theory: cumulative representation of uncertainty. *Journal of Risk and Uncertainty, 5*, 297–323.

The Capital Asset Pricing Model 22

In the previous chapter, we studied the behavior of individuals facing uncertainty, in particular the way in which "mean-variance" (MV) investors best allocate their savings between the various assets available on the market.[1] They need, as data input for solving their problem, the set of asset expected returns and their variance-covariance matrix. This chapter presents a model that expresses expected returns at equilibrium, assuming that the market portfolio is efficient. This Capital Asset Pricing Model (CAPM) was developed in the 1960s by W. Sharpe (1964), J. Treynor (1965), J. Lintner (1965) and I. Mossin (1966), and radically changed the way of thinking in market finance.[2] Its central message is that for any financial asset, *the relation between risk and return is positive and linear.* The CAPM has since experienced many applications, has been subjected to innumerable empirical tests and remains to this day a dominant paradigm despite continual attacks, both theoretical and empirical.

Section 22.1 is devoted to the derivation and interpretation of the CAPM, Sect. 22.2 to its main applications, Sect. 22.3 to some of its extensions, Sect. 22.4 to criticisms that have been made against its validity, and Sect. 22.5 to econometric tests of the model.

22.1 Derivation of the CAPM

Like any model, the CAPM is based on simplifying assumptions that are outlined in Sect. 22.1.1. We demonstrate the model rigorously in Sect. 22.1.3. However, to facilitate its intuitive understanding in the special case where there is a risk-free

---

[1] Portfolio theory was initially developed with reference to stocks. However, the analyses that follow are valid for any financial asset with random payoffs.

[2] For the history of the evolution of ideas in modern finance, the reader will consult the excellent work of Peter L. Bernstein, *Capital Ideas; The Improbable Origins of Modern Wall Street*, The Free Press, A Division of Macmillan, Inc. New York, 1992.

© The Author(s), under exclusive license to Springer Nature Switzerland AG 2022   929
P. Poncet, R. Portait, *Capital Market Finance*, Springer Texts in Business and Economics, https://doi.org/10.1007/978-3-030-84600-8_22

asset, we first derive two intermediate results (see Sect. 22.1.2) concerning the efficient portfolios studied in the previous chapter.

22.1.1 Hypotheses

To determine the relation between expected return and risk of a financial security at market equilibrium,[3] we will aggregate the behavior of individual investors and formulate the following hypothesis (H), which is added to those given in the previous chapter (especially that of perfect markets):

(H): The market portfolio is mean-variance efficient.[4]
The market portfolio is denoted by M and defined as the portfolio comprising *all* the risky assets present in the economy (but not the possibly existing risk-free asset),

Remarks
- The efficient frontier, hence the notion of efficiency, is relative to: (i) expectations (or estimates) concerning the means and variances-covariances of returns (μ, V); (ii) the investment horizon (*Hor*). Efficiency is therefore defined with respect to a frame of reference (μ, V, *Hor*) including probabilities and a given horizon. In the following, we will consider estimates constructed from objective probabilities that may be those used by the representative (or average) investor or those that result from the best available information set or those that are derived from econometric constructions explained later on. In addition, we will consider a standard horizon which can also be interpreted as that of the representative investor. This set of beliefs and a standard horizon will constitute the reference frame from which the efficiency frontier will be constructed and the hypothesis (H) formulated.
- It is often assumed, in order to justify the CAPM, that all investors behave in accordance with the mean-variance paradigm, and that they have the same reference frame, i.e. the same horizon of investment as well as homogeneous expectations (they all use as inputs to their portfolio optimization problem the same vector of expected returns and the same variance-covariance matrix). We explain in the following paragraph why these assumptions *lead* to the hypothesis (H) and are a priori more restrictive than the latter.

---

[3] We will later justify the fact that the CAPM is a model of partial equilibrium, and not of general equilibrium. It suffices here to point out that the supply of securities (by the firms) is in fact ignored and that the CAPM relationship results from optimization conditions concerning the sole demand.
[4] In the sequel, we simply say that it is efficient.

22.1 Derivation of the CAPM

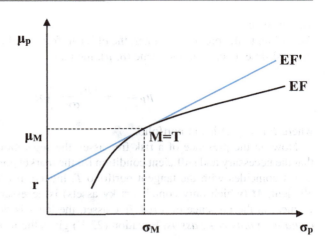

Fig. 22.1 The Capital Market Line (CML)

22.1.2 Intermediate Results in the Presence of a Risk-Free Asset

22.1.2.1 Tobin's Separation Theorem

We have seen in the previous chapter that the set of Markowitz efficient portfolios can be obtained from the combination of any two efficient portfolios yielding different returns (Black's two-fund separation theorem). A particular, useful case is obtained in the presence of a risk-free asset. In this case, indeed, the efficient frontier is the half-line passing through the point of coordinates $(0, r)$ and the point T of coordinates (σ_T, μ_T) representing the portfolio tangent to the hyperbola of Markowitz (see the previous chapter, as well as Fig. 22.1 and Sect. 22.1.2.2). This portfolio T is composed only of risky assets and is efficient within the frame (μ, V, Hor) used to build the efficient frontier. Tobin's two-fund separation theorem states that any efficient portfolio is a combination of the risk-free asset and the tangent portfolio T.

22.1.2.2 The Capital Market Line
The previous result is stated in the following proposition:

Proposition 1
Under the hypothesis H (M is efficient) and in the presence of a risk-free asset, there is a linear relation between the expected return and the risk (measured by the standard deviation) of efficient portfolios. If P is efficient, then:

$$\mu_P = r + \left(\frac{\mu_M - r}{\sigma_M}\right)\sigma_P \qquad (22.1)$$

Justification

According to the previous chapter, the efficient frontier of portfolios in the presence of a risk-free asset is, in the plane (σ, μ), the half-line rT whose equation writes:

$$\mu_P = r + \left(\frac{\mu_T - r}{\sigma_T}\right)\sigma_P \qquad (22.1')$$

where P is any efficient portfolio.

Now, in the presence of a risk-free asset, the separation into two funds implies that the necessary and sufficient condition for the market portfolio M to be efficient is that it coincides with the tangent portfolio T. Indeed, if (H) is true, that is, if M is efficient, M (which only contains risky assets) is necessarily the tangent portfolio T since, in the presence of a risk-free asset, the latter is *the only efficient portfolio containing only risky assets*. Equation (22.1') gives the result.

Remarks

– Note again that the efficient frontier and the tangent portfolio T are a priori specific to each investor since they depend on expectations and the investment horizon. However, *in the event that investors share the same expectations and the same horizon* (they use the same frame (μ, **V**, *Hor*)), they all build the *same* efficient Markowitz frontier and thus *all* hold the same portfolio of risky assets characterized as *the* (unique) tangent portfolio T constructed from the given reference frame. Their portfolios, therefore, differ only in the respective weights allocated to T and the risk-free asset. Therefore, the tangent portfolio T is necessarily identical to portfolio M since the latter includes all the risky assets available for trade and, since T is efficient relative to the frame, the hypothesis (H) is necessarily verified. The efficiency of M is therefore in this case a *consequence* of the hypothesis of homogeneous expectations and horizons, and not a hypothesis itself.
– However, we can imagine that M is efficient relative to the reference frame, in accordance with hypothesis (H), although it does not necessarily conform to the beliefs and horizons of *all* investors. The frame can, for instance, be understood as that of an average or representative investor. Hypothesis (H) is, therefore, a priori less restrictive than the hypothesis of homogeneity of beliefs and horizons, often formulated to justify the CAPM.[5]

Equation (22.1), representing the efficient frontier for optimal portfolios, is represented graphically by a straight line called the *Capital Market Line* (CML). It is illustrated in Fig. 22.1.

---

[5]This last hypothesis is obviously highly unrealistic, with a large number of transactions in the market, if not the majority of them, being in practice motivated by differences of appreciation as to the expected return and/or the presumed risk of the traded assets.

22.1 Derivation of the CAPM

The slope of the straight line (1) is called the *market price of risk* for a portfolio. This price is itself equal to the *market risk premium* offered on average by the market $(\mu_M - r)$ divided by the amount of risk borne by the market portfolio (σ_M). It is important to note that the expected (ex ante) premium $(\mu_M - r)$ must be positive to encourage risk-averse investors to finance risky businesses or projects. However, because of the randomness of R_M, the difference $(R_M - r)$ often turns out to be negative ex post, without obviously calling into question the validity of the theory.

Moreover, the composition of the market portfolio M, which includes all the risky securities, is such that the weight m_i of each security i is equal to the ratio of its market capitalization V_i on the sum of all the market capitalizations, i.e., the total market value:

$$m_i = \frac{V_i}{V} = \frac{V_i}{\sum\limits_{i}^{n} V_i}, \text{ so that } \sum\limits_{i}^{n} m_i = 1.$$

22.1.3 The CAPM

We first present the CAPM in its general formulation before presenting it in alternative forms, in particular the standard form that is in fact less general.

22.1.3.1 Statement of the General CAPM

While important, Eq. (22.1) is only valid in the presence of a risk-free asset and for *efficient portfolios* and not for an *individual security* and in the presence of a risk-free asset. The CAPM establishes the relation that must prevail between the risk and expected return of an individual security. This relation is the subject of the following proposition.

Proposition 2 General CAPM, valid with or without a risk-free asset.

(i) Under the hypothesis (H) stated in Sect. 22.1.1 that the market portfolio M is efficient, in the presence or absence of a risk-free asset there are two positive parameters λ and θ such that, for each asset $i = 1, \ldots, N$:

$$\mu_i = \lambda + \theta \sigma_{iM} \tag{22.2}$$

(ii) Conversely, if there are two positive parameters λ and θ such that, for each security $i = 1, \ldots, N$, the relation (22.2) is true, then the hypothesis (H) of efficiency of the market portfolio M is verified.

(iii) The CAPM (22.2) can also be written equivalently:

$$\mu_i = \lambda + \left(\frac{\mu_M - \lambda}{\sigma_M^2} \right) \sigma_{iM} \tag{22.3a}$$

$$\mu_i = \lambda + \beta_i(\mu_M - \lambda) \quad \text{where } \beta_i = \frac{\sigma_{iM}}{\sigma_M^2} \tag{22.3b}$$

Proof

The CAPM can be demonstrated in several ways. The simplest and most general proof derives directly from a fundamental result of the preceding chapter.

Recall that a portfolio P is efficient *if and only if* there are two positive parameters λ and θ such that, for each security $i = 1, \ldots, N$: $\mu_i = \lambda + \theta \sigma_{iP}$ (Proposition 2 of the previous chapter); (i) and (ii) derive directly from this important result.

Moreover, by applying (22.2) to M itself we obtain: $\mu_M = \lambda + \theta \sigma_M^2$, from which we derive $\theta = (\mu_M - \lambda)/\sigma_M^2$.

By replacing θ with this value in (22.2), we get (22.3a and 22.3b).

Remarks

- The key message of the CAPM is that, at all times, the expected return (μ_i) and the risk (σ_{iM} or σ_{iM}/σ_M^2) of any asset are linearly related.
- The slope of the regression line of return R_i on return R_M is equal to σ_{iM}/σ_M^2 and is usually called β_i. This is why the *beta* notation is used in Eq. (22.3b).
- The proposition and its demonstration underscore that the following two propositions are equivalent: "the CAPM is valid" and "the market portfolio is efficient."
- The parameter θ is interpreted as the average risk aversion of the investors intervening in the market.[6] Indeed, the correct measure of the risk of an asset being its covariance with the market portfolio, θ is the increase in the expected rate of return on an asset required by an additional "risk point" provided by this asset (mathematically, the derivative of μ_i with respect to σ_{iM}).
- Relations (22.2) and (22.3a, 22.3b), which hold for any security, are also valid for any portfolio, efficient or not. Indeed, by noting P any portfolio defined by its weights $\underline{x}$, we have:

$$\mu_P = \sum_{i=1}^{n} x_i \mu_i = \sum_{i=1}^{n} x_i \left(\lambda + \theta \operatorname{cov}(R_i, R_M)\right)$$

$$= \lambda + \theta \sum_{i=1}^{n} x_i \operatorname{cov}(R_i, R_M) = \lambda + \theta \operatorname{cov}\left(\sum_{i=1}^{n} x_i R_i, R_M\right)$$

$$= \lambda + \theta \operatorname{cov}(R_p, R_M) = \lambda + \theta \sigma_{PM}.$$

---

[6]This average market aversion is, mathematically, the *harmonic* mean of the relative risk aversion coefficients of individuals weighted by their respective wealth.

22.1 Derivation of the CAPM

We used the fact that $\sum_{i=1}^{n} x_i = 1$ and the linearity property of the covariance operator.

– Parameters λ and θ are the same for every asset or portfolio.

22.1.3.2 Black and Sharpe–Lintner–Treynor–Mossin CAPMs

Historically, the CAPM was developed by Sharpe, Lintner, Treynor, and Mossin in the mid-1960s, in the special case of the existence of a risk-free asset. Black released this assumption in 1972 and proved the so-called *zero-beta* CAPM, which is strictly equivalent to the one we have just described, with a characterization of the two parameters λ and θ.

Proposition 3

Black's zero-beta CAPM

– With or without a risk-free asset, under hypothesis (H) portfolios or assets z whose returns are not correlated with the market portfolio M all have the same expected return μ_z.

– The expected return of every security $i = 1, \ldots, N$, is equal to:

$$\mu_i = \mu_z + \beta_i (\mu_M - \mu_z) \tag{22.4}$$

Proof

Consider a portfolio or a risky asset z whose covariance with M, so its beta, is zero. There exists an infinity of such portfolios.[7] Applying relation (22.3a and 22.3b) to z gives $\mu_z = \lambda$ for all zero-beta portfolios. Replacing λ by μ_z in (22.3a and 22.3b) yields (22.4) directly.

Equation (22.4) can therefore be considered as a simple rewrite of (22.3a and 22.3b) in which the constant λ is interpreted as the expected return of the zero-beta portfolios.

The additional assumption that there is a risk-free asset, whose certain return is denoted r, which is a *special zero-beta*, leads to the standard Sharpe–Lintner–Treynor–Mossin CAPM.

---

[7]Indeed, by noting $\underline{z}$ the weight vector, we have: $\mathrm{cov}(R_z, R_M) = \mathrm{cov}(\underline{z}'R, \underline{m}'R) = \underline{z}'\, V\underline{m}$ where V denotes the variance-covariance matrix of the N returns. As a result, any vector z orthogonal to the vector $V\underline{m}$ is a zero-beta (it therefore belongs to a vector subspace of dimension N-1 and respects the constraint $\underline{z}'1 = 1$, in the absence of a risk-free asset. An infinite number of such portfolios thus exists). It can also be shown that in terms of variance in return, the best of all zero-beta portfolios is an inefficient portfolio as it is located on the lower branch of the Markowitz parabola.

Proposition 4 Standard CAPM.

If there is a risk-free asset with return r, the hypothesis (H) is true if and only if the expected return on any security or portfolio i is equal to:

$$\mu_i = r + \beta_i(\mu_M - r) \tag{22.5}$$

Remark

In an economy where the rate of inflation is stable enough to be considered non-random, the assumption of a risk-free rate may not be a problem. It can be admitted that in several developed countries the State is an issuer with no default risk and that a nominally risk-free asset does exist (e.g., a one-period Treasury Bill). The risk-free nominal return r thus translates into real return also without risk. The standard CAPM can then be interpreted as a valid model for both real and nominal returns. On the other hand, in an economy in which a random inflation rate prevails, a sure nominal risk-free return r becomes random in real terms. In the absence of an inflation-linked government security, no risk-free asset in real terms exists and the standard CAPM would only apply to nominal and not real returns, which is not based on sound economic foundations. Indeed, the mean-variance paradigm cannot be rationalized in nominal terms, as agents cannot be assumed to be permanent victims of monetary illusion. The zero-beta version of Black (4) (as well as, of course, the general version (3)) avoids this difficulty, since there is no need for a risk-free asset. In addition, this version often gives better empirical results than the standard CAPM (see Sect. 22.5).

22.1.3.3 Intuitive Justification of the Standard CAPM

We now propose two intuitive justifications for the standard CAPM (22.5) and then a numerical application.

– The first is based on an observation, already made in the previous chapter, that the correct measure of the risk of a security included in a portfolio P is σ_{iP}/σ_P.

We have indeed:

$$\sigma_P^2 = \sum_i^n \sum_j^n x_i x_j \sigma_{ij} \quad \text{whence} \quad \frac{\partial \sigma_P^2}{\partial x_i} = 2 \sum_j^n x_j \sigma_{ij} = 2\sigma_{iP}.$$

Moreover, we have: $\frac{\partial \sigma_P^2}{\partial x_i} = 2\,\sigma_P \frac{\partial \sigma_P}{\partial x_i}$.

These two equations imply that $\frac{\partial \sigma_P}{\partial x_i} = \frac{\sigma_{iP}}{\sigma_P}$.

The term σ_{iP}/σ_P measures the marginal contribution of the risk of asset i to the total risk of the portfolio. Here, we can write σ_{iM}/σ_M, since the risky portfolio is the market portfolio M. Therefore, the risk premium $\mu_i - r$ must be proportional to the risk σ_{iM}/σ_M: $\mu_i - r = k \frac{\sigma_{iM}}{\sigma_M}$. This last relation, true for any asset, holds in particular

22.1 Derivation of the CAPM 937

for M: $\mu_M - r = k \frac{\sigma_M^2}{\sigma_M}$ or $k = (\mu_M - r)/\sigma_M$, which, in accordance with the equation $\mu_i - r = k \frac{\sigma_{iM}}{\sigma_M}$, gives relation (22.5).

- The second intuitive rationale is based on the distinction, found in Sharpe's market model, between systematic risk and diversifiable risk. It seems common sense that the market, at equilibrium, compensates by a risk premium the inevitable risk only, as the avoidable risk can be eliminated by appropriate portfolio diversification. The only risk that is inevitable, and therefore rewarded, is the systematic risk measured by beta (β_i), since the second risk can be eliminated by diversification. The risk premium, which is added to the risk-free rate r to obtain the expected return on security i, is therefore proportional to β_i:

$$\mu_i = r + \theta\beta_i \tag{22.i}$$

where θ is the price of "a unit" of risk.

This last relation is true for any asset i and for any portfolio, which is a combination of assets. It, therefore, holds in particular for the market portfolio M. The latter has, by construction, a beta equal to one. It follows that:

$$\mu_M = r + \theta\beta_M = r + \theta, \quad \text{which implies } \theta = \mu_M - r.$$

By replacing θ with this expression in Eq. (22.i) giving μ_i, we obtain Eq. (22.5).

Example 1

Consider an economy in which the risk-free interest rate is 6%, the expected return on the market portfolio is 11%, and the standard deviation of the latter is 18%. All these rates are annualized. Assets ABC and DEF have a standard deviation of 35% and 50% and a correlation coefficient with the market of 0.8 and 0.25, respectively. What are their betas and their expected returns at equilibrium?

Security ABC has a covariance with the market equal to $\rho_{iM}\sigma_i\sigma_M$, or $0.8 \times 0.35 \times 0.18 = 0.0504$ while that of security DEF is $0.25 \times 0.50 \times 0.18 = 0.0225$.

Their respective betas are therefore $0.0504/(0.18)^2 = 1.556$ and $0.0225/(0.18)^2 = 0.694$.

Therefore, their respective expected returns, according to the CAPM, should be $\mu_{ABC} = 6\% + 1.556 \times (11\% - 6\%) = 13.78\%$ and $\mu_{DEF} = 6\% + 0.694 \times (5\%) = 9.47\%$.

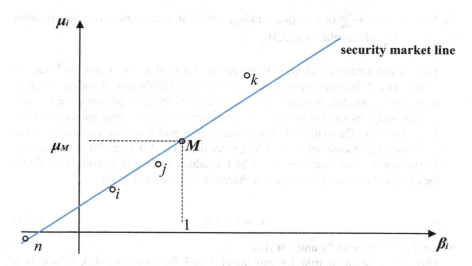

Fig. 22.2 CAPM: Security Market Line

22.1.3.4 Interpretation of the CAPM

If we draw the graph of relation (22.3a and 22.3b) in the plane $[\beta_i \equiv \sigma_{iM}/\sigma_M^2, \mu_i]$, as in Fig. 22.2, we obtain the Security Market Line (SML), passing through the points $(0, \lambda)$ and $(1, \mu_M)$. At market equilibrium, the representative points of all the securities i and all the portfolios must be located on this line: the risk premium $(\mu_i - \lambda)$ offered for each security i is *proportional* to its risk, measured by the beta. Throughout the discussion that follows, the intercept of the SML is denoted by λ, from the general CAPM (22.3a and 22.3b), knowing however that λ is always equal to μ_z and possibly equal to r.

This result explains why a risky asset A with a very high variance may offer a risk premium that is much lower than another asset B with a smaller variance, if its covariance with the market is smaller than that of B. In the same way, a risky asset (such as asset n in Fig. 22.2) offers an expected return lower than λ if its beta is negative, and may even be negative. Indeed, in this case, the asset is a *risk super-diversifier* since its return co-varies negatively with that of the market. This is the case, for example, with gold or gold mine stocks, which are known to have a rate of return very low on average over the long-term (even negative in real terms). These assets are not held for their expected returns, but because their inclusion in the portfolio *reduces* its risk.[8]

---

[8]This is also the case for all puts written on assets with a positive beta, especially puts on a stock index.

22.1 Derivation of the CAPM

Unlike the capital market line, located in the plane (σ_p, μ_p), which is actually only a half-line, the line in (β_i, μ_i) is complete, since there is no theoretical limit to the positive or negative value of a beta.

In Fig. 22.2, the points λ and M are by construction on the SML. So is asset i, indicating that its expected return, and therefore its current price, are correct in view of its risk. In contrast, j and k are not properly valued by the market, j being too expensive (not profitable enough) and k too cheap (too profitable), given their respective risks. The market will be in equilibrium when j, sold because too expensive, and k, bought because cheap, see their prices fall and rise, respectively. As a result, their expected return will increase and decrease, respectively, so that at equilibrium the points j and k will be aligned on the SML.

> **Example 2**
>
> In an economy where on average over the long term the market return exceeds the risk-free rate by 7% and the risk-free interest rate is now 5%, the expected return on the market portfolio must be 12%. Assume that securities A and B have, respectively, an expected return of 15% and 11% and a beta of 1.7 and 0.8. What would investors do in this case and what would be the equilibrium situation?
>
> Despite appearances, asset A would not be profitable enough because $15\% < 5\% + 1.7\,(12\% - 5\%) = 16.9\%$ and this asset would be sold (possibly even shorted). Its price would fall until its expected return reached 16.9%. Conversely, security B would be "too profitable" because $11\% > 5\% + 0.8\,(12\% - 5\%) = 10.6\%$. It would therefore be bought and its price would increase until its expected return reaches 10.6%.

A direct consequence of the CAPM is important both theoretically and practically. It will be recalled that the risk of an individual security i is measured by σ_{iP}, its covariance with the investor's portfolio P in which it is inserted, and therefore depends on it. Thus, if for an investor A this covariance is positive, the risk of i is positive and should justify a positive risk premium, even though if, for an investor B, σ_{iP} is negative, the risk premium of i should be negative. The measure of risk, and therefore the expected return, of each security appears "*subjective*" in that it depends on the investor's portfolio! However, the CAPM makes this difficulty disappear: The risk of the security i is unambiguously measured by its sensitivity β_i to fluctuations of the market (portfolio), i.e., by its covariance with the latter, and this measure is the *same* regardless of the investor. The risk of a security becomes "*objective*" in this sense,[9] and is priced by the market according to the CAPM. The conceptual and practical importance of this paradigm thus is better appreciated.

---

[9]The demanding reader will question the exact scope of this assertion (see the remarks in Sect. 22.2.2). In the *special* case of homogeneous expectations and investment horizons, all investors hold the same portfolio of risky assets (a fraction of M) and the assertion is obvious. In the more

Remark

The CAPM should not be confused, despite appearances and a common paternity, with Sharpe's market model presented in Sect. 21.2.4.2, which assumes a linear statistical relation between the returns of securities and that of the market portfolio. This last model, although very convenient in practice and making it possible to estimate the beta of securities by simple linear regression (see Sect. 22.2.1) is an ad hoc statistical model without any theoretical basis. The CAPM, on the other hand, is a theoretical model that involves an intercept (μ_z or r) common to all assets. The possible confusion comes from the fact that the two models involve the beta of the asset.

22.1.3.5 The Equilibrium Price of Financial Assets

One can wonder how the relation (22.2) (or (22.3a and 22.3b) constitutes a pricing model since the price of the asset i at the initial time does not appear. This evaluation is in fact implicit.

Let P_i be the equilibrium price of the security i (at $t = 0$), and $\widetilde{P}_i^1$ its random liquidation value at the end of the period (at $t = 1$).[10] It is this value that is the subject of investors' expectations, given the firm's investment and financing policies, which are supposed to be known at the beginning of the period. By definition, the random rate of return of asset i is given by:

$$\widetilde{r}_i \equiv \frac{\widetilde{P}_i^1}{P_i} - 1,$$

and so, using the expectation on both sides:

$$\mu_i = \frac{E\left(\widetilde{P}_i^1\right)}{P_i} - 1.$$

Substitute μ_i by its theoretical value given by the CAPM (22.5) and isolate the unknown P_i:

$$P_i = \frac{E\left(\widetilde{P}_i^1\right)}{1 + r + \theta \sigma_{iM}} \tag{22.i}$$

---

general case of hypothesis (H), each investor may have a frame of reference different from that of the other traders and the market as a whole and thus have their own efficient frontier, and M is then the efficient portfolio of risky assets only for the (fictitious) representative individual. The risk of a security i is then "objective" only from the point of view of this (average) individual. For all investors, risk has a market price determined by aggregating individual demands, as if stemming from the average investor, and conform to the CAPM.

[10] Or, in a multi-period model, its market value at the end of the first period.

22.1 Derivation of the CAPM

where $\theta = (\mu_M - r)/\sigma_M^2$, the market price of risk divided by σ_M, is used to simplify the notation. As we have seen, θ is interpreted as the market price of risk per covariance point.

From (22.i), we deduce:

$$E\left(\widetilde{P}_i^1\right) = P_i\left(1 + r + \theta\mathrm{cov}\left(\frac{\widetilde{P}_i^1}{P_i}, R_M\right)\right) = P_i(1 + r) + \theta\mathrm{cov}\left(\widetilde{P}_i^1, R_M\right),$$

which leads to the following proposition.

Proposition 5

The equilibrium value P_i of an asset generating a random unique flow $\widetilde{P}_i^1$, in the presence of a risk-free asset, is given by:

$$P_i = \frac{E\left(\widetilde{P}_i^1\right) - \theta\mathrm{cov}\left(\widetilde{P}_i^1, R_M\right)}{1 + r}. \tag{22.6}$$

The relation (22.6), the wanted evaluation model, is the analogue of the relation $P_i = P_i^1/(1 + r)$ holding under certainty, where P_i^1 would be the only certain future flow on which asset i has a claim.

The numerator of the right member is the *certainty-equivalent* (in the market sense) of the *random flow* $\widetilde{P}_i^1$ (i.e., the certain flow that has the same value as $\widetilde{P}_i^1$); it is equal to the mathematical expectation of this flow minus a risk premium which itself is the product of the risk of the *flow*, $\mathrm{cov}\left(\widetilde{P}_i^1, R_M\right)$, *by the market price of risk* θ.

This allows using the risk-free rate as the discount rate, as under certainty. One might think that the certainty-equivalent is specific to the investor and depends directly on their utility function. The interest of the CAPM lies precisely in the fact that, by aggregation and confrontation of the market participants, the individual certainty-equivalents disappear to make way for a single number, the certainty-equivalent for the market as a whole. This is the exact counterpart of the fact, commented on above, that the risk of an individual security becomes market priced in the CAPM paradigm.[11] We use again this certainty-equivalent approach when examining physical investments by firms (see Sect. 22.2.2).

Example 3

Consider an economy in which the risk-free interest rate is 5%, the expected return of the market portfolio is 13%, and its standard deviation is 20%. All

(continued)

---

[11] What has just been stated for the certainty-equivalent in the market sense is subject to the same difficulty of interpretation as discussed in footnote 9.

these rates are annualized. The expected value of the EASYVAL share in 1 year is equal to £50 and the covariance between this value and the rate of return of the market portfolio is estimated at 4. What is the certainty-equivalent of the random future value of one EASYVAL share? What is its current equilibrium price?

First, calculate the market price of risk per covariance point, θ. It is equal to $(13\% - 5\%)/(20\%)^2 = 2$. The certainty-equivalent is therefore equal to $50 - (2 \times 4) = £42$. The theoretical price of one EASYVAL share is therefore set at $42/(1.05) = £40$.

We now explain why the CAPM is a model of partial equilibrium and not of general equilibrium.[12] It is indeed only a *condition of* equilibrium *emanating from the demand* for securities by investors, with no consideration given to the supply side.[13] Yet the equilibrium price can only be obtained by characterizing also the supply of securities. For example, in relation (22.6), the supply implicitly *issues the claim to the flow* $\widetilde{P}_i^1$, which is exogenous to the CAPM. To achieve a general equilibrium, it is thus necessary to specify the supply of all securities on the market.

Two polar hypotheses can be considered. According to the first, the supply of securities is supposed *infinitely inelastic* to the demand and is thus fixed and given exogenously. This means that the firms and the State have already taken their physical investment and financing decisions once and for all and issued the corresponding securities on the market regardless of the financial market conditions. These decisions entail a probability distribution of their future *cash flows* (dividends and terminal value, for example, as in the case of EASYVAL). This is the hypothesis implicitly formulated in the preceding paragraph where a single security was considered, and $\widetilde{P}_i^1$, the distribution of its value at the end of the period was an exogenous datum. The change in stock prices (at 0) then leads to a change in the *rates of return* until their expectation complies with the CAPM.

On the other hand, according to the second hypothesis, the supply of securities is *infinitely elastic* to the demand, with firms and the state passively adapting their production capacities, through their investments or physical divestments, to the demand for individual securities. The *rates of return* on productive investments are given exogenously, and do not depend on the level of these investments (constant returns to scale).[14] When the demand for securities increases (decreases), firms issue

---

[12] Note also that if the asset valued by Eq. (22.6) is an element of the market portfolio M, which is the case if it is not a contingent asset with zero net supply, the equation in question is not a closed-form solution characteristic of a general equilibrium, since the asset appears on both sides of the equation.

[13] The discussion in footnote 9 is still relevant here for aggregate demand.

[14] The best-known model adopting this hypothesis is that of Cox, Ingersoll and Ross (1985). It is established in continuous time and therefore essentially multi-periodic, and its mathematical difficulty exceeds the targeted level of this chapter.

(buy back) securities and thus mechanically adjust their level of production, hence their future *cash flows*. It is ultimately investors on the financial market who thus decide the volume of physical capital installed.

22.2 Applications of the CAPM

The applications of the capital asset pricing model are numerous and important. We will examine only three of them, which concern (1) the practical use of the model for the purpose of analyzing the opportunity of investing in a given security or portfolio, (2) the physical investment choices of firms under uncertainty, and (3) the measure of performance of a mutual fund or portfolio.

22.2.1 Use of the CAPM for Financial Investment Purposes

To make the CAPM operational, various calculations must be made.

- First, we need to calculate the period-returns R_{it} of the different securities and R_{Mt} of the market portfolio from the observed prices. Two candidates are a priori conceivable, the arithmetic return $[(P_{i,t} - P_{i,t-1})/P_{i,t-1}]$ and the log return $[\ln(P_{i,t} / P_{i,t-1})]$, where $P_{i,t}$ is the last quoted price of the security i at the date t and ln designates the natural logarithm.[15]

Log returns, unlike their arithmetic analogues, have the important and practical property of being additive over time since the composition of the interests is done by simple addition and not by multiplication as for the arithmetic returns. Also, it is easier to know the law of probability and the statistical properties of an additive process than those of a multiplicative process.

However, log returns present a disadvantage relative to their competitors. The arithmetic return of a portfolio is the average of the arithmetic returns of weighted securities in this portfolio. The same would be true of log returns if the price variations of the securities were infinitesimal. Since in reality the variations observed are finite, calculating the log return of the portfolio as the weighted sum of the individual log returns is only a more or less satisfactory approximation. The shorter the investment period, the more log returns are recommended, as is done in continuous time. In practice, as long as the length of the period does not exceed 1 month, log returns are commonly used.

If a dividend $D_{i,t}$ is distributed between $(t-1)$ and t—which, in Europe, generally occurs only once a year, except obviously for the index chosen as a proxy for the market portfolio M—it is necessary to calculate:

---

[15] The logarithmic returns are very close to the arithmetic returns when the duration of the interval $[t-1, t]$ is small (they are equal at the limit). See Sect. 8.2.1.

$$R_{i,t} = \ln\left((P_{i,t} + D_{i,t})/P_{i,t-1}\right) \quad i = 1, \ldots, n, M.$$

- The second step is to estimate the betas of the different assets, by performing, with historical data, a regression of the $R_{i,t}$ on the $R_{M,t}$, for each $i = 1, \ldots, n$. This raises three types of problems: the frequency of the observations, the length of the total estimation period, and the choice of the market portfolio M. We will come back later to these important questions (see Sect. 22.4.2). For the first two, note that it is often considered that the number of observations must be greater than 30 for the estimate to be sufficiently reliable, and that it is preferable to use weekly rather than daily data to (partially) avoid statistical noise. As far as the market portfolio is concerned, when it comes to calculating the beta of securities, a market index is most often used in practice, such as the S&P 100 or 500, the Euro Stoxx 50, the MSCI World or an S&P global.
- Then one has to estimate the average market risk premium $(\mu_M - r)$. Although theoretically an anticipated premium, the historical average of the difference between the market return and the risk-free rate (e.g., the one-year money market rate) is usually calculated on historical data. This empirical average depends on countries and circumstances but is in the range of 4% to 8%.
- Lastly, the expected return is calculated using the CAPM:

$$\mu_i = r + \beta_i(\mu_M - r)$$

where r, the currently observed risk-free rate, β_i and $(\mu_M - r)$ have been estimated as indicated above.

Example 4

Over the last 10 years, the rate of return on the Euro Stoxx 50, with re-invested dividends, has been 5% higher than the 1-year money market rate. By performing the regression of stock X's rate of return on the relative variations of the Euro Stoxx 50 $[R_x(t) = \alpha_x + \beta_x R_M(t) + \varepsilon(t)]$ from monthly observations over the past 5 years, one finds an estimated coefficient β_X of 0.85. Today, the one-year rate is 3.5%. What is the expected return μ_X required today by the market for an investment in X?

The required premium on R_M, i.e. $(\mu_M - r)$, is 5%. Applying the CAPM to X, we find that currently its theoretical expected return should be: $\mu_X = 3.5\% + 0.85 \times 5\% = 7.75\%$.

A first use of the model consists in comparing this theoretical value μ_i with an anticipation of return a_i developed on the basis of a financial analysis for example, in order to judge the opportunity of an investment, a withdrawal, or even a short sale. This amounts to placing the representative point of the asset in the space (β_i, μ_i) using as expectation a_i. If the point is on the market line, then $a_i = \mu_i$, and the price of i is

22.2 Applications of the CAPM

correct. Otherwise, one has to sell or buy the asset according to whether a_i is lower or higher than μ_i.

Example 5

Market conditions and econometric estimates are those of the previous example. Consider a company Y whose share has just yielded a dividend of 9 €, sum that the market estimates to increase at the average annual rate of 6% to infinity (to simplify calculations). The share Y, whose beta is estimated at 1.20, now quotes 300 €. Should you buy it?

The expected return of this share compatible with the market equilibrium is: $\mu_Y = 3.5\% + 1.2 \times 5\% = 9.5\%$.

However, the expected return of an investment in Y securities is only 9% (6% capital gain due to growth +3% as a dividend, see the *Gordon-Shapiro model* in Chap. 8). Buying Y shares is therefore not recommended. It is too expensive, at 300 €. The reader will check that its equilibrium price is 257.14 €.

Another use is to calculate the historical average of the theoretical returns $\mu_{i,t}$ and to compare it with the average of the returns actually observed over the same period, in order to assess the past performance of the examined security.

These analyses also apply to portfolios, the subscript P replacing the subscript i, and in particular to mutual funds. The beta of the portfolio, as we saw in the previous chapter, is the weighted sum of individual betas[16]:

$$\beta_P = \sum_i^n x_i \beta_i \tag{22.7}$$

This property of the CAPM is useful, e.g., to build portfolios of a given beta. All that is required to measure the systematic risk of a portfolio is to know the betas of the securities that make it up.

The model is thus doubly useful with regard to mutual funds and other managed funds: on the one hand, it allows their managers to select the securities that make up the portfolio, and on the other hand, helps them evaluate the performance of the fund (see Sect. 22.2.3).

---

[16]This additivity property results directly from the linearity of the covariance operator. Indeed:
$$\beta_P \equiv \frac{\sigma_{pM}}{\sigma_M^2} = \frac{\text{cov}\left(R_M, \sum x_i R_i\right)}{\sigma_M^2} = \frac{\sum x_i \text{cov}\left(R_M, R_i\right)}{\sigma_M^2} = \sum x_i \beta_i.$$

22.2.2 Physical Investments by Firms

The analysis above does not concern only financial investments but also applies to the evaluation of investment decisions by firms under uncertainty. The two methods, the NPV (Net Present Value) and the IRR (Internal Rate of Return), used under certainty, cannot be used as such when future cash flows are random. However, the CAPM allows us to recover these methods in two ways.

The first is the so-called *certainty-equivalent* method. It makes it theoretically possible to calculate the NPV of a project from the certainty-equivalent (in the market sense) of the flows F_{jt} that it generates (see Sect. 22.1.3.5):

$$NPV_j = \sum_{t=0}^{T} \frac{E(F_{jt}) - \theta \text{cov}(F_{jt}, R_{Mt})}{(1+r)^t} \tag{22.8}$$

where θ and r have been assumed constant.[17]

The selection criterion using the NPV under certainty then applies under uncertainty. In particular, a positive NPV_j is necessary (and usually sufficient) for the project to be acceptable.

The second method is the *adjusted* (for risk) *discount rate* method, which is similar to the IRR method. The implicitly used model is:

$$NPV_j = 0 = \sum_{t=0}^{T} \frac{E(F_{jt})}{(1 + IRR_j)^t} \tag{22.9}$$

where IRR_j is the Internal Rate of Return computed from the series of expected cash flows $E(F_{jt})$. It must be at least equal to μ_j, the sum of the risk-free rate and a risk premium dependent on the project's cash flows. This required average return, also called the *cost of capital* of the project, maybe estimated by applying the CAPM.

The fundamental difference with the certainty case is that comparing the IRR of project j to the risk-free rate is no longer sufficient. Its β_j must indeed be taken into account. In other words, one must consider the plane (β, μ) and see how the point (β_j, IRR_j) representing the project is positioned relative to the security market line. If it is located on or above the SML, then IRR_j is greater than or equal to the expected return μ_j required by the market for investments of the same risk as j, and the investment can be undertaken. Equivalently, in order to be adopted, the project's NPV must be positive when μ_j is used as the discount rate (see Example 6).

The two proposed methods are in principle equivalent. They generalize to two dimensions (expected return and risk) the one-dimensional methods (return) used under certainty.

---

[17] The demanding reader will notice that the CAPM, derived in a mono-periodic framework, is (abusively) used in a multi-periodic context.

22.2 Applications of the CAPM

Example 6

The company ABC plans to invest in a project whose characteristics are as follows:

- Infinite (estimated) life.
- Expected cash flows after tax: $5700 K per year.
- Covariance between project and market *rates of return*: 0.18.

In addition, the rate of return on 12-month Treasury Bills is 4.5%, the expected market rate of return is 11.5% and its variance is 0.12.

What maximum price will ABC accept to pay for this investment, knowing that it is financed only by equity?

The *simplest* solution is to calculate the *cost of capital* for this project knowing that it will be financed without recourse to debt. This cost is the rate of return required by the shareholders (or the market) taking into account the risk of the project, i.e. $\mu = r + \beta(\mu_M - r) = 4.5\% + \beta \times (7\%) = 4.5\% + (0.18/0.12) \times (7\%) = 15\%$. The present value V of the annual perpetual flows generated by the project is then $5700/0.15 = 38,000$.

The second solution uses the *certainty-equivalent* of the random flows of expectation 5700. The (slight) difficulty arises from the fact that an equation must be posit in which the unknown V intervenes on both sides of the equal sign, as follows:

$$V = \sum_{t=1}^{\infty} \frac{E(F_t) - \theta \mathrm{cov}(F_t, R_{Mt})}{(1+r)^t}$$

$$= \frac{1}{r}\left[E(F) - \theta \mathrm{cov}(F, R_M)\right] = \frac{1}{0.045}\left[5700 - \frac{0.07}{0.12}0.18\,V\right]$$

because the flow F is equal to $R.V$, where R is the random rate of return of the project, which, by the linearity property of the covariance operator, implies that $cov(F, R_M) = cov(R.V, R_M) = V.cov(R, R_M) = 0.18\,V$.

Solving for V, we obtain $V = 5700/0.15 = 38,000$.

The maximum price that ABC Company will agree to pay to make this investment is therefore $38,000 K.

22.2.3 Standard Performance Measures

The relative share of portfolio management by professionals continues to increase and that of direct management by individuals correlatively keeps on decreasing. In view of the considerable number of all kinds of mutual funds offered on the market, the investor needs to get reliable information about their performance. The performance measure, however, proves a difficult exercise for which no single satisfactory solution exists.[18] The two traditional measures that we present here derive directly from the particular assumptions that led to the CAPM and involve the mean-variance paradigm. The first is the *Sharpe ratio* (Sect. 22.2.3.1) and the second is *Jensen's alpha* (Sect. 22.2.3.2). They are both widely used by management professionals, despite their limitations.

22.2.3.1 The Sharpe Ratio

What is the appropriate criterion for ranking two funds (portfolios), one of which has a higher average return than the other but a risk (measured by its standard deviation) also larger? In the context of the (very restrictive) assumptions leading to the CAPM, the solution is simple and is suggested by Eq. (22.1) above, representing the capital market line. Let us rewrite it as follows:

$$\frac{\mu_P - r}{\sigma_P} = \frac{\mu_M - r}{\sigma_M} \equiv RS \tag{22.10}$$

At market equilibrium, all efficient portfolios must have the same ratio of "risk premium to standard deviation." This is the Sharpe ratio (SR) which, applied to the market portfolio M, is none other than the market price of risk for portfolios and the slope of the capital market line.

This theoretical relation is true only ex ante, and for efficient portfolios. But expectation and volatility, being unknown, must both be estimated from data that are available, by definition, only ex post. It is important to understand the nature of the difficulty, recurrent in economics and finance, to empirically test a theoretical model using expectations (e.g., expected returns, variances or covariances), as such unobservable, with actual data which are only particular outcomes of random variables. To avoid confusion, we differentiate between "mathematical expectation" (ex ante) and "mean" (ex post) and attach the sign "^" to an empirical estimate (e.g., $\widehat{m}$ and $\widehat{\sigma}$ denote, respectively, the mean and the standard deviation estimated from a sample).

In practice, the ex post Sharpe ratio is used (for any portfolios or funds). The return appearing in the numerator is calculated ex post and is therefore an *observed mean* and not a mathematical expectation. Similarly, the standard deviation at the denominator is an *estimate*[19] developed from past data. Ex post, Sharpe ratios will generally be different for the various portfolios under scrutiny. The comparison of their Sharpe ratios makes it possible to rank them: the higher the ratio found, the better the performance of the fund is presumed.

In addition, the Sharpe ratio makes it possible to compare the performance of the funds with that of a benchmark supposed to represent the market M (for example, the Dax index for German equities). This allows to know which ones "outperformed"

---

[18] See for example Elton et al. (2010).

[19] Since the risk-free rate varies over time, it is necessary to carry out the calculations not from the rates of return R_p but directly from the excess returns (risk premia) $R_p - r$.

22.2 Applications of the CAPM

Fig. 22.3 Comparison of the performances of two funds

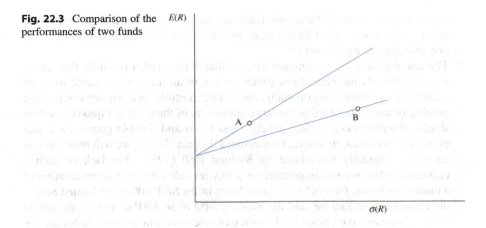

(we also say "beat") the market and which underperformed (were "beaten"). For example, in Fig. 22.3 the fund A, despite its lower return, outperformed portfolio B.

Although very useful, easy to understand and implement, and commonly adopted by professionals, the Sharpe ratio (SR) suffers from drawbacks that seriously limit its scope. We highlight four shortcomings.

1. Let us note beforehand that the program of maximizing ex ante SR (under a risk constraint, i.e. a standard deviation constraint) is equivalent to that of maximizing an expected return under the constraint of a variance (or standard deviation) (quadratic Markowitz optimization). Indeed:

$$\underset{x}{Max}[E(Rx)] \text{ s.t.}: \sigma(Rx) = k \Leftrightarrow \underset{x}{Max}\left[\frac{E(R_x) - r}{k}\right] \text{ s.t.}: \sigma(Rx) = k$$

$$\Leftrightarrow \underset{x}{Max}\left[\frac{E(R_x) - r}{\sigma(R_x)}\right] \text{ s.t. } \sigma(Rx) = k.$$

The two approaches are therefore strictly equivalent and their validity is based on the validity of the mean-variance paradigm. In particular, if the return distributions are not symmetrical, and investors are sensitive to them (preference for a right-sided asymmetry leaving few outcomes much lower than the average), SR is not a satisfactory measure of performance.[20]

2. When the risk premia are negative (which, even if false ex-ante, often happens *ex-post*), the comparison of the SRs of the funds does not make sense. Suppose, indeed, that two funds have the *same* SR and that this one is negative. The

---

[20]Goetzmann et al. (2002) show how to manipulate the Sharpe ratio, upwards or downwards, without any particular management skills.

representative points of these two funds are on the same half-line starting at the point of coordinates $(0, r)$. In this case, the portfolio with the lowest risk is also the one offering the highest return.

3. The calculation of the benchmark SR, i.e. that of the market portfolio that serves as a benchmark, theoretically requires the use of an index constructed from *all* existing risky assets. Unfortunately, this market portfolio is unknown because the number of assets involved is too large, and many of them are not quoted, such as shares of unlisted companies, jewels, most of art and durable goods, most real estate, etc. It cannot, therefore, be measured. In Sect. 22.4.1, we will return to this criticism, originally formulated by Richard Roll (1977). For lack of such a universal index, we use in practice as proxy an index deemed representative of a market such that, for stocks, the Dow Jones or the S&P 500 in the United States, the Euro Stoxx in Europe, and the MSCI World or an S&P global for the whole world. However, the choice of the index of reference can strongly influence the outcome of the comparison between the performance of a fund and that of the "market portfolio".

4. In most cases, a traditional fund explicitly refers to a benchmark, i.e. an index or portfolio (of securities or indices) whose manager is expected to replicate the performance and, if possible, improve it. This benchmark differs in general from the market portfolio. We will see in Chap. 25 how to analyze "benchmarked" management and how a performance criterion closely related to the Sharpe Ratio, called the Information Ratio, is defined and used.

22.2.3.2 Jensen's Alpha

The relevant measure of risk for an investor's total wealth is the standard deviation of the investor's return. The Sharpe ratio is therefore indicated when it applies to a well-diversified portfolio representing the bulk of the individual's wealth. On the other hand, when the investor considers a fund or portfolio (more or less well diversified) that constitutes only a part of their wealth, the relevant risk measure is its sensitivity to the market portfolio, i.e., its beta.

In the latter context, Jensen's alpha measures the ability of the manager of the analyzed fund to select securities that will show an abnormally high return and to ignore (or short sell) the others. This capacity is known as *stock picking* in the case of equities and, more generally, *security picking*. In other words, for a given systematic risk (linked to the market), i.e., a given beta, is the return generated by the portfolio correct or not? Jensen's alpha, which measures the difference between the expected return on the portfolio and its expected theoretical return calculated by the CAPM, answers this question. It is formally defined (ex ante) by:

$$\alpha_P \equiv \mu_P - r - \beta_P(\mu_M - r) \tag{22.11}$$

Jensen's alpha has a linearity-additivity property that simplifies its use and implies that the alpha of a portfolio is equal to the weighted average of the alphas

22.2 Applications of the CAPM

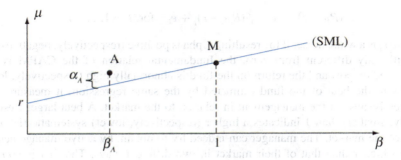

Fig. 22.4 Jensen's alpha

of its components. Indeed, we can easily show that we have,[21] if x_i is the weight of the risky asset i in the portfolio P ($i = 1, \ldots, n$):

$$\alpha_P = \sum_{i=1}^{n} x_i \alpha_i$$

Since any performance measure is by definition carried out ex post, the excess returns ($\mu_P - r$) and ($\mu_M - r$) are in fact the average excess returns observed for the analyzed fund and the market portfolio. It is clear that if the manager has no particular ability to analyze and choose, the ex ante alpha of their portfolio will be zero. This should also end up showing on average, ex post (at least before transaction costs and if this average is computed over a period long enough). However, if α_P is positive (respectively negative), the manager will have over- (under-)performed relative to the index M retained as the market portfolio. This is the case of fund A in Fig. 22.4.

In other words, the CAPM implies that, in the plane (beta-expected return), the representative point of the portfolio should, normally (in this framework of analysis), be on the *security market line* (SML), line passing through the points [0, r] and [1, μ_M]. If it is not there, the vertical gap between this point and its projection on the SML is the alpha of the portfolio.

In practice, to take account of the fact that the risk-free rate is variable over time, the ex post Jensen alpha is obtained by simple linear regression of the fund's excess return over that of the market portfolio M:

---

[21] $\mu_P = r + \sum_{i=1}^{n} x_i(\mu_i - r) = r + \sum_{i=1}^{n} x_i(\alpha_i + \beta_i(\mu_M - r)) = r + \sum_{i=1}^{n} x_i \alpha_i + (\mu_M - r)\sum_{i=1}^{n} x_i \beta_i$; Since $\sum_{i=1}^{n} x_i \beta_i = \beta_P$, we have: $\mu_P - r = \sum_{i=1}^{n} x_i \alpha_i + (\mu_M - r)\beta_P$ which, with the definition (11) of α_P, implies that $\alpha_P = \sum_{i=1}^{n} x_i \alpha_i$.

$$(R_P - r)_t = \alpha + \beta(R_M - r)_t + \varepsilon_t \quad \text{for } t = 1, \ldots, T,$$

where ε_t is a white noise. If the resulting alpha is positive (respectively, negative) and significantly different from zero, the fundamental relation of the CAPM is not respected ex post and the return on the fund is abnormally high (respectively, low).

As to the beta of the fund estimated by the same regression, it measures the aggressiveness of the management in relation to the market. A beta larger (respectively, smaller) than 1 indicates a higher (respectively, lower) systematic risk than that of the market. The manager can indeed try to obtain, by active management, a return higher than that of their market in two different ways. The first is *security picking* already mentioned: choosing securities whose Jensen's alpha is presumed positive. The second is *market timing* which consists of trying to predict periods of rise and fall of the market as a whole, to massively enter in the first case and exit (to invest mainly in risk-free assets) in the second. The manager has a real market-timing capacity if the beta of their portfolio is higher in upward periods than in downside periods (see Sect. 25.2.2 for more details).

Let us note in conclusion that the Sharpe ratio and Jensen's alpha do not generally lead to the same ranking of funds since the SR is the slope of a straight line and the alpha is a distance to this straight line. The rankings may differ significantly when comparing funds that are unequally diversified because the Sharpe ratio takes into account the systematic and specific risks (σ) whereas the alpha incorporates only the systematic risk (β). Example 7 illustrates this point. Alpha is, therefore, more relevant than the Sharpe ratio for ranking funds that are intended to be incorporated into larger portfolios in which diversification occurs, the opposite being true otherwise.

Example 7
Consider the three funds M, A, and B whose characteristics are as follow:

- M is the *proxy* of the market portfolio, so has a beta equal to one; its average return was 11% with an empirical standard deviation of 20%, while the risk-free rate was 5%.
- A is a well-diversified portfolio with an average return of 10% and a standard deviation of 21%. Its beta is 1.
- B is a portfolio with little diversification, average return of 11% and standard deviation of 30%. Its beta is 0.8.

The problem is to rank these three funds in order of performance, according to each method.

The Sharpe ratios of M, A and B, respectively, are (11–5)%/20% = 0.3; (10–5)%/21% = 0.238; and (11–5)%/30% = 0.2. Therefore, the ranking according to the Sharpe criterion is M, A, and B.

(continued)

The alpha of M is zero by construction. The alpha of A is $0.10 - [0.05 + 1 \times (0.11 - 0.05)] = -1\%$. The alpha of B is $0.11 - [0.05 + 0.8 \times (0.11 - 0.05)] = +1.2\%$.

The ranking according to Jensen's criterion is consequently B, M, and A, different from the previous one. This difference is due to differences in specific risks. The specific risk of A is low because $\sigma^2(\varepsilon_A) = \sigma^2(R_A) - \beta_A^2 \sigma^2(R_M) = (0.21)^2 - (0.2)^2 = 0.0041$, which represents only 10% of the total variance $\sigma^2(R_A)$ and reflects large diversification. On the other hand, the specific risk of B is high and equal to $\sigma^2(\varepsilon_B) = (0.3)^2 - (0.8 \times 0.2)^2 = 0.064$, which represents 71% of the variance of its return and reflects a low diversification.

22.3 Extensions of the CAPM

We limit ourselves to presenting two extensions of the CAPM. The first concerns Merton's Intertemporal CAPM with stochastic interest rate (Sect. 22.3.1), and the second the international CAPM (Sect. 22.3.2). We do not demonstrate the results.

22.3.1 Merton's Intertemporal CAPM

In a framework of analysis described in the previous chapter, built in continuous and no longer discrete time, Robert Merton (1973) generalized the CAPM in case the risk-free interest rate evolves over time randomly: at time t, the return on this asset is certain and known for the period of infinitesimal duration dt between dates t and $t + dt$, but investors do not know the risk-free rate that will prevail at $t + dt$ and after, because the level of $r(t)$ changes randomly. The fact that the instantaneously risk-free rate obeys a so-called stochastic process (as opposed to deterministic, i.e. known in advance) introduces into the economy an additional source of risk into the economy.

Merton derives *a three-fund separation theorem* (instead of two-fund): the risk-free asset, the market portfolio M and a portfolio whose return is perfectly negatively correlated with the risk-free asset. The latter fund allows the investor to hedge against the risk of unanticipated changes in the level of the risk-free rate. This Intertemporal CAPM writes:

$$\mu_i = r + \lambda_{1i} \left(\mu_M - r \right) + \lambda_{2i} \left(\mu_{NF} - r \right) \tag{22.12}$$

where rates of return are instantaneous continuous rates, μ_{NF} is the expected rate of return of a portfolio perfectly negatively correlated with r, and where:

$$\lambda_{1i} = \frac{\beta_{iM} - \beta_{i,NF} \cdot \beta_{NF,M}}{1 - \rho_{NF,M}^2}, \quad \lambda_{2i} = \frac{\beta_{i,NF} - \beta_{i,M} \cdot \beta_{NF,M}}{1 - \rho_{NF,M}^2}$$

with: $\beta_{x,y} = \frac{\sigma_{x,y}}{\sigma_y^2}$ et $\rho_{NF,M} = \frac{\sigma_{NF,M}}{\sigma_{NF} \cdot \sigma_M}$.

Naturally, if the risk-free rate is not stochastic, the third fund disappears and (22.12) is reduced to the standard CAPM, except that the rates used are instantaneous. Moreover, the framework of this model involving two sources of risk (r and R_M) is a special case of the more general state-variable framework introduced in Chap. 20 where m sources of risk (represented by m Brownian motions driving state variables) lead to m risk premia and would yield an ($m + 1$)-fund separation.

22.3.2 International CAPM

In the same way that diversification makes it possible to improve the risk-return trade-off for domestic investments, it is possible and in general desirable to proceed with the international diversification of the portfolio. All that is required is that the correlation between the various stock market returns is imperfect, which is actually the case.

A complication, however, would be likely to call into question the interest of this diversification, namely the existence of currency risk. Empirically, however, it is not enough to make international diversification inefficient, for two reasons:

- On the one hand, market risk and currency risk do not add up because of their imperfect correlation (close to zero in fact). If, for example, the standard deviation of the French market portfolio is 20%, the standard deviation of the Euro/Dollar exchange rate variations is 10% and the correlation between the two is negligible, the standard deviation of the rate of return of the French market in dollars (e.g., relevant for a US resident) is not 30% but only $(0.2^2 + 0.1^2)^{0.5} = 22.4\%$, higher than 20% but not by much. In some cases, the correlation between market risk and currency risk may even be negative, which may make overall risk less than strict stock market risk.
- On the other hand, the overall risk is itself reduced by the fact that the internationally diversified portfolio involves several economies not perfectly correlated.

Empirically, international diversification allows, given the expected return, to reduce the overall risk of portfolios due to the imperfect correlations between financial markets.

An international version of the CAPM[22] is obtained by following the same approach as for the domestic model; for any asset i from country p, we have:

---

[22] See Solnik (1974), Grauer et al. (1976) and Stulz (1981) for different versions of this model. An excellent summary, which remains for the most part relevant, is Adler and Dumas (1983).

$$\mu_{i,c} = r_c + \beta_i \left(\mu_{WM} - r_F \right) \tag{22.13}$$

where β_i is the international systematic risk of security i, i.e. calculated in relation to the *global* (world) market portfolio WM, r_c is the rate of the risk-free asset in the country of asset i, r_F is the *average global risk-free* rate, and *all* rates of return are expressed in the national currency of country c of asset i. The weights used to calculate μ_{WM} and r_F are identical and proportional to the market capitalization of each country.

In practice, estimation problems make the empirical tests of the international CAPM rather difficult. This does not, however, destroy the obvious interest in internationally diversifying the portfolio.

22.4 Limits of the CAPM

We will only mention two of the main theoretical and practical limits of the CAPM, namely the question of the efficiency of the market portfolio (Sect. 22.4.1) and that of the stability of the betas (Sect. 22.4.2).

22.4.1 Efficiency of the Market Portfolio and Roll's Criticism

For the CAPM to be valid, it is necessary and sufficient, as we have seen, that the market portfolio M be efficient, i.e., located on the Markowitz frontier. Therefore, the efficiency of M and the validity of the CAPM are two inseparable hypotheses, to be empirically tested jointly. This led Richard Roll (1977) to formulate the following criticisms:

- In fact, the market portfolio M is not observable because it includes in principle, as we have seen, all risky assets, including those that are not exchanged in an effective market (as most bank loans and human capital), or those whose market is opaque (such as art, luxury goods, or, to a lesser extent, real estate), or those for which it is impossible to know the value, in the absence of reliable statistical data (shares of small and medium-sized enterprises, durable goods, collections, etc.) Not being observable, its return is *a fortiori* not measurable and *it is therefore impossible to determine if it is efficient or not.*
- Being unobservable, for empirical purposes M must be represented by a portfolio or an index. Results of empirical tests and model-related performance measures then are likely to be sensitive to the choice of the stock market index P adopted as a proxy of M.
- The linear relation between the risk (measured by the beta with respect to P) and the expected return of the assets comes, as we have seen, from the efficiency of the proxy: if a given ex-post efficient (in the mean-variance sense) portfolio P is chosen as the market portfolio, and if the individual betas β_i are calculated from

this portfolio P, then we necessarily obtain a linear relation between the expected return μ_i and the risk β_i, in apparent accordance with the CAPM.

If, on the other hand, the portfolio P adopted as a proxy is not efficient ex-post, we can find a non-linear relation between μ_i and β_i, and the resulting over- or underperformance of each security i with respect to the CAPM norm is in fact due to the inefficiency of P, which may be an inadequately chosen proxy.

To summarize, the CAPM can be wrongly not rejected if P is ex-post efficient while the real market portfolio M is not. On the other hand, the CAPM can be wrongly rejected if P is not efficient ex post while M in fact is.

These criticisms of Roll make it necessary to be extremely cautious when interpreting the results of econometric tests of the CAPM (see Sect. 22.5).

22.4.2 Stability of Betas

Another difficulty in applying the model is that of beta instability over time (measured against the same market index). The betas appear, in fact, relatively volatile, which poses a difficult problem of measurement and application. Two observations are relevant in this respect:

- On the one hand, individual betas are more stable if they are calculated from bi-weekly or monthly returns than if they are from daily or weekly returns, because of large statistical noise at high frequency;
- On the other hand, betas of *well-diversified portfolios* are much more stable than individual betas. The first reason for this empirical observation is as follows: for an individual security, the fraction of systematic risk $\beta_i^2 \sigma_M^2$ in the total variance $\left(\sigma_i^2 = \beta_i^2 \sigma_M^2 + \sigma_{\varepsilon_i}^2 \right)$ is generally quite low and the share of diversifiable risk consequently quite high. This is indicated by a fairly low value, except for the few "heavyweights" included in the index, of the coefficient of determination R_i^2 of the regression (equal to the square of the correlation coefficient ρ_{iM} between R_i and R_M).[23] Therefore, it is plausible that the precision of the beta measurement is all the more questionable as the (systematic) risk that is measured represents a small fraction of the total risk. The second reason, linked to the former, is that a business is a complex, living entity that is constantly changing because of its investment opportunities and its financing policy (involving more or less financial leverage). Thus the measure of β_i is generally unreliable for an individual security, the noise being very important: the confidence interval (for example at

---

[23] Since $\beta_i = \sigma_{iM}/\sigma_M^2$ and $R_i^2 = \rho_{iM}^2 = \left(\sigma_{iM}^2/\sigma_M^2 \sigma_i^2 \right)$, it follows: $R_i^2 = \beta_i^2 \sigma_M^2/\sigma_i^2$ (R_i^2 here designates a coefficient of determination). Therefore, R_i^2 measures the proportion of systematic risk in total risk. Moreover, for given β_i and σ_M^2, the coefficient of determination is lower when the total variance σ_i^2 is larger. It is obviously equal to one only if $\sigma_{\varepsilon_i}^2 = 0$.

95%) for beta is generally very wide, typically (0.2–1.8) for an estimated beta of one.

On the other hand, for a portfolio, the diversification effect implies, as we have seen, that the diversifiable variance $\sigma_{\varepsilon_P}^2$ tends to zero and that the systematic variance $\beta_P^2 \sigma_M^2$ represents the bulk of the overall risk σ_P^2 (in a one-factor model). The coefficient of determination R_P^2 of the regression, therefore, tends in theory to one, and the reliability of the measurement increases with the number of securities included in the portfolio. As the number of securities entered in the portfolio increases and the weight of each of them decreases, the confidence interval for the beta of the portfolio narrows.

22.5 Tests of the CAPM

Since the development of the CAPM in the mid-1960s, countless empirical studies have been published on its validity, robustness, use in market efficiency tests, sensitivity to a change in the definition of the risk-free asset and, above all, the market portfolio, and the stability of the betas (see the previous section on this last point). We will limit the analysis to the examination of the main problems encountered in the tests.

CAPM tests face several challenges.[24] The first is that the variables μ_i and μ_M that it involves are defined ex ante, and not ex post, since they are expectations that, as such, are not directly observable. Among the statistical methodologies that have been proposed to circumvent this obstacle, the most classic is to *link* the CAPM to the market model. They write, respectively, for the period from $t-1$ to t (a time index must be used because the model test uses time series):

$$\mu_{it} = r_t + \beta_i \left(\mu_{Mt} - r_t \right) \text{ (CAPM)} \tag{22.i}$$

$$R_{it} = \alpha_{it} + \beta_i R_{Mt} + \varepsilon_{it} \text{ (Market model)} \tag{22.ii}$$

As (22.ii) implies:

$$\mu_{it} = \alpha_{it} + \beta_i \mu_{Mt},$$

by combining this last relation with (22.i), we obtain:

$$\alpha_{it} = (1 - \beta_i) r_t.$$

By incorporating this expression of α_{it} in (22.ii), it follows:

---

[24] Since the standard CAPM and its Black's version using the zero-beta portfolio are tested with similar methods, no special treatment of the second is provided. It will, however, be referred to occasionally.

$$R_{it} = r_t + \beta_i(R_{M\,t} - r_t) + \varepsilon_{it} \qquad (22.14)$$

with $E(\varepsilon_{it}) = 0$, and ε_i not correlated with R_M nor with $\varepsilon_j, j \neq i$.

The relation (22.14) is called the ex post CAPM because it involves the observable ex post returns (at t) and not the expectations developed ex ante (at $t - 1$) and unobservable ($\mu_{it} = E_{t-1}(R_{it})$ and $\mu_{Mt} = E_{t-1}(R_{Mt})$). This ex post CAPM (22.14) makes it possible to test the following different propositions:

- A *linear* relationship statistically links expected return to beta; it is called the empirical CAPM and writes:

$$R_{i,t} = \Upsilon_{0t} + \Upsilon_{1t}\,\beta_i + \varepsilon_{it} \qquad (22.15)$$

- The coefficients Υ_{0t} and Υ_{1t} of the empirical CAPM are identical for all securities i and, according to the standard CAPM, should be respectively equal to r_t and $(R_{Mt} - r_t)$;
- Beta is the only factor explaining the expected return of a risky security or portfolio. The introduction of additional explanatory variables such as the variance of R_i, dividends, firm growth rate or size, etc., should therefore not improve the quality of the model and estimates.

Most of the classical empirical studies[25] of the CAPM have proceeded as follows:

- As a first step, individual betas are estimated, using regressions (time series) of individual returns on an index supposed to reflect the successive returns of the market. For example, each regression is carried out over 5 years or more from historical monthly return.
- In a second step, the securities are grouped in portfolios or classes, each class comprising from 10 to 20 assets sorted in descending order of beta. Such grouping eliminates many of the statistical problems associated with errors in estimating betas for individual securities and their instability (see Sect. 22.4).
- Finally, the two coefficients Υ_{0t} and Υ_{1t} are estimated by instantaneous cross-sectional regression and then compared to the estimates of the respective theoretical values r_t and $(R_{Mt} - r_t)$.

With a few exceptions, classical empirical studies before the Fama and French publications (from 1992 onwards) led to the following conclusions:

- The coefficient Υ_{0t} of the regression (22.15) is significantly higher than the risk-free rate r_t and the slope Υ_{1t} is significantly lower than $(R_{Mt} - r_t)$: the low (high)

---

[25] See, for example, Blume and Friend (1973), Black et al. (1972), Fama and MacBeth (1973), and Banz (1981). For studies using a more advanced methodology, see Cochrane (2009).

22.5 Tests of the CAPM

beta securities, therefore, have a higher (lower) return than the standard CAPM. The empirical market line, therefore, has a stronger intercept and a weaker slope than the theory predicts. This empirical observation is more in line with Black's zero-beta CAPM.[26]

- Beta is the main factor that explains the returns on securities.
- Model (22.14), linear in beta, explains the data as well as more complex models.
- On average over a long period, the return on the market is much higher than the risk-free rate (the equity premium is around 7% for the SP500 over the period 1926–2018).[27] Fama and French and some following studies question these results, however, as we will see in Chap. 23 (Sect. 23.2.3).

In summary, the empirical market line (Eq. 22.15) with Υ_0 and Υ_1 estimated and not forced to be respectively equal to r and $(\mu_M - r)$) is an acceptable empirical model and a useful tool for the practitioner.

However, it should not be ignored that the tests presented above of the theoretical CAPM (Eq. 22.i) face, besides the problem of the non-observability of expectations, the other following difficulties:

- The stability over time of betas and risk premia [$(\mu_M - r)$ or $(\mu_M - \mu_z)$] is not assured, as we have seen. This instability makes the use of standard linear regression techniques (ordinary least squares) problematic and the interpretation of the results difficult. More powerful tests should be used, such as those involving GMM (Generalized Method of Moments).[28]
  Since the market portfolio M is not observable, an index P which is only a (small) part of it is used as a proxy. The risk of error is twofold (see the previous section): accept the CAPM when it is false because the portfolio P used instead of M is efficient; or reject the CAPM while it is correct because P is inefficient.
- The diversifiable risks ε_i are empirically (more or less) serially correlated, and are most often heteroscedastic (with non-constant variances), which makes the estimators obtained by simple linear regression inefficient. If the residuals are correlated (this is checked by the Durbin and Watson test), the model is in fact poorly specified because it lacks (at least) an explanatory variable. If they are heteroscedastic, the generalized least squares method must be used.
- Due to measurement errors in the rates of return, individual beta estimators obtained by simple or even generalized linear regression are inefficient and biased.

---

[26] In Black's CAPM the expected return μ_z replaces the risk-free rate r. Hence, $\Upsilon_{0,t}$ is theoretically equal to $\mu_{z,t}$ and $\Upsilon_{1,t}$ to $(\mu_{m,t} - \mu_{z,t})$. If μ_z is greater than the presumed rate r (which in face does not exist) then: $\Upsilon_0 > r$ and $\Upsilon_1 < (\mu_M - r)$.

[27] In CAPM applications, $(\mu_M - r)$ is often chosen between 5% and 8%, and $\sigma(r_M)$ between 15 and 20%.

[28] See, e.g., Cochrane (2009).

The technical answers to these problems, when they exist, generally appear more or less satisfactory, except for the non-observability of the market portfolio, an insurmountable obstacle in theory, but in part circumvented by the use of a market index encompassing several types of assets (equities, bonds, money market securities, real estate, and so on).

22.6 Summary

- The Capital Asset Pricing Model (CAPM) is an equilibrium relation between expected return and risk. It rests on the hypothesis that *the market portfolio M is mean-variance efficient. M* is defined as the portfolio comprising *all* the risky assets present in the economy (but *not* the possibly existing risk-free asset).
- Tobin's two-fund separation states that all efficient portfolios are combinations of the asset yielding the risk-free rate r (if it exists) and of any efficient portfolio, and that the *efficient frontier is the half-line in the* (σ, μ) space tangent to Markowitz's hyperbola, passing through the point $(0, r)$ and the point T (σ_T, μ_T) representing the tangent portfolio. Moreover, T is the *only* efficient portfolio containing *only* risky assets. Hence, since M is assumed efficient, M *is the tangent portfolio and all the efficient portfolios are located on the half-line* $(0, r) - M$ (σ_M, μ_M).
- Consequently, in the presence of a *risk-free asset*, the relation between the expected return and the risk (measured by the standard deviation) of the efficient portfolios is *linear*: $\mu_P = r + \left(\frac{\mu_M - r}{\sigma_M}\right) \sigma_P$ (*Capital Market Line,* CML).
- The slope of the CML, $(\mu_M - r)/\sigma_M$, is interpreted as the *market price of risk* for a portfolio.
- The CML is only valid for *efficient portfolios* and not for *individual assets*. The CAPM establishes the relation that must prevail between the relevant measure of risk and the expected return of any asset or portfolio.
- The CAPM states that, if M is efficient, whether a risk-free asset exists or not, the *linear relation:* $\mu_i = \lambda + \beta_i (\mu_M - \lambda)$ holds for *any* individual security or portfolio i, where λ is a positive parameter (common to all i) and $\beta_i = \sigma_{iM}/\sigma_M^2$.
- The graph of this linear relationship in the plane (μ, β) is the *Security Market Line* (SML) passing through the points $(0, \lambda)$ and $(1, \mu_M)$. The representative points of all securities and portfolios are located on this line.
- In general, λ *is the expected return of any zero-beta asset*. In presence of a risk-free asset, $\lambda = r$, and the *standard* CAPM obtains: $\mu_i = r + \beta_i (\mu_M - r)$.
- An intuitive explanation of the CAPM is based on the distinction, found in Sharpe's market model, between *systematic risk* and *diversifiable risk*. At equilibrium, the market compensates by a risk premium the (inevitable) systematic risk only, as the diversifiable risk can be eliminated by portfolio diversification. Since the systematic risk is measured by beta, the risk premium, which is added to the risk-free rate r to obtain μ_i, is proportional to β_i.

Suggestions for Further Reading   961

- *The equilibrium value P_i of an asset generating a random flow $\widetilde{P}_i^1$, in the presence of a risk-free asset, is given by:* $P_i = \dfrac{E\left(\widetilde{P}_i^1\right) - \theta\,\mathrm{cov}\left(\widetilde{P}_i^1, R_M\right)}{1+r}$. The numerator on the r.h.s. is the certainty-equivalent of the random flow $\widetilde{P}_i^1$.
- To implement the CAPM, one needs to estimate (i) the different β_i, by a regression of the past observed $R_{i,t}$ on $R_{M,t}$, and (ii) the market risk premium ($\mu_M - r$), which is theoretically an *anticipated premium*, but is estimated by an *historical average* of the difference between the market return and the risk-free rate. The "theoretical" expected returns μ_i can then be calculated using the CAPM and compared to anticipated returns, developed through financial analysis for example, in order to assess the opportunity of an investment.
- Two standard measures of portfolio performance are grounded on the CAPM: the *Sharpe Ratio* and *Jensen's alpha.*
- *The Sharpe ratio* (SR) of a portfolio P relates its risk premium to its standard deviation: $\mathrm{SR} = (\mu_P - r)/\sigma_P$.
  All efficient portfolios should have the same SR, including M, equal to the slope of the CML. The higher its SR, the better is the performance of P. A SR is *estimated* from historical data, then available only ex post.
- *Jensen's alpha* is the difference between the expected return on the portfolio and its expected theoretical return calculated by the CAPM: $\alpha_P \equiv \mu_P - r - \beta_P(\mu_M - r)$.
- The *Sharpe ratio* measures a deviation from the CML while the *alpha* is a deviation from the SML. The SR is appropriate for a well-diversified portfolio or fund representing the bulk of the investor's wealth while the alpha is more relevant for a portfolio (diversified or not) that constitutes only a part of the investor's wealth.
- The *intertemporal CAPM* (yielding multi-beta risk premiums), and an *international* version of the CAPM (where M is the *global* market portfolio) are important theoretical and practical extensions of the CAPM.
- The CAPM suffers from several theoretical and practical limits. In particular, the *market portfolio M is not observable*; hence, it is *impossible to assess whether it is efficient or not* and tests of the CAPM are in fact tests of the efficiency of the stock market index adopted as a *proxy* for M. In addition, the betas appear relatively unstable over time, which poses difficult problems of measurement and implementation.

Suggestions for Further Reading

Books

Bodie, Z., Kane, A., & A. Marcus, 2007, 7th ed, Investments, Irwin.
Cochrane, J. H. (2009). *Asset pricing* (2nd ed.). Princeton University Press.
*Dumas, B., & Luciano, E. (2017). *The economics of continuous-time finance*. MIT.

Elton, E., Gruber, M., Brown, S., & Goetzmann, W. (2010). *Modern portfolio theory and investment analysis* (8th ed.). Wiley.

*Merton R., 1999, Continuous time finance, Basil Blackwell.

Mossin, J. (1973). *Theory of financial markets*. Prentice Hall.

Sharpe, W. (2000). *Portfolio theory and capital markets*. McGraw Hill.

Solnik, B. (2000). *International investments* (4th ed.). Addison-Wesley.

Articles

Adler, M., & Dumas, B. (1983). International portfolio choice and corporation finance: A synthesis. *Journal of Finance, 38*, 925–984.

Banz, R. (1981). The relationship between return and market value of common stock. *Journal of Financial Economics, 9*.

Black, F. (1972). Equilibrium with restricted borrowing. *Journal of Business, 45*, 444–454.

Black, F. (1993). Beta and return. *Journal of Portfolio Management, 20*(1), 8–18.

Black, F., Jensen, M., & Scholes, M. (1972). The capital asset pricing model: Some empirical tests. In M. Jensen (Ed.), *Studies in the theory of capital markets*. Praeger.

Blume, M., & Friend, I. (1973). A new look at the capital asset pricing model. *Journal of Finance*.

Cox, J., Ingersoll, J. E., & Ross, S. A. (1985). A theory of the term structure of interest rates. *Econometrica, 53*, 385–408.

Fama, E. F., & MacBeth, J. D. (1973). Risk, return and equilibrium: Empirical tests. *Journal of Political Economy, 81*, 607–636.

Gibbons, M. R. (1982). Multivariate tests of financial models. *Journal of Financial Economics, 10*.

Goetzmann, W. N., Ingersoll, J. E., Siegel, M. I., & Welch, I. (2002). Sharpening the sharpe ratio. *NBER working paper*.

Grauer, F., Litzenberger, R., & Stehle, R. (1976). Sharing rules and equilibrium in an international capital market under uncertainty. *Journal of Financial Economics*.

Grinblatt, M., & Titman, S. (1989). Portfolio performance evaluation: Old issues and new insights. *Review of Financial Studies, 2*, 393–416.

Henriksson, R. D., & Merton, R. C. (1981). On market timing and investment performance of managed portfolios (II): Statistical procedures for evaluating forecasting skills. *Journal of Business, 54*.

Jensen, M. (1969). Risk, the pricing of capital assets, and the evaluation of investment performance. *Journal of Business, 42*(2), 167–247.

Lintner, J. (1965). The valuation of risky assets and the selection of risky investments in stock portfolio and capital budgets. *Review of Economics and Statistics, 47*, 13–37.

Merton, R. C. (1973). An intertemporal capital asset pricing model. *Econometrica, 41*, 867–888.

Mossin, J. (1966). Equilibrium in a capital asset market. *Econometrica, 34*, 768–783.

Roll, R. (1977). A critique of the asset pricing theory's tests, part I: On past and potential testability of the theory. *Journal of Financial Economics, 1*, 129–176.

Sharpe, W. (1964). Capital asset pricing: A theory of market equilibrium under conditions of risk. *Journal of Finance, 19*, 425–442.

Solnik, B. (1974). An equilibrium model of the international capital market. *Journal of Economic Theory, 8*, 500–524.

Stulz, R. (1981). A model of international asset pricing. *Journal of Financial Economics, 9*.

Tobin, J. (1958). Liquidity preference as behavior toward risk. *Review of Economic Studies*, 65–86.

Treynor, J. (1965, January–February). How to rate management of investment funds, *Harvard Business Review*.

Treynor, J., & Mazuy, K. (1966). Can mutual funds outguess the market? *Harvard Business Review, 43*.

Arbitrage Pricing Theory and Multi-factor Models

23

Despite its limitations, the CAPM remains in finance the paradigm as a pricing model. However, in the late 1970s, S. Ross (1976) then R. Roll and S. Ross (1980) developed a competing model, the Arbitrage Pricing Theory (APT). The first motivation of the authors was to free the model from the restrictive hypotheses leading to the mean-variance paradigm and concerning either the investors' utility functions, supposedly quadratic, or the distributions of the rates of return, supposedly Gaussian. In fact, the derivation of the APT does not require a restriction on the former, except implicitly risk aversion, nor does it require the assumption of normality of returns. A second motivation was to find a response to the criticism of the CAPM by R. Roll, explained in the previous chapter, concerning the unobservable, and therefore not measurable, market portfolio, a feature that makes any robust empirical test of the model extremely difficult. A third motivation was to enrich the model by allowing explicit consideration of *several* risk factors common to most securities, thus inducing various non-diversifiable risks that justify several components in the risk premium (whereas the CAPM includes only one).

The CAPM, as we have seen, does not rely on the assumption of a return generating process in line with Sharpe's market model (although the latter model is compatible with it). On the contrary, the APT is based on the assumption that returns are generated by common risk factors. As a result, there is often a confusion between the APT model (and its derivatives) and multi-factor models. We adopt the following respective meanings. Multi-factor models are statistical models that postulate a certain *return generating process* (like the market model which is a special case). They are by nature inter-temporal. The APT, like the CAPM, is an evaluation model, i.e., *an explanation of the risk premia* that must prevail in absence of arbitrage opportunities or equilibrium, respectively, at a given point of time.

Since the APT, unlike the CAPM, is based on a multi-factorial model, we will begin this chapter with a presentation of multi-factor models (Sect. 23.1). Section 23.2 derives and interprets APT as a model for assessing risk premia. Section 23.3 addresses the issue of its practical implementation and examines the most important of its applications. Section 23.4 finally proposes a comparison

© The Author(s), under exclusive license to Springer Nature Switzerland AG 2022
P. Poncet, R. Portait, *Capital Market Finance*, Springer Texts in Business and
Economics, https://doi.org/10.1007/978-3-030-84600-8_23

between the CAPM and the APT, regarding their theoretical status and the difficulties encountered in econometric tests, then a brief comparison of the respective merits of multi-factor models.

23.1 Multi-factor Models

Multi-factor models are pragmatic and not theoretical. They aim, aside from any consideration of equilibrium or of no-arbitrage, to explain empirically the structure of the correlations between the returns on risky assets, assuming a priori that several *common factors* influence them. After presenting these models (Sect. 23.1.1), we will discuss their main practical uses (Sect. 23.1.2).[1]

23.1.1 Presentation of Models

Empirically, common factors other than the market itself may exist that lead to systematic correlations between the rates of return on securities. They can be represented either by indices representing industries, economic sectors, or microeconomic characteristics, by macroeconomic indices, or even by linear combinations of returns determined from a factor analysis. We will come back to the determination of the factors in Sect. 23.3.1 devoted to the implementation of these models.

The objective of multi-factor models is to best explain the structure of correlation of the returns on different assets and to separate systematic risks from diversifiable risks. Formally, these models postulate that the process generating random returns of individual securities is correctly described by the equation:

$$R_i(t) = \mu_i + \sum_{k=1}^{m} b_{ik}F_k(t) + \varepsilon_i(t) \text{ for each } i = 1, \ldots, n \text{ and each period } t, \quad (23.1)$$

where μ_i is the expectation of R_i, the b_{ik} represent the sensitivities[2] of the return on asset i to the m common centered factors F_k, ε_i is a white noise of zero average which has the following properties of non-correlation (in cross sections and time series):

$$\text{Cov}(\varepsilon_i, \varepsilon_j) = \text{cov } (\varepsilon_i, F_k) = 0 \text{ and cov } (\varepsilon_i(t), \varepsilon_j(t')) = 0 \text{ for every } i, k, j \neq$$
$$= i \text{ and } t' \neq t.$$

The common factors F_k have zero mean and, in general, are more or less correlated. However, it is possible to convert a set of correlated factors into another

---

[1] See for more complete developments Elton et al. (2010).
[2] Also called saturation or factor loading.

23.1 Multi-factor Models

set of orthogonal factors.[3] We will assume the F_k to be uncorrelated to ease the analysis. For empirical purposes (these models have an essentially empirical aim), the coefficients μ_i and b_{ik} are supposed to be constant, as are the variances $\sigma^2_{ei(t)}$ and $\sigma^2_{Fk(t)}$ which will allow in the sequel to omit the index t in some equations.

We check that by taking the mathematical expectation of each side of Eq. (23.1), we obtain the wanted equality $\mu_i = \mu_i$. Equation (23.1) means that the deviation of the return on security i from its average is due, on the one hand, to a risk specific to the security, and, on the other, to the deviation of the common factors from their expectations. For example, if the inflation rate is one of the factors and its expected value is 4%, given the economic policy of the concerned country, while the inflation rate actually realized is 5%, it is the unforeseen 1% difference that counts in the formula.

The variances of the securities' returns then write:

$$\sigma_i^2 = \sigma_{\varepsilon_i}^2 + \sum_k^m b_{ik}^2 \sigma_{F_k}^2 \tag{23.2}$$

Then, the covariance between the rates of return of two assets i and j is, under the assumptions above:

$$\sigma_{ij} = \sum_k^m b_{ik} b_{jk} \sigma_{F_k}^2 \quad \forall i, \forall j \neq i \tag{23.3}$$

When $m = 1$, we recover Sharpe's market model (see Chaps. 21 and 22) which reads:

$$R_i = \mu_i + b_i F + \varepsilon_i \quad \forall i = 1,...,n \tag{23.4}$$

Note that, up to a scale factor, F is interpreted as the centered market return ($R_M - \mu_M$).

Indeed, by summing the returns on securities weighted by market weights m_i, we obtain:

$$R_M = \sum_{i=1}^n m_i R_i = \sum_{i=1}^n m_i \mu_i + \sum_{i=1}^n m_i b_i F + \sum_{i=1}^n m_i \varepsilon_i = \mu_M + b_M F + \sum_{i=1}^n m_i \varepsilon_i.$$

Since the last term tends to zero when n tends to infinity, we have: $F = \frac{1}{b_M}(R_M - \mu_M)$ and therefore, by letting $b'_i = b_i / b_M$, we obtain:

$$R_i = \mu_i + b'_i (R_M - \mu_M) + \varepsilon_i \quad \forall i = 1,...,n. \tag{23.5}$$

---

[3] See Appendix 1 for a proof.

23.1.2 Portfolio Management Models in Practice

Multi-factor models are mainly used for purposes of (i) estimating the variances-covariances of asset returns, (ii) risk analysis and management, and (iii) simulation and portfolio composition.

(i) The problem of estimating the variance-covariance matrix, although more complicated than in the one-factor Sharpe model, is still very simplified compared to the general case, since instead of having to measure $n(n-1)/2$ covariances (in addition to the n variances), it is enough to estimate the ($n \times m$) sensitivities b_{ik} and the m variances of the common factors, i.e., $m(n+1)$ variables. For $n = 500$, and $m = 10$, it is therefore "enough" to estimate 5010 parameters instead of 124,750 covariances! For the Sharpe model ($m = 1$), it would be sufficient to estimate 501 parameters (500 betas and the variance of the market).

(ii) A multi-factorial model makes it possible to analyze in detail the risk of a portfolio by breaking it down into its elementary sources (factors). In particular, it allows a better understanding of the differences in the sensitivity of the portfolio to the various risks in relation to the objectives that had been posted, particularly in the case of benchmarking. Then the decision to increase or reduce certain risks is made easier. Moreover, the breakdown of the overall risk between systematic risks, which should be earning a premium on average, and the specific risk that is not, is crucial information, the general principle being to avoid unrewarded risks.

(iii) A multi-factor model also allows estimating the evolution of a future portfolio by simulating the behavior of the various risk factors. It can therefore be used as a tool for active portfolio management (as opposed to passive management that merely replicates a given index or benchmark). By simulating the behavior of the different risk factors and modifying the b_{pk} sensitivities of the portfolio (by changing its composition), one obtains the portfolio that reacts according to the investor's or manager's desires.

23.2 Arbitrage Pricing Theory

The Arbitrage Pricing Theory (APT) is sometimes presented as a more general model than the CAPM in that it frees itself from the restrictions necessary to obtain the mean-variance criterion. In fact, the APT is a model based on a factorial model and a no-arbitrage (AAO) assumption, but it is not in its original form an equilibrium model.[4] Recall that equilibrium implies AAO but that the reciprocal is false

---

[4]Jarrow (1988) has shown how to obtain a version of the APT, not presented in this book, as a partial equilibrium model, at a cost of additional assumptions about the preferences of economic agents.

(in general). An example is the case of the quoted price of an asset under rationing (of supply or demand), a situation in which the (constrained) equilibrium is done by the quantity and not the price. The AAO prevails (investors cannot buy or sell the asset(s) necessary to carry out the arbitrage) but the price is not that of the classic Walrasian equilibrium, since either buyers or sellers cannot perform the desired transactions at quoted prices. The original APT is a less demanding model in some respects (utility functions) than the CAPM, but it requires other assumptions about the structure of returns expressed by the multi-factor model. These models thus do not have the same theoretical status and therefore are not direct competitors.

After the presentation of the assumptions and notations used (Sect. 23.2.1), we derive the model and provide a financial interpretation (Sect. 23.2.2) and then study the relationship and the possible compatibility between the APT and the CAPM (Sect. 23.2.3).

23.2.1 Assumptions and Notations

The APT is based on the assumption of AAO on the one hand and a purely statistical assumption on the other hand. First, in AAO, a certainly positive terminal value portfolio must have a positive initial value (otherwise the market would allow an arbitrage). Secondly, a limited number of *common systematic factors* are assumed to affect the estimated rate of return on all risky financial assets. The dual purpose of the model is to identify these factors and to provide the equation linking the expected return of a security to these common sources of risk.

There are several ways to derive the model. We do not give the most general, but we present a more accessible justification, based on linear algebra. In addition, we provide the discrete mono-periodic version, as for the CAPM.[5]

The statistical decomposition of the return of a risky security i *for a given period* t is obtained by the orthogonal projection on the factors, which leads to Eq. (23.1) that we rewrite to ease the reading:

$$R_i = \mu_i + \sum_{k=1}^{m} b_{ik} F_k + \varepsilon_i$$

where F_k is the unanticipated variation of the risk factor k, therefore of zero mean, the F_k are neither correlated with each other nor correlated with ε_i. The latter represents the specific risk of security i, of zero mean also *by construction*.[6] Note that this decomposition is not an assumption because it is obtained automatically from orthogonal centered factors with: $\mu_i = E(R_i)$ and $b_{ik} = \text{cov}(R_i, F_k)/\sigma^2_{Fk}$.

---

[5] For the continuous-time version of the APT, see Sects. 19.2 and Sect. 20.3.

[6] Formally, we have: $E(\varepsilon_i) = \text{cov}(\varepsilon_i, \varepsilon_j) = E(F_k) = \text{cov}(F_k, F_m) = \text{cov}(F_k, \varepsilon_i) = 0$, $\forall i, j \neq i$, and $\forall k$, $m \neq k$.

The crucial assumption underlying the model is that the risks ε_i are not correlated with each other and are therefore diversifiable: the correlation between the individual returns R_i is entirely due to (and explained by) the influence of the common factors F_k. This allows the law of large numbers to apply and the specific risk of a well-diversified portfolio to be neglected. Other assumptions usually underlying multi-factor models (notably no serial correlation, homoscedasticity of residuals, or constant coefficients) are not necessary for the validity of the theoretical model, but are generally necessary when it comes to implementing it and estimating the coefficients.

The number of marketable securities n is assumed large and the number of common factors m is presumed small as compared to n. These factors may be, for example, particular portfolios based on existing individual securities or macroeconomic magnitudes exogenous to the model. The proper choice of factors depends in particular on the orthogonality of ε_i (at best an approximation) and the value of the coefficient of determination of the regression. We will come back later on to this question, which is crucial for the practical implementation of the model. It should be noted, however, that the theory does not indicate the nature of these factors and that, as a result, the market M as a whole could be one of them.

Let P be any portfolio characterized by the weights $x_1,.., x_n$. The sensitivities b_{pk} of P to the different factors are equal to:

$$b_{pk} = \sum_{i=1}^{n} x_i b_{ik} \quad \forall\, k = 1, ..., m\,;\ \text{with}\ \sum_{i=1}^{n} x_i = 1. \tag{23.6}$$

This result stems directly from the fact that a sensitivity is a covariance (normed by a variance) and that a covariance is a (bi)-linear operator.

23.2.2 The APT

We will proceed in two stages, the model being derived first in the very simplified framework of a single common source of risk, and from a very simple numerical example (Sect. 23.2.2.1), then in the general case, and more rigorously, for individual securities and any number of sources of risk, but lower than the available securities (Sect. 23.2.2.2). The derivation of the model will require the construction of a particular portfolio, called *arbitrage portfolio*, with zero initial value and (approximately) risk free.

23.2.2.1 Simplified Approach

(a) *One source of risk*

Suppose that there is only one common source of risk so that Eq. (23.1) describing the generation of returns reduces to Eq. (23.4) that we recall to ease the reading:

23.2 Arbitrage Pricing Theory

$$R_i = \mu_i + b_i F + \varepsilon_i \quad \text{for each } i. \tag{23.7}$$

If investors hold well-diversified portfolios, their residual risk may be considered negligible in practice (because the ε_i are not correlated with each other). The return of such a portfolio p then writes:

$$R_p = \mu_p + b_p F,$$

and is completely characterized by parameters $\mu_p = \sum_{i=1}^{n} x_i \mu_i$ and $b_p = \sum_{i=1}^{n} x_i b_i$.

Example 1

Let us introduce the model by an example.

Suppose there are two well diversified portfolios A and B whose characteristics are given below:

Portfolio	Expected return μ_p (%)	Sensitivity b_p
A	12	1.0
B	10	0.8

It is easy to determine the equation linking the return on a portfolio to its sensitivity. By writing for each portfolio A and B the equation with two unknowns (λ_0 and λ_1):

$$\mu_p = \lambda_0 + \lambda_1 b_p,$$

and using the data of the table (for μ_p and b_p), we solve a system of two equations with two unknowns whose solution is unique in AAO (if A and B have a different risk, measured by their sensitivity, their expected return must be different, which is the case here). It follows:

$$\lambda_0 = 2 \text{ and } \lambda_1 = 10.$$

The desired equation, a line in the plane (expectation, sensitivity to the common factor), is therefore:

$$\mu_p = 2 + 10\, b_p \tag{23.8}$$

It follows, since any linear combination of points on a line (the sum of the weights making one) lies on the line, the representative point of any portfolio built from portfolios A and B is located on the line described by Eq. (23.8). For example, a portfolio C made up of 75% of A and 25% of B yields on average $0.75 \times 12(\%) + 0.25 \times 10(\%) = 11.5(\%)$ and, as its sensitivity is $0.75 \times 1 + 0.25 \times 0.8 = 0.95$, we also have $\mu_c = 2 + 10 \times 0.95 = 11.5(\%)$. The reader

(continued)

will check that the portfolio D composed by 150% of A and −50% of B has an expectation of 13(%), also in accordance with (Eq. 23.8).

Suppose a well-diversified portfolio E has an expected return that is not in line with Eq. (23.8); for example, it has a sensitivity of 0.9 but yields an average of 13 (%) instead of $2 + 0.9 \times 10 = 11$ (%). We then build an *arbitrage portfolio* H. This one requires, on the one hand, *no* initial down payment, the short sales, which are supposed allowed, financing the purchases, and, on the other hand, is *constructed* in such a way that its systematic risk b_h is zero. Moreover, as it is formed from well diversified portfolios, its specific risk ε is negligible. Its total risk can therefore be considered as zero. Let this portfolio H consist of portfolio E purchased (for example for $1000) and portfolio F (defined later) sold short (also $1000). The net investment is therefore nil. F is composed equally of A and B, and therefore has an expected return of 11% ($= (12+10)/2$) and a sensitivity of 0.9 ($= (1+0.8)/2$)). The sensitivity b_H of the portfolio H is therefore zero (0.9−0.9). The final value of H is (without risk) $1000 (1 + 0.13) − 1000 (1 + 0.11) = \$20$. This risk-free gain without down payment is of course a violation of the AAO.

In general, the APT indicates that the relationship between the expected return on a security or portfolio and its systematic risk, as measured by its sensitivity b, is *linear*: All portfolios must respect Eq. (23.8) whose general writing for a single common source of risk is:

$$\mu_p = \lambda_0 + \lambda_1 b_p, \tag{23.9}$$

the two scalars λ_0 and λ_1 being the same for all securities and portfolios.

Assume that a nonlinear relationship (regardless of the convexity of the representative curve) is observed between the expected return of the portfolios and their sensitivity to this factor, in accordance with Fig. 23.1, where are distinguished the four well-diversified portfolios A, B, C, and D. An appropriate linear combination of portfolios A (bought) and B (short sold) gives a portfolio X, which is insensitive to the common factor and therefore (approximately) risk-free, returning r_x. In the same way, a portfolio Y positively composed of C and negatively of D has an (approximately) risk-free return r_y. It is then sufficient to buy Y and sell X to obtain, (practically) without risk, an arbitrage gain equal to $(r_y - r_x)$.

Such a situation is not compatible with AAO. The only one that is (in the absence of information and transaction costs) is that for which points A, B, C, and D are aligned in the plane (b_p, μ_p).

(b) *Several sources of risk*

An analysis similar to the one that leads to Eq. (23.9) yields in the presence of m common sources of risk the following linear relationship that must prevail in AAO:

23.2 Arbitrage Pricing Theory

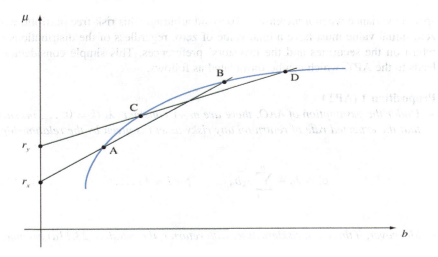

Fig. 23.1 Arbitrage opportunity

$$\mu_p = \lambda_0 + \sum_{k=1}^{m} \lambda_k b_{pk} \qquad (23.10)$$

Just as securities and portfolios must be located, because of Eq. (23.9), on a line in the plane (b_p, μ_p), more generally they must be located in the $((m + 1)$-dimensional) *hyperplane* defined by the sensitivities to the m factors and the expected return. In such a context, $(\mu_p - \lambda_0)$ is interpreted as the risk premium applicable to portfolio p, the λ_k are the market prices of risk and the b_{pk} the systematic risks of p.

The relation (23.10) is the APT for well-diversified *portfolios*. What about *individual* securities i? The answer is the same, i.e., Eq. (23.9) (or 23.10). Although a more rigorous demonstration is required (see Sect. 23.2.2.2 below), the intuition is the following. The difference between a well-diversified portfolio p and an individual asset i is that the latter, in addition to its systematic risk(s), presents a specific risk ε_i that is not negligible (see Eq. 23.1). However, the market does not reward it with a risk premium since it is diversifiable. The *expected* return μ_i should therefore depend only on the systemic risk(s) b_{ik} according to Eqs. (23.9) or (23.10) with $p = i$.

23.2.2.2 A More Rigorous Justification for APT

Since the previous justification is essentially intuitive, a more rigorous demonstration of APT is now provided. It is based on the construction of an arbitrage portfolio. Remember that an arbitrage portfolio has zero initial value, thus does not require a down payment, since short sales finance purchases. In addition, its systematic risks (vis-à-vis the common risk factors) are zero, because it is *constructed* in such a way that each of the b_{pk} given by Eq. (23.6) is equal to zero. Lastly, its diversifiable risk is (almost) zero because the number of securities (n) entering into its composition is assumed large (the law of large numbers ensures the convergence to zero of the specific variance when n increases). To avoid arbitrage, this risk-free portfolio with zero initial value must have a final value of zero, regardless of the distributions of return on the securities and the investors' preferences. This simple consideration leads to the APT, which can be formulated as follows.

Proposition 1 (APT)

- *Under the assumption of AAO, there are $m +1$ constants λ_k ($k = 0, \ldots, m$) such that the expected rate of return on any risky asset i is given by the relationship:*

$$\mu_i = \lambda_0 + \sum_{k=1}^{m} \lambda_k b_{ik} \qquad \text{for } i = 1, \ldots, n. \qquad (23.11a)$$

- *Moreover, if there is a riskless asset with return r, the relation ($23.11a$) becomes:*

$$\mu_i = r + \sum_{k=1}^{m} \lambda_k b_{ik} \qquad \text{for } i = 1, \ldots, n. \qquad (23.11b)$$

*Proof
Consider a portfolio of zero initial value, i.e., such that:

(i) $\sum_{i}^{n} x_i = 0$ or, in vector notation, $\underline{x}'.\underline{1} = 0$,

  where x_i is the amount in dollars invested in security i.

The change in value of this portfolio (also equal to its terminal value) is, given Eq. (23.1), equal to:

$$\sum_{i}^{n} x_i(1 + R_i) - \sum_{i}^{n} x_i = \sum_{i}^{n} x_i R_i = \sum_{i}^{n} x_i \mu_i + \sum_{k}^{m} \left(\sum_{i}^{n} x_i b_{ik} \right) F_k + \sum_{i}^{n} x_i \varepsilon_i$$

For this portfolio to be risk-free, it must first be insensitive to each of the factors k, and secondly, its diversifiable risk $\varepsilon_p = \sum_{i}^{n} x_i \varepsilon_i$ must be diversified away. Therefore, we must first choose the x_i such that:

(ii) $\sum_{i}^{n} x_i b_{ik} = 0 \quad \forall k$, or, in matrix form: $\underline{x}'.\underline{B}_k = 0$, $k = 1, \ldots, m$

  where $\underline{B}_k$ is the kth column of the ($n \times m$) matrix $\mathbf{B}$ of the b_{ik}, which has rank m.

It is then necessary that the diversifiable risk of the portfolio $\sum_{i}^{n} x_i \varepsilon_i$ be negligible, which is possible under the law of large numbers; it is asymptotically zero if the weight of each security tends to zero with the number n of securities:

23.2 Arbitrage Pricing Theory

(iii) $\lim |x_i| = 0$ when $n \to \infty$, for $i = 1, \ldots, n$.

A portfolio with all three properties (i)–(iii) is called an *arbitrage portfolio*.

Such a portfolio requires a zero investment and has a (quasi-) certain terminal value equal to $\sum_{i=1}^{n} x_i \mu_i$; the non-arbitrage condition (AAO) implies that this terminal value is zero, i.e.:

(iv) $\sum_{i=1}^{n} x_i \mu_i = 0 \Leftrightarrow \underline{x}' . \mu = 0$, for all $\underline{x}$ respecting (i), (ii), and (iii).

Consider now the vector subspace of $\mathbb{R}^n$ generated by the $(m + 1)$ vectors $\{\underline{1}, \underline{B}_1, \ldots, \underline{B}_m\}$, which we denote V, and its orthogonal we denote $V^\perp$. The conditions (i) and (ii) ($\underline{x}'.\underline{1} = 0$ and $\underline{x}'.\underline{B}_k = 0$ for $k = 1, \ldots, m$) imply that *any* arbitrage portfolio $\underline{x}$ belongs to the subspace $V^\perp$ and that *any* vector of $V^\perp$ is an arbitrage portfolio.[7] Moreover, according to condition (iv), the vector of expected returns μ is orthogonal to any vector $\underline{x}$ that is representative of an arbitrage portfolio, therefore to *all* vectors of $V^\perp$: Therefore it belongs to V. It can thus be written as a linear combination of $\{\underline{1}, \underline{B}_1, \ldots, \underline{B}_m\}$, i.e. there exists $m + 1$ constants $(\lambda_0, \lambda_1, \ldots, \lambda_m)$, such that:[8]

(v) $\mu = \lambda_0 \underline{1} + \sum_{k=1}^{m} \lambda_k \underline{B}_k$

or, more explicitly, for the *i*th line of (v), which is a particular asset *i*:

$$\mu_i = \lambda_0 + \sum_{k=1}^{m} \lambda_k b_{ik} \qquad i = 1, \ldots, n. \tag{23.11a}$$

It will be noted that the preceding analysis, as well as result (Eq. 23.11a), constitute only a generalization of the argument developed and the result obtained in the simple case of two portfolios and one source of risk (Sect. 23.2.2.1 above).

If there is a risk-free asset with return r, the constant λ_0 is equal to r. Indeed, a risk-free asset, numbered zero, is by definition insensitive to each of the factors k: $b_{0k} = 0$, for all k. Its return, given by Eq. (23.11a), is therefore: $\mu_0 = r = \lambda_0$. The wanted Eq. (23.11b) is the Eq. (23.11a) with $\lambda_0 = r$.

---

[7] Up to the asymptotic condition (iii); this restriction bestows to APT its approximate or asymptotic character.

[8] Result (v) is only approximate in a (real) economy where the number of risky assets is (necessarily) finite. This is because, even though, *on average*, the specific risk of a security *i* has zero return, this does not strictly prohibit that for certain securities this is not the case. Also, if the correlations between the residuals ε_i are not strictly zero, the APT does not necessarily hold exactly for each asset *i*, even if the number n of securities is infinite. In both cases (not mutually exclusive), there may be small errors for most securities and big errors for a small number of securities. See, for example, Ingersoll (1984).

The relationship between expected return and risk remains linear, as for the CAPM. The return required ex ante by the market for an investment in security i is equal to the risk-free rate plus the risk premia (which can be positive, negative, or zero, according to the sign of the sensitivities b_{ik}).

The $\lambda_k b_{ik}$ are analyzed as the *risk premia* associated with the *common factors k,* and each λ_k as the *market premium per unit of risk k, or market price of the risk k.* One can also give a more concrete interpretation of these λ_k.

Let us denote δ_k the expected return of a risky portfolio of sensitivity b_{kk} equal to 1 to the factor k and of sensitivity equal to 0 to all the other factors $k' \neq k$. Applying the APT to this portfolio, which can be called the *factorial portfolio k,* yields:

$$\delta_k = r + \sum_{k=1}^{m} \lambda_k b_{kk} = r + \lambda_k$$

The expected return of a portfolio (or asset), only sensitive to the common source of risk k (with a sensitivity equal to one) must exceed the risk-free rate by the amount λ_k. This implies: $\lambda_k = \delta_k - r$ and leads to the next proposition.

Proposition 2: Alternative writing of APT *Equation (23.11b) giving the APT in the presence of a risk-free asset can be rewritten:*

$$\mu_i - r = \sum_{k=1}^{m} b_{ik}(\delta_k - r) \qquad i = 1, \ldots, n. \qquad (23.12)$$

The expected excess (in relation to the risk-free rate) return on a risky asset in AAO is the sum of the expected risk premia of the factor portfolios weighted by the sensitivities of the asset under consideration to the various sources of risk.

Equation (23.12) is generally the empirically tested version of the APT. It is important to emphasize that, although there are m common risk factors exhibited empirically in the economy, it is possible that only m' factors, with $m' < m$, have a "price," i.e., are paid a risk premium λ_k statistically different from zero. We will come back to this question.

Finally, note that, like for the CAPM and the simplified analysis presented in Sect. 23.2.2.1 above, Eqs. (23.11a), (23.11b), and (23.12) apply to both individual assets and portfolios since these are linear combinations of individual securities.

By construction, the sensitivities b_{ik} are equal to: $b_{ik} = \frac{\text{cov}(R_i, \Delta_k)}{\text{var}(\Delta_k)}$, where Δ_k represents the return on factorial portfolio k of expectation δ_k.

These sensitivities are therefore interpreted as betas. Moreover, if one carries out a multiple linear regression between the return on a security and the common factors, the estimators of b_{ik} are given by the previous formula.

If there is empirically only one common factor justifying a non-zero risk premium, then the APT is formally identical to the CAPM and this common factor is interpreted, as we have seen (in Sect. 23.1) as the market portfolio. It should be noted

23.2 Arbitrage Pricing Theory

that the one-factor APT is written like the CAPM (see Sect. 23.2.3 below for their compatibility). Remember, however, that these two models are based on different assumptions. The second is a model of *equilibrium* giving an *exact* relation (on the theoretical level) whereas the first is only an *arbitrage* model delivering moreover only an *approximate* relation (the number of risky assets available on the market being finite).

Example 2

Assume that the random process generating the rates of return on stocks is the following:

$$R_i = \mu_i + b_{i1}F_1 + b_{i2}F_2 + \varepsilon_i \quad \text{for all } i = 1, \dots, n,$$

where ε_i is a white noise and F_1 and F_2 are the two common orthogonal and centered risk factors.

There is a risk-free asset yielding $r = 8\%$. In addition, two well-diversified equity portfolios, A and B, have the following characteristics:

Portfolio	Expected return (%)	b_{i1}	b_{i2}
A	10.0	1.9	−0.6
B	14.0	0.9	1.4

(a) We are looking for the equation of the plane that describes the expected rates of return for risky assets that are compatible with the APT.

To do this, we use Eq. (23.11b) where λ_1 and λ_2 are the unknowns and solve the following system of two equations with two unknowns:
(Portfolio A): $10 = 8 + 1.9\,\lambda_1 - 0.6\,\lambda_2$
(Portfolio B): $14 = 8 + 0.9\,\lambda_1 + 1.4\,\lambda_2$
The solution is then: $\lambda_1 = 2$ and $\lambda_2 = 3$.
The wanted APT thus writes:
(i) $\mu_i = 8 + 2\,b_{i1} + 3\,b_{i2}.$

(b) Assume now that there are two other well-diversified portfolios C and D. The first, C, has sensitivities equal to 1.1 and 0.3 to risk factors 1 and 2, respectively, and has an expected return of 12.5%. The second, D, has sensitivities equal to 2.3 and 0.7 to the risk factors, and has an expected return of 13.5%. What should a wise investor do if such a situation arises?

Apply APT (i) to these portfolios to calculate their expected theoretical returns. It follows:
$\mu_C = 8 + 2 \times 1.1 + 3 \times 0.3 = 11.1.$
$\mu_D = 8 + 2 \times 2.3 + 3 \times 0.7 = 14.7.$

(continued)

> Since Portfolio C earns an average of 12.5% instead of the required 11.1%, it is relatively cheap (too profitable). The opposite situation prevails for portfolio D: its expected return is 13.5% but, given its high risk, it should be 14.7%. Such a situation is in fact a violation of the APT, and thus provides investors with arbitrage opportunities (C will be bought and D short sold).

23.2.3 Relationship with the CAPM

Although the APT is a model that involves, in general, several factors, and not a single factor (the market portfolio) like the CAPM, and in addition the assumptions underlying these two models are different, CAPM and APT may be *compatible*.

Let us note again that the CAPM and the single factor APT are *formally* identical: in the absence of a risk-free asset, compare Eqs. (22.3a) and (22.3b) of the previous chapter with Eqs. (23.11a) and (23.11b) where $m = 1$ and $\lambda_1 \equiv \mu_M - \lambda$; in the presence of a risk-free asset, compare Eq. (22.5) of the previous chapter with Eq. (23.12), where $m = 1$ and $\lambda_1 = \delta_1 - r \equiv \mu_M - r$, equations that are compatible if $\delta_1 = \mu_M$.

More generally, in the presence of several common sources of risk, the two models can be compatible. Indeed, the CAPM does not assume that the only source of covariation between asset returns is the market. It can be shown (see Appendix 2 for a proof and a numerical example) that CAPM and APT are compatible if the risk premia λ_k turn out to obey the CAPM, i.e., more specifically, if the factor portfolios satisfy the CAPM:

$$\delta_k - r = \lambda_k = \beta_{\Delta k} (\mu_M - r), \text{ with } \beta_{\Delta_k} = \frac{\text{cov}(R_M, \Delta_k)}{\text{var}(R_M)}.$$

This compatibility is, however, not guaranteed. If the common sources of risk can be interpreted as (or replicated exactly by) portfolios, it prevails if the factor portfolios are in compliance with the CAPM. If risk factors are macroeconomic variables such as monetary and fiscal policies, climatic conditions, or exchange rates, such a compatibility would be totally fortuitous.

23.3 APT Applications and the Fama-French Model

We address the issue of the practical implementation of the model based on a multi-factor model (Sect. 23.3.1) before applying it to the selection of portfolios (Sect. 23.3.2). Finally, we examine the three-factor model of Fama and French (Sect. 23.3.3).

23.3.1 Implementation of Multi-factor Models and APT

At the operational level, there are two ways to implement the model: the so-called *endogenous* method because it only uses data related to the universe of securities on which the model is applied, and the so-called *exogenous* method that relies on economic considerations to decide, on a priori grounds, what are the common risk factors influencing asset returns, such as exchange rates and monetary and fiscal policies.

23.3.1.1 The Endogenous Method

The procedure to be used to isolate endogenous common factors, i.e., factors that are combinations (portfolios) of the individual securities under scrutiny, is as follows:

- The universe of securities considered relevant (in number n) is selected and the logarithmic returns on the securities are calculated on a given period (typically 2 years in daily data, and 10–30 years in monthly data).
- The empirical matrix of variances–covariances corresponding to these time series of returns is estimated.
- A factor analysis (with the maximum likelihood criterion, for example) or principal component analysis (with, e.g., the same criterion)[9] is used to identify *simultaneously* the common factors (which are here linear combinations of securities, so portfolios) and the sensitivities b_{ik}; in more formal terms, we estimate the model:

$$\underline{X} \equiv \underline{R} - \underline{\mu} = \mathbf{B}.\underline{F} + \underline{\varepsilon} \Leftrightarrow X_i \equiv R_i - \mu_i = \sum_{k=1}^{m} b_{ik} F_k + \varepsilon_i, \quad \text{for } i$$

$$= 1, \dots, n \tag{23.13}$$

where $\underline{X}$ is the $(n \times 1)$ vector of return "surprises," $\mathbf{B}$ is the $(n \times m)$ matrix of the sensitivities b_{ik}, $\underline{F}$ is the $(m \times 1)$ vector of the common centered factors, and $\underline{\varepsilon}$ is the $(n \times 1)$ vector of residual error terms.

---

[9]The purpose is to give a simplified structure to the matrix $\mathbf{V}$ while impoverishing as little as possible the information it contains. Factor analysis assumes that the covariances between the returns are fully explained by the common factors so that the residuals are white noises. Its objective is to reproduce as much as possible the empirical *covariances*. Principal component analysis does not assume that the ε_i are pure residuals and therefore that the factors capture the entire covariability of returns. It aims at explaining (retaining) the largest possible share of the *variance* of returns. Moreover, the advantage of the maximum likelihood method, by far the most used in practice, is that it provides a precise test (the Chi-square test) of the number of factors to be retained, for a given desired accuracy. Its disadvantage lies in the fact that, to give an analytic expression to the likelihood, it reintroduces the hypothesis of (multi-) normality of the returns (from which the Chi-square test results).

The common factors once isolated and decorrelated, the matrix $\mathbf{V}$ of variance-covariance of the returns decomposes in:

$$\mathbf{V} = \mathbf{B}.\mathbf{W}.\mathbf{B}' + \mathbf{E} \tag{23.14}$$

where $\mathbf{B}'$ is the transpose of $\mathbf{B}$, $\mathbf{W}$ is the diagonal $(m \times m)$ matrix of factor variances, $\mathbf{B}.\mathbf{W}.\mathbf{B}'$ is, therefore, the matrix $(n \times n)$ of systematic (risks) variances, explained by common sources of risk, and $\mathbf{E}$ is the diagonal matrix $(n \times n)$ of diversifiable variances (residual risks) $\sigma^2(\varepsilon_i)$.

Since $\mathbf{W}$ and $\mathbf{E}$ are diagonal matrices, we have:

$$\text{cov}(R_i, R_j) = \sum_{k=1}^{m} b_{ik} b_{jk} w_k^2 \quad \text{and} \quad \text{cov}(R_i, R_i) = \sum_{k=1}^{m} b_{ik}^2 w_k^2 + \sigma^2(\varepsilon_i).$$

- We finally use the estimated b_{ik} to verify, by cross-sectional regressions, that the different individual returns are well explained by the model, and to measure the size and the significance of the risk prices λ_k ($= \delta_k - r$, assuming the existence of a risk-free asset) associated with the factors:

$$\underline{\mu} - r.\underline{1}_n = \mathbf{B}.\underline{\lambda} \tag{23.15}$$

where $\underline{1}_n$ is the unit $(n \times 1)$ vector and λ the $(m \times 1)$ vector of the risk prices.

This test of Eq. (23.15) [by the cross-sectional regression: $\mu_i = \lambda_0 + \sum_k^m \lambda_k b_{ik} + z_i$, where the parameters to be estimated are the λ_k] constitutes the true test of the APT and makes it possible to know to which sources of risk the market attributes a price (reward) not equal to zero. It should be noted that the procedure just described, like the one proposed below, poses in practice delicate technical problems whose examination goes beyond the scope of this book.

23.3.1.2 The Exogenous Method

Alternatively, one can select a priori, preferably from economic theory and/or financial theory, microeconomic or macroeconomic factors deemed likely to influence asset returns. The same procedure as above is applied except that it is not necessary to use factor analysis or principal components to identify the factors, since they are given a priori exogenously. It is nevertheless necessary to orthogonalize them (we provide a method in Appendix 1).

The intuition underlying models based on sectoral or microeconomic factors is that the covariations between the different securities are explained, in addition to the systematic influence of the market, by their belonging to the same industry or sector: for example, financial institutions are specifically sensitive to the gap between lending and borrowing interest rates, which makes their industry special, while oil companies are very sensitive to the gap between the world price of crude oil and that

23.3 APT Applications and the Fama-French Model

of other sources of energy. Sometimes factors are microeconomic elements derived from corporate balance sheets such as debt ratio, dividend distribution rate, liquidity ratio, firm size (measured by total assets or market capitalization), or EBITDA (Earnings Before Interests, Taxes, Depreciation, and Amortization).

Similar analyses underpin models based on macroeconomic indices, but here the common sources of risk are very general (and may include a global market index), such as macroeconomic elements relevant to monetary and fiscal policies, economic growth and level of inflation, interest rates, and exchange rates.

For example, Chen, Roll and Ross (1986) used the following four factors for the USA (these are unanticipated components, i.e., innovations):

- The rate of inflation.
- The growth rate of Industrial Production by volume.
- The change in the risk premium (measured by the difference between the rate on US Treasury bonds and the rate of *junk bonds*, bonds issued by companies with a high risk of default).
- The variation of the slope of the range of rates.

Haugen (1999) used, for the USA, the following six factors (these are again innovations):

- The yield of Treasury bills.
- The difference between short- and long-term Treasury bonds yields.
- The difference between the yield of Treasury bonds and that of private bonds with the same (long) maturity.
- The growth rate of consumer prices.
- The growth rate of industrial production by volume.
- The dividend/price ratio for the S&P 500 index.

Whatever the factors selected, an important practical problem is that of estimating *unanticipated* variations (shocks, or innovations). Remember that in Eq. (23.1), each factor F_k must have an expected value equal to zero. This estimate is actually difficult because the expectations of economic agents are not directly observable. We must therefore adopt a model for the formation of expectations. For instance, expectations may be assumed to be *extrapolative*, i.e., the anticipation of a variable for the date $(t+1)$ depends only on its past observations until the date t.

23.3.2 Portfolio Selection

According to the CAPM, well-diversified portfolios with the same beta (measured against the same market index) are equivalent in terms of risk/return. However, they may have very different sensitivities to the various factors included in the APT. If risk is multidimensional, the correlations between returns are better explained by a multifactorial model than by Sharpe's market model. Investors' knowledge of

980   23 Arbitrage Pricing Theory and Multi-factor Models

portfolio sensitivities can be useful information for the development and management of portfolios. For example, some of the *alternative investment* styles, of which hedge funds are a special case, which will be discussed in Chap. 25, are based in part on the estimation of these sensitivities.

This can be done in a manner similar to that which, in the context of the market model, results in Jensen's alpha based on the excess return on the market portfolio, that on the considered portfolio and the beta of the latter. For each common risk factor k, its excess return is calculated and multiplied by the sensitivity of the portfolio b_{pk}, which makes it possible to calculate a normative return, and, by comparison with the observed return, to estimate an analogue of the Jensen's alpha of the portfolio.[10]

23.3.3 The Three-Factor Model of Fama and French

Challenging the famous mono-beta model, Fama and French (1992) showed that for the USA from 1928 to 1990, it was not so much the beta that explained the average return on a stock (or its average risk premium) than its *book-to-market* ratio (the book value of the equity over its market capitalization). This ratio makes it possible to distinguish *value stocks* from *growth stocks*. The former has a high ratio, the latter a low ratio. This is because the market values not only current investments but also predictable future investments. When these have a high expected net present value, market capitalization is high, so that the book-to-market ratio is low. From 1928 to 1997, the average return on value stocks actually exceeded that of growth stocks by 440 basis points (a huge gap). The instant impact of this research was due to the fact that there seemed to be no relationship between this return gap and the (negligible) difference in beta between these two groups of stocks.

More generally, Fama and French (1995, 1997, 2006) have found that it takes three factors, *book-to-market ratio*, *size* (in relative terms, measured by *market capitalization*), and the *market* itself, to explain the observed returns. Formally, the expectations of excess returns (risk premia) $\widehat{\mu}_i$ on stocks obey:

$$\widehat{\mu}_i = b_{iM}\widehat{\mu}_M + b_{iT}\widehat{\mu}_T + b_{iB}\widehat{\mu}_B \qquad (23.16)$$

where $\widehat{\mu}_M (= \mu_M - r)$ is the expected risk premium of the selected market index, $\widehat{\mu}_T$ is the expected difference in return between the stocks of small and large companies, and $\widehat{\mu}_B$ is the expected difference in return between high book-to-market ratio stocks and those with low book-to-market ratio.

For the period 1963–1994, the estimate of the (annualized) risk premium on the American market is 5.2%, that relating to the size effect is 3.2% and that relating to the book-to-market ratio is 5.4%. The value of these premia, however, varies

---

[10] See the Sect. 22.2.3 on performance measures in the previous chapter.

23.3 APT Applications and the Fama-French Model

according to the period studied; in particular, they are weaker both for much longer records (e.g., 1928–2000) and for shorter and more recent periods (e.g., 1992–2002).

Moreover, if the stocks are grouped into economic sectors, the sensitivities to the factors of the various sectors appear very different (and often of opposite sign), except the sensitivity to the market (which remains between 0.79 and 1.21).

The empirical status of the Fama and French model among financial economists, however, is not as clear as the previous results might suggest. On the one hand, some authors have highlighted biases that taint the validity of the methodology used. On the other hand, the model does not apply as well to other advanced economies. In addition, the size effect seems to have, at least partially, disappeared. Lastly, it is unlikely that the number, nature, and magnitude of the common risk factors will remain unchanged over time, given the dynamic nature of the economic and financial environment. It is therefore up to the designers of the models and the quantitative managers to adapt to them.

On a more theoretical level, it remains to provide convincing explanations of the empirical presence of factors other than the market itself. Fama and French have suggested that the HML factor[11] may be related to the risk of default by some companies due to their particular vulnerability to adverse economic conditions (corporate distress) and therefore reflects the risk premium required by investors to finance such firms. Authors, however, do not unanimously agree with this explanation. According to some researchers, companies in great difficulty would in fact have lower returns than other firms, although their market sensitivities and the two factors of Fama and French are stronger. This is obviously incompatible with the assumption that the size and the book-to-market ratio capture the compensation of the risk of default.[12]

Other interpretations of the presence of these factors are also possible. The SMB[13] and HML factors would contain, for example, relevant information about the future level of the Gross Domestic Product, and this information would be essentially independent of that conveyed by the stock market. These factors could be indicators (*proxies*) for innovations (unanticipated components) of state variables in an inter-temporal CAPM. These variables, linked to monetary and fiscal policies, for example, would influence the expected returns on securities in a differentiated way, which would explain why the market is not a sufficient explanatory variable. For example, the level of the nominal interest rate, a measure of the slope of the yield curve, a measure of the average default spread, and the aggregate dividend rate would constitute such explanatory variables. Or else, HML could be a proxy for the growth rate of physical capital invested globally in the economy.

---

[11] This factor is a measure of the expected difference in profitability between high book-to-market ratio stocks and low ratio stocks (High minus Low).

[12] See for example Campbell, Hilscher and Szilagyi (2008).

[13] This factor is a measure of the expected difference in profitability between the stocks of small and large companies (Small minus Big).

Carhart (1997) added with some success a fourth factor to the Fama-French model, the *momentum*, which is the tendency of the price of an asset to continue to rise (or to fall) once it started doing it. This model has become popular among practitioners.

23.4 Econometric Tests and Comparison of Models

The number of empirical work on the APT, a more recent model than the CAPM, is significantly smaller than the latter, but is still significant. The detailed analysis of these econometric studies, sometimes complex, is beyond the scope of this book. Only the main problems encountered in the tests will be mentioned (Sect. 23.4.1) and the major conclusions that can be drawn, on the one hand, from the comparison between the CAPM and APT (Sect. 23.4.2), and, on the other hand, from the comparison between multifactorial models (Sect. 23.4.3).

23.4.1 Tests of the APT

Like the tests of the CAPM, the tests of the APT first involve time series in order to uncover the common factors on the one hand and the sensitivities of the assets on the other. Then, equipped with these estimates, the linear relationship between the expected asset returns and their sensitivities to the factors must be verified in the cross section, which reveals the common sources of risk that are actually rewarded by the market. We have already analyzed (see Sect. 23.3.1 of the previous section) the problems related to the implementation of the model.

Let us only mention here that, in addition to the difficulties presented in Sect. 23.3.1, common to the CAPM and the APT, the second faces two other problems:

- As regards the endogenous factor procedure, the number of found common factors depends (positively) on the size of the set of securities studied, the statistical method used (factor analysis or principal components), and the type of criterion used (maximum likelihood or others); it can also depend on the time interval on which returns are calculated.
- It is most often difficult to give an economic interpretation to the endogenous common factors, except the market portfolio itself.

Note that, despite hundreds of APT-related publications, drawing general conclusions is difficult, as the whole set of procedures allows for arbitrary choices that lead to bias and complications.[14] Results may be plagued by data mining or data snooping (grinding data until it yields a desired or expected result).

---

[14] For a literature review, see for instance Cochrane (2009).

23.4.2 Empirical and Practical CAPM-APT Comparison

The two competing paradigms have both advantages and disadvantages, and it is difficult to discern a consensus among both theoreticians and professionals.

The CAPM has for it that it has been the subject of countless studies and has proved until recently fairly robust (especially in its zero-beta form) with respect to the many tests it has been submitted to, that it constitutes an exact, not asymptotic, equilibrium relation, that it is very simple to understand and implement, once chosen the index considered as representative of the market, and that it does not suffer the drawback of identifying the source of risk (the market portfolio).

The APT, on the other hand, assumes the return generating process but allows for several common sources of risk, and does not assign any particular role to the market portfolio (although, as we have seen, it can be one of the common factors).

Empirically, the rigorous econometric tests of both models are difficult to establish. In particular, while the CAPM does not pose any problem in *identifying* the source of risk, which is unique, it suffers, on the other hand, from a serious problem of *estimating* it since it is by nature unobservable. Moreover, it has been noted that, at least in recent decades, the CAPM is unable to explain the cross-sectional differences in risk premia on portfolios ranked by size and book-to-market ratio. In particular, "value" stocks (with a high book-to-market ratio) have high return despite relatively low betas.[15] On the other hand, if the APT does not pose (too many) problems in *estimating* the sources of risk, since these variables are in principle measurable, it suffers from a serious problem of *identifying* the factors (nature and number), both in the endogenous method (what is the significance of the factors highlighted?) and in the exogenous method (what factors should be postulated on a priori grounds?).

However, the bottom line remains that the assumed linear relationship between expected return and risk seems *robust*, with one or more sources of risk. In practice, the adoption of one model rather than the other depends essentially on the objective pursued. For example, the investor who wants to know the sensitivities of the portfolio to changes in macroeconomic magnitudes will adopt the APT. Those

---

[15] The relative failure of the traditional CAPM, even in its conditional version, has given birth to a new generation of theoretical models. This failure is known as "puzzles in pricing theory." In a nutshell, traditional models predict a much too high riskless interest rate, a much too low equity risk premium, and too low stock market volatility. One avenue is to develop valuation models based on physical investment or production-based (production-based asset pricing theory) rather than terminal wealth. See, for example, Cochrane (1996). Another is to use a consumption-based pricing theory à la Breeden (1979) augmented by various other sources of risk such as housing, labor, or political factors. Another is to use more sophisticated (than Von Neumann-Morgenstern's) investor preferences, such as internal (see, e.g., Constantinides 1990) or external (see, e.g., Santos and Veronesi 2010) habit formation, recursive utility [see, e.g., Epstein and Zin (1989) and Skiadas (1998)]. Another is to assume that investors have heterogeneous expectations and update them in a Bayesian way to account for information flows over time (e.g., Basak 2005). The risk created by changes in the expectations of the other investors justifies a (novel) risk premium. Dumas and Luciano (2017) provide a rigorous and lucid treatment of these extensions.

wishing to appreciate the expected return-risk tradeoff of their portfolio compared to that of the market will use the CAPM.

23.4.3 Comparison of Factor Models

In practice, as we have seen, the models generally used are single- or multi-factor models. A few firms, mostly of American origin, specialized in the design and sale of software including multi-factor models for risk analysis and quantitative portfolio management.

The potential number of these models being huge, and their relatively new systematic use by professionals, it is difficult to make a clear judgment on their main merits and drawbacks. Two main questions concerning them are worth asking: what advantage do they have with respect to a completely general model that does not assume any restriction on the matrix V of the variances-covariances? And what is their performance compared to Sharpe's single-factor model (which is a special case)?

Concerning the first question, the estimation of the unconstrained matrix V (including $n(n - 1)/2$ arbitrary covariances) requires, given the very large number of its elements, very long data histories (several decades) so that the estimation errors are not too large and the uncovered correlations not purely fortuitous. The major disadvantage is that changes in the characteristics of assets resulting from changes in investment or financing policies by firms over very long periods are not taken into account. This difficulty is such as to make the estimation of the correlation structure rather hazardous.

On the other hand, multi-factor models, requiring significantly shorter histories (typically 1 year in daily data, 5 years in weekly data or 10 years in monthly data), offer an undeniable operational advantage from this point of view. In addition, they have the following other advantages:

- The economic or financial logic that in principle underpins them makes them quite realistic and understandable by the users (at least when the factors are exogenous).
- They take into account changes in the fundamental characteristics of the securities as long as microeconomic factors are accounted for, such as the debt ratio or the liquidity ratio.
- The risk being decomposed into different elements, its analysis is finer.

The answer to the second question, concerning the respective merits of the Sharpe model and multi-factor models, is more nuanced. It is necessary to distinguish the explicative power of the models concerning the historical structure of the correlations from their capacity to predict the future structure of these correlations.

It is obvious that taking into account several sources of risk, rather than one, improves the explanation of covariations between assets. There are, however,

23.5 Summary

several problems, due in particular to the stability of the sources of risk, the way factors are constructed and estimated, and the econometric methods used.

23.5 Summary

- The Arbitrage Pricing Theory (APT) does not rely on the CAPM restrictive hypotheses sustaining the mean-variance paradigm (quadratic utilities, or normality of returns) but rests on the assumption that *returns are generated by several common risk factors* yielding a more realistic and complex structure of the correlation of the returns and to several risk premia.
- In a *multi-factor model (MFM)*, the returns on the n individual securities are assumed correctly described by:

$$(\text{MFM}) \quad R_i(t) = \mu_i + \sum_{k=1}^{m} b_{ik} F_k(t) + \varepsilon_i(t)$$

for each $i = 1, \ldots, n$ and each period t; μ_i is the expectation of R_i, the b_{ik} represent the sensitivities of the return on asset i to the m *common centered factors* F_k, ε_i is a white noise of zero average and properties of serial and cross non-correlation: $\text{Cov}(\varepsilon_i, \varepsilon_j) = \text{cov}(\varepsilon_i, F_k) = \text{Cov}(\varepsilon_i(t), \varepsilon_i(t')) \, 0$ for every $i, j \neq i, k$ and $t' \neq t$. The factors F_k have zero mean and can be orthogonalized, in which case $b_{ik} = \text{cov}(R_i, F_k)/\sigma^2(F_k)$, $\sigma_i^2 = \sigma_{\varepsilon_i}^2 + \sum_k^m b_{ik}^2 \sigma_{F_k}^2$ and the covariances $\sigma_{ij} = \sum_k^m b_{ik} b_{jk} \sigma_{F_k}^2 \;\; \forall i, \forall j \neq i$, are only due to the influence of the common factors F_k, source of m systematic risks. When $m = 1$, we obtain Sharpe's model (Chap. 21).
- In a MFM, the *specific risk* (weighted average of ε_i's) of a *well-diversified portfolio vanishes* by virtue of the law of large numbers. Hence, intuitively, the specific risk ε_i of any individual asset *should not be rewarded* by a risk premium since it is diversifiable away: the expected return μ_i should depend only on the m systematic risks b_{ik} through m risk premia proportional to the b_{ik}'s.
- The APT validates the previous intuitive conclusions: under the assumptions of AAO and MFM, there are $m + 1$ constants λ_k ($k = 0, \ldots, m$) such that the expected rate of return on any risky asset i is given by:

$$(\text{APT}) \quad \mu_i = \lambda_0 + \sum_{k=1}^{m} \lambda_k b_{ik}$$

where λ_k *is the market price of risk* k, and $\lambda_0 = r$, if there is a riskless asset yielding r.
- The formal justification of APT rests on the construction of an *arbitrage portfolio* P whose loose definition is:

(i) It is (almost) *risk-free* since its systematic risks are zero (it is *constructed* in such a way that each of its b_{pk} is zero) and its diversifiable risk is (almost) zero by virtue of a (quasi) perfect diversification obtained with a "large" number of securities; (ii) it has *zero initial value* (short sales finance purchases). The AAO condition is that the terminal value of such an arbitrage portfolio is zero, which yields the APT.

- The APT in the presence of a risk-free asset can alternatively be written: $\mu_i - r = \sum_{k=1}^{m} b_{ik}(\delta_k - r),$

δ_k being the expected return of a "factor portfolio" of sensitivity b_{kk} equal to 1 to the factor k and sensitivity 0 to all the other factors and the MPR $\lambda_k = \delta_k - r$. The global risk premium on a risky asset in AAO is thus the sum of the expected risk premia of the factor portfolios weighted by the sensitivities of the asset to the various sources of risk. It is under this form that APT is often empirically tested and used in practice.

- In presence of only one factor (R_M), the APT and the CAPM relations are identical. But the APT involves, in general, several factors which explain better the systematic risk than one factor only. Besides, the assumptions underlying these two models are different. In spite of these differences, CAPM and multi-factor APT *may be compatible.*

- Two ways to implement the model can be followed: the *endogenous* determination of the factors resting on a statistical analysis of the data related to the security returns (for instance factor analysis), and the *exogenous* choice of the factors on the basis of a priori economic and financial considerations (like the rate of inflation, the growth rate of Industrial Production, or the level and slope of the yield curve).

- When the APT is applied to portfolio selection and performance measurement, the excess return on each risk factor k is first calculated and then multiplied by the sensitivity of the considered portfolio b_{pk}, which allows computing a normative return, and, by comparison with the observed return, estimating an analogue of Jensen's alpha.

- Fama and French, for instance, have found that it takes three factors, *book-to-market ratio, market capitalization,* and the *market* itself to explain returns, which leads to three corresponding risk premia. Carhart added a fourth factor, the *momentum.*

- The practical use of APT and MFM suffer from complications in *identifying* the factors (nature and number), both in the endogenous method (what is the significance of the factors statistically highlighted?) and in the exogenous method (what factors should be postulated on a priori grounds?). However, the bottom line on APT tests remains that the assumed linear relationship between expected return and risks seems relatively *robust.*

Appendix 1: Orthogonalization of Common Factors

Equation (23.1) of the text: $R_i = \mu_i + \sum_{k=1}^{m} b_{ik} F_k + \varepsilon_i, \forall i = 1,...,n$, describes, for a given period t, the return of security i. The econometric test of this relation involves time series and it is appropriate to rewrite (Eq. 23.1) in the more explicit manner for any asset i:

$$R_{it} = \mu_i + \sum_{k=1}^{m} b_{ik} F_{kt} + \varepsilon_{it} \text{ for } t = 1, \ldots, T. \tag{23.17}$$

where the expected value of R_i and the sensitivities b_{ik} are assumed to be constant over time.

This equation involves common factors F_k (in fact unanticipated changes in risk factors k) that are not correlated. This is not a limiting assumption as it is always possible to obtain orthogonal factors from correlated factors. Although there are other, more complex, methods of orthogonalization, we propose a procedure that is easy to understand and implement based on linear regression. We illustrate it from a two-factor model ($m = 2$), the generalization to m sources of risk being only sketched out because it is obvious. So assume we have:

$$R_{it} = \mu_i + a_{i1} G_{1t} + a_{i2} G_{2t} + \varepsilon_{it} \tag{23.18}$$

where G_1 and G_2 are centered but correlated and the sensitivities are denoted by a_{ik}.

To decorrelate the factors, let us define $F_1 = G_1$ (the first factor will be unchanged) and make the following regression:

$$G_{2t} = c_1 F_{1t} + h_t \tag{23.19}$$

where the residual h is, by construction, uncorrelated with F_1. It should be noted that this regression must be performed without a constant because the factors are of mean equal to zero in Eqs. (23.17) and (23.18). In the case where the Eq. (23.17) is written without the constant μ_i, as in Eq. (23.9) of the text, the factors are not of mean equal to zero and regression (Eq. 23.19) must be performed with a constant c_0.

By defining $h = G_2 - c_1 F_1 \equiv F_2$, the second factor, F_2, is decorrelated from the first, F_1.

By replacing G_2 in Eq. (23.18) by its value given by Eq. (23.19), G_1 by F_1, h by F_2, and by rearranging the terms, we obtain:

$$R_{it} = \mu_i + (a_{i1} + c_1 a_{i2}) F_{1t} + a_{i2} F_{2t} + \varepsilon_{it}$$

which, with the definitions $(a_{i1} + c_1 a_{i2}) \equiv b_{i1}$ and $a_{i2} \equiv b_{i2}$, gives the wanted Eq. (23.17) (with $m = 2$).

In the case of three factors, the following regression would be carried out:

$$G_{3t} = d_1 F_{1t} + d_2 F_{2t} + j_t$$

and we would define $j = G_3 - d_1 F_1 - d_2 F_2 \equiv F_3$. And so on for the m factors.

Appendix 2: Compatibility of CAPM and APT

Recall that the APT and the MEDAF are written respectively:[16]

(a) $\mu_i = r + \sum_k^m \lambda_k b_{ik}$

 for $i = 1, ..., n$.

(b) $\mu_i = r + \beta_i(\mu_M - r)$

 for $i = 1, ..., n$.

Recall also that the factorial portfolio k $(k = 1, ..., m)$ is such that:

$$b_{kk} = \frac{\text{cov}(F_k, \Delta_k)}{\text{var}(F_k)} = 1 \text{ and } b_{kk'} = \frac{\text{cov}(F_{k'}, \Delta_k)}{\text{var}(F_{k'})} = 0 \text{ for } k' \neq k,$$

where Δ_k represents the return on this factorial portfolio and $\delta_k = E(\Delta_k)$.

We have stated in the body of the text that CAPM and APT are compatible if factorial portfolios satisfy the CAPM:

$$\delta_k - r = \lambda_k = \beta_{\Delta k} (\mu_M - r), \text{ with } \beta_{\Delta k} = \frac{\text{cov}(R_M, \Delta_k)}{\text{var}(R_M)}.$$

Proof

Assume that the risk premia λ_k (remember that they can be interpreted as the excess return expected from the factorial portfolios, $\delta_k - r$) obey the CAPM. It then follows:

(c) $\delta_k - r = \lambda_k = \beta_{\Delta k} (\mu_M - r)$
 for $k = 1, ..., m$.

By replacing the value of λ_k given by (c) in (a), we obtain:

$$\mu_i = r + \sum_k^m \beta_{\Delta k}(\mu_M - r)b_{ik} = r + (\mu_M - r)\sum_k^m \beta_{\Delta k}b_{ik}.$$

---

[16]We adopt here, for each model, the version with a risk-free asset. The proof is analogous for the version without such an asset.

Now just set the beta of the asset i as $\beta_i = \sum_k^m \beta_{\Delta k} b_{ik}$ to recover the CAPM.

Example 3

Consider the conditions of the example of Sect. 23.2.2.2 (Example 1).

We want to find the values of the four variables (betas) which make the expected rates of return on the A and B portfolios compatible with the CAPM, knowing that $(\mu_M - r) = 4\%$. The four variables are the betas (in the sense of the CAPM) of the two common factors F_1 and F_2 (then considered as portfolios) and the betas (in the sense of the CAPM) of the two portfolios A and B. Apply the CAPM to F_1 and F_2:

$$\lambda_1 = 2 = \beta_{\Delta 1} (\mu_M - r) = 4\beta_{\Delta 1}$$

$$\lambda_2 = 3 = \beta_{\Delta 2} (\mu_M - r) = 4\beta_{\Delta 2}$$

From this we deduce that $\beta_{\Delta 1} = 0.5$ and $\beta_{\Delta 2} = 0.75$. Then apply the CAPM to A and B:

$$\mu_A = 10 = r + \beta_A (\mu_M - r) = 8 + 4\beta_A$$

$$\mu_B = 14 = r + \beta_B (\mu_M - r) = 8 + 4\beta_B$$

From which we deduce that $\beta_A = 0.5$ and $\beta_B = 1.5$.

The proof of the compatibility of the two models is that we recover the betas of A and B from the betas of the factorial portfolios and the sensitivities of A and B to the common factors:

$$\beta_A = 0.5 = 1.9\beta_{\Delta 1} - 0.6\beta_{\Delta 2} = 1.9 \times 0.5 - 0.6 \times 0.75.$$

$$\beta_B = 1.5 = 0.9\beta_{\Delta 1} + 1.4\beta_{\Delta 2} = 0.9 \times 0.5 + 1.4 \times 0.75.$$

Suggestions for Further Reading

Books

Alexander, G. J., Sharpe, W. F., & Bailey, J. V. (2001). *Fundamentals of investments* (3rd ed.). Prentice Hall.

Cochrane, J. H. (2009). *Asset pricing* (2nd ed.). Princeton University Press.

*Dumas, B., & Luciano, E. (2017). *The economics of continuous-time finance*. The MIT Press.

Elton, E., Gruber, M., Brown, S., & Goetzmann, W. (2010). *Modern portfolio theory and investment analysis* (8th ed.). Wiley.

Haugen, R. (1999). *The inefficient stock market: What pays off and why*. Prentice Hall.

*Jarrow, R. A. (1988). *Finance theory*. Prentice Hall International Editions.

*Merton, R. (1999). *Continuous-time finance*. Basil Blackwell.

Articles

Basak, S. (2005). Asset pricing with heterogeneous beliefs. *Journal of Banking and Finance, 29*, 2849–2881.

Breeden, D. T. (1979). An intertemporal asset pricing model with stochastic consumption and investment opportunities. *Journal of Financial Economics, 7*, 265–296.

Campbell, J., Hilscher, J., & Szilagyi, J. (2008). In search of distress risk. *Journal of Finance, 63*(6), 2899–2939.

Carhart, M. (1997). On persistence in mutual fund performance. *Journal of Finance, 52*(1), 57–82.

Chen, N. F., Roll, R., & Ross, S. (1986). Economic forces and the stock market. *Journal of Business, 59*(3).

Cochrane, J. (1996). A cross-sectional test of an investment-based asset pricing model. *Journal of Political Economy, 104*, 572–621.

Constantinides, G. (1990). Habit formation: A resolution of the equity premium puzzle. *Journal of Political Economy, 98*, 519–543.

Epstein, L. G., & Zin, S. E. (1989). Substitution, risk aversion and the temporal behavior of consumption and asset returns: A theoretical framework. *Econometrica, 57*, 937–969.

Fama, E. F., & French, K. R. (1992). The cross-section of expected stock returns. *Journal of Finance, 47*, 427–465.

Fama, E. F., & French, K. R. (1995). Size and book-to-market factors in earnings and returns. *Journal of Finance, 50*, 131–155.

Fama, E. F., & French, K. R. (1997). Industry costs of equity. *Journal of Financial Economics, 43*, 153–193.

Fama, E. F., & French, K. R. (2006). The value premium and the CAPM. *Journal of Finance, 61*, 2163–2185.

Ingersoll, J. E. (1984). Some results in the theory of arbitrage pricing. *Journal of Finance, 39*(4), 1021–1039.

Ross, S. A. (1976). The arbitrage theory of capital asset pricing. *Journal of Economic Theory, 13*(3), 341–360.

Roll, R., & Ross, S. A. (1980). An empirical investigation of the arbitrage pricing theory. *Journal of Finance, 35*(5), 1073–1103.

Santos, T., & Veronesi, P. (2010). Habit formation, the cross section of stock returns and the cash flow risk puzzle. *Journal of Financial Economics, 98*, 385–413.

Shanken, J. (1992). The current state of arbitrage pricing theory. *Journal of Finance, 47*(4), 1569–1574.

Skiadas, C. (1998). Recursive utility and preferences for information. *Economic Theory, 12*, 293–312.

Strategic Portfolio Allocation

24

Quantitative portfolio management was born with the work of Markowitz (introduced in Chap. 21), which clarified and formalized the concepts of diversification, separation, and efficiency. Remember that Markowitz model is based on two simplified assumptions:

- The investor takes an initial position (at time 0) and makes no transactions until a final date (at time T), at which time he evaluates the results of his investment decision. It is, therefore, a static model that only allows *buy and hold* strategies.
- The investor arbitrates (at time 0) the expected return and the variance of her investment over the period 0 to T: at a given expected return, she minimizes the variance, thus constituting a so-called efficient portfolio. This is based on the mean-variance criterion.

However, the assumption of lack of portfolio review is unrealistic in the long run (more than 1 year). Strategies involving portfolio revision between *time 0 and time T* are called dynamic strategies, as opposed to static strategies or Markowitz's "buy and hold" strategies. Dynamic strategies are the subjects of this chapter and the next one.

Dynamic portfolio management covers a wide variety of short- and long-term management styles that can be classified according to different criteria.

First, we distinguish passive management from active management; the former seeks to replicate an index as perfectly as possible, while the latter differs more or less distinctly from it.[1]

We also distinguish strategic asset allocation from tactical asset allocation. Strategic allocation concerns a small number of asset classes (typically stocks, bonds, and monetary securities) and depends on the investors' risk aversion and wealth;

---

[1] Passive management is essentially static, except under certain circumstances, as described in the following chapter.

© The Author(s), under exclusive license to Springer Nature Switzerland AG 2022
P. Poncet, R. Portait, *Capital Market Finance*, Springer Texts in Business and Economics, https://doi.org/10.1007/978-3-030-84600-8_24

991

while tactical allocation focuses on individual securities, constitutes a bet on the investor's ability to take advantage of opportunities or to predict market reversals and does not depend on their utility function. In addition, the horizon of strategic allocation is long (several years), while the horizon of tactical allocation is short (less than 6 months).

In fact, a rational asset allocation process comprises two successive and articulated stages: the first stage is strategic allocation, during which investors determine the investments by large asset classes; the next stage is tactical allocation, during which investors choose different individual securities and determine their respective weightings, under the constraints previously defined at the strategic allocation stage. By this logic, we divide the dynamic portfolio allocation into two chapters: this chapter is dedicated to strategic allocation and the next chapter to tactical allocation.

The different approaches of strategic allocation will be presented according to the following structure:

- The so-called common sense management rules (allocation according to age, risk profile, etc.), which concerns strategic allocation will be discussed in Sect. 24.1.
- The various forms of portfolio insurance, which are based on more or less automatic rules and are designed to preserve the value of capital in bearish conditions while partially benefiting from bullish conditions, will be discussed in Sect. 24.2.
- Finally, dynamic optimization model (Merton's model), which theoretically allows portfolios to be adapted at any time to the investors' own objectives, will be discussed in Sect. 24.3.

24.1 Strategic Asset Allocation Based on Common Sense Rules

Strategic asset allocation, contrary to tactical asset allocation which will be discussed in the next chapter, does not focus on individual securities but focuses on a small number of large asset classes. In addition, strategic asset allocation takes into consideration each investor's characteristics (especially his/her age and his/her risk aversion level) and most often concerns long-term investments (several years).

In this section, we define by three rules a strategic asset allocation.

The first rule dictates the distribution of weights in (typically) each of the *three main categories* of financial securities: monetary assets (≤ 1 year), bonds (fixed-rate loans with a duration of 5–10 years) and stocks (stock market index or predetermined portfolio). The composition of the portfolio should obviously depend on the investor's objectives (which further depend on their wealth and risk aversion) as well as their investment horizon (T).

24.1 Strategic Asset Allocation Based on Common Sense Rules

The second rule determines the evolution of these weights (buying or selling risky securities) over time (between 0 and T) or as a function of the investor's age.

The third rule dictates the changes within a portfolio with the evolution of market conditions (interest rates, stock market index).

Although a consensus has been reached on the first two rules based on common sense, diverse opinions exist as to the third one.

24.1.1 Common Sense Rules

When dealing with strategic allocation problems, investment managers often comply with two qualitative principles known as "common sense," which are related to the first two rules: risk aversion profile on the one hand and asset allocation according to age on the other hand. We describe these two consensus rules before presenting various justifications, some of which are well-founded, and others are misleading.

24.1.1.1 Consensual Rules Based on Common Sense and Reactions to Market Evolutions

We outline two commonly accepted portfolio management principles.

(a) *First principle of common sense: risk profile*

"Aggressive" investors who demand a high expected return will mainly invest in stocks (in return, they accept a high level of risk).[2] Those who wish to take a lower risk than that of an index (stock) while demanding a substantial return will also buy some bonds, and those who prioritize security will mainly invest in bonds and monetary assets. In any case, the ratios of bond/stock and monetary assets/total portfolio should increase with the investor's risk aversion. A specification of these rules leads to the notion of "*risk profile.*"

(b) *Second principle of common sense: modification of asset allocation over time or asset allocation according to an investor's age*

When their investment horizon is far away, investors should first invest massively in stocks (time 0), then gradually reduce the proportions of stocks to make room for bonds and then monetary assets, until finally (time T) they obtain a portfolio composed essentially of monetary assets. This second rule of common sense, which determines the optimal allocation according to the investor's age, is called "*conventional wisdom*" in the literature.

---

[2] Possibly with leverage (debt) that can be obtained through forward transactions.

24.1.1.2 Attempts to Rationalize Common Sense Rules, Puzzles, and Errors in Reasoning

In fact, the principles of common sense which were just presented *are difficult to justify* theoretically from standard static models (Markowitz; *see* Chap. 21).

(a) *Attempts to justify the first rule (risk profile)*

It should be noted at the outset that the principle according to which the bond/stock ratio should increase with investor's risk aversion level and which governs the composition of *profiled funds* is in contradiction with the Tobin-Black-Markowitz' two-fund separation theorem. In fact, according to the latter theorem, all efficient portfolios are obtained by combining risk-free assets and a (single) fund composed of risky assets (the tangent portfolio). The proportion of risk-free assets depends on the investor's risk aversion, but in the risky fund, the proportions of different securities are fixed rather than dependent on risk aversion.

Therefore, when the monetary asset is risk-free (assumed to be free of interest rate and counterparty risks), the bond/stock ratio is the one prevailing in the risky funds (identical for all investors) and should be independent of the investor's attitude toward risk.

This contradiction between the standard portfolio theory and the consensual behavior of fund managers, as highlighted by Canner, Mankiw and Weil (1997), was considered as *a puzzle, as it could not be explained* by a simple loosening of Markowitz' static model's assumptions. The followings are some representative examples among the various attempts of explanation, all of which have proved to be unsatisfactory:[3]

- Behavior that does not comply with the mean-variance criterion.
- Absence of risk-free assets due to inflation rate risk which makes the real return of monetary asset uncertain.
- Inclusion of illiquid assets.
- An investment horizon T greater than the duration of the monetary asset.

In fact, the puzzle *can only be resolved by considering medium- or long-term investments* ($T > 1$ year), for which the money market fund becomes risky due to the uncertainty of short-term rate in the future, *by allowing revisions between time 0 and time T and by using a dynamic optimization model.* The technical explanation of the puzzle is beyond the scope of this book.[4]

(b) *Attempts to justify the second rule (allocation according to investor's age) and falsity of the time diversification argument*

---

[3] See Canner, Mankiw and Weil (1997).

[4] See Bajeux, Jordan and Portait (2001) for this explanation.

24.1 Strategic Asset Allocation Based on Common Sense Rules

Recall that, according to the second consensual rule, an investor should first allocate a large proportion of their investment to stocks, then gradually reduce stocks' proportions to make room for the bonds and finally increase the proportions of monetary assets as the investment horizon gets near.

This allocation rule, which implies that more risk can be taken in a long investment horizon, is often justified by the principle of *time diversification*. This diversification could intuitively be explained as follows: the annualized return of a portfolio is calculated as the average of successive annual returns. For simplicity, assuming that successive annual returns are identically and independently distributed (random-walk hypothesis), the expected annualized return is then equal to the average annual return. As for the risk of this annualized return, it is lower than the risk of each year's annual return due to the independence assumption ("bad years can be compensated by good years"). In addition, the time diversification effect increases with the investment horizon (over a long period, the risk can be eliminated because of the law of large numbers). In fact, thanks to time diversification, effective annualized return should differ little from its annual average in the long run.

In a more rigorous way, let us note $R_{0,n}$ as the *annualized* return over the period which covers year 1, year 2, ...year n; and denote R_i the return of year i, so that:

$$(1 + R_{0,n})^n = (1 + R_1) \ldots (1 + R_i) \ldots (1 + R_n).$$

Denoting $r_x = \ln(1+R_x)$, we have: $r_{0,n} = (r_1+r_2+\ldots+r_n)/n$.

If the R_i ($i = 1, 2, 3,\ldots n$) are independently and identically distributed (i.i.d.), the r_i ($i = 1, 2, 3,\ldots n$) will also be. Let us denote μ the mean and σ the standard deviation of r_i. We have:

$variance(r_{0,n}) = \sigma^2/n$, so that when n approaches infinity, r_{0n} tends to μ (perfect time diversification).

When the R_i ($i = 1, 2, 3,\ldots n$) are autocorrelated, diversification is only partial.

In fact, the hypothesis of *time diversification is fundamentally questionable.* Its weakness is essentially due to the fact that *a rational investor does not base their investment choices on the annualized return.*

To explain the fallacious aspect of this principle of time diversification, we consider an investor whose preferences comply with the mean-variance criterion. His objective is to minimize the variance of his terminal wealth (under the constraint of a given expected wealth) and not to minimize the variance of the *annualized return*. In that respect, while it is true that the variance of the annualized return decreases with the investment horizon, the risk associated with the terminal wealth increases with the investment horizon. Far from producing a time diversification effect, the plurality of periods adds successive risks, thus increasing the total risk. Think about a game in which we toss a coin N times, giving each toss a gain of 1 € if "heads" appears and a loss of -1 € otherwise. If N is large, we have: (i) an average gain of 0 which is almost certain (the variance is equal to $1/N$) and (ii) an expected terminal wealth of 0 but extremely random (its variance is equal to N).

However, the reasoning of "adding successive risks" is itself fallacious, as the utilities related to different horizons are not comparable.[5]

There exist nevertheless some valid explanations for the second rule. We present two of them in the following paragraphs.

We can first find an explanation simply in the evolution of the investor's risk aversion. It is plausible that younger investors are willing to take more risks than their elders. As a result, in accordance with the second rule, the investor will decrease their relative investment in the risky asset as time elapses.

A second convincing justification involves human capital.[6] Human capital, which is significant when people are young and weak when they are old, is subject to less risk than financial capital. Therefore, a high allocation on stocks (risky) can offset a high allocation on human capital (less risky) for young people. Over time, as human capital decreases, allocation on stocks should also decrease to make room for less risky assets in order to maintain the optimal level of risk.

24.1.2 Reactions to the Evolution of Market Conditions and of the Portfolio: Convex and Concave Strategies

The rule that dictates portfolio changes following market evolutions (interest rates, stock market index) depends on the management's objectives and style. In fact, investment managers' reactions to the changes in portfolio value (X) and economic variables are not univocal. For example, following a change in the stock index, which generally results in a change of the same direction in the portfolio value (X), two opposite types of behavior are observed:

- **The first type of behavior consists in buying risky securities (stocks)** following a price increase and selling them following a decrease. Portfolio insurance is part of the family of so-called *momentum* (we act in the same direction as the market) or *convex* strategies. Selling risky assets in bearish conditions helps to cushion the effect of a prolonged fall in stock prices and buying risky assets in the event of successive stock price increases accentuates their positive effect, which makes the terminal value of the portfolio a convex function (of the benchmark stock market index). We will see that portfolio insurance strategies are characterized by such a convexity (*see* Sect. 24.2).
- **The second type of behavior, in contrast, consists in selling risky securities** following a price increase and buying them following a decrease. These strategies are called *contrarian* (one acts against the direction of the market) or *concave* strategies. Portfolios with constant weight distributions are part of this family. As

---

[5] This is why it is tempting to reason in terms of annualized returns which seem (falsely) to allow for the comparison of investments of different horizons.

[6] See Bodie, Merton and Samuelson (1992). To simplify, human capital is defined as the present value of all expected labor income until retirement.

a matter of fact, when the stock price of a security increases faster than that of the portfolio average, its relative weight mechanically increases. To maintain constant weights, one should therefore sell the securities whose relative prices increase and buy the ones whose relative prices decrease.

24.2 Portfolio Insurance

Portfolio insurance is a dynamic allocation strategy that allows to limit a portfolio's loss of value in the event of a market downturn, while still enabling investors to benefit from an increase. As the guarantee against the decline is obviously not free of charge, the profit in the event of an increase will generally represent only a fraction of the gains realized by the market.

In its most basic sense, portfolio insurance takes the form of *Stop Loss* strategies implemented since 1920s. More sophisticated techniques developed by American academics[7] in particular date back to the 1970s. The original idea was to use options, bought or "synthesized" from an appropriate dynamic position involving the underlying stock and the risk-free asset. These techniques are called OBPI (*Option Based Portfolio Insurance*).

As OBPI encountered many problems which will be further discussed, new methods appeared, including the CPPI (*Constant Proportion Portfolio Insurance*), which no longer uses options but some mechanical assets allocation rules.

In practical terms, Portfolio Insurance is suitable for an automatic or quasi-automatic management, with the orders to be executed being triggered by computers (*Program Trading*).

In conceptual and formal terms, Portfolio Insurance is part of strategic allocation because it only involves two asset classes in its standard form. In addition, it belongs to the family of quantitative management techniques aimed at obtaining a certain profile for the terminal value of a portfolio. Given a risky asset, a particular portfolio or an index of value S (underlying), we want to determine a dynamic strategy involving the underlying and a risk-free asset such that the portfolio value at time T can be written:

$$V(T)/V(0) = f(S(T)/S(0))$$

where the deterministic function $f(.)$ reflects the desired profile.

For example, we may have $f(S(T)/S(0)) = \text{Max}\ (S(T)/S(0), k)$, where k is the minimum performance threshold (100%, for example) which is fixed at the begin-

---

[7] H. Leland and M. Rubinstein developed Portfolio Insurance based on the theoretical and practical plans within the framework of their corporation LOR (Leland, O'Brien and Rubinstein) which was to experience a rapid and spectacular development then a sudden drop due to the crash of October 1987, which highlighted the limits of the method. Other academics (Perold, Sharpe, etc.) have also contributed to the development of Portfolio Insurance.

ning and called the floor. It should be noted that such a profile, which is actually that of a call at maturity, is convex. The convexity of the profile $f(.)$ is a characteristic of insurance strategies.

Portfolio insurance techniques are often required to have the following properties:

C1—A minimum value or return (floor) must be ensured, as in the function $f(.) = Max\left(\frac{S(T)}{S(0)}, k\right)$.

C2—The probability of bursting this floor should be very low, 0 if possible.

C3—In case of an increase in the underlying asset (represented by $S(T)$) above the floor price, the captured rate of return should be a *predictable* percentage of that of the underlying asset.

The methods of *Stop Loss,* OBPI, and CPPI, which we will explore successively, own these properties, at least in theory.

24.2.1 The *Stop Loss* Method

The simplest of the Portfolio Insurance methods, *Stop Loss*, applies the following rules:

- We observe the evolution of the value of a hypothetical portfolio $M(t)$, initially normalized to 100 and only composed of monetary assets.
- In reality, the funds are invested in the desired risky asset S, the Dow Jones, for example, for a value of $S(0)$.
- S will be entirely liquidated to buy M as soon as $S(t) < M(t)$, and M will also be entirely liquidated to buy S once $S(t) > M(t)$, just as illustrated in Fig. 24.1. The values of $M(t)$ therefore represent the threshold of *stop-loss* buy or sell orders.
- In principle, we end up with Max $(S(T), M(T))$, the insured floor being here the value of the money market portfolio (*see* Fig. 24.1).

It is clear that such a strategy theoretically owns the characteristics (C1, C2, and C3) mentioned above. However, this strategy in practice will suffer from potentially high transaction costs since the entire portfolio is liquidated at each transaction. If S (t) fluctuates around $M(t)$ with a high volatility, the costs will become prohibitive. As a result, not only the outcome will be affected, but it will also be very dependent on the path of $S(t)$ (*path dependency*), which is an undesirable characteristic compared to strategies whose result is path-independent (e.g., the value of a European option) and only depends on the final value $S(T)$ of the risky underlying.

Therefore, in practice it is necessary to set an interval around the trajectory of $M(t)$ and only sell (buy) when $S(t)$ falls below (rises above) the lower (upper) limit of the

24.2 Portfolio Insurance

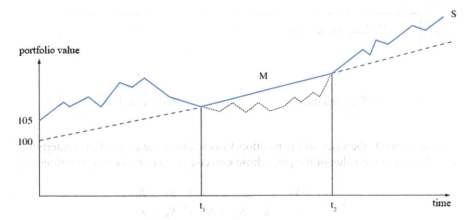

At t_1 S is liquidated to buy M; at t_2 M is liquidated to buy S.

Fig. 24.1 Evolution of the value of the portfolio under management according to the *stop-loss* method

interval. This strategy is less expensive in terms of the transaction costs but it makes the characteristics C1, C2, and C3 invalid.

24.2.2 Option-Based Portfolio Insurance

The objective of Option-Based Portfolio Insurance (OBPI) is to ensure a minimal value for the portfolio at a given maturity while still allowing the investor to benefit from any potential increase in the underlying security. Are relevant to this technique:

- The purchase (or replication) of a *put option* to hedge a long position in the underlying asset.
- The purchase (or replication) of a *call option* to take advantage of the increase in the underlying security, combined with a default-less zero-coupon bond that ensures the capital guarantee.

24.2.2.1 Portfolio Insurance with Long Puts or Replicated Puts
(a) **Basic set-up**

The position includes an underlying security, which benefits from its potential value increase, and a put option with maturity T. The put will appreciate in case the underlying security price decreases, thus providing a hedge.

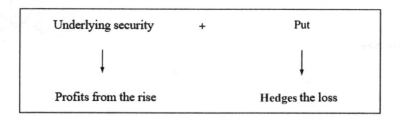

At maturity T, the value of the position V_T consists of the price of the underlying security S_T and the value of the put, whose exercise price is K. V_T can be written as:

$$V_T = \begin{cases} S_T & \text{if } S_T > K \\ S_T + K - S_T = K & \text{if } S_T \leq K \end{cases}$$

Thus, $V_T = Max\,[S_T, K]$.

The exercise price K of the put, therefore, constitutes the floor below which the value of the position V_T cannot fall. To obtain such a result, the required investment at $t = 0$ is $S_0 + P_0$. The price of the put P_0 is the premium paid to ensure that the portfolio value will not fall below the floor K. It's obvious that the higher the floor K is, the more expensive the premium is.

Above, it has been implicitly assumed that the put is purchased by the insured fund. However, there are some cases where it is difficult and/or expensive to find a put with the desired characteristics. In fact:

- The options offered on the markets are most often American options and therefore unnecessarily expensive compared to the European options (the European options are sufficient since what only matters are the values at maturity).
- The available exercise prices and dates of maturity may not be suitable for the investors' objectives; in particular, options with long maturities are rarely listed, even if they exist, they are often illiquid and abnormally expensive.
- Certain management or regulatory constraints limit the proportion of options contained in the portfolio (especially for mutual funds).
- Finally, the counterparty risk that many option sellers present should not be overlooked.

When these disadvantages or obstacles are prohibitive, it is preferable, through dynamic management involving the underlying security and a monetary asset, to *build a synthetic put*, i.e., to replicate it. The method of replication (as we are now going to explain) is similar to the one devised by Black and Scholes for the construction of the arbitrage portfolio underlying their model.

24.2 Portfolio Insurance

(b) *Replication of the put*

Assume $P(S,t)$ to be the theoretical price of a put at time t (thus $T - t$ is the distance from the maturity date) of exercise price K, the price of the underlying security being S.

A small variation dS implies a small variation dP *of opposite sign such that:* $dP = \frac{\partial P}{\partial S}.dS = \delta_p.dS$ ($\frac{\partial P}{\partial S} \equiv \delta_p$ is the delta of the put, between -1 and 0).

Note that δ_p is negative (because a price increase of the underlying security implies a decrease in the value P of the put) and that it depends on S and t.

A position combining B euros of short-term money market asset and a *short* position of $|\delta_p|$ units of the underlying (i.e., selling $\delta_p S$ *euros* of the underlying) behaves like a put with respect to variations of S. This position will then be written by specifying its two components: $\{B, \delta_p S\}$. Its value W is equal to the sum of the values of its components:

$$W(S, t) = B + \delta_p S \quad \text{and} \quad \frac{\partial W}{\partial S} = \delta_p = \frac{\partial P}{\partial S}$$

Note that, if the position $\{B, \delta_p S\}$ reacts as a put with respect to variations of S, it does not react as a put with the passage of time, except for a particular value of B. In fact, B generates interest which is proportional to its value. For this portfolio to react as a put with the passage of time, we must always keep $W=P$, which means: $W = B + \delta_p S = P$; thus : $B = P - \delta_p S$.

Therefore, a position at time t is composed of

$$\{P - \delta_p S \text{ euros in monetary assets}, \quad \delta_p S \text{ euros in underlying assets}\}$$

acts as a put in a short time interval $(t, t+dt)$ during which only small variations dS are possible. When these proportions are respected at each moment t between 0 and T, the portfolio behaves like a put purchased at time 0 with a maturity T.

Remarks
1. $\delta_p S < 0$, which means that the position on the underlying security is short.
2. $P(t,S)$ and $\delta_p(t,S)$ vary: the passage of time and the price change dS of the underlying asset, therefore, imply the changes of the two components of the portfolio which could duplicate the put option. Hence, the portfolio duplicating the put option should be subject to continuous review.
3. The value of the *put* $P(t,S)$ as well as the value of its delta $\delta_p(t,S)$ can only be calculated from a valuation model (Black-Scholes model, for example).
4. The synthetic put can alternatively be obtained by selling forward/future contracts on the underlying security rather than selling the underlying security itself.
5. The method is based on two assumptions: the path of the underlying asset price is assumed to be continuous (only small variations of S are possible in a short time

interval); its volatility σ is assumed to be constant according to the Black-Scholes model.

These two assumptions, unrealistic in practice, limit the validity of the method and raise implementation problems, which we will discuss later.

(c) *The composition of the portfolio hedged by the replication of a put*

The hedged portfolio hence includes a synthetic put for each underlying security that is in long position. Its composition is the sum of these two components:

$P—\delta_p\,S$ euros in monetary assets and $S + \delta_p\,S = S\,(1 + \delta_p)$ euros in the underlying asset. Maintaining these proportions requires continuous revisions.

Example 1

Consider an initial situation characterized by:

–

An underlying security with price $S = 100$ €.

–

An *at-the-money put*, with maturity T equal to the chosen horizon, with $P = 3$ €, and delta $= \delta_p = -0.5$.

The initial portfolio is therefore a multiple of:

$1 + \delta_p = 0.5$ securities $= 50$ € of securities S,
$P - \delta_p.S$ € of B $= 53$ € of B.

Total value: $V_0 = 103$ €; Proportions: $\alpha_S = \frac{50}{103}$; $\alpha_B = \frac{53}{103}$.

Assume that S increases to 103 €, and as a result, δ_p increases from -0.5 to -0.45 and P decreases to 1.5 €:

Thus V *increases by* 1.50 € (3–1.5) to 104.5 €.

The portfolio should therefore be revised so that its composition is a multiple of:

$$1 + \delta_p \text{ securities } S = 0.55 \text{ securities } S = 56.65 \text{ €}$$
$$P - \delta_p.S \text{ € of } B = 1.5 + 46.35 \text{ € of } B = \underline{47.85 \text{ €}}$$
$$V = 104.50 \text{ €}$$

Thus, the portfolio revision involves:
– Buying 0.05 securities S (5.15 €)
– Selling 5.15 € of B (53 − 47.85).

24.2.2.2 Portfolio Insurance with Calls

The position includes:

- A zero-coupon bond with maturity T which guarantees the capital.
- A call that benefits investors in case of a value increase in the underlying security.

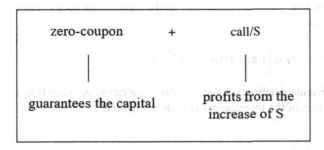

As in the case of the set-up involving the put, the option can either be purchased or replicated. Since the replication of the call is based on the same principles as the put,[8] we will only explain the set-up involving the purchase of the option.

The terminal value V_T of the position, which includes a zero-coupon bond valued at K euros at time T and a call on S with exercise price K maturity T, is the sum of the values of its two components:

$$V_T(S_T) = Max[S_T - K,\ 0] + E = Max(S_T, K).$$

We can see the identity of the result obtained with that of the set-up "underlying security + put" analyzed in the previous paragraph, which is explained by the redundancy of the assets involved in the two set-ups (put-call parity): a call can be synthesized by a put with the same maturity and the same exercise price K, a long position in S and a loan of amount $K/(1+r)^T$ (here r represents the discount rate of the zero-coupon loan with a maturity of T discounting back to time 0).

24.2.2.3 A Special Case: Guaranteed Capital Fund

The [underlying security + put option] or [zero-coupon security + call option] set-up makes it easy to construct a capital guaranteed capital fund. These two set-ups being actually equivalent, we will explore only the second ([zero-coupon security + call option]) in the following paragraphs. If the nominal value of the zero-coupon is equal to V_0 (thus the purchase price B_0 is lower than V_0) and the call option (the initial price of which is $V_0 - B_0$) is *at-the-money*, the fund will have the following two properties:

---

[8] At each moment, a portfolio which duplicates a call includes:
$\begin{cases} C - \delta_C S & \text{euros in monetary assets} \\ \delta_C S & \text{euros in underlying asset} \end{cases}$

- The fund value V_T at horizon T is necessarily greater than or equal to the nominal value of the zero-coupon security, i.e., the initial value V_0.
- In addition, thanks to the call option, the yield $\frac{V_T-V_0}{V_0}$ is a fraction k of $\frac{S_T-S_0}{S_0}$ if the latter yield is positive.

Therefore, for an investment V_0 of 100 € at time 0, the guaranteed capital fund will have a value of V_T at time T given by:

$$V_T = Max\left[100, \ 100\left(1 + k\frac{\Delta S}{S}\right)\right]$$

where k is a positive coefficient sometimes called *gearing* (gear ratio), between 0 and 1, which measures the portion of the increase in the underlying asset captured by the insured portfolio.

Example 2

Assume that $k = 0.6$ and that the CAC40 index is the underlying asset. The value of the fund will be 100 € in all circumstances where the value of the CAC40 drops between 0 and T $\left(k\frac{\Delta S}{S} < 0, \quad \text{thus} \quad V_T = 100\right)$. In favorable economic conditions, the fund will have a growth rate equal to 60% of CAC40's growth rate (e.g., if CAC40 increases by 20% between time 0 and time T, the value of the fund will increase by 12%).

To obtain such a result, we use the available 100 € to buy the following securities at time 0:

- A zero-coupon with a value of 100 € at time T, which costs $\frac{100}{(1+r)^T}$ €.
- $\left(100 - \frac{100}{(1+r)^T}\right)$ € of a call option on the CAC40.

Now consider an investment of 1 million € over 2 years and assume that:

- A zero-coupon with a value of 1 million € at $T = 2$ years now costs 0.907 million € (if the 2-year discount rate is 5%).
- An at-the-money call option on the CAC40 (for a 1 million € underluing value) which expires in 2 years costs 0.1550 million € (the volatility of the underlying index is 19%). Then we can buy the zero-coupon which guarantees the capital as well as $\frac{1-0.907}{0.155} = 0.60$ call options per 1 million €, thus obtaining a *gearing* $k = 0.60$.

Remarks

1. The performance of such a fund is all the better (in terms of *gearing* and/or the capital guaranteed) because:
   - interest rates are high (as the zero-coupon security is cheaper).[9]
   - the volatility is low (because the price of the call is lower).
   - the period between 0 and T is long (since the amount allocated to the purchase of the call $\left(100 - \frac{100}{(1+r)^T}\right)$ increases with T more quickly than its cost, for realistic values of r).
2. In practice, such a fund is built with *bullet* coupon bonds rather than zero-coupons. The nominal value of the bonds guarantees the capital, the coupon sequence allows the fund to buy a call on CAC40 (the premium of the call is equal to the current value of the coupon sequence, net of management fees. Therefore, the coupon sequence should be sold on the bond market in exchange for cash to buy the call).

24.2.3 CPPI Method

We introduce the CPPI method (*Constant Proportion Portfolio Insurance*) before exploring its properties and possible extensions.

24.2.3.1 Presentation of the Method

This method, which is very flexible in its use, is not based on option theory and does not require the investors to set an investment horizon T.

The portfolio value $V(t)$ can be decomposed in two different ways:

- In the first way (which is the traditional one), $V(t)$ can be broken down into $S(t)$, the risky part (called exposure), and $M(t)$, the risk-free monetary part (here we can choose from several monetary instruments with different durations): $V(t) = S(t) + M(t)$; the price of the risky security will be denoted $s(t)$ and will be distinguished from the amount invested in this security denoted $S(t)$.
- In the second way, $V(t)$ can be broken down into P, the guaranteed floor (assumed to be constant for the moment) and $C(t)$ the variable cushion defined as the difference between $V(t)$ and P, thus $V(t) = P + C(t)$.

The ratio $S(t)$ over $C(t)$, denoted by $m(t)$, is called *multiple*. Its target value m^* is fixed and strictly higher than one. Similarly, the inverse of $m(t)$, denoted by $\lambda(t)$, is called *management ratio*, the target ratio being the constant λ^*:

$$m(t) = S(t)/C(t); \quad \lambda(t) = C(t)/S(t)$$

---

[9] Despite the fact that the price of the call is a positive function of the interest rate.

The investor first chooses the floor P as well as the amount of initial investment $V(0)$. She obtains the initial cushion by computing the difference between $V(0)$ and P: $C(0) = V(0) - P$; then she gets the amount initially invested in the risky asset: $S(0) = m(0) * C(0)$; finally, she obtains the amount allocated to the monetary assets:

$$M(0) = V(0) - S(0).$$

Example 3

Assume that $V(0) = 1000$ and $P = 900$. Then $C(0) = 100$. Further assume that $m^* = 4$, thus $\lambda^* = 0.25$; Hence $S(0) = m^* C(0) = 400$. By making the difference with the initial total value (1000), we obtain the risk-free part invested in monetary assets (600).

In general, at any moment t, we have:

$$S(t) = V(t) - M(t) = m(t)C(t) = m(t)(V(t) - P), \text{thus}$$
$$S(t) = m(t)(P - M(t))/(m(t) - 1).$$

In practice, as prices fluctuate, the effective multiple and the ratio constantly deviate from their target values and the manager must choose a third parameter, called *tolerance*, to apply to the multiple (or equivalently, the management ratio), which will determine when the portfolio should be readjusted. The manager will only execute a transaction if, after a market fluctuation, the multiple which stems from the position deviates from its target value, from above or below, by a percentage greater than the *tolerance* or, equivalently, if $\lambda(t)$ falls outside of the interval $(\lambda_{min}, \lambda_{max})$.

It is important to note that when the price of the risky security $s(t)$ increases, this will cause an increase of ΔS *in the overall value of the risky component of the portfolio*, $M(t)$ will be unchanged and $C(t)$ will increase by the same amount ΔS since P is fixed. Thus, the multiple $m(t)$ decreases and $\lambda(t)$ increases. If the price $s(t)$ rises (falls) above (below) the tolerance threshold, the portfolio must be rebalanced through buying (selling) risky securities and simultaneously selling (buying) risk-free assets so that $m(t)$ and $\lambda(t)$ will return to their target value.

Example 4

Consider again example 3 as a starting point ($V(0) = 1000$; $C(0) = 100$; $m = 4$; $S(0) = 400$) and assume that the tolerance parameter is 20%, which implies that the thresholds of intervention for $m(t)$ are 3.2 and 4.8 (4 plus or minus 20%) and thus those for $\lambda(t)$ are 0.2083 and 0.3125.

Now assume that the price s increases rapidly by 10% and S reaches 440; thus the portfolio value is now 1040 (here we neglect the interest earned on the risk-free monetary part). The cushion $C(t)$ *is* 140, then the multiple becomes $440/140 = 3.14 < 3.2$ and the management ratio becomes $0.3182 > 0.3125$.

(continued)

24.2 Portfolio Insurance

The investor brings m back to 4 and λ to 0.25 through buying risky securities for an amount of 120 and selling monetary assets for the same amount. Indeed, we will have $440 + 120 = 560 = 4 \times 140$. Thus, the portfolio includes 560 invested in risky assets and 480 invested in risk-free assets. Following is the general equation to be resolved to obtain the amount of purchases/sales of the risky asset, denoting respectively $S(t^-)$ and $S(t^+)$ the value of the risky assets just before and just after the portfolio revision:

$$S(t^+) - S(t^-) = m\,[V(t) - P] - S(t^-) = 4 \times (1\,040 - 900) - 440$$
$$= 560 - 440 = 120.$$

24.2.3.2 Properties of the CPPI Strategy

(a) *General properties of the CPPI strategy*

The properties of the value V of the portfolio managed according to CPPI strategy comply with the following principles:

- Even in the event of a *continuous* decrease in $s(t)$, $V(t)$ remains greater than or equal to the floor P as the value of the cushion should be positive or at least zero.
- In case of a sudden drop in price s such that the investment manager does not have time to make the necessary adjustments, the cushion allows to absorb a decrease that is less than or equal to the lower limit λ_{min} of the management ratio (0.2083 in our example).
- The higher λ_{min}, the greater is the probability of not bursting the floor in case of a crash; however, the multiple m will be lower and thus the likelihood of benefiting from a significant market increase will also be lower. As always, an investor should therefore find a compromise between risk and expected return, the optimal situation depending on the investor's utility function.
- At time T, the profile of the portfolio value $V(T)$ is a convex function of the price s (T) (approximately like a call option) according to Fig. 24.2 (blue curve). The degree of convexity is an increasing function of the adopted multiple m^*.

Remarks
- The *Buy and Hold* strategy is a special case of CPPI in which the floor is zero and m^* is equal to one. The function $V_T(s_T)$ is then represented by the bisector (dotted line) in Fig. 24.2.
- Figure 24.2 shows that CPPI "outperforms" the underlying asset in extremely favorable market conditions. This is due to the fact that such conditions lead CPPI investors to be borrowers (hold a short position in monetary assets) and thus create a leverage effect.

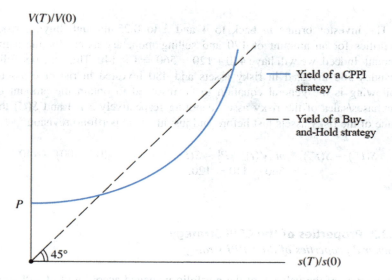

Fig. 24.2 Performances of a portfolio managed under a CPPI strategy

(b) *An analytical solution in a special case*

An analytical expression of the value of the portfolio managed under CPPI can be obtained simply under the following assumptions:

- The multiple m is always kept constant, by means of continuous transactions.
- The price $s(t)$ of the risky security follows the geometric Brownian motion:

$$\frac{ds}{s} = \mu dt + \sigma dW.$$

We emphasize that it is necessary to clearly distinguish the price s of the security from the amount S invested in this security.

The floor P is revised according to the interest rate r which is supposed to be constant:

$$dP = P(t)\, r\, dt, \quad P(t) = P(0)\, e^{rt}.$$

Under these assumptions, we first demonstrate that the cushion follows a geometric Brownian motion. In fact, since $V(t) = S(t) + M(t)$, we have:[10]

---

[10]The dynamics of the underlying *security price* $s(t)$ and the value invested in the security $S(t)$ are different since the number of the underlying security included in the portfolio changes. However, in the infinitesimal interval separating t and $t+dt$, we have $dS/S = ds/s$ ($= \mu\, dt + \sigma\, dW$), due to the self-financing condition imposed on the portfolio; see Chaps. 10 and 11 concerning self-financing strategies.

24.2 Portfolio Insurance

$$dV = dS + dM = S(t)(\mu dt + \sigma dW) + M(t)rdt,$$

and by expressing the dynamics of C:

$$d(V - P) = dC = dS + dM - dP = S(t)(\mu dt + \sigma dW) + M(t)rdt - P(t)rdt,$$

which implies, since $S = mC$ and $M = V - S = C + P - S = C(1 - m) + P$:

$$dC = Cm(\mu dt + \sigma dW) + C(1 - m)rdt.$$

Thus the cushion follows a geometric Brownian motion:

$$dC = C[r + m(\mu - r)] \, dt + Cm\sigma \, dW \Leftrightarrow C(t)$$
$$= C(0) \, \exp \left[\left(r + m(\mu - r) - (m\sigma)^2/2 \right) t + m\sigma \, W(t) \right].$$

Since $V(t) = C(t) + P(t)$, the log-normality of the cushion directly implies that the portfolio value can be written as the sum of the floor and a log-normal variable:

$$V(t) = P(t) + (V(0) - P(0)) \, \exp \left[\left(r + m(\mu - r) - (m\sigma)^2/2 \right) t + m\sigma \, W(t) \right].$$

One checks verified that the portfolio value is consistently higher than the value of the floor, bearing in mind that this result depends on the assumed continuity of the $s(t)$ trajectories.

The terminal portfolio value $V(T)$ can also be expressed as a function of the price $s(t)$ of the risky security ($s(t) = s(0) \, e^{(r - \sigma^2/2)T + \sigma \, W(T)}$) by highlighting the relationship between the two overall returns $\frac{V(T)}{V(0)}$ and $\frac{s(T)}{s(0)}$:

$$\frac{V(T)}{V(0)} = \frac{P(T)}{V(0)} + \left(1 - \frac{P(0)}{V(0)} \right) \left(\frac{s(T)}{s(0)} \right)^m e^{-(m-1)(r+m\sigma^2/2)T}$$

From this analytical relationship, we find all the characteristics of the CPPI yield profile ($\frac{V(T)}{V(0)}$ as a function of $\frac{s(T)}{s(0)}$) described in Fig. 24.2 (convexity, outperformance compared to the underlying in favorable market conditions, etc.).

It should be noted that the return on the CPPI portfolio, the minimum value of which is $\frac{P(T)}{V(0)}$, does not only depend on the yield $\frac{s(T)}{s(0)}$ of the underlying asset, but also depends on the parameters σ and r *which are actually random*. It should also be noted in particular that the return on the CPPI portfolio decreases with the volatility σ of the risky asset price (because $m > 1$).

24.2.3.3 Extensions of the CPPI Method

The CPPI method is extremely flexible and can be adapted in many ways.

The simplest extension is to raise the floor P (we then write it $P(t)$):

- Either in a predetermined way, as presented in Sect. 24.2.3.2 (b);
- Or according to the evolution of the chosen monetary rate;
- Or else according to the evolution of $S(t)$, to increase the guarantee in case of a significant rise in the risky asset (we will come back to this point later when we discuss ratchet products).

These extensions do not pose any particular problem and make the method flexible and attractive.

Another possibility is to have several risky assets instead of just one and allocate the risky part of the portfolio among these assets by using the *Tactical Asset Allocation technique* (see the next chapter). We can also consider changing the multiple m (or the management ratio λ) over time to increase the monetary component as the investment horizon approaches for example. It is also possible to assume a more complex process for the evolution of risky assets, for example, by making their volatility stochastic.[11] Finally, the risky component can also be a bond portfolio which is sensitive to interest rate movements.

Despite their undeniable attractiveness, the methods of Portfolio Insurance based on replication of options or CPPI have limitations.

- Portfolio Insurance strategies involve, as we have seen before, buying the risky security when the market rises and selling it when the market falls, which seems to be uncomfortable.
- With a given tolerance, readjustments are more frequent when the underlying is volatile. As each transaction generates some cost, the overall net return is lower when this volatility (observed during the insurance period) proves to be significant.
- That the actual performance of the portfolio deviates from the theoretical one can be explained by the sometimes discontinuous nature of the underlying asset price, always discontinuous interventions of the investment manager, and the existence of transaction costs.
- The random evolution of the risk-free interest rate can also be problematic.
- In addition, it should be noted that in case of a sudden drop in the asset price, the required proportions for the option replication will not be met and thus, the floor will be burst. In this respect, CPPI is better than option replication because the floor is guaranteed as long as the sudden drop of the security price does not exceed, in percentage terms, λ_{min}. In fact, the Portfolio Insurance techniques that do not involve the actual purchase of options hedge against adverse price fluctuations which are potentially severe, but not against true crashes. In addition, we should not overlook the counterparty risk in the event of an option purchase (except on an organized market, due to the margin call procedure).
- Changes in the volatility of the underlying price are a serious problem in case of option replication. In the case of an underestimation of the volatility, the

---

[11] See Prigent and Bertrand (2003), for example.

24.2 Portfolio Insurance

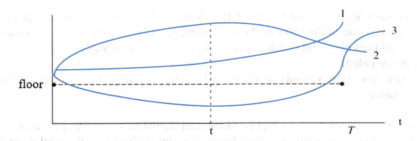

Fig. 24.3 Possible trajectories for the insured portfolio value at time T.
Note: Trajectory 3 is possible for OBPI method, but impossible for CPPI method; trajectory 2 can be excluded by a ratchet CPPI.

theoretical price of the synthesized option is too low and the portfolio is under-guaranteed. In contrast, in case of an overestimation, the portfolio is over-guaranteed, which leads to underperformance when the underlying asset price increases. An advantage of the CPPI method over its competitor is that it does not require the estimation of the volatility (which obviously does not mean that the results of the cushion method are independent of the realized volatility).

– It should also be noted that the fraction of the increase in the underlying price captured in the case of CPPI method is difficult to quantify precisely, ex ante. In particular, we noted in Sect. 24.2.3.2 (b) (when presenting the analytical solution obtained in a simple case) that the return on the CPPI portfolio not only depends on the return on the underlying, but also depends on the parameters σ and r. However, these parameters are in fact stochastic. It is therefore perilous to promise an indexation on the evolution of the underlying price. Remember that the parameters σ and r do not explicitly intervene in the expression of the return of an option-based Portfolio Insurance. In this respect, OBPI is better than CPPI, especially in the case of actual option purchases.

– Note that the OBPI method allows the trajectory of the portfolio value to potentially fall below the floor and rebound to at least the floor level at time T (as trajectory 3 of the Fig. 24.3 demonstrates) whereas the trajectory remains (in principle) constantly above the floor with the CPPI method. Therefore, the CPPI method offers the advantage over its competitor of not requiring the choice of an investment horizon (as, in principle, the floor is never burst).

– Finally, it should be noted that the results of these methods are, at least *in practice* if not in theory, *path-dependent*, in particular, due to the existence of transaction costs as well as the discontinuous nature of transactions.

24.2.4 Variants and Extensions of the Basic Methods

Different variants of the basic OBPI and CPPI methods described above could be implemented.

We have already noted that the CPPI floor can be revised upwards, especially with the increase in the value of the underlying security. Such a *ratchet (or cliquet) fund* facilitates entry and exit at any time (for example, the trajectory of the Fig. 24.3 can be excluded).

There also exist variants of OBPI and especially the guaranteed capital fund which offers:

- A better *gearing* and/or a higher guaranteed capital in return for entry fees.
- A distribution of coupons (albeit lower than those offered by the market) at the price of a reduction in *gearing* and/or guaranteed capital.
- A better *gearing* and/or a higher guaranteed capital in return for a *cap* on the performance (so-called *capped* product).

In addition, other more complex financial products, the set-up of which involves the use of exotic options in some circumstances, have undergone strong development in recent years. These financial products include ratchet products, products with maximum or minimum value, tiered funds, and various *promised performance funds* whose real prospects are often less attractive than they appear.

- *Product with ratchets (cliquets).* For example, the actual performance can be calculated based on the *maximum*:
  - Of the four underlying security values at the end of each trimester, if it is a 1-year product.
  - Of the four underlying security values at the end of each year, if it is a 4-year product.
- *Product in which the increase in the risky asset is acquired if it exceeds a certain level (or barrier).* For example, consider the case of a product whose index is based on a market index of initial value 3000. A barrier of 3200 will allow to benefit from a new minimum of 3200 provided that the index reaches this level at any time between 0 and T.
- *Products indexed on the best performance of N indices*: $V_T = \text{Max}\ [I_1(T),...., I_N(T)]$.
- The overall performance is often conditional on a minimum performance obtained by *each* element of a basket, for example:

$$V_T = \text{Max}\ [I_1(T),\ ...,I_N(T)] \qquad \text{if}\quad I_j(T) > kI_j(0) \quad \text{for } j = 1,..,N;$$
$$V_T = V_0 \qquad\qquad\qquad\qquad\qquad \text{otherwise.}$$

A thorough analysis of the indexation conditions thus often shows how highly improbable it is.

However, the existence of formulas for valuing exotic options, the ensuing hedging techniques, and the development of simulation methods have facilitated the development of such products.

24.2.5 Portfolio Insurance, Financial Markets Volatility and Stability

As we have seen, there are two possibilities for investors who wish to follow portfolio insurance strategies. The first one consists in buying options, for example puts, to hedge a long position in shares, stock indices, or bonds. The second one is to replicate an option or to implement a CPPI strategy.

Note that for the first method—buying put options—the seller of the puts almost always hedges in delta-neutral position, i.e., constructs synthetic puts options to cover their short position. Therefore, the consolidation of the positions of the seller and the buyer of the puts creates a portfolio identical to the one obtained when the investor replicates himself the puts (the second method).

Therefore, in both cases, Portfolio Insurance automatically implies the sale (purchase) of the underlying security[12] in response to a decrease (increase) in its price. *A priori*, this mechanism increases the price volatility and is a factor of market instability to the extent that it tends to amplify and self-sustain price movements. This downward spiral argument is invoked by the critics of Portfolio Insurance, who blamed it for both the 1987 and 1929 stock market crises (because of *stop-loss orders* in the latter case).

However, in the absence of important news concerning the fundamental elements which determine stock prices in the medium and long terms, this downward spiral logic is questionable. In fact, investors who base their decisions on fundamentals can be expected to react in the opposite direction to investors who apply Portfolio Insurance, thus offsetting the effects of the latter.

These questions were asked in concrete terms, especially in Brady's report (1988) following the October 1987 crisis, and were seriously analyzed in a few scientific articles.[13] Brady's report suggests that derivative markets and Portfolio Insurance played a role in increasing volatility and causing security prices to fall sharply. According to some theoretical models, Portfolio Insurance could increase the volatility of the financial market, weakly when the market is efficient and well-informed, significantly when the market lacks transparency and is incorrectly informed about the relative importance of the insurance strategies. In the latter case, poorly informed operators may attribute sell (buy) orders related to Portfolio Insurance transactions to investors who have just received unfavorable (favorable) inside information. Knowing that they are poorly informed, they lower (raise) their expectations regarding the fundamentals and increase the amplitude of the stock market movement initiated by Portfolio Insurance through their own transactions.

In any case, investors who legitimately wish to limit their losses could not be prevented from reducing or liquidating their losing positions, nor they are to blame.

---

[12] Or the sale of forward (futures) contract on the underlying asset, which in turn involves, through *cash-and-carry* transactions, the sale of underlying asset.

[13] See Brennan and Schwartz (1989) and Genotte and Leland (1990) in particular.

24.3 Dynamic Portfolio Optimization Models

The management styles described in the previous sections do not rely on the investor's specific preferences to provide a *tailor-made* portfolio. In these approaches, the investor chooses a management style with more or less precise objectives, depending on their preferences and in particular their risk aversion, among a wide range of possibilities; in that way, we can say it is more like *ready-to-wear*. An alternative approach is to build a portfolio that optimizes the expected utility of the investor. The static Markowitz model presented in Chap. 21 belongs to such an approach, but it excludes any possibility of portfolio revision. This limitation, which is prohibitive when it comes to make a long-term investment, motivated more recent work on portfolio optimization with possible revisions.

In particular, as part of a fundamental scientific work that partly won him the Nobel Prize in Economics in 1997, Robert Merton (1969, 1971, 1973) developed an optimization model in which the investor continuously carries out transactions between time 0 and time T (their investment horizon). It is a continuous-time *dynamic* model, based on stochastic dynamic programming, whose objective is not necessary to acquire an efficient portfolio in the sense of mean-variance but more generally[14] to maximize the expected utility of the terminal portfolio value. Merton was the pioneer in continuous-time finance and his model allows generalizing different results obtained in the static case (such as Tobin's and Black's two-fund separation principle). However, practical applications in the field of quantitative portfolio management are limited by the great complexity of stochastic dynamic programming which, with the exception of very special cases, does not allow for an analytical expression of the optimal strategy. However, a method subsequent to Merton's stochastic dynamic optimization method, developed by Cox and Huang (1989) and Karatzas, Lehoczky and Schreve (1987), allows to determine the optimal dynamic strategy relatively simply in some cases.

24.3.1 Dynamic Strategies: General Presentation and Optimization Models

24.3.1.1 Presentation of the Problem and Notations

Generally speaking, "dynamic portfolio strategy" is understood as a rule that allows an investor to determine at any time the allocation of their wealth to each available financial securities: instantly risk-free assets, various risky assets (bonds, stocks, real estate, etc.). This allocation obviously depends on the investor's objectives (utility function), the time elapsed (t between 0 and T), the investor's wealth (portfolio value $X(t)$ attained at time t), and a vector $\underline{Y}(t)$ of variables called state variables

---

[14]Efficient portfolios in this sense are obtained by maximizing the expectation of a quadratic utility function. This function is only a special case of a class of utilities called HARA (*Hyberbolic Absolute Risk Aversion*); *see* Chap. 21.

24.3 Dynamic Portfolio Optimization Models

characterizing the market conditions (especially interest rate and stock market index). A strategy can therefore be represented by a function f which gives, at any time t, the weights $\underline{x}(t)$ *assigned to the different assets*:

$$\underline{x}(t) = f\left(t, X(t), \underline{Y}(t); U, T\right).$$

This makes it clear that the function f *depends on the horizon T and the preferences* (utility U) of the investor and/or the investment manager, and also on the management rules given a priori.

We study the problem of an investment management which starts at $t = 0$ and ends at $t = T$. Markets are assumed to be continuously open and transaction costs to be negligible.

We use the notations and the analytical framework of Chap. 21, which we adapt to the case of continuous time. Readers should therefore get familiar with this analytical framework as well as stochastic calculus and change of probability measure and numeraire (*see* Chap. 19). In addition, some developments are based on the results and concepts of financial theory which will be presented briefly.

There are $N+1$ securities, denoted by $0, 1, \ldots, N$, traded on the market. Security 0 is a risk-free short-term asset (for any t, the investment horizon is dt) while securities $1, \ldots, N$ are risky. $S_i(t)$ represents the price of the security i at time t. We write:

$$\frac{dS_0}{S_0}(t) = r(t)dt; \quad \frac{dS_i}{S_i}(t) \equiv dR_i(t) \quad i = 1, \ldots, N$$

where $r(t)$ is the short-term rate (stochastic) and $dR_i(t)$ is the random return of security i between t and $t + dt$.[15]

The vector of the return of N risky securities is denoted by $d\underline{R}$ and is assumed to be governed by the multidimensional Itô process:

$$d\underline{R} = \underline{\mu}(t)dt + \Sigma(t)d\underline{W}. \tag{24.1a}$$

The variance-covariance matrix $\mathbf{V}(t)$ of the general term v_{ij} ($v_{ij}\, dt = Cov\,(dR_i, dR_j)$) writes: $\mathbf{V}(t) = \Sigma(t)\Sigma'(t)$.

We assume that $\mathbf{V}(t)$ is invertible, which is the same as assuming that the N risky assets are not redundant (*see* Chap. 21).

At a given time t, a portfolio is defined, up to a one scale factor, by the weights $x_1(t), \ldots, x_N(t)$ of the N risky securities. The weight $x_0(t)$ of the instantly risk-free asset is equal to $x_0(t) = 1 - \sum_{i=1}^{N} x_i(t)$, or $x_0(t) = 1 - \underline{1}'\underline{x}(t)$ ($\underline{1}$ is the unitary vector of $\mathbb{R}^N$, the prime denotes the transposition and $\underline{1}'\underline{x}$ is a scalar).

---

[15] For dR_i to represent the return of i, any potential dividend it distributes must be reinvested in i and incorporated into S_i as if it were the share of a capitalization mutual fund composed solely of security i.

The portfolio value $X(t)$ is associated with the strategy of portfolio $\underline{x}(t)$ $(t \in [0, T])$ and the initial investment $X(0)$. The strategy will be denoted by $(\underline{x}, X)$ to indicate the weights and value. We only consider self-financing strategies, i.e., strategies that do not involve contributions nor withdrawals of funds between 0 and T. The relative change in the value of a self-financed portfolio then equals its return, equal to the weighted sum of the returns of its components, written as:

$$\frac{dX}{X} = x_0(t)r(t)dt + \underline{x}'(t)d\underline{R}$$
$$= [1 - \underline{1}'\underline{x}(t)]r(t)dt + \underline{x}'(t)d\underline{R} \qquad (24.1b)$$
$$= r(t)dt + \underline{x}'(t)(d\underline{R} - r(t)\underline{1}dt)$$

The reader will notice the analogy between this formula and the one used to represent the portfolio return in discrete time, in the presence of a risk-free asset (*see.* Chap. 21).

We study the problem of the choice of a self-financing strategy by an investor who tries to maximize the expected utility of their portfolio's terminal value $U(X(T))$, the program of which writes:

$$(P) \quad \underset{\underline{x}}{Max} E\{U(X(T))\}$$

$$s.t. \begin{cases} \dfrac{dX}{X} = x_0(t)r(t)dt + \underline{x}'(t)d\underline{R} & (C_1) \\ X(0) \equiv X_0 = 1 & (C_2) \end{cases}$$

By making $d\underline{R}$ explicit using Eqs. (24.1a) and (24.1b), the first constraint (C_1) can also be written as:

$$\frac{dX}{X} = \left[r(t) + \underline{x}'(t)\left(\underline{\mu}(t) - r(t)\underline{1} \right) \right] dt + \underline{x}'(t)\underline{\Sigma}(t)d\underline{W} \qquad (C_1')$$

The second constraint (C_2) indicates that the initial investment is equal to X_0 which, without loss of generality, can be normalized to 1. Remember that the terminal value $X(T)$ obtained from an initial investment of 1 is equal to 1+ rate of return over the entire period $(0, T)$.

The program (P), the control variable of which is the strategy $\underline{x}(t)$, can be solved using stochastic dynamic programming with Merton's method or, after a simple adaptation, with that of Cox and Huang or Karatzas, Lehoczky, and Shreve. However, before discussing the two models, we will discuss the principle and problems of dynamic programming and then consider the special case of a logarithmic investor (also called Bernoulli investor) which is simple but important.

24.3.1.2 Dynamic Programming: Notations, Problems, and Principle

We will denote $E_t\{.\}$ the conditional expectation obtained from the information available at time t and $J(t, X(t), \underline{Y}(t))$ the maximum expected utility (max $E_t\{U(X(T))\}$) that an investor could obtain, between time t and T, by following an optimal policy during this period. This maximum expected utility, which depends on t, the wealth level $X(t)$ and the state of the economy $\underline{Y}(t)$, is called the *indirect utility function*.

At any time t, the operator chooses the composition of their portfolio maintained between time t and $t + dt$. It is important to note that in most cases, the investor could only take into consideration the consequences of their choice till time $t+dt$, through a simple maximization of $E_t\{U(X(t+dt))\}$. In fact, *the optimal policy between time t and t + dt depends on the events which are likely to occur in the future between time t+dt and T, and depends on the investor's reactions to these events.* This remark is very general in nature; for example, if a person wishes to travel from A to B as quickly as possible, the choice between two alternative routes for the first part of the journey obviously depends on the expected hazards and congestion in the second part of the journey. In the context of portfolio selection, the policy of maximizing $E_t\{U(X(t+dt))\}$ at any time t "without thinking about what happens next" is described as *myopic and suboptimal*, except for the logarithmic utility function *alone*, as we will see in the next subsection.

In general, the optimal policy is based on the following principle, which is the core of dynamic programming:

– Whatever situation is attained at time $t + dt$ (characterized by $X(t+dt)$ and $\underline{Y}(t+dt)$), the investor is assumed to follow an optimal policy between time $t+dt$ and T, which will allow to obtain an expected utility (calculated with E_{t+dt}) equal to $J(t+dt, X(t+dt), \underline{Y}(t+dt))$.
– At time t, the optimal portfolio $\underline{x}^*(t)$ maximizes $E_t\{J(t+dt, X(t+dt), \underline{Y}(t+dt))\}$, where $X(t+dt)$ obviously depends on $\underline{x}^*(t)$.

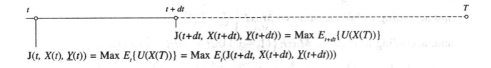

24.3.2 The Case of a Logarithmic Utility Function and the Optimal Growth Portfolio

We denote $(\underline{h}, H)$ the portfolio which solves the logarithmic investor's optimization program. Remember that (*see.* Chap. 21) the risk aversion coefficient (Arrow-Pratt coefficient) is constant and equal to 1 for any logarithmic utility.

More specifically, by definition and under the program (P), the logarithmic portfolio $(\underline{h}, H)$ solves:

$$\underset{\underline{x}, X}{MaxE}\{\ln[X(T)]\}$$

$$s.a. \begin{cases} \dfrac{dX}{X} = rdt + \underline{x}'(d\underline{R} - rdt.\underline{1}) \\ X(0) = 1 \end{cases}$$

Such a portfolio is called the *numeraire portfolio* (Long 1990), *optimal growth portfolio* (Merton 1992), or *logarithmic portfolio*. We have discussed it in Sect. 19.6, under the name of P-numeraire and here we analyze it from a new perspective.

The solution to this program is simple because of the myopia that characterizes the logarithmic utility function. To understand that the logarithmic investor does not have to worry about what will happen in future periods in making their portfolio choices in the upcoming period, let us denote $1, 2, \ldots, N$ as the periods and focus on the terminal value $\ln(X(N))$ of the portfolio.

Let us start with the identity: $X(N) = X(0)\frac{X(1)}{X(0)} \frac{X(2)}{X(1)} \cdots \frac{X(N)}{X(N-1)} = X(0)R_1 \cdot R_2 \ldots R_N$ (where R_i refers to the portfolio return during the period i). Therefore:

$\ln(X(N)) = \ln(X(0)) + \sum\limits_{t=1}^{N} \ln(R_t)$, and maximizing $E\{\ln(X(N))\}$ is equivalent to maximizing $\sum\limits_{t=1}^{N} E\{\ln(R_t)\}$.

As the choice of portfolio at time t does not affect the rate of return that could be obtained at time $t+1, .., N,$[16] maximization is therefore obtained simply by choosing, at each step t, the portfolio which ensures the maximization of $E\{\ln(R_t)\}$ (without taking into account the $\ln(R_{t'})$ relative to the subsequent periods t', the expected value of which will be maximized at the appropriate time t').

This myopia greatly facilitates the calculation of the weights $\underline{h}(t)$ assigned to risky securities.

In fact, $\underline{h}(t)$ solves $\underset{\underline{x}(t)}{MaxE}\{\ln(X(t+dt))\} \Leftrightarrow \underset{\underline{x}(t)}{MaxE}\{d \ln(X(t))\}$

and according to Itô's lemma, $\Leftrightarrow \underset{\underline{x}(t)}{MaxE}\left\{\frac{dX}{X} - \frac{1}{2}\left(\frac{dX}{X}\right)^2\right\}$

and, according to (C_1): $\Leftrightarrow \underset{\underline{x}(t)}{MaxE}\left\{(1 - \underline{x}'\underline{1})rdt + \underline{x}'d\underline{R} - \frac{1}{2}\text{variance}(\underline{x}'d\underline{R})\right\}$

$\Leftrightarrow \underset{\underline{x}(t)}{Max}\left\{\underline{x}'\left(\underline{\mu} - r\underline{1}\right) - \frac{1}{2}\underline{x}'\mathbf{V}\underline{x}\right\}$ obtains if: $\frac{d}{dx}\left[\underline{x}'\left(\underline{\mu} - r\underline{1}\right) - \frac{1}{2}\underline{x}'\mathbf{V}\underline{x}\right] = 0$,

therefore, for $\underline{x} = \underline{h}$ such that:

---

[16]Except for an investor who can influence security prices because of their size, hence future rates of return.

$$\underline{h} = \mathbf{V}^{-1}\left(\underline{\mu} - r\underline{1}\right). \tag{24.2}$$

The portfolio $(\underline{h}, H)$ has many properties. We discuss four of them.

– At any given time t, it is efficient at horizon $t+dt$ (*see.* proposition 4 of Chap. 21 characterizing efficient portfolios). It is therefore *homothetic to the tangent portfolio.*
– It maximizes the expected growth rate G of the wealth level at any time horizon T. In fact, as $\frac{X(T)}{X(t)} = e^{G(T-t)}$, we have $G = \frac{1}{T-t}[\ln(X(T)) - \ln(X(t))]$, and at any time t, the strategy that maximizes $E\{\ln(X(T))\}$ also maximizes $E(G)$. Because of this property, it is named *optimal growth portfolio.*
– In a market free of arbitrage opportunities, it can be demonstrated (*see.* Chap. 19) that the prices $X(t)/H(t)$ of self-financing securities or portfolios, denominated in the numeraire H, are *martingales under the historical probability:*

$$\frac{X(0)}{H(0)} = X(0) = E\left(\frac{X(T)}{H(T)}\right). \tag{24.3}$$

$H(T)$ is therefore the *appropriate stochastic deflator* (or *stochastic discount factor*), which allows to calculate the value of a random future flow using a simple expectation. It involves an alternative valuation formula equivalent to that obtained by changing the historical probability to the (risk-neutral) probability Q^{17} with asset 0 used as numeraire. We have analyzed this portfolio $(\underline{h}, H)$ in a deeper and more formal way in Chap. 19, which discusses mathematical finance. The choice of the numeraire H avoids the change in probability.

– Finally, it can also be shown that the optimal growth portfolio $(\underline{h}, H)$ "dominates over the long term" all other portfolios in the following sense:
For any attainable and self-financing portfolio strategy, different from $(\underline{h}, H)$, *whose terminal value is* $X(T)$, *the probability of the event* $[H(T) > X(T)]$ *tends toward* 1 *when T tends towards infinity.*

This property is less attractive than it seems. It is confusing and has given rise to a criterion, called Kelly's criterion, which is in favor among some practitioners. According to this criterion, *any* investor with a *distant investment horizon* should follow the logarithmic strategy since the probability that the value of their portfolio will be higher than the value obtained by following a different strategy is, asymptotically, equal to one. However, we can prove that Kelly's criterion leads to extremely

---

[17] Which means: $X(0) = E\left(\frac{X(T)}{H(T)}\right) = E^Q\left(\frac{X(T)}{S_0(T)}\right)$.

risky portfolios; it is only valid for logarithmic investors whose risk aversion is very low, and for whom the investment horizon does not matter since they are myopic.[18]

24.3.3 The Merton Model

In a series of fundamental articles, Merton (1969, 1971, 1973) proposed a general method[19] based on stochastic dynamic optimization techniques to solve the program (P).

24.3.3.1 General Presentation of the Model and the General Form of the Solution

In addition to the general assumptions described in Sect. 24.3.1, Merton's model assumes that the interest rate r and the security price follow diffusion processes that depends on m state variables $\underline{Y}(t)$ which are also governed by diffusion processes, so that we have:

$$d\underline{Y}(t) = \underline{\alpha}(t, \underline{Y})dt + \Omega(t, \underline{Y})d\underline{W}; \quad dr = \phi(t, Y)dt + \underline{q}'(t, \underline{Y})d\underline{W};$$
$$d\underline{R} = \underline{\mu}(t, \underline{Y})dt + \Sigma(t, \underline{Y})d\underline{W} \tag{24.4}$$

where $\underline{W}(t)$ represents a standard p-dimensional Brownian motion, $\Omega(.)$ is a $(p \times m)$ matrix, $\Sigma(.)$ is a $(n \times m)$ matrix, and $\underline{\alpha}$ and $\underline{q}$ are m- and n-dimensional vectors, respectively.

The derivation of this model is complex and beyond the scope of this book. It only leads to precise analytical relationships in very specific cases concerning both processes and utility functions. Nevertheless, we obtain the general form for the optimal weights, which write:

---

[18] We can indeed show that $Lim_{T \to \infty} Proba [H(T) > X(T)] = 1$ does not imply $E(U(H(T)) > E(U(X(T))$ for any U and for T large enough. Thus, Kelly's criterion is incompatible with the principle of maximizing expected utility, and is unfounded in this respect.

[19] In fact, the Merton model addresses a more general case than the problem of simple portfolio optimization, since it also includes optimal consumption/saving trade-offs. The objective function is written as:

$MaxE \left\{ \int_0^T u(c(t), t)dt + U(X(T)) \right\}$ where $c(t)$ represents the consumption density from which the consumer-investor derives, between t and $t+dt$, a utility $u(c(t),t)dt$. The function U is interpreted as the utility of the bequest. The dynamics of the portfolio value (constraint C_1) writes: $dX = X\{rdt + \underline{x}'(d\underline{R} - rdt\underline{1})\} - cdt$.

24.3 Dynamic Portfolio Optimization Models

$$\underline{x}^*(t, X, \underline{Y}) = \tau(t, X, \underline{Y}) \ \underline{h}(t, X, \underline{Y})$$

$$+ \sum_{i=1}^{m} \xi_i(t, X, \underline{Y}) \ \underline{h}_i(t, X, \underline{Y}) \quad \left(\text{and } x_0 = 1 - \underline{x}^* \underline{1}\right) \tag{24.5}$$

where:

$\tau = -\frac{J_X}{XJ_{XX}} Z$ represents the *relative risk tolerance* (*cf.* Chap. 21) calculated at time t from the indirect utility function;

$\underline{h} = \mathbf{V}^{-1} (\mu - r\underline{1})$ is the optimal growth portfolio discussed before;

ξ_i depends on the indirect utility function and reflects its sensitivity to the state variable Y_i; the state variables influence investment prospects,[20] especially expected return and risk, which further influence utilities that could be derived in the future and thus the indirect utility function; the evolution of a state variable may be favorable or unfavorable in this sense;

$\underline{h}_i$ is a portfolio whose return is perfectly correlated with the variation dY_i of the ith state variable during the interval t, $t+dt$ (and uncorrelated with the dY_j); the position on $\underline{h}_i$ is a hedge protecting the investor from the risk of an unfavorable evolution of Y_i.

24.3.3.2 Principle of Separation into $m + 2$ Funds and Interpretation

Merton's solution leads directly to the following separation principle.

Theorem of separation into $m + 2$ funds *The optimal portfolios for all investors are the dynamic combinations of the same $m+2$ funds below:*

Risk-free asset 0; optimal growth portfolio $\underline{h}$; m funds $\underline{h}_1, .., \underline{h}_m$, the ith fund $\underline{h}_i$ being constructed in such a way that its return is locally (which means between time t and $t+dt$) and perfectly correlated with the variations of the ith state variable Y_i.

Note that the composition of each of the $m+1$ separating funds $\underline{h}, \underline{h}_1, ...,\underline{h}_m$ is continuously readjusted (according to t and $\underline{Y}(t)$) and that the weights investors assign to the $m+2$ funds separators 0, $\underline{h}, \underline{h}_1, ...,\underline{h}_m$ depend on agents' preferences and are also continuously modified.

Let us emphasize the interpretation of this fundamental result. The portfolio of any investor who maximizes a concave expected utility includes:

- An allocation in the logarithmic portfolio $\underline{h}$ which, along with asset 0, would be the only one present if the investor were myopic. This allocation is proportional to the investor's risk tolerance: it is therefore homothetic to the tangent portfolio or obtained as a combination of the tangent portfolio and the risk-free asset (it is, therefore, efficient at horizon $t+dt$ in the mean-variance sense).

---

[20] The *investment opportunity set* is defined as the set formed by the coefficients $\alpha(t, \underline{Y})$; $\Omega(t, \underline{Y})$; $\phi(t, \underline{Y})$; $\underline{q}'(t, \underline{Y})$; $\mu(t, \underline{Y})$; $\Sigma(t, \underline{Y})$.

- m allocations in the hedge portfolios $\underline{h}_1, \ldots, \underline{h}_m$, intended to protect the investors who are not myopic from economic developments that would darken the prospects for future investment.
- a position (short or long) on monetary asset 0 that "accommodates" the allocations to other $m+1$ separating funds, so that the sum of the weights is always equal to 1: $x_0 = 1 - \underline{x}^*\underline{'}\underline{1}$.

24.3.3.3 Special Cases

Case n° 1: The rate r is constant and no state variable affects the coefficients $\underline{\mu}$ and Σ which are also constant.

Thus:
- The weight of the logarithmic portfolio is constant: $\underline{h} = \Sigma\Sigma'^{-1}(\mu - r\underline{1})$.
- Since the number of the state variables m is zero, the separation into two funds prevails: all optimal portfolios are the combinations of the risk-free asset and the logarithmic portfolio.
- The risky components of all optimal portfolios have an identical structure (they are homothetic to $\underline{h}$, therefore to the tangent portfolio), as in the static case studied by Markowitz. The optimal weights of risky assets are $\underline{x}^* = \tau(X(t))\ \underline{h}$ and are therefore proportional to the investor's relative risk tolerance.

Case n° 2: A single state variable, the interest rate r(t), affects the coefficients μ and Σ.

The model is written as:

$$dr = \phi(t,r)dt + \underline{q}'(t,r)d\underline{W}; \quad d\underline{R} = \underline{\mu}(t,\underline{r})dt + \Sigma(t,\underline{r})d\underline{W}$$

Here, $m = 1$ and a separation into *three* funds prevails: asset 0; logarithmic portfolio; portfolio perfectly (positively or negatively) correlated with changes in the interest rate $r(t)$ (a bond portfolio).

24.3.4 The Model of Cox-Huang and Karatzas-Lehoczky-Shreve

In related and independent work, different authors (Pliska, Karatzas-Lehoczky-Shreve, and Cox-Huang) have proposed an alternative method to the method of Merton, which helps to solve the problem of dynamic optimization (P) when the markets are efficient. We introduce the notion of completeness (discussed in more detail in Chap. 19) before presenting the model.

24.3.4.1 The Notion of a Dynamically Complete Market

Let us consider all the attainable (achievable) self-financing strategies (there are an infinite number of them). Each strategy allows to reach a random terminal value; for example, $(\underline{x}, X)$ reaches $X(T)$. Conversely, let us consider a random variable X_T which belongs to the space L^2 of the random variables having a first and second finite moments $X_T \in \mathbb{L}^2$ if $E\left(X_t^2\right) < \infty$.

24.3 Dynamic Portfolio Optimization Models

X_T is achievable if it is equal, with certainty, to the terminal value of a self-financing strategy $(\underline{x}, X)$. If all the elements of L^2 are achievable, the market is said to be complete.

We should understand that the ability to revise one's portfolio at any time enhances the investor's possibilities to the point of making the completeness assumption plausible. In fact, the investor chooses among an infinite number of funds (each strategy being interpreted as a fund) rather than chooses among a finite number of securities (static case). The market which is made complete thanks to the possibility of revisions is described as *dynamically complete*. For example, a dynamically complete market can be obtained if the N prices of the risky assets follow diffusion processes driven by N Brownian motions, if these assets are not redundant and if there exists an instantly risk-free monetary asset (*see*. Chap. 19).[21]

24.3.4.2 The Model

The solution to the optimization program (P) can be obtained in two steps.

Step 1

It consists in determining the optimal pay-off $X_T{}^*$ by solving the following program (P_1):

$$(P_1) \qquad \underset{X_T}{Max} \; E\{U(X_T)\}$$

subject to two constraints:

(CR_1) X_T is attainable through a self-financing strategy.

(CR_2) X_T requires an initial investment of 1.

(P_1) thus seeks the optimal terminal value (not the weights $\underline{x}(t)$) that can be obtained at the price of an initial unit stake (budgetary constraint).

Note that if the strategy $(\underline{x}^*, X^*)$ solves the program (P) of Sect. 24.3.1.1, it is self-financing (since it respects C_1) and requires an investment equal to 1 (constraint C_2) and maximizes expected utility. Its terminal value $X^*(T)$ is therefore necessarily a solution of (P_1). Conversely, if X_T^* is the solution of (P_1), X_T^* is attainable through a self-financing strategy that satisfies C_1 and C_2, maximizes the expected utility and is therefore necessarily a solution of (P).

In summary, step 1 consists in solving (P_1) to obtain the optimal pay-off X_T^*. However, both constraints CR_1 and CR_2 *must be expressed in mathematical form to solve* (P_1).

---

[21] No dR_i return can be replicated between t and $t+dt$ by a portfolio containing the N assets $0, \ldots, i-1, i+1, \ldots, N$ (*see* Chap. 21). In a diffusion model, this assumes that the variance-covariance matrix $\mathbf{V}(\tau) = \mathbf{\Sigma\Sigma}'$ is invertible, which is true if and only if $\mathbf{\Sigma}$ is of full rank (N).

In incomplete markets, the constraint CR_1, according to which X_T is attainable, is difficult to formulate. In the following, we will therefore consider the case of (dynamically) complete markets for which the constraint CR_1 is always satisfied.

With regard to the budgetary constraint CR_2 (obtaining X_T requires an initial unitary investment), two possibilities are offered:

- Use the risk-neutral probability Q associated with asset 0 (risk-free asset), and write that the values of the self-financing portfolios are martingales under this probability:

$$X_0 = 1 = E^Q\left(\frac{X_T}{S_0(T)}\right)$$

- Use the optimal growth portfolio as numeraire to obtain martingale prices under the historical probability. It is the latter path that we are taking by rewriting the program (P_1) as follows:

$$\underset{X_T}{Max}E\{U(X_T)\} \quad \text{s.c.}: \quad E\left(\frac{X_T}{H(T)}\right) = X_0 = 1 \qquad (P_1)$$

This program *is interpreted as a static program under a single budgetary constraint:* each attainable wealth X_T is associated with the fund that reaches it; it is a matter of choosing among an infinite number of funds at $t = 0$ the one that maximizes expected utility at the price of a unit investment.

(P_1) can be easily solved, regardless of the utility function and the form of the processes governing rates and prices, by the Lagrange multiplier method. To simply obtain the solution, we will consider the case of a discrete space Ω of the states of the world, the state ω occurring with probability $p(\omega)$. The optimization program which allows to determine the optimal value $X_T^*(\omega)$ in each state of the world ω writes simply, from the Lagrangian, as a function of the multiplier λ:

$$\underset{X_T(\omega)}{Max}\left[\sum_{\omega'\in\Omega}p(\omega')U(X_T(\omega')) - \lambda\sum_{\omega'\in\Omega}p(\omega')\frac{X_T(\omega')}{H_T(\omega')}\right]$$

$$\Leftrightarrow \left[\frac{d}{dX_T(\omega)}\sum_{\omega'\in\Omega}p(\omega')U(X_T(\omega')) - \lambda\sum_{\omega'\in\Omega}p(\omega')\frac{X_T(\omega')}{H_T(\omega')}\right] = \quad \text{for any }\omega,$$

that is: $U'(X_T^*(\omega)) = \frac{\lambda}{H_T(\omega)}$ (after elimination on both sides of the term $p(\omega')$). The last relation should prevail for any state of the world ω, whence:

24.3 Dynamic Portfolio Optimization Models

$$U'(X_T^*) = \frac{\lambda}{H_T}.\tag{24.6}$$

In addition, the constraint $(E\left(\frac{X_T^*}{H(T)}\right) = 1)$ gives the value of λ.

The justification of condition (24.6) which has just been presented, valid in the context of a discrete space Ω of states of the world, extends to the case of an infinite space Ω which is not countable but probabilistic.[22]

Condition (24.6) almost immediately gives the optimal pay-off form for any utility function U and any price and rate processes (provided that the market is complete), as shown in the following examples.

> **Example (5) of logarithmic utility**
>
> The optimal pay-off X_T^* is derived from the condition (24.6): $U'(X_T) = \frac{d}{dX_T}$ $\ln(X_T) = \lambda/H_T$, whence:
>
> $X_T^* = (1/\lambda)H_T$. The constraint $(E\left(\frac{X_T^*}{H(T)}\right) = 1)$ implies $\lambda = 1$, thus $X_T^* = H_T$.
>
> We therefore recover that the strategy $(\underline{h}, H)$ maximizes the logarithmic expected utility.
>
> **Example (6) of iso-elastic (or power) utility**
>
> Consider the utility function $\frac{1}{1-\gamma}X_T^{1-\gamma}$, characterized by a constant risk aversion coefficient γ.
>
> The condition (24.6) writes: $(X_T^*)^{-\gamma} = \lambda/H_T$ and leads to the optimal pay-off $X_T^* = (H_T/\lambda)^{1/\gamma}$.

Step 2

Although the optimal pay-off X_T^*, which is the solution of (P_1), is interesting itself, it is not the optimal strategy $(\underline{x}^*, X^*)$ desired, because the rule that allows us to choose the weights $\underline{x}^*$ is not explicit. We should also solve the problem (P_2) which allows to determine the weights $\underline{x}^*$ to attain X_T^*.

This is an exercise similar to that of replicating an option in which the purpose is to determine the portfolio strategy which attains the terminal value of the option (e.g., *Max* $(0, S_T - K)$ in the case of a call with exercise price K). This value depends on the processes that are supposed to govern asset prices.

If the solution of program (P_1) is easy, that of (P_2) is often difficult but, in any case, the decomposition of the optimization program into two steps generally facilitates its solution.

---

[22] The demonstration is based on the calculus of variations and not on the simple calculation of a Lagrangian.

Example 7

In this example, we assume that the prices of the risky assets follow the geometric Brownian motions:

$$d\underline{R} = \underline{\mu}\, dt + \Sigma d\underline{W}; \quad dS_0 = S_0\, rdt,$$

the coefficients μ and Σ as well as the interest rate r being constant.

The weight of the log-optimal portfolio $(\underline{h}, H)$ is also constant: $\underline{h} = V^{-1}(\underline{\mu} - r\,\underline{1})$ with $V = \Sigma\Sigma'$, and $H(t)$ follows a geometric Brownian motion. We use the notation: $\sigma^2(dH/H) = \sigma_H^2$.

Let us determine an optimal strategy for an investor whose risk aversion coefficient γ is constant. Define $c = 1/\gamma$. As we have seen in the previous example, such an investor follows a strategy that leads to a portfolio whose terminal value is proportional to $(H_T)^c$.

Let us consider a portfolio x *which allocates, at each moment,* a weight c to the portfolio H and $1 - c$ to the risk-free asset and show that its terminal value X_T is indeed proportional to $(H_T)^c$. We have:

$$dX/X = c\, dH/H + (1 - c)\, r\, dt,$$

according to the Itô's lemma:

$$d\ln(X) = c\, d\ln(H) + k\, dt, \text{with}\, k = (1-c) + \sigma_H^2 - \sigma_X^2 = (1-c) + \sigma_H^2(1-c^2).$$

Thus:

$$\ln(X_T) = c\, \ln(H_T) + kT, \text{and} : X_T = e^{kT}(H_T)^c,$$

which shows that portfolio x does solve the problem of the investor who has a constant relative risk aversion. It should be noted that this result had already been obtained with the Merton model (special case n° 2).

24.4 Summary

- In contrast with the static "buy and hold" Markowitz framework, dynamic asset allocation (AA) involves frequent portfolio revisions. Dynamic AA covers a wide variety of styles.
- Strategic AA is concerned with a few large asset classes (such as stocks, bonds, monetary assets) and depends on the investor's characteristics (risk aversion, age, and investment horizon). It relies on three rules:

24.4 Summary

- The risk profile rule dictates the allocations to each of the three main asset classes: the higher is the investor's risk aversion, the smaller is the share of stocks or stock indices and the larger the weight of monetary assets.
- The second rule determines the evolution of these weights according to the time remaining before the investment horizon or the investor's age: the proportions of stocks should be gradually decreased to make room for bonds and then monetary assets.
- The third rule dictates portfolio rebalancing according to the evolution of market conditions.
- The previous first two rules, based on common sense, can be (not so easily) justified on theoretical grounds, while different opinions concerning the third rule lead to various strategies:
  - One family of strategies consists in buying risky securities (stocks) following price increases and selling them following decreases. Portfolio insurance is part of this family of so-called momentum or convex strategies.
  - The second family, contrary to the previous one, consists in selling risky securities following a price increase and buying them following a decrease. They are called contrarian or concave strategies. Portfolios with constant weights are part of this family, since to maintain constant weights, one must sell (buy) the securities whose relative prices increase (decrease).
- Portfolio insurance is a convex dynamic strategy that allows limiting a portfolio's loss in the event of a market downturn, while still enabling investors to get some benefit from a market increase.
- Portfolio insurance may take the form of Stop Loss, OBPI (Option Based Portfolio Insurance), or CPPI (Constant Proportion Portfolio Insurance). It is suitable for automatic execution (Program Trading).
- Under OBPI, the manager either purchases (or replicates) a put option to cover a long position in the underlying asset, or purchases (or replicates) a call option to take advantage of the increase in the underlying asset, in combination with a default-free zero-coupon bond that ensures the capital guarantee.
- Under CPPI, the portfolio whose total value is $V(t)$ contains risky assets for an amount $S(t)$ and risk-free assets. A guaranteed floor P is predetermined, and the variable cushion is defined as $C(t) = V(t) - P$. Under basic CPPI, the multiple $m = S(t)/C(t)$ is maintained theoretically constant so that the risky component $S(t)$ tends to 0 when $C(t)$ gets close to 0, i.e., when the portfolio value gets close to the floor. By following this management rule, the portfolio value $V(t)$ can never become smaller than the floor. The CPPI method is flexible and can be adapted in several ways, for instance by raising or indexing the floor.
- In the previous approaches, the investor chooses a management style with more or less precise objectives. A more theoretical approach is to build a portfolio that optimizes the investor's expected utility.

- In continuous-time portfolio optimization, an investor chooses continuously the weights on the different assets. The optimal vector of weights $\underline{x}(t)$ is obtained by maximizing an expected utility U (or stream of utility). In general, at any time t, $\underline{x}(t)$ is a function of the value X(t) of the portfolio and of the state of the economy characterized by a vector of exogenous state variables $\underline{Y}(t)$: $\underline{x}(t) = f(t, X(t), \underline{Y}(t); U, T)$.
  Besides, the vector of the N risky returns, $d\underline{R}$, is assumed to be a multidimensional diffusion process: $d\underline{R} = \underline{\mu}(t)dt + \Sigma(t)d\underline{W}$, implying a variance-covariance matrix of instantaneous returns $\mathbf{V}(t) = \Sigma(t)\Sigma'(t)$.
- The continuous time portfolio optimization problem can be solved, either by dynamic programming techniques (as in Merton's model), or following an alternative approach assuming that the markets are dynamically complete (i.e., any random terminal wealth X_T is attainable by a feasible dynamic strategy). Simple solutions can be obtained in some special cases like (i) and (ii) below:
  - (i) Log-utility function. The optimal vector of risky assets is $\underline{h} = \mathbf{V}^{-1}(t)(\underline{\mu}(t) - r(t)\underline{1})$ where $\underline{1}$ is the N-dimensional unit vector and $r(t)$ the risk-free rate. This log-optimal portfolio $\underline{h}$, also called numeraire portfolio or optimal growth portfolio, has many remarkable properties:
    $\underline{h}$ is myopic, meaning logarithmic investors do not worry about what will happen in future periods in making their portfolio choices for the upcoming period;
    $\underline{h}$ is instantaneously efficient and therefore homothetic to the tangent portfolio.
    $\underline{h}$ maximizes the expected growth rate of wealth at any time horizon.
    When $\underline{h}$ is used as numeraire, the value of self-financed securities or portfolios are martingales under the historical probability.
  - (ii) Constant interest rate r and parameters μ and Σ, implying that $\underline{h} = \mathbf{V}^{-1}(\underline{\mu} - r\underline{1})$ is constant. Then:
    The two-fund separation prevails, all optimal portfolios being combinations of the risk-free asset and the logarithmic portfolio $\underline{h}$, the weight on $\underline{h}$ being inversely proportional to the investor's relative risk aversion.
    The risky components of optimal portfolios of all investors thus have an identical structure (they are homothetic to $\underline{h}$, therefore to the tangent portfolio), as in the static case studied by Markowitz (Chap. 21).

Suggestions for Further Reading

Books

Bodie, Z., Kane, A., & Marcus, A. (2010). *Investments* (9th ed.). Irwin.
Elton, E., Gruber, M., Brown, S., & Goetzman, W. (2010). *Modern portfolio theory and investment analysis* (8th ed.). Wiley.
*Merton, R. (1999). *Continuous-time finance*. Basil Blackwell.

Articles

Bajeux-Besnainou, I., Jordan, J. V., & Portait, R. (2001). An asset allocation puzzle: Comment. *American Economic Review, 91*, 1170–1179.

Bajeux-Besnainou, I., Jordan, J. V., & Portait, R. (2003). Dynamic asset allocation for stocks, bonds and cash. *Journal of Business, 76-2*, 263–287.

Bajeux-Besnainou, I., & Portait, R. (1998). Dynamic asset allocation in a mean-variance framework. *Management Science, 44*(11), 79–95.

Bertrand, P., & Prigent, J. L. (2003). Portfolio insurance strategies: A comparison of standard methods when the volatility of the stock is stochastic. *International Journal of Business, 8*(4), 461–472.

Bertrand, P., & Prigent, J. L. (2005). Portfolio insurance strategies: OBPI versus CPPI. *Finance, 26*, 5–32.

Black, F., & Jones, R. (1987). Simplifying portfolio insurance. *Journal of Portfolio Management*, 48–51.

Black F. and A. Perold. Theory of constant proportions portfolio insurance. *Journal of Economic Dynamics and Control*, 1992, vol. 16, n° 3–4, 403-426.

Brennan, M., & Schwartz, E. (1989). Portfolio insurance and financial market equilibrium. *Journal of Business, 62*, 455–472.

Bodie, Z., Merton, R. C., & Samuelson, W. F. (1992). Labor supply flexibility and portfolio choice in a life cycle model. *Journal of Economic Dynamics and Control, 16*, 427–449.

Canner, N., Mankiw, N. G., & Weil, D. N. (1997). An asset allocation puzzle. *American Economic Review, 87*(1), 181–191.

Cox, J., & Huang, C. F. (1989). Optimal consumption and portfolio policies when asset prices follow a diffusion process. *Journal of Economic Theory, 49*, 33–83.

Genotte, G., & Leland, H. E. (1990, December). Market liquidity, hedging and crashes. *American Economic Review*.

Karatzas, I., Lehoczky, J., & Shreve, S. (1987). Optimal portfolio and consumption decisions for a 'small investor' on a finite horizon. *SIAM Journal of Control and Optimization, 25*, 1157–1186.

Leland, H. E., & Rubinstein, M. (1988). The evolution of portfolio insurance. In D. L. Luskin (Ed.), *Portfolio insurance: A guide to dynamic hedging*. Wiley.

Long, J. B. (1990). The numeraire portfolio. *Journal of Financial Economics, 26*, 29–69.

Merton, R. C. (1969). Lifetime portfolio selection under uncertainty: The continuous time case. *Review of Economics and Statistics, 51*, 247–257.

Merton, R. C. (1971). Optimum consumption and portfolio rules in a continuous time model. *Journal of Economic Theory, 3*, 373–413.

Merton, R. C. (1973). An intertemporal capital asset pricing model. *Econometrica, 41*, 867–888.

Merton, R. C., & Samuelson, P. A. (1974). Fallacy of the log-normal approximation to portfolio decision-making over many periods. *Journal of Financial Economics, 1*, 67–94.

Nguyen, P., & Portait, R. (2002). Dynamic mean variance efficiency and asset allocation with a solvency constraint. *Journal of Economics Dynamics and Control*, 11–32.

Perold, A., & Sharpe, W. (1988). Dynamic strategies for asset allocation. *Financial Analysts Journal, Jan–Feb*, 16–27.

Pliska, S. R. (1986). A stochastic calculus model of continuous trading: Optimal portfolios. *Mathematical Operations Research, 11*, 371–382.

Samuelson, P. A. (1969). Life time portfolio selection by dynamic stochastic programming. *Review of Economics and Statistics, August*, 239–246.

Samuelson, P. A. (1989). The judgement of economic science on rational portfolio management: Timing and long-horizon effects. *The Journal of Portfolio Management, 16*, 4–12.

Benchmarking and Tactical Asset Allocation 25

A *benchmark, usually an index,* is often used to analyze the risk, return, and performance of portfolios. We have already distinguished between passive management, which merely duplicates a benchmark index, and active management, which tries to "beat" it by deviating more or less from it.

Active management generally has two phases. The first, described as strategic, was discussed in the previous chapter. It determines general policy and, as such, is concerned only with broad asset classes and long-term horizons. But once the large classes and trends have been determined, the manager most often proceeds with Tactical Asset Allocation, which concerns the weighting of the various asset classes (which may differ, in the short term, from the strategic weightings) as well as the choice of individual securities within the same class. Tactical allocation is a bet on the manager's ability to detect under- and over-valued assets and/or predict market reversals. The tactical allocation horizon is thus generally short, typically in the order of 3–6 months.

In this chapter benchmarking, which aims at replicating an index, is examined in Sect. 25.1.1. Section 25.1.2 deals with the active management of a portfolio more or less linked to a benchmark, explicitly or implicitly. Generally speaking, active management aims to achieve a performance objective based on the tactical or technical ability of the manager, within the framework of guidelines defining a management style. Alternative management (of which hedge funds are a special case), which has undergone many developments in recent years, is the subject of Sect. 25.1.3.

25.1 Benchmarking

This management style uses an index as a benchmark and aims to replicate its performance. The index is then referred to as the benchmark. It aims at replicating the index as accurately as possible, while so-called active management (see Sect. 25.1.2) will give managers more latitude and bet on their ability to beat the index.

© The Author(s), under exclusive license to Springer Nature Switzerland AG 2022
P. Poncet, R. Portait, *Capital Market Finance*, Springer Texts in Business and
Economics, https://doi.org/10.1007/978-3-030-84600-8_25

25.1.1 Definitions and Classification According to the Tracking Error

We will use the following definitions:

- The monitoring gap, Δ_t, is the difference of returns between the fund or portfolio x and its benchmark B for a period t: $\Delta_t \equiv R_{x,t} - R_{B,t}$.
- The « distance » between a portfolio and its benchmark[1] is estimated from the *tracking error* (TE below), defined as the standard deviation of the monitoring gap[2]:

$$TE_t \equiv \sigma(\Delta_t) \tag{25.1}$$

This definition of TE is probabilistic and theoretical. Its empirical estimator, called *empirical tracking error*, is calculated ex post from a history (path) that most often consists of N weekly returns.

The empirical weekly tracking error is given by the following formula:

$$Tw = \sqrt{\frac{1}{N-1} \sum_{i=1}^{N} \left(\Delta i - \overline{\Delta}\right)^2}$$

where Δ_i refers to the monitoring gap observed in week i and $\overline{\Delta}$ the mean of the N gaps.

The annualized TE then results from the formula: $Ta = Tw\sqrt{52}$.

These calculations are similar to those used to estimate a historical volatility (which is technically a standard deviation) and to annualize it.

Actually, benchmarking is a relatively heterogeneous category due to the wide disparities between the maximum tracking errors to which managers are implicitly or explicitly constrained.

A classification accepted in different countries and by different regulatory authorities distinguishes, in a continuum of indexed management styles, the following categories:

- *Index Funds* duplicating the index as accurately as possible; they follow a so-called passive strategy whose objective is to minimize the tracking error. Funds in this category have an annualized empirical TE of less than 1%.

---

[1] The second moment of the difference between two random variables constitutes a distance between them in the mathematical sense.

[2] Our definition conforms to that adopted by most asset managers. TE then may be also called "tracking error volatility." However, the reader should be aware that sometimes TE is defined as the difference Δ_t and not its standard deviation.

25.1 Benchmarking

- Enhanced index funds, or tilted index funds, which, while closely linked to the index, are trying to beat it. Their profitability may differ from that of the index up to a TE of 2%.
- Index-based active management by Mutual Funds (UCITS, Undertakings for Collective Investments in Transferable Securities, in the European Union) which actively manage performance and risk against an agreed-upon benchmark (see Sect. 25.1.2).

A fourth type of fund must be added to this classification, consisting of Mutual Funds whose performance is not linked to that of an index but which use the latter, *ex post*, as a simple element for comparing their management; they are therefore not included in our definition of benchmarking. In this section, we present passive benchmarking (pure index funds), before studying, in Sect. 25.1.2, active benchmarking in its various forms.

25.1.2 Pure Index Funds and Trackers

The success of index funds that replicate the index is based on the relative efficiency of markets that has been observed many times, and therefore on the confirmed difficulty of beating the market. As discussed in a previous chapter, empirical studies show that very few funds beat the market over a long period of time, and that, on average, the performance of funds is equal to that of the market, less transaction and management costs. So isn't the wisest management attitude to be limited to duplicating a global index? It is up to the investor to choose the risk level in accordance with his or her own objectives and to build a portfolio that respects it, by combining an equity index fund with bond and money-market funds (see the strategic allocation discussed in the previous chapter).

25.1.3 Replication Methods

- In principle, the replication of the index is obtained simply by building a portfolio that includes, at all times, the same assets as the index and with identical weights. The advantage of such a management is that it does not require any special skills and, in theory, involves few transaction costs. Indeed, the replication of an index weighted by market capitalizations is theoretically obtained by a *buy and hold* strategy (it is sufficient to initially weight the portfolio as the index it seeks to duplicate). In fact, equity issues, dividend payments, the fact that the composition of the index is modified from time to time, and the management of unpredictable cash inflows (contributions to the fund) or outflows (withdrawals) force the manager of a so-called *passive* index fund to carry out transactions, some of which may pose problems (liquidity of the securities and so on)[3] and are inevitably a source of tracking error.

This is why benchmarking often uses methods that do not require the creation of a portfolio that accurately reflects the index and thus avoids or limits the above-mentioned difficulties. This type of management uses derivatives (synthetic replication). They may also be satisfied with imperfect replication, by building a portfolio containing well-chosen securities, in a smaller number than the index components (statistical replication). They then lead to higher TEs than strict replication and may therefore not fall into the category of pure index funds, characterized by a low TE.

– *Synthetic* replication consists in using derivatives, mainly index futures contracts, to replicate the index. The principle is simple and based on the well-known redundancy between a futures contract, its underlying asset, and the risk-free asset (see Chap. 9). *The index is replicated by purchasing the contract combined with an investment in the money market*; the latter is equal to the spot value of the replicating portfolio and its maturity coincides with that of the futures contract. As futures contracts are generally liquid and the number of instruments in the replicating portfolio is considerably reduced, this method eliminates many of the above difficulties.

– *Statistical* replication consists of an approximate reproduction of the index, using a sub-basket that contains a smaller number of assets. The M securities in the replication portfolio can be selected in several ways: either on the basis of their weight (M securities that weigh the most heavily in the index, which contains $N > M$), or by sampling (the sample of M securities is the most accurate possible reflection of the index in terms of capitalizations, sectors, styles, etc.), or finally by statistical optimization that determines the portfolio of M securities that is most correlated with the index (or the one that minimizes TE).

25.1.4 Trackers or ETFs

Trackers, also known as Exchange Traded Funds (ETFs), which appeared in Europe in 2000, are similar to index funds in that they attempt to replicate an index with a minimum TE. They differ in that they are continuously quoted and are therefore similar to securities held directly. As with the latter, no entry or exit fees are charged. However, there are brokerage fees, as well as management fees (e.g. 0.3%) that compensate trackers' issuers. The minimum stake is low (their price represents a small fraction of the index, e.g., one thousandth). Overall, trackers are an interesting alternative to index funds, which explains their fast growth.

---

[3] The low liquidity of some of the securities in the index may hinder the transactions necessary for replication; some indices contain too many securities to allow portfolios built in practice to be, at all times, perfect clones, especially when the index fund is small. In this case, rounding problems may also arise.

25.2 Active Tactical Asset Allocation

The objective of a management that compares its performance to a benchmark to which it is more or less closely linked differs from that of pure index funds: it is no longer a question of minimizing a TE, but rather of maximizing an expected return under a more or less strict constraint on a given TE. This management style is a fairly broad class which comprises:

– Extended benchmarking. The portfolios managed in this way are closely linked to the index (the TE is less than 2%), and their beta relative to the index is generally very close to 1, but they can marginally deviate from the benchmark by different weightings on certain securities.

– Active index-based management. The portfolios managed in this way may deviate, to varying degrees, from the benchmark by appropriate management choices (replication mode, overweighting of certain securities, use of derivatives, arbitrage, etc.) and their beta may significantly differ from 1. The techniques used to try to beat the index are those of tactical allocation (security picking and market timing), as described in Chap. 21 and discussed below.

In addition, a third group includes active management that does not refer to a benchmark, but has more or less well-defined characteristics (styles) such as "European mid caps," "diversified bonds," "aggressive yield", "reasoned profitability" or "Socially Responsible Investment (SRI)." The assessment of these funds, which in fact have an implicit benchmark, must be conducted using the same methods as those used for extended or active index funds. Their benchmark is simply more difficult to determine. We, therefore, include these three groups in the same tactical asset allocation analysis and discuss active management in general.

We first present the formal problem of an active manager who competes with a benchmark and its solution (Sect. 25.1). We will then analyze the performance of active management (Sect. 25.1.2).

25.2.1 Modeling and Solution to the Problem of an Active Manager Competing with a Benchmark

Let us consider the problem of a manager seeking to maximize the expected return of her portfolio while respecting a maximum TE compared to a benchmark. The latter is defined by the weights $\underline{b}$ and its random return is denoted by R_B. The other notations are those of Chap. 21: $\underline{R}$, μ and $\mathbf{V}$ refer respectively to the (random) vector of returns of the N risky securities, their expectations, and the $N \times N$ matrix of their variances-covariances. A portfolio x is defined by the weight vector $\underline{x}$ in the N risky assets; in the absence of a risk-free asset, the sum of the weights on risky assets is equal to 1: $\underline{x}'\underline{1} = 1$.

The program of an active index manager writes:

$$\underset{x}{\text{Max}} \ [E(Rx)] \ s.t. : \sigma(RxR_B) = k$$

The constant k represents the maximum TE allowed; it is interpreted as the degree of freedom left to the manager and we will call it *tolerance*. By denoting $\Delta_x = R_x - R_B$ and noting that, once chosen, R_B is "exogenous" for the manager (she does not have control over it), this program is equivalent to maximizing the expectation of the difference Δ_x under a TE constraint:

$$\underset{x}{\text{Max}} \ E(\Delta_x) \ s.t. : \sigma(\Delta_x) = k \tag{PI}$$

Proposition 1 Two-fund separation and efficiency condition; Roll (1992)
(i) The optimal portfolio of the active index manager that solves the program (PI) is a combination of the benchmark and a portfolio q identical for all index managers[4]:

$$\underline{x}^* = \underline{b} + \hat{\theta}\underline{q}. \tag{25.2}$$

The vector of weights $\underline{q}$ defining q is independent of the benchmark and such that:

$$\underline{q} = \boldsymbol{V}^{-1}(\underline{\mu} - m\underline{1}). \tag{25.3}$$

$\hat{\theta}$ *is a positive scalar proportional to the tolerance k and m denotes the expected return on the minimum variance portfolio.*
More precisely:
$m = r$, *if the manager can trade the risk-free asset;*
$m = \dfrac{\mu'\mathbf{V}^{-1}\mathbf{1}}{\mathbf{1}'\mathbf{V}^{-1}\mathbf{1}}$ *if the manager only trades on risky assets.*

$\hat{\theta} = \dfrac{k}{\sigma_q}$ *is twice the inverse of the Lagrange multiplier of the constraint on the T;*
$\sigma^2{}_q = \left(\underline{\mu} - m\underline{1}\right)' \boldsymbol{V}^{-1} \left(\underline{\mu} - m\underline{1}\right)$ *represents the variance of the return on portfolio q.*
(ii) This portfolio is efficient, in the sense of Markowitz's mean-variance criterion (MV), if and only if the benchmark itself is MV efficient.

Proof
(i) It is similar to that of propositions 1 and 4-*iv* of Chap. 21, related to the optimal portfolio as defined by Markowitz, in the absence and presence of risk-free assets.

Let us call $\underline{y}$ the weight differences between the portfolio $\underline{x}$ and its benchmark: $\underline{y} = \underline{x} - \underline{b}$. It is these weight differences $\underline{y}$ that are the source of the tracking error.

First, consider the case of portfolios that involve the risk-free asset (0).

It should be noted that:

---

[4]If they have the same horizon and expectations ($\underline{\mu}$ and **V**).

25.2 Active Tactical Asset Allocation

$$\Delta_x = R_x R_b = [(\underline{1}\underline{x}'\underline{1})\ r + \underline{x}'R][(\underline{1}\underline{b}'\underline{1})\ r + \underline{b}'R] = r\ (\underline{b}'\underline{x}')\underline{1} + (\underline{x}'\underline{b}')R = \underline{y}'(Rr\underline{1}).$$

The program (PI) is then rewritten:

$$\underset{y}{\mathrm{Max}}\ \underline{y}'(\mu r\underline{1})\ \ s.t.:\underline{y}'\mathbf{V}\underline{y} = k^2$$

which is expressed, from the Lagrangian, as: $\underset{y}{Max}\ \underline{y}'\left(\mu - r\underline{1}\right) - \frac{\theta}{2}\underline{y}'\mathbf{V}\underline{y}.$

where $\frac{\theta}{2}$, the Lagrange multiplier, is interpreted, as in Chap. 21, Section 21.1.3, as a risk aversion coefficient (but here the risk is to deviate from the benchmark).

The solution is $\underline{y}^* = \widehat{\theta}\mathbf{V}^{-1}(\mu - r\ \underline{1})$, with $\widehat{\theta} = \frac{1}{\theta}$ which proves (25.2) and (25.3), in the presence of a risk-free asset.

Let us now consider the case of no risk-free asset. Since $\underline{b}'\underline{1} = \underline{x}'\underline{1} = 1$, it comes: $\underline{y}'\underline{1} = 0$; the weight difference vector $\underline{y}$ is then interpreted as a portfolio y with a zero weight sum (zero net investment) and $\Delta_x = R_x - R_b = \underline{y}'R$.

The program (PI) must contain the constraint on the sum of the weights of y and writes:

$$\underset{y}{\mathrm{Max}}\ \underline{y}'\mu\ \ s.t.:\underline{y}'\mathbf{V}\underline{y} = k^2\ \ \text{and}\ \ \underline{y}'\underline{1}$$

$$= 0 \Longleftrightarrow \underset{y}{\mathrm{Max}}\ \underline{y}'\mu - \frac{\theta}{2}\underline{y}'\mathbf{V}\underline{y} - \lambda\underline{y}'\underline{1}$$

where $\theta/2$ and λ represent the Lagrange multipliers of the two constraints.

The solution is: $\underline{y}^* = \widehat{\theta}\mathbf{V}^{-1}(\mu - \lambda\ \underline{1})$; the two constraints make it possible to determine θ and λ:

$\underline{1}'\ \underline{y}^* = 0$ implies: $\lambda = \frac{\underline{1}'\mathbf{V}^{-1}\mu}{\underline{1}'\mathbf{V}^{-1}\underline{1}}$. But (see Chap. 21) $\frac{\mathbf{V}^{-1}\underline{1}}{\underline{1}'\mathbf{V}^{-1}\underline{1}}$ is the weight vector of the

minimum variance portfolio; therefore $\frac{\underline{1}'\mathbf{V}^{-1}\mu}{\underline{1}'\mathbf{V}^{-1}\underline{1}} = m$ is its expected return, which

proves (2) and (3) in the absence of a risk-free asset.

In both the presence and absence of a risk-free asset, the constraint on the tracking error is written:

$$\underline{y}^{*'}\mathbf{V}\underline{y}^* = k^2 = \widehat{\theta}^2(\mu - m\underline{1})'\mathbf{V}^{-1}(\mu - m\underline{1}) = \widehat{\theta}^2\sigma_q^2,\quad \text{hence}\quad \widehat{\theta} = \frac{k}{\sigma_q}.$$

(ii) According to Proposition 4-iv in Chap. 21, in the presence of a risk-free asset, a portfolio $\underline{x}^*$ is MV-efficient if and only if it is homothetic to $\mathbf{V}^{-1}(\mu - r\ \underline{1})$, i.e. if there is a scalar λ such that $\underline{x}^* = \lambda\ \mathbf{V}^{-1}(\mu - r\ \underline{1})$. However, according to (i), such a relationship prevails if and only if the benchmark $\underline{b}$ is itself homothetic to $\mathbf{V}^{-1}(\mu - r\ \underline{1})$, and therefore efficient, which proves (ii). This result, which is very intuitive, is due to Richard Roll. In the absence of a risk-free asset, the proof follows the same logic based on Black's theorem on two-fund separation (see Chap. 21). It is intuitive and will be proved in 3.2 that, when B is inefficient, a combination of q and B, even if inefficient, dominates B in the M-V sense, since it has a higher Sharpe ratio than B.

Finally, it should be noted that (PI) is equivalent to:

$$\operatorname*{Max}_{x}\left[\frac{E(\Delta)}{k}\right] = \operatorname*{Max}_{x}\left[\frac{E(\Delta)}{\sigma(\Delta)}\right] \quad \text{s.t.} : \sigma(\Delta) = k.$$

Definition *The ratio to be maximized is called the information ratio and is denoted IR:*

$$IR = \frac{E(\Delta)}{\sigma(\Delta)} \tag{25.4}$$

IR quantifies the performance of the manager and, as just defined, it is an ex ante measure. It has the merit of combining in a single criterion the two elements, expectation and risk (measured by the tracking error). The IR is a generalization of the Sharpe ratio (see Chap. 22), the latter being the information ratio of a portfolio whose benchmark is the risk-free rate.

25.2.2 Analysis of the Performance of Active Portfolio Management: Empirical Information Ratio, Market Timing, and Security Picking

The ex-post performance of a portfolio P is commonly estimated on the basis of the regression of excess returns $R_{P,t} - r_t$ on $R_{B,t} - r_t$:

$$R_{P,t} r_t = \alpha + \beta \ (R_{B,t} r_t) + \varepsilon_t \tag{25.5}$$

where r_t represents the risk-free rate with a maturity equal to the investment horizon, α, which is *assumed constant, refers to the performance (it is the equivalent of Jensen's alpha, and it would be Jensen's alpha if B were the market portfolio) and the white noises ε_t are independently and identically distributed. We explained this regression when we examined the CAPM tests (Chap. 22). We will revisit here some of these analyses by distinguishing cases where the beta coefficient is equal to 1 and different from 1.*

25.2.3 Beta Coefficient Equal to 1

This is the case of a fund that effectively monitors the evolution of its *benchmark* (there is no market timing, i.e., no attempt to anticipate periods when the benchmark

25.2 Active Tactical Asset Allocation

provides a return higher than the risk-free rate and those when it is the opposite[5]). The manager then only *manages the alpha*, i.e. tries to improve the average return of his benchmark by relying exclusively on his ability to select the right securities (security picking). The relationship (25.5) simplifies in this case as:

$$R_{P,t} = \alpha + R_{B,t} + \varepsilon_t, \text{ hence :}$$
$$R_{P,t} - R_{B,t} \equiv \Delta_t = \alpha + \varepsilon_t \tag{25.6}$$

The deviation Δ_t is therefore broken down into an expected gain α (over-performance) due to the choice of a portfolio different from the benchmark and a white noise ε_t of zero expectation representing the risk induced by this selection.

The *ex-ante* IR, which the manager of a benchmarked fund must maximize, then writes:

$$IR = \frac{E(\Delta)}{\sigma(\Delta)} = \frac{\alpha}{\sigma(\varepsilon)}$$

The regression of R_P on R_B yields the estimators $\widehat{\alpha}$ and $\widehat{\sigma}(\Delta) = \widehat{\sigma}(\varepsilon)$. The latter, the empirical or ex post standard deviation of the tracking error, is the square root of $\widehat{\mathrm{var}}(\Delta)$, with:

$$\widehat{\mathrm{var}}(\Delta) = \frac{1}{T-1} \sum_{t=1}^{T} (\Delta_t - \overline{\Delta})^2 \quad \text{and} \quad \overline{\Delta} = \frac{1}{T} \sum_{t=1}^{T} \Delta_t$$

We deduce from this the *empirical ratio of information*, an *ex-post* performance measure:

$$\widehat{IR} = \frac{\overline{\Delta}}{\widehat{\sigma}(\Delta)} = \frac{\widehat{\alpha}}{\widehat{\sigma}(\varepsilon)} \tag{25.7}$$

It is worth noting that $\widehat{IR} = \frac{\widehat{\alpha}}{\widehat{\sigma}(\varepsilon)} = \frac{T(\alpha, N-1)}{\sqrt{N}}$, where N is the number of observations in the regression and $T(\alpha, N-1)$ is a random variable which has a Student's t-distribution centered on α, with $(N-1)$ degrees of freedom.

The information ratio is therefore directly related to the Student's test on the empirical mean $\widehat{\alpha}$, by means of which the statistical significance of the hypothesis $\widehat{\alpha} = 0$ (i.e. the hypothesis that the manager is "in the middle") can be measured by assuming that the returns are Gaussian. This, therefore, makes it possible, in principle, to assess whether the manager's performance was achieved by chance or due to their skills.

---

[5] See Sect. 2.2.2 and the appendix to this chapter.

Example 1

Let us consider a manager whose portfolio shows a beta of 1 compared to its benchmark and a monthly empirical alpha of 0.25% (approximately 3% per year). The empirical standard deviation of this average gain is 2.5%. Suppose that, by chance, the value of these estimators is not biased and therefore corresponds to their ex ante value. The manager is indeed efficient but, as this fact is not known *a priori*, the H_0 hypothesis is that she is just average. How many years are needed to reject, at the confidence level p, the hypothesis that the manager has no real management skill, i.e. the null hypothesis $\alpha = 0$? We get:

$$\hat{IR} = \frac{\hat{\alpha}}{\hat{\sigma}(\varepsilon)} = \frac{0.25\%}{2.5\%} = 0.1 \text{ (monthly)} = \frac{t_p}{\sqrt{N}}, \text{hence,} \quad t_p = 0.1\sqrt{N}.$$

For a confidence level p of 95%, the value of t_p is 1.96 (for 98%, it would be 2.33). It, therefore, takes N $= (19.6)^2 = 384$ months, or 32 years, to affirm with 95% confidence that this manager is performing better than her benchmark! However, it only takes 8 years if $\hat{\sigma}(\varepsilon) = 1.25\%$.

This example, in which it takes 32 years to reject the hypothesis that the manager is average ($\alpha = 0$) while she is good ($\alpha > 0$), clearly shows that it is necessary to put into perspective statements such as "it is impossible to beat the market." These are based on tests of the null hypothesis "all fund managers are average" which is almost impossible to reject. On the other hand, it would also take 32 years to reject the hypothesis that the manager is average ($\alpha = 0$) while she is poor (monthly empirical alpha of -0.25%). These simple computations show that the mutual funds' charts regularly published in specialized newspapers must be read with the greatest caution.

Finally, it should be noted that: $\frac{\widehat{\text{var}(\Delta)}}{\text{var}(R_B)} = (1 - R^2)$.

where R^2 is the coefficient of determination of the regression of R_P on R_B.

25.2.4 Beta Coefficient Different from 1

The method of assessing the stock-picking ability presented above and based on the relation (25.6) only makes sense if the beta of the fund has been maintained equal to one during the studied period. If the manager has significantly changed the composition of his portfolio over time and therefore his beta (in a market timing attempt), the relation (25.5) must be used to conduct the analysis and the statistics from the regression are much more difficult to interpret and are generally biased. Let us consider the case of a portfolio whose beta relative to the index may differ from 1 for opportunistic reasons and let us take up the more general Eq. (25.5):

25.2 Active Tactical Asset Allocation

$$R_{P,t} - r_t = \alpha + \beta_t \ (R_{B,t} - r_t) + \varepsilon_t.$$

By subtracting $R_{B,t} - r_t$ from the two sides of this relation, we obtain an expression for the monitoring gap, $R_{P,t} - R_{B,t}$, as the sum of two components, one of which is generated by market timing and the other by security picking:

$$R_{P,t} - R_{B,t} \ = \Delta_t = \ \underbrace{(\beta_t - 1)(R_{B,t}r_t)}_{\text{Market timing}} + \ \underbrace{(\alpha + \varepsilon_t)}_{\text{Security picking}} \tag{25.7}$$

This leads to the following breakdowns that highlight the respective contributions of market timing and security picking to excess profitability on the one hand and tracking error on the other:

$$E(\Delta_t) = \ \underbrace{(\beta_t - 1 \ ) \ E(R_{B,t} - r_t)}_{\text{Market timing}} + \ \underbrace{\alpha}_{\text{Security picking}} \qquad \text{and} \qquad \sigma^2(\Delta_t)$$

$$= \underbrace{(\beta_t - 1)^2\sigma^2(R_{B,t} - r)}_{\text{Market timing}} + \underbrace{\sigma^2(\varepsilon_t)}_{\text{Security picking}}$$

Two definitions of the Information Ratio (ex ante) are possible here:
$IR_1 = \alpha / \sigma(\varepsilon_t)$ *measures the effectiveness of security picking;*
$IR_2 = E(\Delta_t)/\sigma(\Delta_t)$ *measures overall performance.*
Ex post, the corresponding IRs are calculated from regression estimates, with the empirical IR_1 being directly related to the t-Student of α:

$$\hat{IR}_1 = \frac{\hat{\alpha}}{\hat{\sigma}(\varepsilon)} = \frac{T(\alpha, N - 1)}{\sqrt{N}}.$$

The market timing ability of a management centered on a beta equal to 1, but which may deviate from it in periods when it anticipates with a high probability a return on the benchmark higher than the risk-free rate r (its beta will then be >1), or lower than r (its beta will then be <1), can be tested using a regression such as:

$$R_{P,t} - R_{B,t} = \alpha + \beta_1 \ \text{Max}(0, R_{B,t} - r_t) + \beta_2 \ \text{Max}(0, r_t - R_{B,t}) + \varepsilon_t$$

The ability to anticipate increases results in a coefficient of $\beta_1 > 0$ and the ability to predict decreases by a coefficient of $\beta_2 > 0$.

A similar method has been proposed by Henriksson and Merton (1981). It is based on the regression:

$$R_{P,t} - R_{B,t} = \alpha + \beta_3(R_{B,t} - r_t) + \delta(R_{B,t} - r_t)D + \varepsilon_t$$

where D is a dummy variable, equal to one when the risk premium $(R_{B,t} - r_t)$ is positive or zero and zero when it is negative. If the manager's expectations regarding the benchmark's profitability are correct, the value of δ is positive. In this case, the managed fund has a total beta equal to $(\beta_3 + \delta)$ in favorable market conditions, which is greater than β_3, the beta chosen in adverse circumstances.

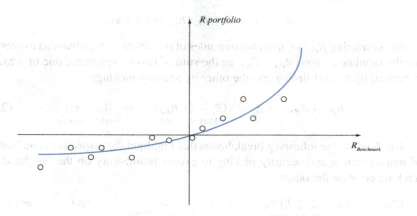

Fig. 25.1 Effective market timing capacity

Alternatively, the profitability gap of the fund can be analyzed as the sum of the excess profitability of the benchmark and the value of a put that is positive only if this excess return is negative:

$$R_{P,t} - R_{B,t} = \alpha + \beta_4 (R_{B,t} - r_t) + \gamma \, \text{Max} \, [0, -(R_{B,t} - r_t)] + \varepsilon_t$$

where γ is positive if the manager demonstrates a real market timing capacity.

Another method was proposed by Treynor and Mazuy (1966). If the market timing attempt is successful, the fund depreciates less than its benchmark when it falls and appreciates like it (or even more) when it rises. Therefore, the curve representing the relationship between the fund's returns and those of the benchmark, in the event of successful market timing, has an upward concavity [see Fig. 25.1; the points represent the combined returns of the sub-periods].

To obtain such a curve, it can be assumed that the returns on the fund are generated by a quadratic function such as:

$$R_{P,t} = \alpha + \beta \, R_{B,t} + \delta \, R^2_{B,t} + \varepsilon_{P,t}$$

Therefore, if the delta coefficient is significantly different from zero, the curve is convex or concave (its second derivative is equal to 2δ), so the *market timing* is effective and allows to improve (respectively, deteriorate, if $\delta < 0$) the performance of the benchmark if it is done correctly (resp., in the wrong direction).

Finally, it should be noted that many practitioners adopt an alternative approach to decompose the $R_P - R_B$ gap between market timing and security picking and describe this breakdown as *performance attribution*. The appendix presents this approach and provides an illustration.

25.2.5 Information Ratios, Sharpe Ratio, and Active Portfolio Management Theory

We have had several opportunities to discuss the thesis, widespread in academic circles and among some practitioners, that active management is irrelevant in a world of efficient markets. Indeed, it may seem illusory, in such a world, to look for alpha-positive securities: the dominant strategy is simply to hold a proportion of the market portfolio (which is nothing but the tangent portfolio T), and to supplement it with the risk-free asset. The following developments can be considered as a theory of active management, which gives meaning to it, even in the context of efficient markets. They are grounded in part on the model of Treynor and Black (1973) and are based on the unobservable nature of the tangent portfolio T and on the fact that any proxy B of the latter (here, the benchmark), is probably inefficient, and so can be dominated.

25.2.6 The Construction of a Maximum IR Portfolio from a Limited Number of Securities

We first consider a portfolio P, which is compared to a benchmark B which is generally not efficient. The portfolio consists of m securities or sub-portfolios with $1 \leq m \leq N$, where N refers to the number of different securities that can be traded on the market.

$IRP = \frac{E(R_P - R_B)}{\sigma(\varepsilon)}$ is the IR of the P portfolio considered. The IR of the sub-portfolio or security j is denoted IR_j, and its alpha and specific risk are respectively denoted α_j and $\sigma_{\varepsilon\,j}$.

The following proposition characterizes the maximum achievable IR from m securities and the weights that lead to this maximization.

Proposition 2

(i) The weights x_1^*, ..., x_m^* (constituting the portfolio P* of m securities) that maximize the IR information ratio are proportional to the ratios $\frac{\alpha_j}{\sigma_{ej}^2} = \frac{IR_j}{\sigma_{ej}}$:

$$x_j^* = \frac{\frac{\alpha_j}{\sigma_{ej}^2}}{\sum_{k=1}^{m} \frac{\alpha_k}{\sigma_{ek}^2}} \quad \text{for} \quad j = 1, \ldots, m; \tag{25.8}$$

(ii) The square of the IR of P* is equal to the sum of the squares of the IR of its components:

$$(IR_P)^2 = \sum_{j=1}^{m} (IR_j)^2. \tag{25.9}$$

Proof

Let us consider a portfolio P characterized by weights $x_1, \ldots, x_m$ and let us start from the relations:

$$\alpha_P = \sum_{j=1}^{m} x_j \alpha_j \quad \text{and} \quad \sigma_{\varepsilon_p}^{\ 2} = \sum_{j=1}^{m} x_j^2 \sigma_{\varepsilon_j}^2. \quad \text{Consequently} : (IR_P)^2 = \frac{\alpha_P^2}{\sigma_{\varepsilon_P}^2}$$

$$= \frac{\left(\sum_{j=1}^{m} x_j \alpha_j\right)^2}{\sum_{j=1}^{m} x_j^2 \sigma_{\varepsilon_j}^2}.$$

The weights $x_1^*, \ldots, x_m^*$ that maximize $(IR_P)^2$ (and therefore IR_P) set the m derivatives $\frac{\partial IR_P^2}{\partial x_j}$ to 0:

$$2(\sigma_{\varepsilon P*})^2 \alpha_j - 2(\alpha_{P*})^2 x_j^* (\sigma_{\varepsilon j})^2 = 0; \text{ hence:}$$

$$x_j^* = \frac{\sigma_{\varepsilon P*}^2}{\alpha_{P*}} \frac{\alpha_j}{\sigma_{\varepsilon_j}^2}. \tag{25.10}$$

Since the sum of the weights is equal to 1, $\sum_{j=1}^{m} x_j^* = \frac{\sigma_{\varepsilon P*}^2}{\alpha_{P*}} \sum_{j=1}^{m} \frac{\alpha_j}{\sigma_{\varepsilon_j}^2} = 1$, we have:

$\sum_{j=1}^{m} \frac{\alpha_j}{\sigma_{\varepsilon_j}^2} = \frac{\alpha_{P*}}{\sigma_{\varepsilon P*}^2}$. This last relation, associated with (25.10), gives (25.8).

In addition, $\alpha_{P*} = \sum_{j=1}^{m} x_j^* \alpha_j$; therefore, pursuant to (25.10): $\alpha_{P*} = \frac{\sigma_{\varepsilon P*}^2}{\alpha_{P*}} \sum_{j=1}^{m} \frac{\alpha_j^2}{\sigma_{\varepsilon_j}^2}$;

hence:

$\frac{\alpha_{P*}^2}{\sigma_{\varepsilon P*}^2} = \sum_{j=1}^{m} \frac{\alpha_j^2}{\sigma_{\varepsilon_j}^2}$, which is (25.9).

The relationship (25.9) means that the IR can be increased without any limits other than the number of existing securities with $\alpha \neq 0$ and the ability, as well as the time needed, to detect them. If $\overline{IR}$ refers to the "average" IR ($\overline{IR}^2 = \frac{1}{m} \sum_{j=1}^{m} (IR_j)^2$), (25.9) can be written as:

$$IR_P = \overline{IR} \sqrt{m}.$$

A direct application of the latter formula may lead to the conclusion that the IR increases as the square root of the number of securities with $\alpha \neq 0$ that the manager (or team of managers) is likely to detect. This result has been called, somewhat emphatically, *the fundamental law of active management*. To mitigate the scope of such a law in the reader's mind, we insist on the following caveats:

- the manager may make mistakes and take long positions in securities with a negative α;
- the additivity of the squares of the IR is based on the assumption of an optimal composition of the portfolio;
- the search for securities has a cost and is subject to decreasing returns to scale ($\overline{IR}$ decreases with m); at the optimum, the financial analysis effort must stop when its marginal gain (related to the marginal IR_j and the value of the managed portfolio) is equal to its marginal cost.

25.2.7 The Construction of a Portfolio That Dominates the Benchmark (Higher Sharpe Ratio)

A second important proposition establishes a relation between the information ratio of a portfolio, its Sharpe ratio, and that of the benchmark. This proposition allows:

- To legitimize the maximization of the information ratio as a management objective.
- To justify the role of active management in a potentially efficient market.
- To provide a method and indicate an operational process leading to an *achievable second best* (the *first best* being unreachable, in practice, for reasons of cost and accessibility of information).

Let us consider a portfolio P (generally inefficient) and benchmark B (also generally inefficient). Let us consider among the portfolios obtained by combinations of P and B (in blue in Fig. 25.2) the one that maximizes the Sharpe ratio, which we call A.

We denote by S_A and S_B, respectively, the Sharpe ratios of portfolios A and B. The following proposition characterizes Portfolio A and its Sharpe ratio S_A as a function of IR_P and S_B.

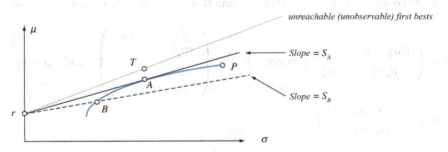

Fig. 25.2 Sharpe ratio maximization

Proposition 3

Let A be the portfolio combination of portfolios P and B that maximizes the Sharpe ratio.

(i) The square of the Sharpe ratio of A is equal to the sum of the squares of the Sharpe ratio of B and the IR of P:

$$S_A{}^2 = S_B{}^2 + IR_P{}^2 \qquad (25.11)$$

(ii) Assuming that P is the combination of m sub-portfolios or securities that maximizes the IR (portfolio P* described in proposition 2), the square of the Sharpe ratio of A is equal to:

$$S_A{}^2 = S_B{}^2 + \sum_{j=1}^{m} (IR_j)^2 \qquad (25.12)$$

Proof

Portfolio A is the portfolio corresponding to the point of tangency of the half-line starting from (0, r) with the curve representing the combinations of P and B (see Fig. 25.2). The composition of this tangent portfolio (weight x on A, 1–x on B) is given by Proposition 4-vi of Chap. 21: $\dfrac{\mathbf{V}^{-1}\left[\underline{\mu} - r\underline{1}\right]}{\mathbf{1}'\mathbf{V}^{-1}\left[\underline{\mu} - r\underline{1}\right]}$, *which implies:*

$$\mu_A - r = \frac{[\underline{\mu} - r\underline{1}]'\mathbf{V}^{-1}[\underline{\mu} - r\underline{1}]}{\mathbf{1}'\mathbf{V}^{-1}[\underline{\mu} - r\underline{1}]}; \qquad \sigma_A^2 = \frac{[\underline{\mu} - r\underline{1}]'\mathbf{V}^{-1}[\underline{\mu} - r\underline{1}]}{(\mathbf{1}'\mathbf{V}^{-1}[\underline{\mu} - r\underline{1}])^2};$$

$$S_A{}^2 = \frac{(\mu_P - r)^2}{\sigma_P^2}, \text{ hence :}$$

$$S_A{}^2 = (\underline{\mu} - r\underline{1})'\mathbf{V}^{-1}(\underline{\mu} - r\underline{1}). \qquad (25.13)$$

In the context of two risky securities P and B with $R_P - r = \alpha_P + \beta_P (R_B - r) + \varepsilon_P$:

$$\mathbf{V} = \begin{pmatrix} \beta_P^2 \sigma_B^2 + \sigma_{\varepsilon_P}^2 & \beta_P \sigma_B^2 \\ \beta_P \sigma_B^2 & \sigma_B^2 \end{pmatrix}, \text{ hence } \quad \mathbf{V}^{-1} = \begin{pmatrix} \dfrac{1}{\sigma_{\varepsilon_P}^2} & -\dfrac{\beta_P}{\sigma_{\varepsilon_P}^2} \\ -\dfrac{\beta_P}{\sigma_{\varepsilon_P}^2} & \dfrac{\beta_P^2}{\sigma_{\varepsilon_P}^2} + \dfrac{1}{\sigma_B^2} \end{pmatrix} \quad \text{and} \quad \underline{\mu} - r\underline{1}$$

$$= \begin{pmatrix} \alpha_P + \beta_P(\mu_B - r) \\ \mu_B - r \end{pmatrix}.$$

(25.13) then implies: $S_A{}^2 = \left(\underline{\mu} - r\underline{1}\right)'\mathbf{V}^{-1}\left(\underline{\mu} - r\underline{1}\right) = \dfrac{\alpha_P^2}{\sigma_P^2} + \dfrac{(\mu_B - r)^2}{\sigma_B^2} = S_B{}^2 + IR_P{}^2$, which is (25.11); (25.12) derives from (25.11) and Proposition 2 (Eq. 25.9).

25.2.8 Synthesis, Interpretation and Application to Portfolio Management

The above proposals constitute a theory of active portfolio management, in its *stock picking* component, and provide a method and framework for a management procedure that can be decentralized. The theory is based on the Markowitz model and a possibly efficient market in which the tangent portfolio T is not necessarily known, particularly because of its complexity (see Roll's critique). Performances are assessed using an index B which, in general, is not efficient (it is not a combination of T and the risk-free asset). The likely inefficiency of the benchmark legitimates theoretically and gives practical credibility to the objective of active management which is to beat the index.

The first step is to detect alpha-positive securities. It should be noted that, since alpha is calculated in relation to the benchmark, such securities can theoretically exist even in an efficient market, due to the inefficiency of the benchmark. It should also be noted that the assessment of alpha requires a specific study (e.g., financial analysis) and that, at the operational level, the number of securities studied is necessarily limited. This step can be decentralized, with m managers building m portfolios with the highest possible IRs, independently. In a second step, these m securities or portfolios are aggregated according to Proposition 2 (weight proportional to the estimated $IR_j/\sigma_{\varepsilon j}$) in order to maximize the overall IR. The result is a P* portfolio. In a third step, the latter is combined with the benchmark in order to maximize the Sharpe ratio. The result is a portfolio A whose Sharpe ratio is higher than that of B if (and only if) the IR of P* (therefore its alpha) is positive. Relations (25.11) and (25.12) clearly show that at the level of the first two steps, the objective that must be assigned to the managers involved is to maximize the IR. Indeed, since the benchmark is chosen, and therefore S_B^2 is given, the maximization of the Sharpe ratio requires that of the IR_j and IR_P. Ultimately, the added value of active management, therefore, lies in its ability to detect alpha-positive securities (positive IR) compared to the benchmark and to combine them in an optimal way.

25.3 Alternative Investment Management and Hedge Funds

Alternative investment management appeared in the United States after the Second World War, with the emergence of hedge funds (which are a special case of the type of management). Despite varying fortunes and some spectacular bankruptcies such as those of LTCM in 1998 or Amaranth in 2006, hedge funds experienced a strong expansion before the sub-prime crisis of 2007–2008. After a general description of alternative investment[6] (Sect. 25.1), we will define the main management styles

---

[6]Note that we use here the terms "alternative investment" or "alternative management" in the restrictive sense that *excludes* non standard financial investment such as real estate, commodities or art works, which are also often termed "alternative" investments.

encountered (Sect. 25.1.2), then attempt to assess the interest of this type of portfolio management for investors (Sect. 25.2), which poses serious problems due to the particular difficulties of measuring its performance (Sect. 25.3).

25.3.1 General Description of Hedge Funds and Alternative Investment

The regulations applied to traditional mutual funds are obviously not intended to eliminate all the risks inherent in financial investments, but essentially to protect uninformed investors against risks that are excessive relative to their assets and/or which they may not be aware of. For more sophisticated or wealthier than average investors, the over-cautious regulation of mutual funds may be too restrictive in that it prohibits them from certain transactions and thus deprives them of certain opportunities. This is why the legislator has provided for more flexible regulation of funds dedicated to large and/or experienced investors.

Hedge funds, which are private companies, cannot make public offerings and are only accessible to wealthy and/or well-informed, so-called accredited partners.[7] In return, they can take long or short positions, play on financial leverage and trade in derivatives at will; in addition, they are subject to less restrictive reporting than traditional funds.

Hedge funds are typically organized in limited partnerships, in which investors are the limited partners and managers are the general partners. As general partners, fund managers usually invest a significant proportion of their personal wealth in the fund and are personally, jointly, and severally liable for the company's debts, while limited partners have a liability limited to a contractually agreed amount. Investors pay (performance-based) fees that can be much higher than those of traditional mutual funds.[8] Limited partnerships continue to be the dominant structure although, recently, some hedge funds have opted for limited liability company statutes. In addition, for tax reasons and to avoid national regulatory frameworks that are not well suited to alternative management, many hedge funds are based offshore, in places such as the Cayman Islands, Bermuda, the Virgin Islands, etc. The fear of under-regulated finance since the sub-prime crisis has put hedge funds under fire from critics, hence a need to strengthen global control.

Alternative management is often the realm of relatively light and specialized structures, focused on their expertise in asset allocation and subcontracting operations such as accounting, portfolio valuation, custody and settlement, securities delivery, order placement, or marketing. Among the hedge fund's privileged partners, the prime broker plays an essential role. The latter is a financial services

---

[7]Note however that a limited number of non-accredited investors may be given access and that the general public can invest in funds of (alternative) funds.

[8]The manager's compensation may be as high as 15 to 25% of returns exceeding a given threshold (*high water mark*).

provider that can perform all of the above functions, as well as financing and securities lending, thereby facilitating short selling.

The hedge fund universe is characterized by a very wide diversity of management styles. Some of these funds were created simply to allow stable management, free from the problems of managing cash inflows and outflows, but traditional otherwise. On the other hand, those in the alternative investment management to which this section is dedicated take advantage of the possibilities offered by a softer regulation. Like Georges Soros's Quantum Fund, some have in the past posted annual return series of more than 40%, obtained by taking short or long positions in various products and currencies.

Although their diversity makes it illusory to determine a common denominator, most alternative investment styles share one or more of the following characteristics: a low correlation with other asset classes; the search for absolute performance without reference to a benchmark; leverage effects that increase risks and expected returns; the implementation of sophisticated, dynamic trading techniques, often designed to exploit certain niches or even market inefficiencies.

25.3.2 Definition of the Main Alternative Investment Styles

There are many possible classifications of hedge funds and alternative investments. We base our presentation on the styles defined by Credit Suisse, one of the leading producers of hedge fund indices. We group (somewhat arbitrarily) these different styles into four main families: long-short strategies; directional trading strategies; arbitrage or relative value strategies; and event strategies.

(i) Long-short equity strategies combine long positions in stocks considered undervalued and short positions in stocks presumed overvalued, in order to take advantage of any opportunities and of high leverage (sales financing purchases) and to reduce certain risks (market or sector risks). Four management styles fall within this category.

– *Equity Market Neutral*

This strategy seeks to exploit inefficiencies in equity markets. It usually involves being simultaneously long and short on equity portfolios of the same size in a given country. These non-directional portfolios are designed to be either beta-neutral or neutral against certain currencies, or both. Leverage is often used to improve returns (at the price of increased risk).

– *Long/Short Equity*

This strategy also involves long and short equity positions but, unlike the previous strategy, it is not market neutral. Managers have the option of alternating value and growth strategies (see the Fama and French model in Chap. 23), small-cap and large-cap equities, and net long and net short positions. They can use futures and options to hedge their positions. The allocation can be geographical (as in long/short US or European equity strategies), or sectoral (as in long/short technology or health sector strategies). Long/short equity funds are generally much less diversified than traditional funds.

– Dedicated Short Bias

Portfolios of this type may include long and short positions in most equities and derivatives, but they maintain a net short position. The short bias of the portfolio must be consistently positive for it to be classified in this category. The capital raised by net short sales is invested in the money market.

– Emerging Markets

This strategy consists of investing in emerging markets (equities or bonds). Because many of these do not allow short selling and do not offer the possibility of taking short positions using futures or other derivatives, these strategies mainly involve long positions. Emerging markets include Latin American countries, Eastern Europe, Russia, Africa, and some Asian countries. The weighting in these different countries is modified according to market conditions and the manager's expectations. Some funds are specialized by region.

(*ii*) Relative value strategies seek to take advantage of price or rate differentials that are considered abnormal and expect these anomalies to be resolved. Two management styles defined by CSFB/Tremont correspond to this general definition.

– Convertible Arbitrage

These are transactions involving a company's convertible bonds. A typical investment is long on the convertible bond and short on the shares of the same company or on calls written on these shares. The objective is to take advantage of the relative undervaluation of the bond, while protecting the portfolio from market and counterparty risks through the short sale of the shares or the calls. Some managers may also hedge interest rate risk and/or counterparty risk on the bond portion of the convertible bond (e.g., by using CDS, credit default swaps; see Chap. 30).

– Fixed Income Arbitrage

This strategy seeks to exploit valuation anomalies between different fixed-rate securities while neutralizing the risk of distortion of the yield curve. Most managers try to achieve low volatility returns. This category includes interest rate swap arbitrage, sovereign and non-sovereign bond arbitrage, forward rate curve arbitrage, and mortgage-backed securities arbitrage (see Chap. 30).

(*iii*) "Event-driven" strategies seek to take advantage of price movements generated by an economic event affecting one or more companies. CSFB/Tremont identifies three styles among these strategies.

– Risk arbitrage

This involves taking long and short positions simultaneously in two companies involved in a merger/acquisition. Typically, risk arbitrageurs are long on the target company and short on the acquiring company. The main risk is that the merger/acquisition operation will be delayed or fail.

– Distressed securities

These are positions in the securities of companies in distress, the bet being that their position may improve or, on the contrary, become desperate.

– Regulation D

This type of transaction is specific to the US market and concerns investments in small-cap public companies. It often involves the purchase of convertible securities which are offered to the managers on a confidential basis by companies in dire need of financing.

25.3 Alternative Investment Management and Hedge Funds

– High yield

These are investments in bonds (often called junk bonds) of companies with low ratings, high yield rates, and high upside potential (but also bankruptcy potential!).

(*iv*) Finally, trading strategies consist of directional positions based on precise forecasts regarding certain markets. Two management styles can be distinguished.

– Global Macro

The "global macro"managers have long and short positions in the main international capital and derivatives markets. These positions reflect their views on the general trend of stock markets, interest rates, exchange rates, and commodity prices and rely most often on market timing. Their expectations are based on the economic and political situation of the various countries, the world supply and demand for raw materials, etc. Portfolios may include all cash or derivative instruments.

– Managed Futures

This strategy involves trading in futures on financial instruments, commodities, and exchange rates. These managers are commonly referred to as Commodity Trading Advisors (CTAs). Their management methods are systematic or discretionary. The first ones have a quantitative and often automated approach, while the second ones adopt a more qualitative approach.

Finally, it should be noted that there are funds of hedge funds that represent nearly half of alternative investments. Such funds of funds invest in the alternative funds of other managers. As a result, portfolios are more diversified and have lower volatility than those of a simple hedge fund. The manager may focus on funds within the same style, or may also diversify her portfolio into different styles. The minimum investment in a fund of funds may be lower than that required by a hedge fund, which allows for better diversification for the same total invested amount.

25.3.3 The Interest of Alternative Investment

The interest in hedge funds and alternative investment is partly due to the diversification of risks they allow, and partly to their innovative feature and complementarity to traditional management.

The ability to diversify, linked to the often low correlation with standard markets, appears all the more attractive as international or sectoral diversification is lacking at the very moment when the investor needs it most, namely in periods of sharp market declines; indeed, the correlation between securities and that between the equity markets of different countries increase sharply in situations of sudden decline.

In fact, detailed analysis of hedge fund returns indicates that they are heterogeneous in terms of correlation with other asset classes. Some strategies such as market neutral, convertible arbitrage, fixed income arbitrage, or short bias are in normal market conditions weakly correlated with the performance of the S&P 500 for example (correlations typically below 0.5 in absolute value); they can therefore be used effectively as risk reduction instruments. On the other hand, other strategies such as emerging markets, event driven or global macro lead to portfolios that are highly correlated with the market, and are therefore inadequate diversification instruments, although they often offer high returns.

On a theoretical level, the interest of alternative investment can be assessed in terms of market completion.[9] In theory, alternative portfolios consisting of long and short positions in existing assets are redundant (replicable) assets and, since they do not constitute a new asset class, cannot claim to complete the market and improve the risk/return trade-off for investors. In practice, however, some of the strategies used by the managers are sufficiently sophisticated (and require sufficiently sophisticated technical means of access and processing of information and optimization software) so that the individual investor, who also faces higher transaction costs, cannot actually implement them.[10] In addition, we have already seen that the ban or practical impossibility of short selling, which affects most individual investors, reduces the slope of the market line because it restricts all reachable portfolios and thus moves the efficient frontier in the wrong direction (to the southeast). Alternative portfolios, free of such a constraint, make it possible to avoid this movement.

25.3.4 The Particular Difficulties of Measuring Performance in Alternative Investment

Although the objective of alternative investment is to achieve absolute, not relative, performance, in practice it is difficult to avoid a performance measure that does not use a benchmark. It is then necessary to first build a good index that is representative of a homogeneous management style (failing that, the average return of comparable funds), and then to compare the return of a given fund with that of the relevant index.

At least a dozen hedge fund index providers compete with each other and differ in their definitions and index construction methods.

In addition to the difficulties inherent in the construction of stock market indices, presented in Chap. 8, which are increased in the world of alternative management, there are specific obstacles and pitfalls.

First of all, information on the assets managed by hedge funds is generally not available in real time and the absence of any obligation to publish results explains why the databases are neither reliable nor exhaustive (in particular, rates of return are only available on a monthly basis).

Moreover, in a world where competitive advantage is based on sophisticated and confidential management techniques, managers are reluctant to disclose their strategies and index providers must generally be satisfied with using the managers' self-proclaimed styles.

---

[9] See Chaps. 19 and 24 for the definition of a *complete* market.

[10] In the Black–Scholes world, the market is complete so that any option has been seen (in Chap. 11) to be redundant, hence replicable and accurately priced. In the real world, however, the option is not useless as the exact replication strategy is impaired by market imperfections such as transaction and information costs, discontinuous trading, and presence of price jumps.

25.3 Alternative Investment Management and Hedge Funds

In addition, as the terminology of hedge fund strategies is not uniform, the same strategy may be referred to under different names, depending on the index provider.

Finally, there are many biases in databases and indices, including survivor bias and selection bias. The latter are classic but they particularly affect alternative investment indices.

As a result of these difficulties, the overall funds considered for a given management style vary widely from one index provider to another, the returns of competing indices for the same management style differ significantly (the return differential may reach 20%) and their correlation is generally low (often smaller than 45%).

In addition to these problems related to data quality and reliability, the two main reasons why performance measurement is particularly difficult are related to the definition of risk on the one hand and the instability of the distributions of the returns of alternative investment on the other hand:

– The standard deviation is a good measure of risk only in the Gaussian case. However, the distributions of most returns are not Gaussian and actually exhibit left asymmetry and too high a kurtosis. Now, the Sharpe ratio systematically overestimates a fund's actual performance when the distribution of its returns is not Gaussian, because it neglects the impact of asymmetry and kurtosis. For example, it is easy to increase the value of a fund's Sharpe ratio without particular management skills, by distorting the distribution of its profitability to move it away from normality and increase its asymmetry coefficient. To do this, it is sufficient to sell call options on the fund's representative index. This truncates the allocation on the right because if the index rises, and with it the value of the fund, calls are exercised against the position. The distribution queue on the left is not affected because, if the index drops, calls are not exercised. Very few investors appreciate this type of negative (left) asymmetric distribution, which can cause very large losses without the possibility of very large gains. However, this distribution benefits from an attractive Sharpe ratio. The reason for this is that, over a wide range of the index's value around its initial value, the fund's profitability is good or very good, due in particular to the receipt of call sales. Profitability can be catastrophic in the event of a sharp drop or crash (since the sold calls do not hedge a (long) portfolio). However, since these events are quite rare, the mean distribution is relatively high and its standard deviation relatively low. The Sharpe ratio is therefore higher. However, this improvement is only achieved at the cost of increased asymmetry.

– In addition, the assumption of identically and independently distributed returns underlying traditional performance measures is particularly challenged for hedge funds. Indeed, they are often positively self-correlated, due in particular to the illiquid nature of some of the assets used. In addition, they are most often not identically distributed because of the very strong opportunism that characterizes these managers' style.

Finally, it should be noted that the difficulties encountered are all the more acute because managers seek results that are less correlated with those of a benchmark market (or index).

A few specific performance measurement methods for hedge funds have been proposed to overcome some of the shortcomings of traditional measures. We will only mention here the application of style analysis. The idea behind style analysis (formalized by Sharpe in 1994) is to estimate the composition of a fund (alternative or not), i.e. its style, using econometric methods, knowing the evolution of the elementary indices likely to compose it. For example, a traditional mutual fund invested in equities can be composed on average of 70% US equities (a stock index can be used), 25% US bonds (a bond index can be used), and 5% US monetary assets. It should be noted, however, that around these averages, the composition may change significantly according to the manager's forecasts, and sometimes even include different asset classes. The simplest method is to regress the periodic returns on the considered fund, denoted by R_P (dependent variable), on those of the (n) potential indices chosen a priori, denoted by R_i, $i = 1,\ldots, n$ (explanatory variables):

$$R_P = \alpha + \sum_{i=1}^{n} w_i\, R_i + \varepsilon_P$$

under the constraints $w_i \geq 0$, where R_P and R_i are daily or weekly rates of return, the w_i are the weights of the relevant indices in the fund under study (or exposures to different styles) and ε_P is the regression's residual.

Each weight w_i measures the average sensitivity of the fund to changes in the corresponding index i. Depending on the values of these weights, the hedge fund is characterized by a certain style, which makes it possible to assess the consistency between the stated objectives and the manager's actual decisions. The origin ordinate α measures the part of the fund's average return that is due to the manager deviating more or less from the average weights. If the weights w_i are constrained to have a sum equal to one, α measures the over- or under-performance in relation to the displayed style depending on whether it is positive or negative. The coefficient of determination R^2 (between zero and one) of the regression measures the part of the variance of the periodic return of the fund explained by the composite index (weighted by the weights w_i). A coefficient close to one (respectively, zero) indicates a very good (respectively, very poor) correlation between the return on the fund and that on the composite index.

For example, the nine Crédit Suisse indices (Convertible Arbitrage, Dedicated Short-Term Bias, Event-Driven, Global Macro, Long Short Equity, Emerging Markets, Fixed Income Arbitrage, Market Neutral, Managed Futures) representative of the most common types of strategies can be used to determine the distribution of a hedge fund's activity. The advantage of this variant is that these nine style indices are real hedge fund portfolios and not exogenously given factors.

25.4 Summary

- A *benchmark,* usually an *index,* is often used to analyze the risk, return and performance of portfolios.
- Let $\Delta_t \equiv R_{x,t} - R_{B,t}$ be at period t the deviation of the return on a portfolio x to the benchmark B. The « distance » between portfolio x and its benchmark is

25.4 Summary

- estimated from the *tracking error* (TE), defined as the standard deviation of Δ_t:
$TE_t \equiv \sigma(\Delta_t)$.
- The previous definition of TE is theoretical. Its estimator, called *empirical tracking error*, is calculated ex post from a history that most often consists of N past weekly returns and is given by: $Tw = \sqrt{\frac{1}{N-1} \sum_{i=1}^{N} \left(\Delta - \overline{\Delta} \right)^2}$.
- A variety of indexed management styles exist: some Index Funds or Exchange Traded Funds (ETF) *replicate the index as accurately as possible* by minimizing the tracking error (TE < 1%); more active management tries to "beat" the benchmark by deviating more or less from it, allowing a higher TE.
- In principle, the replication of the index is obtained by building a portfolio that includes, at all times, the same assets as the index and with identical weights. In practice, to avoid too many transactions, the manager of a *passive* index fund uses a sub-basket, or derivatives (synthetic replication), leading to *imperfect replication* (higher TE).
- An *active manager who maximizes the expected return while respecting a maximum* TE (mean-TE criterion) chooses a particular combination of the benchmark and the portfolio q whose vector of weights is $q = \mathbf{V}^{-1}(\mu - m\,\underline{1})$, m being the expected return of the minimum variance portfolio ($m = r$ in presence of a risk-free rate r).
- This fund q is identical for all index managers and benchmarks (two-fund separation) and is homothetic to the tangent portfolio. The combinations of q and the benchmark are efficient iff the benchmark B is efficient. If B is inefficient, these combinations have a Sharpe ratio higher than that of B.
- Following the *mean-TE criterion is equivalent to maximizing the Information Ratio* defined by: $IR = E(\Delta)/\sigma(\Delta)$.
- The IR generalizes the Sharpe ratio, which is a special IR when B is the market portfolio. IR subsumes the manager's *performance* in a single criterion that accounts for both expected return and risk (measured by the tracking error). As previously defined, it is a theoretical and *ex ante* measure.
- The *ex-post* performance of a portfolio P is commonly estimated by regression of excess returns: $R_{P,t} - r_t = \alpha + \beta\,(R_{B,t} - r_t) + \varepsilon_t$, ε_t being a white noise and α a *performance* (identical to *Jensen's alpha* if B is the market portfolio).
- A β_t different from 1 may reflect an attempt at *market timing*. A manager has a skill in market timing if, more often than not, $(\beta_t - 1)(R_{B,t} - r_t)$ is *positive*.
- When β is kept equal to 1 (no market timing), $R_{P,t} = \alpha + R_{B,t} + \varepsilon_t$. The ability of an active manager to outperform the benchmark relies on selecting the right securities (*security picking*) and should be reflected by a *positive* α but also by deviations ε_t from the benchmark. Besides, $TE = \sigma(\varepsilon)$, and $IR = \alpha/\sigma(\varepsilon)$. These last equations, which are theoretical, lead to standard *estimators of α and IR*.
- In particular, the *empirical ratio of information*, an *ex-post* performance measure, is: $\hat{IR} = \frac{\overline{\Delta}}{\hat{\sigma}(\Delta)} = \frac{\hat{\alpha}}{\hat{\sigma}(\varepsilon)}$. The information ratio is therefore directly related to the Student's test on the empirical mean $\hat{\alpha}$.

- The weights on *m* given assets (or sub-portfolios) that maximize the IR of the resulting portfolio P* are *proportional* to the ratios $IR_j/\sigma(\varepsilon_j)$ of these assets.
- The square of the IR of P* is equal to the sum of the squares of the IRs of its components. Then the so-called «*fundamental law of active management*» states that the IR of a portfolio increases as the square root of the number of securities with non-zero alphas that the manager is able to detect.
- Active management may still be *relevant* in *efficient* markets since the market portfolio is *unobservable* and the *selected benchmark may be inefficient,* thus could be dominated.
- *Hedge funds* are mainly dedicated to sophisticated or wealthy investors seeking to avoid over-cautious regulation.
- Hedge funds involve a *wide variety of investment styles* sharing one or more of the following characteristics: *low correlation* with other asset classes; search for *absolute* performance without reference to a benchmark; *leverage* that increases risks and expected returns; sophisticated, dynamic *trading techniques*; the exploitation of some *niches* or even market *inefficiencies*.
- The different hedge funds and alternative investment strategies may be classified into *four wide families*: long-short; directional trading; arbitrage or relative value; event-driven.
- The interest in hedge funds and alternative investment is due to the *diversification of risks* they allow (linked to the often low correlation with standard asset classes), their flexibility, and their innovativeness. From a theoretical viewpoint, the interest of alternative investment can be assessed in terms of *market completion*.
- Measuring the performance of hedge funds raises specific issues related to: the construction of *relevant* indices; the definition of risk; the *instability of the distributions* of alternative investment returns. A few specific performance measurement methods for hedge funds have been proposed, in particular, those grounded on *style analysis* (whereby the composition of a fund, i.e., its style, is determined by econometric methods).

Appendix

Breakdown of the Tracking Error and Performance Attribution

The purpose of performance measurement is to assess the quality of past investment decisions in terms of return and risk, to assess its persistence for forecasting purposes, and to compare the results of the managed funds. A breakdown of the tracking error, referred to as performance attribution, aims to identify and explain the sources of a fund's over- or underperformance in relation to its explicit or implicit benchmark, i.e. to attribute them to specific decisions of the manager. The type of breakdown considered here, therefore, assumes that the fund which is analyzed

Appendix

1057

follows a strategy of the same nature as a correctly identified index (benchmarked management).

In the (restrictive) sense used by practitioners, the attribution of performance only concerns the average return observed. For them, the fundamental element of assessment of the fund's performance is the difference observed *ex post* between the average profitability of the fund and that of its benchmark over the analysis period. Performance attribution (in the narrow sense) is then the breakdown of this gap between market timing (opportunistic weighting of the different asset classes) and security picking (selection of individual securities).

The individual securities constituting the fund to be analyzed are grouped into broad asset classes, as for its benchmark. These classes may represent different types of securities (equities, bonds, money markets, etc.), different sectors (oil equities, banking, mass distribution, technology, etc.) or different countries or geographical areas (Eurozone, America, Pacific, etc.). We assume that there are n asset classes accessible to the manager, denoted by i ($i = 1, \ldots, n$). Moreover, we will denote:

– x_i: weighting (relative value with respect to the total value of the portfolio) of the class i in the fund to be evaluated.
– y_i: weighting of the class i in the benchmark.
– R_i: average return of class i in the fund.
– B_i: average return of class i in the benchmark.
– R: total average return of the fund.
– B: total average return of the benchmark.

By construction, we have:

$$\sum_i x_i = \sum_i y_i = 1 \tag{25.14}$$

$$\sum_i x_i R_i = R \tag{25.15}$$

$$\sum_i y_i B_i = B \tag{25.16}$$

The over-(under)performance (tracking error) can be broken down into three elements measuring the market timing effect, the stock selection effect and the interaction between the two.

The market timing effect is defined by:

$$MT = \sum_i (x_i - y_i)(B_i - B) \tag{25.17}$$

It measures the manager's ability to over–/under-weight ($x_i >$ or $< y_i$) the most (/ least) profitable asset classes compared to the benchmark.

The effect of selecting particular securities in a given asset class is defined by:

$$ST = \sum_i y_i(R_i - B_i) \tag{25.18}$$

It measures the impact on the profitability of the portfolio of the choice of the particular assets constituting each class i and therefore the manager's ability to choose the most profitable securities (winners) and to abandon the least profitable ones (losers).

The interaction is defined by:

$$I = \sum_i (x_i - y_i)(R_i - B_i) \tag{25.19}$$

It measures the combined impact of market timing and security picking on the return on the portfolio. For example, suppose that the manager over-weighted the classes for which her picking was better than that of the benchmark, and under-weighted those for which her picking was worse than that of the benchmark. In this case, all (n) elements of the sum defining I are positive and I is therefore positive. As there are 4 possible cases (x_i higher or lower than y_i, and R_i higher or lower than B_i), I can be positive, negative or nil.

Finally, by adding the 3 components (25.17) to (25.19), and using the three Eqs. (25.14) to (25.16), it is easy to verify that we obtain:

$$MT + ST + I = R - B = \text{global over-(under) performance.} \tag{25.20}$$

Example 2

Let us consider a fund whose composite benchmark index is: 25% Eurozone equities; 30% Japanese equities; 25% Eurozone money market equities; and 20% US bonds. All data are converted into euros using the prevailing exchange rates. The managed fund is analyzed over the past year and we assume for simplicity that the weights of the asset classes in the portfolio have not changed significantly during the period. The data on weights and observed returns are gathered in Table 25.1. Table 25.2 presents the results of the performance attribution.

Table 25.1 Data

Class	Weight in the portfolio x_i	Weight in the benchmark y_i	Return on class i in the portfolio R_i	Return on class i in the benchmark B_i
Eurozone Stocks	35%	25%	12.0%	12.5%
Japanese Stocks	40%	30%	14.0%	13.0%
Eurozone Money Market	15%	25%	6.0%	8.0%
US bonds	10%	20%	7.0%	6.0%
Total	100%	100%	$R = 11.40\%$	$B = 10.225\%$

Suggestion for Reading

Table 25.2 Performance attribution

Class	Market Timing effect $(x_i-y_i)(B_i-B)$	Security picking effect $y_i(R_i-B_i)$	Interaction effect $(x_i-y_i)(R_i-B_i)$
Eurozone Stocks	0.2275%	−0.125%	−0.05%
Japanese Stocks	0.2775%	0.30%	0.10%
Eurozone Money Market	0.2225%	−0.50%	0.20%
US bonds	0.4225%	0.20%	−0.10%
Total	1.15%	−0.125%	0.15%

The portfolio and the benchmark achieved an average return of 11.40% and 10.225% respectively over the year. The fund's over-performance was therefore 1.175%. The latter breaks down into 1.15% due to market timing effect, −0.125% due to security picking effect, and 0.15% due to the interaction of the two effects. The manager, therefore, achieved her performance from successful market timing, but her security picking capacity was poor.

This method of attributing performance (in the narrow sense of return, without considering risk), frequently used in practice, has the advantage of simplicity, once the fund benchmark has been precisely defined. It can also be transposed to reveal other components of total return, such as its performance against an exchange rate for international portfolios. Its main weakness is that it does not take into account the relative risks of the fund and its benchmark. It should therefore be complemented by a similar analysis of relative risk. Indeed, if market timing and security picking have led to an outperformance in terms of average return but also to a greater risk-taking, the conclusion as to the manager's real performance is ambiguous.

Suggestion for Reading

Books

Bodie, Z., Kane, A., & Marcus, A. (2010). *Investments* (9th ed.). Irwin.
Elton, E., Gruber, M., Brown, S., & Goetzman, W. (2010). *Modern portfolio theory and investment analysis* (8th ed.). Wiley.

Articles

Agarwal, V., & Naik, N. Y. (2004). Risks and portfolio decisions involving hedge funds. *Review of Financial Studies, 17,* 63–98.
Ammann, M., & Zimmermann, H. (2001). The relation between tracking error and tactical asset allocation. *Financial Analyst Journal, 57,* 32–43.
Fung, W., & Hsieh, D. A. (2001). The risk in hedge funds strategies: Theory and evidence from trend followers. *Review of Financial Studies, 41,* 313–341.
Fung, W., & Hsieh, D. A. (2002). Hedge fund benchmarks: Information content and bias. *Financial Analysts Journal, 58*(1), 22–34.

Henrikson, R. D. (1984). Market timing and mutual funds performance: An empirical investigation. *Journal of Business, 57*, 73–96.

Henriksson, R. D., & Merton, R. C. (1981). On market timing and investment performance of managed portfolios (II): Statistical procedures for evaluating forecasting skills. *Journal of Business, 54*, 513–533.

Jorion, P. (2003). Portfolio optimization with tracking-error constraints. *Financial Analyst Journal, 59*, 70–82.

Merton, R. C. (1981). On market timing and investment performance: An equilibrium theory of value for market forecasts. *Journal of Business, 54*(3), 363–406.

Roll, R. (Summer, 1992). A mean/variance analysis of tracking error. *The Journal of Portfolio Management*, 13–22.

Sharpe, W. (Winter, 1992). Asset allocation: Management style and performance measurement. *The Journal of Portfolio Management*, 7–19.

Part IV

Risk Management, Credit Risk, and Credit Derivatives

The fourth part, made up of Chaps. 26–30, focuses on risk management. It emphasizes general methods of analysis, and then assesses default and credit risks and the instruments to hedge them (credit derivatives). It is of an intermediate conceptual and mathematical difficulty, comparable to that of the third part.

Chapter 26 first presents the basic version of the so-called Monte Carlo simulation method, and then proposes various extensions (copulas, regressions for the application to American derivatives, etc.). The method is applied both to the valuation of assets and to the assessment of market risks.

Chapter 27 is devoted to the most widely used risk measure, namely the Value-at-Risk (VaR), its uses, limits, and extensions, such as the Expected Shortfall and the extreme value method. The conditions that must be satisfied by a risk measure to be coherent are also briefly commented.

Chapter 28 provides an empirical and theoretical analysis of credit and/or counterparty risks, essential elements of concern in the management of financial institutions. Both so-called intensity and structural models are presented and discussed.

Chapter 29 applies techniques inherited from VaR to the analysis and management of credit risk (Credit-VaR) and presents the notion of economic capital.

Chapter 30 briefly presents some aspects of banking regulation (Basel 2 and 3) concerning credit risk. It also provides an analysis and valuation of the main types of credit derivatives (mainly Credit Default Swaps) that investors and financial institutions use to hedge against, or to take a position on, credit events affecting the issuers of debt assets. It then presents the notion and role of securitization in credit risk management. It finally introduces and analyses the CVA (and more generally the xVA) mechanisms that have been designed to cope with the counterparty risk affecting derivatives.

Monte Carlo Simulations

26

This chapter presents in a concrete and practical manner the main operational techniques related to Monte Carlo simulations. The reader in search of a more in-depth and rigorous presentation may consult the various specific literature dedicated to this topic.[1] As an introduction, we present the problem to be solved, the notations to be used as well as the methodology.

Let us consider a portfolio or a security whose value is denoted by $V(t, \underline{Y}(t))$ at time t; this value is a function depending both on time and on m random factors denoted $\underline{Y}(t) = (Y_1(t), Y_2(t), \ldots, Y_m(t))$. It may for instance represent an option whose value depends on two random factors, namely the underlying price $S(t)$, and the interest rate $r(t)$ (in this case $m = 2$, $Y_1(t) = S(t)$ et $Y_2(t) = r(t)$). Here, we aim to determine, at least empirically, the probability distribution of $V(t, \underline{Y}(t))$ or certain moments of this distribution. Depending on the situation, it will be necessary to either determine $V(t, \underline{Y}(t))$ in the whole interval $(0, T)$—trajectories of V are of interest—or only its value $V(T, \underline{Y}(T))$ at time T.

Monte Carlo simulation is a probabilistic technique, which is used when the probability distribution of $V(t, \underline{Y}(t))$ fails to be represented with a closed formula and its empirical distribution shall hence be simulated. As a first step, we generate a large sample from the probability distribution of $\underline{Y}(t)$ and we map each element of this sample onto the corresponding value $V(t, \underline{Y}(t))$, hence building a "simulated empirical" distribution $V(t, \underline{Y}(t))$ when repeating this procedure. Depending on the problem we face, Monte Carlo simulation will either require the generation of a large number M of paths $\underline{Y}_t]_{t=t_1,t_2,\ldots,t_N}$ from which are derived the corresponding paths $V(t, \underline{Y}_t)]_{t=t_1,t_2,\ldots,t_N}$ or the generation of M values of $(\underline{Y}(T), V(T, \underline{Y}(T)))$ at a single time T.

The reader will notice that Y_t denotes the specific outcome of the random variable $Y(t)$: this specific notation permits to distinguish a random draw $\underline{Y}_t$ from the random variable $\underline{Y}(t)$ itself.

---

[1] Readers should consult Glasserman (2004) or Fusai and Roncoroni (2008) for a rigorous and complete discussion.

© The Author(s), under exclusive license to Springer Nature Switzerland AG 2022
P. Poncet, R. Portait, *Capital Market Finance*, Springer Texts in Business and Economics, https://doi.org/10.1007/978-3-030-84600-8_26

The Monte Carlo methodology relies upon the generation of a large number of random draws from a given probability distribution. We describe in Sect. 26.1 the methodologies to be used to perform such random samplings. We then explain in Sect. 26.2 how to perform Monte Carlo simulation with a single risk factor Y, before addressing in Sect. 26.3 the simulation with multiple risk factors (generally correlated). Section 26.4 presents some observations related to the efficiency of the simulation framework and introduces some techniques to improve this efficiency. Finally, Sect. 26.5 proposes a solution to the problem posed by American options.

26.1 Generation of a Sample from a Given Distribution Law

Let us briefly examine the generic problem of sampling from a given probability distribution before addressing how to sample from a normal distribution.

26.1.1 Sample Generation from a Given Probability Distribution

Let us consider a given random variable X whose cumulative distribution function is denoted by F_X, assumed to be known, and from which we aim to generate a random sample. Recall that $F_X(x) = \text{Proba}(X \leq x)$ takes on values in $(0,1)$.

Let us consider the sample $\underline{a} \equiv (a_i)_{i=1, 2, \ldots, N}$ drawn from a random variable A distributed according to a *uniform law* in the interval $(0, 1)$ whose distribution function is $F_A(y) = y$ for $y \in (0,1)$.

Such a sample is fairly easy to obtain, for instance by using a random number table or a built-in random number generator function available in various vendor solutions (Excel, Matlab, Mathematica) or in the programming language libraries (Python, C++, Java, etc.).

Let us define $x_i = F_{X^{-1}}(a_i)$ where $F_{X^{-1}}$ is the inverse distribution law of X. We hence build a new sample $\underline{x} \equiv (x_i)_{i=1, 2, \ldots, N}$.

The sample $\underline{x}$ is generated from the distribution law of X since the random variable $F_{X^{-1}}(A)$ has the same distribution as X; in fact:

$$\text{Proba}[F_{X^{-1}}(A) \leq x] = \text{Proba}[F_X(F_{X^{-1}}(A)) \leq F_X(x)] \text{ (because } F_X \text{ is increasing)}$$
$$= \text{Proba}[A \leq F_X(x)]$$
$$= F_X(x) \text{ (based on the cumulative distribution function of a uniform law)}$$

We can thus build a random sample, drawn from any given distribution F, from a random sample drawn from a uniformly distributed variable.

26.1.2 Construction of a Sample Taken from a Normal Distribution

Monte Carlo simulations are often built upon random samples generated from normal probability distributions. The above methodology permits building a sample from any given cumulative distribution function F, hence in particular from a normal distribution. Those Gaussian samples may also be obtained by using the so-called Box-Muller technique. This technique is more accurate and simpler to implement than the one relying upon the normal inverse cumulative distribution, which does not have a closed form. Let A_1 and A_2 be two independent and uniformly distributed random variables in (0, 1) and let us posit:

$$U = \sqrt{-2 \ln (A_1)} \cos (2\pi A_2).$$

One can prove that U defined as such follows a standard normal distribution (beware that, in these notations, A stands for the uniform distribution while U is the normal distribution). Thus, one can easily generate N numbers from a normal distribution starting with $2N$ numbers generated from a uniform distribution.

Many vendor softwares are equipped with built-in normal distribution sample generators, which in practice makes rarely necessary the use of the Box-Muller technique.[2] Besides, it is useful and desirable to link each drawing u_i to its "antithetic" - u_i to double the size of the generated sample, reduce the computation burden and ensure that its average value is exactly zero.

26.2 Monte Carlo Simulations for a Single Risk Factor

Let us first address the case of a single risk factor denoted by Y. The value of a security, $V(t, Y(t))$, is a function of Y. We will distinguish two situations. The former situation addresses the problem where the simulation of various paths $V(t, Y(t))$ between time 0 and T (dynamic simulation) is required. The latter addresses the simpler problem where the simulation of different values $V(T, Y(T))$ at a single time T is required (static simulation).

26.2.1 Dynamic Paths Simulation of $Y(t)$ and $V(t, Y(t))$ in the Interval $(0, T)$

Let us begin with the simulation of $Y(t)$. This simulation is grounded upon the stochastic equation which should drive the dynamic of $Y(t)$ over time. This dynamic is modeled in discrete or continuous time depending on the situation.

---

[2]For example, Excel analytics module allows to generate random outcomes distributed according to different distribution laws, including the normal law. This module, albeit imperfect, may prove to be useful for simple applications.

In continuous time, we use for instance the following diffusion process:

$$dY = \mu(t, Y(t))dt + \sigma(t, Y(t))dW, \tag{26.1}$$

where dW denotes the incremental variation of a standard Brownian motion. $\mu(.)$ (resp. $\sigma(.)$) are two known functions which represent the drift (resp. the volatility) of the process.

We split the interval $(0, T)$ into N evenly spaced intervals of common length $\Delta t = T/N$. N is chosen to be sufficiently large so that this common length remains short (usually a working day or a week).

We define $t_i \equiv i\Delta t$ so that the N periods of equal length Δt are simply denoted by $(t_0, t_1), (t_1, t_2), \ldots, (t_{i-1}, t_i), \ldots, (t_{N-1}, t_N)$. We let $t_0 = 0$ and $t_N = T$.

We simulate $Y(t)$ and $V(t, Y(t))$ on dates $t = t_1 \ldots, t_N$. We hence need to discretize the above Eq. (26.1) which leads to:

$$Y(t_j) - Y(t_{j-1}) = \mu(t_{j-1}, Y(t_{j-1}))\Delta t + \sigma(t_{j-1}, Y(t_{j-1}))\sqrt{\Delta t}\, U(j), \tag{26.2}$$

where $U(j)$ (for $j = 1, \ldots, N$) are independent standard normal random variables. U_j denotes the outcome of $U(j)$ and Yt_j the corresponding value of $Y(t_j)$.

In some cases, the dynamics of $Y(t)$ is directly given by a discrete time process of type (26.2) which permits computing $Y(t_j)$ from the pair $(Y(t_{j-1}), U(j))$.

Any specific path $Y_t]_{t=t_1,\ldots,t_N}$ is computed by generating a sequence of N independent random draws $U_j]_{j=1,\ldots,N}$ from a standard normal distribution and by applying recursively Eq. (26.2).[3] Those random samples U_j may be directly obtained by using an appropriate software (for instance the built-in analytics functions in Excel) or coded by relying upon the Box-Muller formula (see Eq. (26.1)). Besides, each trajectory $Y_{t_j}]_{j=1,\ldots,N}$ corresponds to a trajectory $V(t_j, Y_{t_j})]_{j=1,\ldots,N}$.

In a nutshell, the simulation of *a trajectory* comprises the following steps in sequence of order: (*i*) drawing N values $U_j]_{j=1,\ldots,N}$ from a standard normal law; (*ii*) computing from those values, using Eq. (26.2), the array $Yt_j]_{j=1,\ldots,N}$ constituting one specific path of $Y(t)$; (*iii*) computing the corresponding path $V(t, Y(t))$, in accordance with the scheme below:

$$U_j]_{j=1,\ldots,N} \rightarrow Y_{t_j}]_{j=1,\ldots,N} \rightarrow V(t_j, Y_{t_j})]_{j=1,\ldots,N}.$$

The above procedure will be repeated as much as needed in order to obtain a set of different trajectories: the number of simulations depends on the problem and the methodology to be used and ranges from many thousands up to several millions of simulations.

---

[3] Y_1 is computed with (26.2) from Y_0 (known) and the first random draw U_1; Y_1 thus permits computing Y_2 using (26.2) and U_2, and so on step by step, Y_n is computed from Y_{n-1} and U_n using (26.2).

Example 1

Let us consider a stock s with market price S and a portfolio composed of this stock and derivatives written on s, such as options. We assume a Black–Scholes (BS) universe in which the value $V(t, S(t))$ of this position depends on time t and only one random factor $Y(t) \equiv S(t)$; the function $V(t, S)$ is assumed to be known (option prices are for example given by the BS formula below).

In accordance with BS we assume that $S(t)$ follows the geometric Brownian motion:

$$dS/S = \mu dt + \sigma dW \Leftrightarrow \qquad (26.3\text{a})$$

$$S(t) = S(0)\,e^{\left(\mu - 0.5\sigma^2\right)t + \sigma W(t)} \qquad (26.3\text{b})$$

where $W(t)$ is a standard Brownian motion, μ and σ are known annualized constant parameters, and time is measured in years.

We perform simulations with a weekly time step. The weekly drift and volatility parameters are written respectively $\mu/52$ and $\sigma/\sqrt{52}$.

Suppose that the listed share price is $S(0) = 100\ \text{€}$ on date 0 and that its annualized expected rate of return and volatility are equal to $\mu = 12\%$ and $\sigma = 36\%$ respectively. In terms of weekly parameters, this implies:

drift $= 0.12/52 = 0.0023$; volatility $= 0.36/\sqrt{52} = 0.050$. We can therefore rewrite (26.3b) with these parameters, to express the share price at the end of the nth week, noted in this example $S(n)$:

$$S(n) = 100e^{0.00105n + 0.05\left(U_1 + U_2 + \ldots + U_n\right)}, \qquad (26.4)$$

where the sum of n independent standard normal variables $U_1 + U_2 + \cdots + U_n$ also follows a centered normal distribution with a variance equal to n corresponding to $W(n) = W(1) + (W(2) - W(1)) + \cdots + W(n) - W(n-1)$.

A trajectory $S_n]_{n = 1, \ldots, N}$ is obtained using (26.4) by a sequence of N draws $U_n]_{n = 1, \ldots, N}$. Table 26.1 below contains the first 6 and last 6 trajectories obtained in a simulation of 2000 trajectories over 10 weeks (trajectories $i = 1$ to 6 and 1995 to 2000). A trajectory corresponds to a row; the first table contains 12 sequences of $U^i_n]_{n = 1, \ldots, 10}$ and the second the corresponding 12 sequences $S^i_n]_{n = 1, \ldots, 10}$.

Note that the trajectories go in "antithetic" pairs; thus, the first two trajectories 1 and 2 are linked by the relationship $\underline{U}^2_n = -\underline{U}^1_n$ for $n = 1, \ldots, 10$ (for example $\underline{U}^1_1 = 0.054991$ and $\underline{U}^2_1 = -0.054991$). More generally, from each ith odd trajectory obtained from the 10 draws $U^i_n]_{n = 1, \ldots, 10}$ we calculate the $(i + 1)^{th}$ even trajectory by $U^{i+1}_n = -U^i_n$ for $n = 1, \ldots, 10$. We take advantage of the fact that the $-U^i$ sample is as legitimate as the U^i sample because the standard normal distribution is symmetric with respect to 0. This hence gives 2000 trajectories for

(continued)

Example 1 (continued)

$S(t)$ from only $1000 \times 10 \ U^i_n$ draws (and not 2000×10).[4] This method, known as the "antithetic variables" method, saves computation time and ensures that the empirical mean of the U draws is exactly equal to 0.

Let us now consider a portfolio containing the stock s and derivatives, for example, options, written on s. At each moment t, the value of this portfolio is a function $V(t,S(t))$ assumed to be known. Even if the function V is complex and the distribution of $V(t, S(t))$ has no analytical expression, trajectories of $V(t, S(t))$ can be constructed from those of $\underline{S}(t)$ using the function $V(t, S)$.

26.2.2 Simulations of $Y(T)$ and $V(T, Y(T))$ at Time T (Static Simulations)

We demonstrated above how to compute trajectories $\left(Y_{t_j}, V\left(t, Y_{t_j}\right) \right)_{j=1,\ldots,N}$.

In various cases, especially when it comes to evaluating European options with a maturity T or assessing a VaR (Value at Risk; see Chap. 27) with a maturity T, solely the knowledge of the empirical distribution of $V(T, \underline{Y}(T))$ proves to be useful. It would then be useless to simulate the intermediary values of $\underline{Y}(t)$ for $t < T$ if, as shown in the above examples, simulations of $\underline{Y}(T)$ can be directly computed.[5]

Example 2

Let us consider again the previous example where the only risk factor $Y(t)$ is the price $S(t)$ which follows the geometric Brownian with (annualized) parameters $\mu = 12\%$ and $\sigma = 36\%$.

In contrast to the previous case, we are only interested here in the terminal values $S(T)$ and $V(T, S(T))$. We then use, preferably to (26.4), the following Eq. (26.5):

$$S(T) = 100 \, \mathrm{e}^{\left(\mu - \sigma^2/2\right)T + \sigma\sqrt{T}U} \tag{26.5}$$

where U is a standard normal random variable.

(continued)

---

[4] Naturally, it is awkward and unnecessary to record $-U^i$ antithetics in the random draw table because it is sufficient to calculate two values of $S(n)$ from a single U^i_n draw. We have written the antithetics in the U^i_j table for the sole purpose of easing the explanation.

[5] In some cases, the simulation of $Y(T)$ cannot be directly obtained and necessitates a step by step simulation of $\underline{Y}_{t_j|j\,=\,1,\,\ldots,\,N}$ by applying Eq. (26.2); we then come back to the preceding problem. This approach prevails when the stochastic differential equation (26.1) fails to be solved analytically in contrast with the simple example of the geometric Brownian motion chosen as a toy example for pedagogical purposes.

26.2 Monte Carlo Simulations for a Single Risk Factor

Table 26.1 $U^i_{n,n=1,\ldots,10}$ for $i=1$ to 2000 (2000 trajectories, including 1000 antithetics, of 10 points each)

	1	2	3	4	5	6	7	8	9	10
1	0.054991	−1.813432	0.044959	0.264168	−0.160376	−1.426192	−2.168090	−0.502937	0.121664	0.397167
2	−0.054991	1.813432	−0.044959	−0.264168	+0.160376	+1.426192	+2.168090	+0.502937	−0.121664	−0.397167
3	−0.767845	−1.069466	−0.572245	1.455346	1.671888	−1.825847	0.632415	−1.849021	0.552977	−1.006028
4	0.767845	1.069466	0.572245	+1.455346	−1.671888	1.825847	−0.632415	+1.849021	−0.552977	+1.006028
5	−0.453447	−1.240337	−0.537091	1.765638	−0.620594	1.733133	3.334717	−0.823488	−1.356143	2.183588
6	0.453447	1.240337	0.537091	−1.765638	0.620594	−1.733133	−3.334717	0.823488	1.356143	−2.183588
⋮	⋮	⋮	⋮	⋮	⋮	⋮	⋮	⋮	⋮	⋮
1995	−1.256008	−0.175427	−0.985433	−0.261634	−0.744326	−1.032652	0.158593	2.504312	0.202607	−1.717544
1996	1.256008	0.175427	0.985433	0.261634	0.744326	1.032652	−0.158593	−2.504312	−0.202607	1.717544
1997	0.650646	0.851408	0.495744	−0.322394	−1.099211	−0.352683	0.792108	0.774435	0.764460	−2.529778
1998	−0.650646	−0.851408	−0.495744	0.322394	1.099211	0.352683	−0.792108	−0.774435	−0.764460	2.529778
1999	−0.453276	−0.119198	0.948919	1.007299	0.303355	0.227346	−2.375655	0.189273	−0.100497	−0.260923
2000	0.453276	0.119198	−0.948919	−1.007299	−0.303355	−0.227346	2.375655	−0.189273	0.100497	0.260923

$S^i_{n,n=1,\ldots,10}$ for $i=1$ to 2000 (2000 10-week trajectories)

	1	2	3	4	5	6	7	8	9	10
1	100.38	91.78	92.08	93.40	92.76	86.46	77.66	75.82	76.36	77.97
2	99.83	109.42	109.29	107.97	108.96	117.13	130.68	134.15	133.48	131.00
3	96.34	91.42	88.93	95.75	104.20	95.21	98.38	89.78	92.40	87.96
4	104.02	109.85	113.16	121.83	112.18	123.04	119.33	131.03	127.59	134.32
5	97.86	92.07	89.73	98.12	95.22	103.95	122.94	118.11	110.48	123.36
6	102.40	109.07	112.16	102.79	106.14	97.43	82.56	86.12	92.26	82.80
⋮	⋮	⋮	⋮	⋮	⋮	⋮	⋮	⋮	⋮	⋮
1995	94.01	93.29	88.90	87.84	84.72	80.54	81.27	92.21	93.24	85.66
1996	106.59	107.65	113.20	114.82	119.30	125.75	124.89	110.31	109.31	119.24
1997	103.42	108.03	110.86	109.20	103.47	101.77	105.99	110.29	114.71	101.19
1998	96.90	92.96	90.78	92.36	97.68	99.52	95.76	92.22	88.85	100.94
1999	97.86	97.38	102.22	107.62	109.38	110.75	98.45	99.49	99.10	97.92
2000	97.24	99.04	103.16	103.14	120.01	126.50	138.08	143.93	153.32	141.23

> **Example 2** (continued)
>
> For $T = 10$ weeks $= 0.1923$ years, $(\mu - 0.5\sigma^2)T = 0.0105$ and $\sigma\sqrt{T} = 0.158$ years, we hence have: $S(T) = 100\, e^{0.0105 + 0.158U}$.
>
> We perform for instance 2000 draws of U, $U_i]_{\,i\,=\,1,\,...,\,2000}$, to which their antithetics are associated, are then obtain 4000 simulations of $S(T)$:
>
> $$S(T)_i = 100\, e^{0.0105+0.158U_i}]_{i=1,...,2,000} \text{ and antithetics } S'(T)_i$$
>
> $$= 100\, e^{0.0105-0.158\,U_i}]_{i=1,...,2,000}.$$
>
> We finally compute 4000 corresponding values $V(T, S(T)_i)$ and $V(T, S'(T)_i)$.

26.2.3 Applications

26.2.3.1 Application 1: Calculation of VaR and ES (*See* Chap. 27)

The computation of the VaR of a portfolio whose value on time T is $V(T, S(T))$ is performed from the simulated values of $V(T, S(T))$ and the loss $V(0, S(0)) - V(T, S(T))$. In Example 2 , the VaR (10 days, 5%) corresponds to the most unfavorable outcome among the 4000 simulated values. Moreover, the arithmetic mean of the 200 largest losses corresponds to the Expected Shortfall: ES(10 days, 5%).

26.2.3.2 Application 2: Evaluation of a European Option

Let us evaluate the price $O(0)$ of a European option with a maturity T whose payoff is $V(T, S(T))$.

This may be a non-vanilla option (otherwise the BS formula would apply) whose payoff $V(T, S(T))$ may be represented by a complex, albeit known, formula, whose expectation cannot be expressed under a closed form. Monte Carlo simulations permit to evaluate in a simple manner this option by discounting with the risk-free rate the empirical average of the payoff simulations performed under a risk-neutral dynamic. We proceed as follows:

- We simulate M ($M > 1000$) values $S_i]_{i\,=\,1,\,...,M}$ of $S(T)$, derived from a risk-neutral dynamic and from M Gaussian draws U_i, for instance by using the equation:

$$S_i = S(0)\, e^{\left(r-\sigma^2/2\right)T+\sigma\sqrt{T}U_i}.$$

Recall that in the risk-neutral world, the expectation of the stock price return is equal to the interest rate r (which differs from μ).

26.2 Monte Carlo Simulations for a Single Risk Factor

- We compute the M corresponding payoff values, namely $V(T, S_i)]_{i=1,\ldots,M}$
- We compute the arithmetic mean of those M payoffs and we discount the obtained result at time horizon T using the interest rate r in order to obtain the option value $O(0)$ at current time 0:

$$O(0) = e^{-rT} \frac{1}{M} \sum_{i=1}^{M} V(T, S_i).$$

Example 3

Let us consider the data from Examples 1 and 2 and assume the continuous rate r to be constant and equal to 4%. We aim to evaluate the value O of a European option with a maturity $T = 10$ weeks, written on a stock with a price $S(t)$ and volatility $\sigma = 0.36$ where $V(T, S(T))$ denotes the option payoff.

$S(T)$ simulations are carried out here using the formula:

$$S(T) = 100 \, e^{(r-\sigma^2/2)T+\sigma\sqrt{T}U} = 100 \, e^{-0.004769+0.158\,U}$$

From a draw of 2000 U_i values and their antithetics $-U_i$ we obtain 4000 $S_i]_{i=1,\ldots,4000}$ values from $S(T)$, therefore an array $V(T, S_j)]_{j=1,\ldots,4000}$ of 4000 values. The value of the option will then be estimated at time $t = 0$ by: $O(0) = e^{-0.00769} \frac{1}{4,000} \sum_{j=1}^{4,000} V(T, S_j)$. In fact, the efficiency of this computation can be improved with the *importance sampling* technique (see Sect. 26.4.3).

26.2.3.3 Application 3: Evaluation of a *Path-Dependent Option*

Now suppose that we aim to evaluate a *path-dependent* option, i.e. one whose value depends on the values taken by the underlying asset price between 0 and T (expiry date of the option). We presented this type of option in Chap. 14. In contrast to applications 1 and 2 above, we cannot solely simulate the variable $S(T)$ but need to also simulate the whole path of $S(t)$ between 0 and T.

The method consists in simulating M trajectories of N points for $S(t)$ (as indicated in paragraph 1 above) and to map each trajectory i $(S^i_1, S^i_2, \ldots, S^i_N)$ onto a corresponding discounted payoff $\Psi(S^i_1, S^i_2, \ldots, S^i_N)$ for $i = 1, \ldots, M$ and evaluate the price of this option by the average as follows:

$$\frac{1}{M} \sum_{i=1}^{M} \Psi(S^i_1, S^i_2, \ldots, S^i_N).$$

In practice, the evaluation of such an option when using this methodology requires to perform an especially large number of simulations (the number of instants N to be considered between 0 and T must be very high) in order to avoid a key element driving the option value being missed due to the adopted discretization scheme.

A simple example will be sufficient to illustrate this issue. Let us evaluate a down-and-in barrier call whose activating barrier stands at 110 (assume that the current stock price is 124 and the strike 125). Let us assume that, for a given trajectory, the minimum stock price that would have been observed under a time step of 1 h is 109.90, which would have activated the option, and that the final stock price ends at 134. The terminal value of the call for this given trajectory would have been $(134-125) = 9$. Unfortunately, adopting a time step of 1 day (corresponding to the close stock price), the "actual" minimum was skipped by the simulation which displays a minimal value of (for instance) 110.25. The terminal value for the call is then 0 (instead of 9) for this path, the call having not been activated. As numerous simulated trajectories may meet this issue, the simulated call value may turn out to be significantly lower than its actual value, all the more so because the chosen time step is large, since the actual minimum (continuous) is always smaller than the simulated minimum. In other examples (typically knock-out options), the option simulated value may in contrast be all the larger than the actual value because the chosen time step is large.

Finally, let us remark that the simple methodology exposed here is not adapted to the evaluation of an *American option*. In fact, the simulation is bound to be performed under a forward induction, i.e., the time evolving from the current date to the option maturity T. The various simulated trajectories (forward looking) are not connected to each other, making it therefore impossible to go backward from the option maturity (date on which the option value is known) as in the Cox-Ross-Rubinstein-like binominal trees. In order to successfully apply the Monte Carlo framework, one should be able to determine, for each corresponding step at each future date t, whether or not it is optimal to exercise the option (when in the money). Doing so requires estimating its time value for the remaining time $(T\text{-}t)$. However, there are now methods that permit to incorporate such estimations into the Monte Carlo framework as explained in Sects. 26.4 and 26.5.

26.2.3.4 Application 4: Evaluation of the Greek Parameters of an Option

Let us assume $V(T, S(T))$ to be the payoff of an option to be evaluated. It is possible, once the value $O_1(0)$ has been estimated, to compute its greeks. In order to do so, we use a slightly different underlying value (for instance 101 instead of 100) to obtain the delta, or in case we are willing to evaluate a different sensitivity such as the vega (resp. the rho), we use a different parameter such as the volatility (resp. the interest rate).

A second simulation is hence performed with this modified value, all other parameters used in the first simulation remaining equal, i.e., the number N of periods, the number M of trajectories, and using the same sequences U_i (and antithetic $-U_i$ as well). We thus come up with a second value $O_2(0)$. The greek parameter is then computed by using the formula:

$$\frac{O_2(0) - O_1(0)}{\Delta X}$$

where ΔX corresponds to the change applied to either the underlying asset price (equals to 1 in our numerical example) or the relevant parameter.

26.3 Monte Carlo Simulations for Several Risk Factors: Choleski Decomposition and Copulas

Let us now consider the more general case of a position whose value $V(t, \underline{Y}(t))$ depends on m risk factors $\underline{Y} \equiv (Y_1, \ldots, Y_m)$ which are most often correlated; we cannot, therefore, simulate the outcomes of the different factors $Y_i(t)$ independently. As before, we distinguish the case where the simulation of the trajectories of $\underline{Y}(t)$ between 0 and T is needed from the case where solely the terminal values $\underline{Y}(T)$ and $V(T, \underline{Y}(T))$ are useful. In addition, we will distinguish the case of a Gaussian vector $\underline{Y}(T)$ from the non-Gaussian case.

Recall that it is not sufficient to have all components Y_i to be Gaussian to ensure $\underline{Y}$ to be a Gaussian vector: it is still necessary (and sufficient) to ensure that all linear combinations of Y_i are Gaussian as well.

The drawing of a Gaussian vector $\underline{Y}(T)$ whose components are correlated is grounded on the Choleski decomposition (see § 26.3.1 below). Copulas, presented in § 26.3.2 and § 26.3.3, allow to simulate the outcomes of a *non-Gaussian vector with correlated components*. The simulation of trajectories of a multi-variate diffusion process is described in § 26.3.4.

26.3.1 Simulation of a Multi-variate Normal Variable: Choleski Decomposition

Recall that, in general, the joint density of the components of a Gaussian vector $\underline{X}$ with m components whose expectation vector (also known as mean vector) is $\underline{\mu}$ and the variance-covariance matrix Σ is as follows:

$$\frac{1}{(2\pi)^{m/2}(\det \Sigma)^{1/2}} \exp\left[-0,5\left(\underline{x} - \underline{\mu}\right)'\Sigma\left(\underline{x} - \underline{\mu}\right)\right].$$

The related cumulative distribution is denoted by $N_m\left(\underline{\mu}; \Sigma\right)$. The problem studied below aims at simulating the drawings of such a law.

Let us first present the simple case of a two-dimensional Gaussian vector $\underline{X} \equiv (X_1, X_2)$ whose components have standard deviations equal to σ_1 et σ_2 respectively, expectations equal to μ_1 et μ_2 and are linked by a correlation coefficient ρ (all these different parameters are assumed to be known).

In the simulation of the pair (X_1, X_2), it is therefore impossible to simulate the two components independently because of their correlation.

To obtain correlated drawings of (X_1, X_2) we posit:

$$X_1 = \mu_1 + \lambda_{11} U_1,$$

$$X_2 = \mu_2 + \lambda_{21} U_1 + \lambda_{22} U_2,$$

where U_1 and U_2 are independent standard normal variables.

To ensure U_1 having a standard deviation of σ_1, we must have $\lambda_{11} = \sigma_1$. To ensure having X_2 a standard deviation σ_2 and being correlated with X_1 with ρ as the correlation parameter, the following two conditions must be met: $\lambda_{21}{}^2 + \lambda_{22}{}^2 = \sigma_2{}^2$ and $\lambda_{11} \lambda_{21} = \rho\sigma_1\sigma_2$.

These two conditions determine the two coefficients: $\lambda_{21} = \rho\sigma_2$ and $\lambda_{22} = \sigma_2\sqrt{(1 - \rho^2)}$.

The joint variate (X_1, X_2) hence rewrites as follows:

$$X_1 = \mu_1 + \sigma_1 U_1;$$

$$X_2 = \mu_2 + \rho\sigma_2 U_1 + \sigma_2\sqrt{1 - \rho^2}U_2.$$

It is then sufficient to draw a sample of the two independent standard normal variables U_1 and U_2 to obtain a sample of (X_1, X_2) distributed as required above. $2N$ drawings of a reduced standard normal provide N drawings for the pair (X_1, X_2), or even $2N$ draws using the antithetics of U_1 and U_2.

This method may be extended to simulate any *m-dimensional* Gaussian vector $\underline{X}$ with correlated components. The Choleski decomposition of the assumed known variance-covariance matrix of $\underline{X}$ is then used. Recall that any symmetric positive definite matrix can be written as the product of a *triangular* matrix and its transpose Λ': $\Sigma = \Lambda\Lambda'$.[6]

The vector $\underline{X}$ then writes as: $\underline{X} = \underline{\mu} + \Lambda\underline{U}$.

The m components of $\underline{U}$ are independent random variables and follow a standard normal law. $2N$ simulations of $\underline{X}$ are obtained from $m \times N$ drawings and their antithetics.

The m components of $\underline{X}$ follow the equations below, whose coefficients λ_{ij} are to be determined:

---

[6]With 2 factors as above, we have $\Lambda = \begin{bmatrix} \sigma_1 & 0 \\ \rho\sigma_2 & \sqrt{1 - \rho^2}\sigma_2 \end{bmatrix}$ and $\Lambda' = \begin{bmatrix} \sigma_1 & \rho\sigma_2 \\ 0 & \sqrt{1 - \rho^2}\sigma_2 \end{bmatrix}$, the product of which indeed is: $\Sigma = \begin{bmatrix} \sigma_1^2 & \rho\sigma_1\sigma_2 \\ \rho\sigma_1\sigma_2 & \sigma_2^2 \end{bmatrix}$.

$$X_1 = \mu_1 + \lambda_{11} U_1$$
$$X_2 = \mu_2 + \lambda_{21} U_1 + \lambda_{22} U_2$$
$$\ldots$$
$$X_i = \mu_i + \lambda_{i1} U_1 + \lambda_{i2} U_2 + \cdots + \lambda_{ii} U_i$$
$$\ldots$$
$$X_m = \mu_m + \lambda_{m1} U_1 + \lambda_{m2} U_2 + \cdots + \lambda_{mm} U_m$$

The λ_{ij} are determined step by step (starting with λ_{11})), from the variance-covariance matrix Σ with generic term σ_{ij}, as follows:

$$\lambda_{11} = \sqrt{\sigma_{11}}, \text{(variance of } X_1).$$

λ_{21} and λ_{22} are the solutions to the two following equations:

$\lambda_{21}^2 + \lambda_{22}^2 = \sigma_2^2$ (variance of X_2)
$\lambda_{11}\lambda_{21} = \sigma_{12}$ (covariance of X_1 and X_2)

The generic terms, λ_{i1}, λ_{i2}, ..., λ_{ii} are determined as solutions to the system of i equations:

$$\sum_{j=1}^{i} \lambda_{ij}^2 = \sigma_i^2, \text{ to comply with the assumed variance of } Xi$$

$$\sum_{j=1}^{k} \lambda_{ij}\lambda_{kj} = \sigma_{ik}, \text{ for } k = 1, \ldots, i-1, \text{ in order to respect the assumed } i-1 \text{ covariances}$$

between X_i and X_k.

An example of the application of the above methodology is presented in § 26.3.4.

26.3.2 Representation and Simulation of a Non-Gaussian Vector with Correlated Components Through the Use of a Copula

The method exposed above allows representing and simulating a Gaussian vector with correlated components. Now suppose that we aim to simulate a vector with m non-Gaussian components $(X_1, X_2, \ldots, X_m)$. We note F_j the unconditional (also called marginal) distribution law of X_j.

Recall that the marginal (or unconditional) distribution $F_j(x) = \text{Proba}(X_j \le x))$ is that of X_j in the absence of any assumptions or knowledge concerning the X_k for $k \ne j$. Recall also that the joint distribution $J(x_1, x_2, \ldots, x_m) = \text{Proba } (X_1 \le x_1$ and $X_2 \le x_2 \ldots$ and $X_m \le x_m)$ differs from the simple product $F_1(x_1) F_2(x_2) \ldots F_m(x_m)$ of the marginal laws, except in the particular case when X_j are independent variables.

Except in very specific cases, such as Gaussian distributions,[7] there is no direct and unambiguous way to represent the joint distribution J as a function of the marginal laws F_j, even when we know the correlation coefficients ρ_{jk} between X_j and X_k for all pairs j,k. However, the use of a copula permits to represent the joint distribution from the marginal distributions. The following analysis focuses on Gaussian copulas which are the only ones used in this book. Non-Gaussian copulas are briefly discussed in § 26.3.3.

The use of a Gaussian copula enables to transform a Gaussian vector $(U_1, U_2, \ldots, U_m)$ into a non-Gaussian vector $(X_1, X_2, \ldots, X_m)$, thus obtaining indirectly a representation of the joint distribution of $(X_1, X_2, \ldots, X_m)$, and to carry out simulations of this vector.

To ease the understanding of Gaussian copulas, let us start by considering the one-dimensional case for a purely pedagogical purpose. Let X be any random variable whose distribution law is Φ. We then have for every real number y between 0 and 1: Proba $(\Phi(X) \leq y) \equiv y$.

Indeed, as Φ^{-1} is an increasing function, we have: Proba $(\Phi(X) \leq y) =$ Proba $(X \leq \Phi^{-1}(y)) = \Phi(\Phi^{-1}(y)) = y$.

In the sequel, N denotes the cumulative distribution function of U assumed to obey the standard normal law. We then have $N(u) = \frac{1}{\sqrt{2\pi}} \int_{-\infty}^{u} e^{-\frac{1}{2}z^2} dz$. We will use the previous remark (by letting $\Phi = N$, $y = F(x)$) which implies that Proba $(N(U) \leq F(x)) = F(x)$.

Let us now expose how we can transform a random variable U following a standard normal distribution into a random variable X following any given distribution F.

The correspondence between the respective distributions of U and X is presented in the following proposition:

Proposition

(a) *Let F be the cumulative distribution function of any given law and U a standard normal random variable. The random variable $X = F^{-1}(N(U))$ has F for distribution law.*

(b) *Conversely, let X be a random variable X whose cumulative distribution function is a given F. The random variable $U = N^{-1}(F(X))$ follows a standard normal distribution.*

Proof

(a) Proba$(X \leq x) =$ Proba$(F^{-1}(N(U)) \leq x) =$ Proba$(N(U) \leq F(x)) = F(x)$ (the last equality is derived from a previous remark). Hence Proba$(X \leq x) = F(x)$ and F is indeed the cumulative distribution function of X.

(b) The proof is derived in a similar way as in (a).

---

[7] In the particular case of Gaussian variables, the joint law is expressed analytically: $exp \frac{1}{(2\pi)^{m/2}(det\Sigma)^{1/2}}$ $[-0.5(\underline{x}\text{-}\underline{\mu})'(\underline{x}\text{-}\underline{\mu})]$.

26.3 Monte Carlo Simulations for Several Risk Factors: Choleski Decomposition... 1077

Corollary

Random draws x_i for any given law F are obtained from random draws u_i of a standard normal law $N(0, 1)$ by using the correspondence: $x_i = F^{-1}(N(u_i))$.

Intuitively, this correspondence, also written $F(x_i) = N(u_i) = p_i$, means that x_i and u_i correspond to the *same p_i-quantile* of their respective distribution. For each draw u_i, which corresponds to a p_i-quantile of the law N ($p_i = N(u_i)$), we associate the p_i-quantile x_i of law F according to the succession of the following relations:

$$u_i \rightarrow p_i = N(u_i) \rightarrow x_i = F^{-1}(p_i).$$

It should be noted that this method is grounded on the same principles as the one described at the beginning of the chapter, which consists in obtaining drawings from F using a sample a_i drawn from a uniform law ($x_i = F^{-1}(a_i)$).

In fact, the method is mostly useful when applied, in a more general form, to multi-variate distributions with correlated components. Indeed, it allows to construct a sample drawn from a multi-variate law $(X_1, \ldots, X_m)$ where the marginal law for each component X_j, denoted by F_j, is given and where the different components are correlated.

First, the method consists in constructing a sample $[\underline{u}_i \equiv (u_i^1, \ldots, u_i^m)]_{i = 1, \ldots, N}$ of N draws from a Gaussian vector $(U_1, \ldots, U_m)$ whose U_j components all follow the same marginal law $N(0,1)$ and are linked to each other by a set of given correlation coefficients $\rho_{jk} = \text{Cor}(U_j, U_k)$, called copula correlations and chosen to match the desired correlation between X_j and X_j. In the following, i denotes an element of the (*m-dimensional*) drawing and j or k a (one-dimensional) component. The sample of $\underline{u}_i$ is, for example, obtained by applying the method described in Sect. 26.3.1 (Choleski decomposition).

In a second step, each $\underline{u}_i \equiv (u_i^1, \ldots, u_i^m)$ is mapped onto an $\underline{x}_i \equiv (x_i^1, \ldots, x_i^m)$ with: $x_i^1 = F_1^{-1}(N(u_i^1)), \ldots, x_i^m = F_m^{-1}(N(u_i^m))$. The sample $[(x_i^1, \ldots, x_i^m)]_{i = 1, \ldots, N}$ thus obtained has components distributed from the desired marginal laws F_j and their correlations are a function of the copula correlations $\rho_{jk} = \text{Cor}[U_j, U_k]$.

It should be noted that the correlation coefficient between X_j and X_k is a function of ρ_{jk} but in general slightly differs from the latter (because $\text{Cor}[F_j^{-1}(N(U_j)), F_k^{-1}(N(U_k))] \neq \text{Cor}[U_j, U_k]$) and that ρ_{jk} copula correlations are often obtained from a factor model assumed to govern the variables U_i.

Example 4

Let us consider two random variables T_1 and T_2 distributed according to exponential laws with their marginal distributions being respectively:

$$F_1(t) = \text{Proba}(T_1 \leq t) = 1 - e^{-\lambda_1 t}; \quad F_2(t) = \text{Proba}(T_2 \leq t) = 1 - e^{-\lambda_2 t}.$$

λ_1 et λ_2 are two positive parameters with values generally lying between 0 and 1.

(continued)

Example 4 (continued)

In Chaps. 28 and 30, we explain that such models are used to represent the probability distributions of the first default dates of two debts, occurring within a time interval dt, with marginal default probabilities equal to $\lambda_1 dt$ and $\lambda_2 dt$ respectively (Poisson process of intensities λ_1 et λ_2). We also assume that the two variables T_1 and T_2 are correlated and we account for this correlation by using a Gaussian copula.

We will therefore associate to T_1 and T_2 two random variables U_1 and U_2 distributed according to $N(0,1)$ and linked together by a correlation coefficient ρ.

We proceed in a two-step process as described below.

- First step: we simulate the outcomes of such a pair of standard correlated Gaussian variates (U_1, U_2). We hence write $U_2 = \rho U_1 + \sqrt{1 - \rho^2}\, V$ with V distributed according to $N(0,1)$ and being independent of U_1. We carry out an equal number M *of* independent draws for each of the two standard Gaussian random variable U_1 and V and deduce, for each of the M outcomes pairs (u_1, v) the corresponding values of $u_2 = \rho u_1 + \sqrt{1 - \rho^2}\, v$. The M pairs (u_1, u_2) thus calculated as well as their antithetics $(-u_1, -u_2)$ form a sample of $2M$ elements derived from a bi-variate standard normal distribution of correlation ρ.
- Second step: we associate each pair (u_1, u_2) with the pair (t_1, t_2) such that:

$$1 - e^{-\lambda 1 t 1} = N(u_1), \text{ i.e. } t_1 = -\frac{1}{\lambda_1} \ln(1 - N(u_1)) = -\frac{1}{\lambda_1} \ln(N(-u_1)); t_2 = -\frac{1}{\lambda_2} \ln(N(-u_2)).$$

(or $t_i = -\frac{1}{\lambda_i} \ln(N(u_i))$ using antithetics).

The pairs (t_1, t_2) computed as explained above are assumed to be drawn from the joint distribution of (T_1, T_2). The correlation coefficient between T_1 and T_2 derived from the copula correlation ρ can be estimated "empirically" using the sample of the simulated $2M$ pairs (t_1, t_2). A too large discrepancy between the copula correlation ρ and the "empirical" correlation would necessitate to re-run the simulation with different values of ρ.

For example, a simulation was performed with the following parameters: copula correlation $\rho = 0.3$; $\lambda_1 = 5\%$; $\lambda_2 = 4\%$. Table 26.2 below shows 4 draws (v, u_1) of this simulation which allow to calculate 4 pairs (u_1, u_2), then 4 pairs (t_1, t_2) representing the remaining time lag (expressed in years).

As an indication, in a simulation of 20,000 draws (a very small number), an empirical correlation coefficient between T_1 and T_2 of 0.264 was obtained from a copula correlation ρ equal to 0.3, which could justify performing a second simulation with a coefficient $= 0.34$ between U_1 and U_2.

In Chap. 30, we will use this example to illustrate the valuation of *CDS first to default* and *CDS second to default*.

26.3 Monte Carlo Simulations for Several Risk Factors: Choleski Decomposition...

Table 26.2 Simulation using a copula

v	u_1	$u_2 = 0.3 * u_1 + 0.954 * v$	$N(u_1)$	$N(u_2)$	$t_1 = -\log(N(u_1))/\lambda_1$	$t_2 = -\log(N(u_2))/\lambda_2$
0.41352791	0.09857445	0.42405282	0.53926198	0.66423632	12.3510756	6.81862148
−0.03622972	−1.75234618	−0.56026481	0.03985711	0.28764938	64.4490913	20.766883
−0.7020958	1.59387127	−0.19159533	0.94451757	0.42402962	1.14161989	14.2991996
−0.6094092	−0.94007646	−0.86336226	0.17358915	0.19396911	35.0212792	27.3342723

26.3.3 General Definition of a Copula, and Student Copulas (*)

Let us first introduce the Gaussian copula in a more formal way.

The correspondence between X_i and U_i is written: $X_i = F_i^{-1}(N(U_i))$ and $U_i = N^{-1}(F_i(X_i))$. Let us denote by $N_m(\ .; \rho)$ the standard normal m-dimensional law whose marginal components are linked by the correlation matrix ρ (copula correlation between each pair of marginal components).

The joint distribution of the m variables $(X_1, \ldots, X_m)$ is written as follows:
$$\text{Proba}(X_1 \leq x_1, \ldots, X_m \leq x_m) = \text{Proba}(U_1 \leq N^{-1}(F_1(x_1)), \ldots, U_m \leq N^{-1}(F_m(x_m)),$$
i.e.:

$$\text{Proba}(X_1 \leq x_1, \ldots, X_m \leq x_m) = N_m \left[N^{-1}(F_1(x_1)), \ldots, N^{-1}(F_m(x_m)); \rho \right]. \tag{26.6}$$

Draws $(x_1, \ldots, x_m)$ of this law are obtained from correlated Gaussian draws $(u_1, \ldots, u_m)$ by using the formula: $x_i = F_i^{-1}(N(u_i))$.

By definition, the Gaussian copula is the function of m variables $N_m[N^{-1}(.), \ldots, N^{-1}(.); \rho]$. In other words it is a function C mapping $[0,1] \times \ldots \times [0,1]$ onto $[0,1]$.

Gaussian copulas are only special cases of a broader family. In general, a copula is a function C mapping $[0,1] \times \ldots \times [0,1]$ onto $[0,1]$ satisfying the requirements of a joint distribution of uniform random variables, namely: i) non-decreasing for each of its m marginal components and ii) such that $C(a_1, \ldots, a_{i-1}, 0, a_{i+1} \ldots, a_m) = 0$ and $C(1, \ldots, 1, a_i, 1 \ldots, 1) = a_i$ (i.e. the marginal distribution of the ith component is uniform).

According to Sklar's theorem, the joint distribution of m random variables $X_1, \ldots, X_m$ following any given marginal distributions $F_1(.), \ldots, F_m(.)$ can be expressed with a copula[8] by the following relationship:

$$F(x_1, \ldots, x_m) = C(F_1(x_1), \ldots, F_m(x_m)). \tag{26.7}$$

Remark that: (i) the random variable $F_i(X_i)$ follows a uniform law, (ii) $C(F_1(.), \ldots, F_m(.))$ does not alter the marginal distributions (because $C(1, \ldots, 1, F_i(x_i), 1 \ldots, 1) = F_i(x_i)$), (iii) in the particular case of a Gaussian copula: $C(a_1, \ldots, a_m) = N_m[N^{-1}(a_1), \ldots, N^{-1}(a_m)])$, which implies:

$$F(x_1, \ldots, x_m) = C(F_1(x_1), \ldots, F_m(x_m)) = N_m \left[N^{-1}(F_1(x_1)), \ldots, N^{-1}(F_m(x_m)) \right],$$

in accordance with Eq. (26.6).

Among the non-Gaussian copulas, let us consider the t-copula (or *Student copula*) whose implementation technique is based on the Gaussian copula.

---

[8] Sklar's theorem ensures that, for any set of random variables $X_1, \ldots, X_m$ whose marginal laws are $F_1(.), \ldots, F_m(.)$ and a joint distribution law $F(.)$, there exists a copula $C(\ldots)$ such that $F(x_1, \ldots, x_m) = C(F_1(x_1), \ldots, F_m(x_m))$ for every $(x_1, \ldots, x_m)$ in $\mathbb{R}^m$. In addition, $C(.)$ is unique if the F_i are continuous functions.

A Student copula with f degrees of freedom is constructed as follows:

We start from m standard random variables $U_1, \ldots, U_m$ with marginal laws $N(0,1)$ and whose joint distribution is $N_m(\,.\,; \rho)$. Let us consider as well a random variable Y independent of all U_i and distributed according to a χ^2 with f degrees of freedom.

The Student copula with f degrees of freedom and correlation matrix ρ is the joint distribution of the m variables $t_f(\sqrt{f/Y}U_i)$, where t_f represents the t-Student, univariate distribution with f degrees of freedom.

The drawings of such a law are obtained as follows:

- A drawing y is made from a χ^2 distribution with f degrees of freedom (the Excel functions, for example, permit to perform such a drawing).
- A Gaussian vector $\underline{u}$ is derived from a multivariate normal distribution with correlated components, for example by using a Choleski decomposition.
- We define the *m-dimensional* vector $\underline{z} = \sqrt{f/y}\,\underline{u}$ whose m components $z_i = \sqrt{f/y}\,u_i$ are correlated and whose marginal law is t_f (Student law with f degrees of freedom).
- *The m* components x_i are obtained by the formula $x_i = F_i^{-1}(t_f(z_i))$ and hence a first vector $\underline{x} = F_1^{-1}(t_f(z_1)), \ldots, F_m^{-1}(t_f(z_m))$.
- The above procedure is repeated N times to obtain a sample of N vectors $\underline{x}$.

We can show that the empirical correlations between the X_i obtained using a Student copula increase over the extreme value ranges (whilst they are constant with a Gaussian copula). This property of the correlation on distribution tails (*tail correlation*) is an advantage of the Student copula over the Gaussian copula, since we observe that the correlations of market variables actually increase during financial crises (see Chap. 27).

26.3.4 Simulation of Trajectories

Simulations of the trajectories of the *m*-dimension random vector $\underline{Y}(t)$ between 0 and T must take into account the correlation structure between its m factors as expressed by their variance-covariance matrix $\Sigma(t)$. This matrix is assumed to be known.

Depending on whether the dynamics of $\underline{Y}(t)$ *is* expressed in continuous (Eq. 26.1) or in discrete (Eq. 26.2) time, the generic term of the matrix $\Sigma(t)$ writes:

$$\sigma'_{ij}(t, \underline{Y}(t)) = \frac{1}{dt}\,cov\big(dY_i, dY_j\big), \text{ for } i,j = 1, \ldots, m \text{ in continuous time;}$$

$$\sigma_{ij}(t, \underline{Y}(t)) = \sigma'_{ij}(t, \underline{Y}(t))\,\Delta t \text{ and } \sigma_i = \sqrt{\sigma_{ii}} = \sigma'_i\sqrt{\Delta t} \text{ in discrete time.}$$

In the above formula, and exceptionally, the primes refer to annualized parameters to distinguish them from the parameters corresponding to periods of time of length Δt.

As Monte Carlo simulations require a discretization of continuous processes, the second formula (in discrete time) is used in practice.

In addition, $\rho_{ij}(t,\underline{Y}(t)) = \frac{\sigma_{ij}}{\sigma_i \sigma_j}$ denotes the correlation coefficient between the variations of the factors i and j;

$\mu(t, \underline{Y}(t))$ refers to the instantaneous drift of $\underline{Y}$ between t and $t + \Delta t (\mu = \mu' \Delta t$ if μ' denotes the annualized drift).

As in § 26.2.1, we decompose the time period $(0, T)$ into N sub-periods $\Delta t = T/N$ denoted (t_{j-1}, t_j) $(j = 0, \ldots, N; t_0 = 0$ and $t_N = T)$. Variations of the first component Y_1 between t_{j-1} and t_j are expressed by using the discretized equation of type (26.2) above:

$$Y_1(t_j) - Y_1(t_{j-1}) = \mu_1(t_{j-1}, \underline{Y}(t_{j-1})) + \lambda_{11}(t_{j-1}, \underline{Y}(t_{j-1})) U_1(j),$$
$$\text{for } j = 1, \ldots, N,$$

where the random variables $U_1(j)$ are standard normal and mutually independent variables.

The variations of all Y_i must be correlated with each other. In order to do so, we use the triangular Choleski decomposition as presented in Sect. 26.3.1 above, which, in this context, leads us to posit:

$$Y_1(t_j) - Y_1(t_{j-1}) = \mu_1(.) + \lambda_{11}(.)U_1(j)$$
$$Y_2(t_j) - Y_2(t_{j-1}) = \mu_2(.) + \lambda_{21}(.)U_1(j) + \lambda_{22}(.)U_2(j)$$
$$\ldots$$
$$Y_i(t_j) - Y_i(t_{j-1}) = \mu_i(.) + \lambda_{i1}(.)U_1(j) + \lambda_{i2}(.)U_2(j) + \cdots + \lambda_{ii}(.)U_i(j)$$
$$\ldots$$
$$Y_m(t_j) - Y_m(t_{j-1}) = \mu_m(.) + \lambda_{m1}(.)U_1(j) + \lambda_{m2}(.)U_2(j) + \cdots + \lambda_{mm}(.)U_m(j)$$

where the $U_i(j)$ are standard normal variables, mutually independent and independent of each other and where, for the sake of simplicity, (.) denotes $(t_{j-1}, \underline{Y}(t_{j-1}))$

Two trajectories of N points for $\underline{Y}$ are obtained from $m \times N$ draws of U and their antithetics.

According to the procedure explained in Sect. 26.3.1, $\lambda_{ij}(.)$ are determined from one parameter to another (starting with λ_{11} from the variance-covariance matrix $\Sigma(.)$ of general term σ_{ij}) and are solutions of the following i equations:

$$\sum_{j=1}^{i} \lambda_{ij}^2 = \sigma_i^2, \text{ to comply with the assumed variance of } Y_i(t_j) - Y_i(t_{j-1});$$

$$\sum_{j=1}^{k} \lambda_{ij}\lambda_{kj} = \sigma_{ik}, \text{ for } k = 1, \ldots, i-1, \text{ in order to comply with the assumed covariance}$$

between the variations of $Y_i(t)$ and $Y_k(t)$.

26.3 Monte Carlo Simulations for Several Risk Factors: Choleski Decomposition... 1083

Example 5. Simulations in a Three-Factor Model (Stochastic Price, Interest Rate, and Volatility)

Consider a portfolio composed of tradable debt securities, a stock (or an index) and options written on this stock (or index) whose value depends on three random factors: the interest rate r, the price of the stock S, and the volatility σ of the stock that influences the dynamics of S and of the option prices. These three factors are assumed to follow the following three-dimensional diffusion process governed by the three independent standard Brownian motions W_1, W_2, W_3:

$$dr = a(b - r(t))\, dt + \lambda_{11}\, dW_1 \tag{r}$$

$$dS/S = (r(t) + \theta)\, dt + \lambda_{21}(.)\, dW_1 + \lambda_{22}(.)\, dW_2 \tag{S}$$

$$d\sigma = c(\phi(S(t)) - \sigma(t))\, dt + \lambda_{31}(.)\, dW_1 + \lambda_{32}(.)\, dW_2 + \lambda_{33}(.)\, dW_3 \tag{σ}$$

Let us briefly explain these three equations (r), (S) and (σ).

- Equation (r): The rate $r(t)$ follows an Ornstein-Uhlenbeck process with a mean-reverting parameter a that attracts it to the (long-term) mean value b. λ_{11} is equal to the annualized instantaneous standard deviation σ_r of the interest rate variation. The three parameters a, b and σ_r are assumed to be constant and known. This Gaussian process is the one underlying the Vasicek interest rate model (*see* Chap. 17).
- Equation (S). The equation of the stock return dS/S assumes a constant risk premium θ that is added to the interest rate so that, at any given point in time, the expected profitability of the stock is equal to $r(t) + \theta$. The coefficients $\lambda_{21}(.)$ et $\lambda_{22}(.)$ must be compatible with the volatility $\sigma(t)$ of the stock, which is distributed according to the process represented by the third equation (σ); this condition writes:

$$\lambda_{21}(.)^2 + \lambda_{22}(.)^2 = \sigma^2(t) \tag{26.c1}$$

In addition, the coefficient $\lambda_{21}(.)$ is chosen so that the (negative) correlation between the interest rate r process and the price S process complies with the correlation parameter set to the value ρ_1. This requirement implies that $\sigma_r \lambda_{21}(.) = \rho_1 \sigma_r \sigma(t)$ and hence leads to a second condition:

$$\lambda_{21} = \rho_1 \sigma(t) \tag{26.c2}$$

The two conditions (26.c1) and (26.c2) therefore determine the following values of λ_{21} and λ_{22}:

(continued)

Example 5 (continued)
$$\lambda_{21} = \rho_1 \sigma(t); \quad \lambda_{22} = \sigma(t)\sqrt{1 - \rho_1^2}$$

– Equation (σ). Volatility $\sigma(t)$ follows a process involving a mean-reverting parameter to a (long-term) value $\phi(S(t))$ that is negatively dependent on the stock price level S. We have already pointed out that we observe in practice that the volatility is all the higher because the stock price is lower and that this phenomenon is theoretically explained by the leverage effect (*see* in particular Chap. 12).

The coefficients λ_{31}, λ_{32}, and λ_{33} impose a correlation with r and S determined by two correlations parameters and by the assumed standard deviation of the volatility changes as presented in the three equations below:

$\lambda_{31} = 0$ (interest rate and volatility are assumed to be uncorrelated);
$\lambda_{32}^2 + \lambda_{33}^2 = k^2$ (k represents the instantaneous standard deviation of the volatility, assumed to be given);
$\lambda_{22}\lambda_{32} = \rho_2\, k\, \sigma$ (the correlation parameter between the relative variation of the stock price S and the volatility is assumed to be equal to ρ_2).

We therefore derive: $\lambda_{32} = \dfrac{\rho_2 k}{\sqrt{1-\rho_1^2}}$ and $\lambda_{33} = k\sqrt{\dfrac{1-\rho_1^2-\rho_2^2}{1-\rho_1^2}}$.

The diffusion process grounded upon the three random factors (Brownian motions) which complies with assumed volatilities and correlation parameters thus is as follows:

$$dr = a(b - r(t))dt + \sigma_r\, dw_1$$

$$dS/S = (r(t) + \theta)dt + \rho_1\, \sigma(t)\, dw_1 + \sigma(t)\sqrt{1 - \rho_1^2}\, dw_2$$

$$d\sigma = c(\phi(S) - \sigma(t))\, dt + \frac{\rho_2 k}{\sqrt{1 - \rho_1^2}}\, dw_2 + k\sqrt{\frac{1 - \rho_1^2 - \rho_2^2}{1 - \rho_1^2}}\, dw_3$$

Albeit this three-dimensional process is presented in continuous time, the Monte Carlo simulation is of course based on a discretization of this process:

$$r(t_j) - r(t_{j-1}) = a(b - r(t_{j-1}))\Delta t + \sigma_r\sqrt{\Delta t}\, U_1$$

(continued)

Example 5 (continued)
$$S(t_j) - S(t_{j-1}) = S(t_{j-1})\left[(r(t_{j-1}) + \theta)\Delta t + \rho_1\,\sigma(t_{j-1})\sqrt{\Delta t}U_1\right.$$

$$\left. + \sigma(t_{j-1})\sqrt{\Delta t}\sqrt{1 - \rho_1^2}U_2\right]$$

$$\sigma(t_j) - \sigma(t_{j-1}) = c\left[\phi(S(t_{j-1})) - \sigma(t_{j-1})\right]\Delta t + \frac{\rho_2 k}{\sqrt{1 - \rho_1^2}}\sqrt{\Delta t}U_2$$

$$+ k\sqrt{\frac{1 - \rho_1^2 - \rho_2^2}{1 - \rho_1^2}}\sqrt{\Delta t}U_3.$$

For a numerical application with weekly time-steps ($\Delta t = 1/52 = 0.01923$), the following parameters could be for instance selected:

$$a = 0.18; b = 0.04; \sigma_r = -0.2; \theta = 0.05; \rho_1 = 0.30; c = 0.5; \rho_2 = -0.2; k = 0.08.$$

We may choose $\phi(S(t))$ equal to a constant (for example 0.3) or a decreasing function of $S(t)$, for example:

$$\phi(S(t)) = 0.15 \times \left(1 + \left(\tfrac{2S(0)}{S(0)+S(t)}\right)^2\right);$$ the "normal" annualized volatility is

thus equal to 0.3 for the current stock price $S(t) = S(0)$, but it is multiplied by 2.5 when $S(t)$ converges to zero and divided by 2 when converging towards infinity.

Two trajectories of 52 weekly points will be obtained from 156 draws of U and their antithetics; the simulation of 1000 trajectories over one year, therefore, necessitates 78,000 draws (counting U_j and its antithetic for a single draw).

26.4 Accuracy, Computation Time, and Some Variance Reduction Techniques

Monte Carlo simulations are often very time-consuming. In fact, in most applications, a compromise is required between two contradictory objectives: *accuracy* and comprehensiveness of the empirical information obtained (which we want maximum), which increases with the number of simulations performed; and *computation time* (which we want minimum). An efficient procedure should comply with a sufficient accuracy obtained at the cost of a limited calculation time.

To define more precisely this notion of efficiency and introduce some techniques to increase it, we assume that we try to estimate, by Monte Carlo simulations, a parameter e of the distribution of $V(T, \underline{Y}(T))$. This may be a quantile of this distribution (e.g., when assessing a VaR), an expectation (e.g., when seeking to value an option), or any moment of the distribution $V(T, \underline{Y}(T))$.

Consider the case of a simple Monte Carlo (as opposed to a Monte Carlo incorporating sophisticated techniques some of which are introduced later on). Let M be the number of simulations that lead to an estimator $\hat{e}$ of parameter e. The standard error of $\hat{e}$ is inversely proportional to $\sqrt{M}$: the number of simulations will therefore have to be quadrupled if we want to double the accuracy of $\hat{e}$ (when it is measured with the standard error of $\hat{e}$). However, several techniques can be implemented to improve the efficiency of simulations.

26.4.1 Antithetic Variables

We have already used the technique of antithetic variables consisting in associating to each draw U_j its opposite $-U_j$. This halves the number of draws and ensures that the empirical mean of the sample drawn from the outcomes of U is equal to zero, which can greatly contribute to the efficiency of the estimator $\hat{e}$.

Similarly, in addition to using the antithetics that set the mean to 0, it can be ensured that the empirical standard deviation of the U_j sample is exactly equal to 1. It is sufficient to divide all U_j by their empirical standard deviation ($\sqrt{\frac{1}{M-1} \sum_{j=1}^{M} U_j{}^2}$) to obtain a perfectly standardized sample.

26.4.2 Control Variate

Another technique that improves the efficiency of simulations is the *control variate* technique.

To ease the explanation, we assume that the parameter e to be estimated represents the expectation of $V(T, \underline{Y}(T))$. Let us also assume that there is a random variable $V'(T, \underline{Y}(T))$ strongly correlated with $V(T, \underline{Y}(T))$ whose expectation e' is close to e (although different) and that e' *is known*.

Rather than estimating e from simulations on V using the standard estimator $\hat{e} = \frac{1}{M} \sum_{i=1}^{M} V(T, \underline{Y}_i)$, we would rather use the information we have of e' and estimate the difference $\delta = e - e'$ using simulations on $V - V'$.

Therefore, in its most basic form, the control variable technique is as follows:

- We compute from N simulations $\underline{Y}_i$: $\hat{\delta} = \frac{1}{M} \sum_{i=1}^{M} (V(T, \underline{Y}_i) - V'(T, \underline{Y}_i))$.

26.4 Accuracy, Computation Time, and Some Variance Reduction Techniques 1087

– The estimator of e will hence be: $\hat{e}_1 = e' + \widehat{\delta}$ (rather than $\hat{e} = \frac{1}{M} \sum_{i=1}^{M} V(T, \underline{Y}_i)$).

It is intuitively understandable that sampling errors affecting $\underline{Y}_i$ will affect $V'(T, \underline{Y}_i)$ values in the same proportions as $V(T, \underline{Y}_i)$ without significantly affecting the difference $\widehat{\delta}$ between the two (errors tend, to some extent, to offset each other). Thus, by using the control variate (V', e'), we obtain an estimator $\hat{e}_1$ that can be significantly more efficient than the standard estimator $\hat{e}$. Intuitively again, the improvement of efficiency depends crucially on the degree of correlation between $V(T, \underline{Y}(T))$ and $V'(T, \underline{Y}(T))$. The choice of an "appropriate" control variate, therefore, is key in this exercise.

Application 5 and Example 6

Assume that a Monte Carlo simulation is used to evaluate a *path-dependent call option (a barrier, an average, a lookback, etc.)*. The estimation of the latter requires "tight" simulations of the entire trajectory of the underlying asset price, which are very time-consuming to calculate (see Application 3, Sect. 26.2, § 26.2.3.3 above).

Assume also that a vanilla European call option is written on the same underlying (as the above *path-dependent* option), listed on the market and evaluable with a BS-type closed-formula. The listed price e' *of* this European option is therefore known and equal to the risk-neutral expectation of the discounted payoff $V'(T, S(T))$: $e' = e^{-rT} E(V'(T, S(T)))$.

The price e of the exotic *path-dependent* option is unknown, differs from e' and is equal to the present value of the risk-neutral expectation of its payoff. This payoff depends on the trajectory of $S(t)$.

Recall that the simple Monte Carlo method consists in simulating M trajectories of N points for $S(t)$, associating to each simulated trajectory i $(S^i_1, S^i_2, \ldots, S^i_N)$ a discounted payoff $\Psi(S^i_1, S^i_2, \ldots, S^i_N)$ and estimating the value of this option by the average $\hat{e} = \frac{1}{M} \sum_{i=1}^{M} \psi(S^i_1, S^i_2, \ldots, S^i_N)$.

The use of the control variate $(e', V'(T, S(T)))$ consists in estimating the deviation $\delta = e - e'$ using the empirical mean $\widehat{\delta} = \frac{1}{M} \sum_{i=1}^{M} \times$ $(\psi(S^i_1, \ldots, S^i_N) - e^{-rT} V(T, S^i_N))$ and using $\hat{e}_1 = e' + \widehat{\delta}$ as the price estimator of the exotic option.

Intuitively, the information contained in the (known) price of the European option and relevant to the estimate its exotic counterpart is used to improve the estimation.

We now briefly present an improved and more complex version of the control variate technique. Let us suppose again that we aim to evaluate the expectation e of the distribution of $V(T, \underline{Y}(T))$ and that the random variable $V'(T, \underline{Y}(T))$, correlated with $V(T, \underline{Y}(T))$, has a known expectation e'. We define $\hat{e}'$ as the estimator of $\frac{1}{M} \times \sum_{i=1}^{M} V'(T, \underline{Y}_i)$ of the known parameter e' and $\hat{e} = \frac{1}{M} \sum_{i=1}^{M} V(T, \underline{Y}_i)$ as the standard estimator of the unknown parameter e whose estimate is aimed to be improved.

Let us consider the family of estimators: $\hat{e}_\alpha = \hat{e} - \alpha(\hat{e}' - e')$, where α is a real number.

Note that the estimator $\hat{e}_1$ described above belongs to this family when we set $\alpha = 1$, because $\hat{e}_1$ writes as follows:

$$\hat{e}_1 = e' + \frac{1}{M} \sum_{i=1}^{M} (V(T, \underline{Y}_i) - V'(T, \underline{Y}_i)) = e' + (\hat{e} - \hat{e}') = \hat{e}_\alpha \text{ with } \alpha = 1.$$

Intuitively, the estimator $\hat{e}_\alpha$ is obtained by correcting the standard estimator $\hat{e}$ by a fraction α of the difference $\hat{e}' - e'$ that comes from the sample error. However, in general, the best estimator $\hat{e}_\alpha$ is not obtained with $\alpha = 1$ but with the parameter $\alpha = \alpha^*$ which minimizes the variance of $\hat{e}_\alpha$, namely: $Var(\hat{e}_\alpha) = Var(\hat{e}) + \alpha^2 Var(\hat{e}') - 2\alpha Cov(\hat{e}, \hat{e}')$.

This minimum is attained for $\alpha^* = \frac{Cov(\hat{e}, \hat{e}')}{Var(\hat{e}')}$ which corresponds to the regression coefficient of $\hat{e}$ on $\hat{e}'$.

This regression coefficient may be obtained from a time series of simulated values of $\hat{e}$ and $\hat{e}'$ using the Monte Carlo method. Thus, the best estimator based on the control variate $(V'(.), e')$ is $\hat{e}_{\alpha^*}$ and not $\hat{e}_1$.

26.4.3 Importance Sampling

To explain this methodology, let us take the example of valuing an option with a payoff $V(T, S(T))$. The number of trajectories resulting in a null value for this option is typically large, or even very large, if the option is deep out-of-the-money. In the case of a European option, these (very) numerous trajectories with a zero terminal value do not contribute to the current price of the option, and computing their outcome (equal to zero) is, therefore, a waste of computing time. For this reason, it seems reasonable to try to obtain trajectories for which the option expires in-the-money.

Let F be the unconditional (risk-neutral) probability distribution of the price $S(T)$ of the underlying at T. Let us take the case of a call and assume to be known analytically the probability p that $S(T)$ is greater than the strike K (this analytical formula may be accurate[9] or be an approximation of a non-analytical formula). Then

---

[9] For example $N(d_2)$ in the Black–Scholes (BS) model.

$G = [F - (1 - p)]/p$ is the probability distribution of $S(T)$ conditional on $S(T)$ being greater than K. The samples are drawn from the law of G and not the law of F when performing the simulations. By multiplying the result (the average of the values obtained for the call), discounted at the risk-free rate, by the probability p, we obtain an estimate of the value of the option.[10] For the same calculation time, this procedure allows to simulate many more useful trajectories (e.g., in-the-money), and thus to significantly improve the efficiency of the estimator.

26.4.4 Stratified Sampling

Stratified sampling is an effective method of drawing samples from a population with relatively *heterogeneous* sub-populations (strata). It consists of drawing independent samples from each sub-population, whose members are relatively *homogeneous* and have been clustered together. This procedure may improve the representativeness of the full sample and reduce the estimation error: an appropriately weighted average (e.g., in proportion to the relative size of sub-populations) of each sub-sample mean displays less variance than the arithmetic mean of a single sample drawn from the entire population.

Applied to the problem of asset valuation, this method splits the probability distribution of the asset price into different intervals, draws samples in each of them and calculates the average. If the number of intervals is relatively high, the size of each interval being hence relatively small, this average may be deemed as representative of the entire interval. When sampling in an interval, this representative value is always considered. For example, in the case of a standard normal distribution $N(0,1)$ split into n equally probable intervals, the representative value for the ith interval is calculated by $N^{-1}((i - 0,5)/n)$ where $N^{-1}(x)$ may be obtained by a simple numerical procedure. For example, if we retain $n = 4$ probability intervals, we obtain $N^{-1}(0,125)$, $N^{-1}(0,375)$, $N^{-1}(0,625)$, and $N^{-1}(0,875)$, the latter term being approximately equal to 1.15. However, it should be noted that in general it is not optimal to choose identical quantiles, as decreasing computation time can be obtained by taking larger quantiles for the distribution tails.

26.5 Monte Carlo and American Options

We already mentioned that Monte Carlo simulations in their standard form described in this chapter are not suitable for evaluating American options (Sect. 26.2, § 26.2.3.3). The reason is that the simulated trajectories are "dissociated" and therefore *backward* induction of the option value (i.e. the calculation of a discounted expectation from T to 0), which is doable in the case of a recombining tree, is not feasible

---

[10] The call value is equal to $C = e^{-rT}\{E[S(T)|\ S(T) \geq K] - K\} \times Proba[S(T) \geq K]$, and, in the BS model, $Proba[S(T) \geq K]$ is equal to $N(d_2)$.

when dealing with dissociated trajectories.[11] Therefore, at a date $t_i < T$, Monte Carlo simulation cannot simply determine the continuation value that must be compared to the intrinsic value in order to decide whether to exercise (kill) the option or keep it alive.

This is why the two competing numerical approaches, namely trees and finite difference methods, have been considered for a long time as the only ones suitable for valuing American options. However, alternative approaches grounded on Monte Carlo simulations have been developed to evaluate these options. The seminal method was proposed by Carrière (1996) and then developed by Longstaff and Schwartz (2001) to become a popular approach for valuing American Options. It combines least square regressions with a Monte Carlo simulation. After an introduction to the general approach to American Option valuation, we describe the regression method and then present and compare the Carrière and Longstaff–Schwartz respective approaches.

26.5.1 General Description of the Problem and Methodology

The idea is to estimate, at each date t_j on which S takes on the value S_j, a *continuation value* $V_c(t_j, S_j)$. This value may be compared to the exercise value $V_e(S_j)$ (or intrinsic value[12]), to decide whether to exercise the option or not.

The first step of the implementation involves the simulation of M risk-neutral trajectories for the underlying: $(S^i_1, S^i_2, ..., S^i_N)_{i = 1, ..., M}$; S^i_j representing the outcome of $S(t_j)$ on path i. To each path i corresponds a trajectory of the exercise values $(V_e(S^i_1), V_e(S^i_2), ..., V_e(S^i_N))$. For the sake of computational purposes, the option is deemed as if it were *Bermuda*, i.e., exercisable exclusively on certain dates, here at the N dates t_j considered in the simulation.

The value of the American option on date t_j, denoted by $O(t_j, S_j)$, is such that:

$$O(t_j, S_j) = \text{Max}\left[V_c(t_j, S_j), V_e(S_j)\right] \tag{26.8}$$

In addition:

$$V_c(t_j, S_j) = e^{-r\Delta t} \, \text{E}\left[O(t_{j+1}, S(t_{j+1})) \mid I_j\right] \tag{26.9a}$$

where $\Delta t = t_{j+1} - t_j = T/N$ is the duration of the simulation step and $\text{E}[\, . \mid I_j]$ is the conditional risk-neutral (or forward-neutral) expectation depending on the information I_j available at time t_j.

---

[11] Indeed, the possibility of exercising the option at any time requires calculating a conditional expectation on each date and for each trajectory, thus performing new drawings of trajectories at each time step. This is impossible in practice, even if there is a finite number of possible exercise dates.

[12] For example, Max $(K - S(t_j), 0)$ in the particular case of an American vanilla put, but the method applies to American options that are not necessarily vanilla.

26.5 Monte Carlo and American Options

In the case of an option whose price depends only on the last value of a single underlying, we write:

$$V_c(t_j, S_j) = e^{-r\Delta t}\, \mathrm{E}\big[O\big(t_{j+1}, S(t_{j+1})\big) \mid S(t_j) = S_j\big] \tag{26.9b}$$

Equations (26.8) and (26.9a) serve as the basis for the different approaches which vary from the way the conditional expectation $\mathrm{E}[O(t_{j+1}, S(t_{j+1})) \mid S(t_j) = S_j]$ is computed. Regardless of the approach chosen (e.g., Carrière or Longstaff–Schwartz) the option price is determined by backward induction (i.e., from date t_N to date 0), in the same vein as in the binomial tree method.

26.5.2 Estimation of the Continuation Value by Regression (Carrière, Longstaff and Schwartz)

Recall that any conditional expectation $\mathrm{E}[\,.\mid I_j]$ is a random variable which, by construction, is a function of the information I_j available at time t_j.[13]

Let us posit for the continuation value defined in Eq. (26.9b) a linear regression form with $q + 1$ regressors: $1, \Phi_1(S_{N-1}), \ldots, \Phi_q(S_{N-1})$[14]:

$$E\left[O\big(t_{j+1}, S\big(t_{j+1}\big)\big)|S(t_j) = S_j\right] = a_j + \sum_{k=1}^{q} \beta_{k,j} \Phi_k(S_j)$$

From a mathematical perspective, the choice of the number q of explanatory variables and the form chosen for the functions Φ_k, as well as the estimation method, should always balance accuracy (which increases with the number of regressors) and speed (which decreases with the number and complexity of regressors). Under its simplest expression, we may assume for instance $q = 2$ (3 regressors), $\Phi_1(S_j) = S_j$, and $\Phi_2(S_j) = (S_j)^2$ (see below, just after Fig. 26.1).

Once chosen the number of regressors as well as the corresponding functions Φ_k, we need to estimate in practice the value $V_c(t_j, S_j)$ at each time t_j.

Let S^i_j be the value at time t_j of the ith simulated trajectory of S. Let M be the number of simulated trajectories.

We start with terminal date t_N; each trajectory has a terminal value for the option $O(S^i_N)$ equal to its payoff.

On date t_{N-1} the continuation value is estimated only in the case where the option is *in-the-money* on t_N. Namely, for those S^i_{N-1} outcomes where the option is *out-of-*

---

[13] More precisely, in Eq. (26.9b), I_j is a vector space generated by the single random variable $S(t_j)$. As such, any linear combination of $S(t_j)$ would be suited to express the conditional expectation in this equation.

[14] To ease the exposition, we omit the almost surely *("a.s.")* restriction on all conditional expectation equalities.

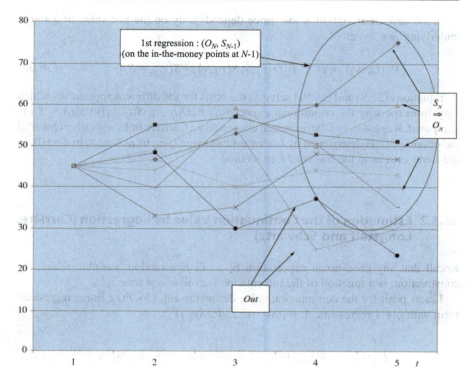

Fig. 26.1 Estimation of the continuation value by regression

the-money the exercise value $V_e(S_{N-1})$ of the option is zero and hence the value of the option equals obviously its continuation value.

To each simulated trajectory i corresponds a pair $(S^i_{N-1}, O(S^i_N))$. We thus have M pairs out of which M' corresponds to situations where the option is *in-the-money* at t_N. These M' pairs allow to perform a cross-sectional regression "explaining" the M' outcomes $O(t_N, S^i_N)$ (where i ranges from 1 to M') using the $q + 1$ regressors $1, \Phi_1(S^i_{N-1}), \ldots, \Phi_q(S^i_{N-1})$. We have:

$$E[O(t_N, S(t_N)) | S(t_{N-1}) = S_{N-1}] = \alpha_{N-1} + \sum_{k=1}^{q} \beta_{k, N-1} \Phi_k(S_{N-1}).$$

In other words, the cross-sectional regression corresponds to the M' options values outcomes $O(t_N, S(t_N))$ which are regressed on their corresponding *outcomes* M of S_{N-1}.

Figure 26.1 illustrates this step of the process for an American call on a futures contract, strike 40 with $M = 8$ and $N = 5$.

26.5 Monte Carlo and American Options

Let us consider 3 regressors ($q = 2$): 1, $\Phi_1(S_{N-1}) = S_{N-1}$ and $\Phi_2(S_{N-1}) = (S_{N-1})^2$, and estimate the corresponding 3 regression coefficients by using the least square regression technique.

We will therefore choose a_{N-1}, $\beta_{1,N-1}$ and $\beta_{2,N-1}$ which minimize the sum:

$$\sum_{i=1}^{M'} \left[O\left(S^i_N\right) - a_{N-1} - \beta_{1,N-1} S^i_{N-1} - \beta_{2,N-1} \left(S^i_{N-1}\right)^2 \right]^2$$

Let us denote by $\widehat{a}_{N-1}$, $\widehat{\beta}_{1,N-1}$ and $\widehat{\beta}_{2,N-1}$ the estimators of the corresponding parameters a_{N-1}, $\beta_{1,N-1}$ and $\beta_{2,N-1}$.

Thus, on date t_{N-1} on the ith trajectory where $S = S^i_{N-1}$, the continuation value $V_c(S^i_{N-1}, t_{N-1}) = [\widehat{a}_{N-1}, + \widehat{\beta}_{1,N-1} S^i_{N-1} + \widehat{\beta}_{2,N-1} (S^i_{N-1})^2] e^{-r\Delta t}$ will be assigned. If the exercise value $V_e(S^i_{N-1}, t_{N-1})$ is larger than $V_c(S^i_{N-1}, t_{N-1})$, the option is early exercised for trajectory i at time t_{N-1}.

26.5.3 Overview of the Carrière Approach

Carrière (1996) starts from the technique described in Sect. 26.5.2 above and iterates the same procedure on date t_{N-2}, t_{N-3}, t_{N-4}, ..., until the initial date $t_0 = 0$.

For the sake of clarity, let us briefly sketch the computation performed on date t_{N-2}. We start from M pairs $(S^i_{N-2}, O(S^i_{N-1}))$ and carry out a cross-sectional regression explaining the outcomes $O(S^i_{N-1})$ (where i ranges from 1 to M) using the same regression model, namely $\Phi_1(S_{N-2}) = S_{N-2}$ and $\Phi_2(S_{N-2}) = (S_{N-2})^2$:

$$E[O(t_{N-1}, S(t_{N-1})) \mid S(t_{N-2}) = S_{N-2}] = a_{N-2} + \beta_{1,N-2} S^i_{N-2} + \beta_{2,N-2} \left(S^i_{N-2}\right)^2.$$

We determine in a similar manner a_{N-2}, $\beta_{1,N-2}$ and $\beta_{2,N-2}$ which minimize the sum:

$$\sum_{i=1}^{M'} \left[O\left(S^i_N\right) - a_{N-2} - \beta_{1,N-2} S^i_{N-2} - \beta_2\left(S^i_{N-2}\right)^2 \right]^2$$

Finally, $\widehat{a}_{N-2}$, $\widehat{\beta}_{1,N-2}$ and $\widehat{\beta}_{2,N-2}$ are the estimators of the corresponding parameters a_{N-2}, $\beta_{1,N-2}$ and $\beta_{2,N-2}$.

What is then the difference with the previous regression? Recall that at time t_{N-1}, we come up with an array of M estimated options values $(O(S^1_{N-1}), ..., O(S^i_{N-1}), ..., O(S^M_{N-1}))$ whose components are either equal to $V_c(S^i_{N-1}, t_{N-1})$ or $V_e(S^i_{N-1}, t_{N-1})$. Recall that for each trajectory $i*$ where $O(S^{i*}_{N-1}) = V_c(S^{i*}_{N-1}, t_{N-1})$, this latter value stems itself from the regression and hence approximated whilst for trajectories $j*$ where $O(S^{j*}_{N-1}) = V_e(S^{j*}_{N-1}, t_{N-1})$ the option value is "exact" as it corresponds to the option pay-off.

In other words, some of the "explained" variables where $O(S^j_{N-1}) = V_c(S^j_{N-1}, t_{N-1})$ are themselves the outcomes of the previous regression at time t_{N-1} and hence subject to the choice of the regression form (here 3 regressors and a simple polynomial function) and the number of simulations (M). They are consequently subject to estimation errors. In these cases, each estimation error made on $V_c(S^j_{N-1}, t_{N-1})$ is then "diffused" backwards to time t_{N-2} then t_{N-3}, t_{N-4}, ..., until the initial date $t_0 = 0$. Intuitively, doing so increases the variance of the estimated option value at time $t = 0$.

Carrière shows that his approach systematically overestimates option values. He also presents some variance reduction techniques which go beyond the scope of this book.

26.5.4 Introduction to Longstaff and Schwartz Approach

Longstaff and Schwartz (2001) build on Carrière's approach and introduce a specific feature to avoid the propagation of the estimation error of the option value while going backward.

Let τ_m[15] be for each simulated trajectory m, the first instance where the early exercise value is larger than the continuation value, namely $V_e(S^m \tau_m, \tau_m) > V_c(S^m \tau_m, \tau_m)$. In other words, at τ_m it is optimal to kill the option early.

We start the algorithm backward by letting $\tau_m = t_N$, i.e. at maturity the exercise value equals the continuation value. Let us determine at time t_{N-1} exactly in a similar manner as explained above the estimators $\widehat{a}_{N-1}$, $\widehat{\beta}_{1,N-1}$ and $\widehat{\beta}_{2,N-1}$ of the corresponding parameters a_{N-1}, $\beta_{1,N-1}$ and $\beta_{2,N-1}$.

Any trajectory i, where $O(S^i_{N-1}) = V_c(S^i_{N-1}, t_{N-1}) > V_e(S^j_{N-1}, t_{N-1})$ (estimated by the results of the regression), corresponds to a situation where the early exercise is not optimal. The optimal exercise date hence remains t_N for the corresponding trajectory i and we set $\tau_i = t_N$ and $O(S^i \tau_i) = V_e(S^i_N, t_N)$ (i.e., the option payoff at maturity).

Any trajectory j, where $O(S^j_{N-1}) = V_e(S^j_{N-1}, t_{N-1}) > V_c(S^j_{N-1}, t_{N-1})$, corresponds to a situation where the early exercise at time t_{N-1} is optimal. We hence set $\tau_j = t_{N-1}$ and $O(S^j \tau_j) = O(S^j_{N-1}) = V_e(S^j_{N-1}, t_{N-1})$.

At time t_{N-2} we have in a similar manner as at time t_{N-1}:

$$\mathrm{E}[O(t_{N-1}, S(t_{N-1})) \mid S(t_{N-2}) = S_{N-2}] = a_{N-2} + \beta_{1,N-2} S^i_{N-2} + \beta_{2,N-2} \left(S^i_{N-2}\right)^2$$

By the law of iterated conditional expectations, we can write:

$$\mathrm{E}\left[\mathrm{E}\left[e^{-r\Delta t} O(t_N, S(t_N)) \mid S(t_{N-1}) = S_{N-1}\right] \mid S(t_{N-2}) = S_{N-2}\right]$$
$$= a_{N-2} + \beta_{1,N-2} S^i_{N-2} + \beta_{2,N-2} \left(S^i_{N-2}\right)^2$$

---

[15] τ_m is by construction a random variable called "stopping time."

26.5 Monte Carlo and American Options

Thus, the explained variables $O(t_{N-1}, S(t_{N-1}))$ are equal to $\mathrm{E}[e^{-r\Delta t} O(t_N, S(t_N)) \mid S(t_{N-1}) = S_{N-1}] = \mathrm{E}[e^{-r\Delta t} O(S^i \tau_i) \mid S(t_{N-1}) = S_{N-1}]= \mathrm{E}[e^{-r\Delta t} V_e(S^j_N, t_N) \mid S(t_{N-1}) = S_{N-1}]$ where early exercise at time t_{N-1} is not optimal.

The explained variables $O(t_{N-1}, S(t_{N-1}))$ are equal to $O(S^j \tau_j) = V_e(S^j_{N-1}, t_{N-1})$ for the trajectories where anticipated exercise at time t_{N-1} is optimal.

If we look closely at both expressions we notice that, in contrast to Carrière's approach, none of the explained variables $O(t_{N-1}, S(t_{N-1}))$ is using any continuation value V_c determined by the regression. In other words, there is no propagation error due to V_c estimation. The only error which is made by the regression bears on the early exercise date and not on the value V_e which is equal to the pay-off on the early exercise date. We then understand intuitively that this algorithm converges faster and the option estimator has less variance.

$O(S_{N-2})$, $O(S_{N-3})$, $\ldots$ are estimated by repeating the same above procedure.

At time t_0, we come up for each trajectory with 0, 1 or several potential situations of early exercise, i.e. occurrences for which $V_e > V_c$. In trajectories that experience such a situation on several dates, the option is presumed to be exercised the first time its exercise value exceeds its continuation value, i.e., on date τ_i for trajectory i. In the absence of an early exercise situation, the option is exercised on date $t_N = T$ if and only if its exercise value is positive on that date.

Figure 26.2 represents, in the case of a call on a futures contract, the four possible scenarios:

- Trajectory #0: no early exercise and an option that expires out-of-the-money $V_e(S_{\tau_0}) = V_e = 0$.
- Trajectory #1: 1 single early exercise on trajectory 1 on point a; we hence set $\tau_1 = t_3$ and $V_e(S_{\tau_1}) = V_e(S^1_3, t_3)$.
- Trajectory #2: 2 early exercise dates when $V_e > V_c$ on trajectory 2 (at t_1 and t_2 on points b and c); we hence set $\tau_2 = t_1$ according to the definition of τ_2 and we have $V_e(S_{\tau_2}) = V_e(S^2_1, t_1)$.
- Trajectory #3: 1 single "degenerate" early exercise date only at maturity on point d $\tau_3 = t_4$ and we have $V_e(S_{\tau_3}) = V_e(S^3_4, t_4)$ (at this stage, the dotted curve "Exercise boundary" plays no role).

More broadly, for M simulated trajectories the option value at time 0 is estimated (by default) by: $\frac{1}{M} \sum_{i=1}^{M} e^{-r\tau_i} V_e(S_{\tau_i})$.

The convergence of this algorithm towards the true price of the American option has been verified, but its speed and accuracy depend on the choice made for the functions Φ_i (the polynomial form being the simplest one[16]), their number q, as well as the number M of simulated paths and N of simulation dates (discretization scheme).

---

[16] Several polynomial forms are possible, such as Legendre and Laguerre polynomials.

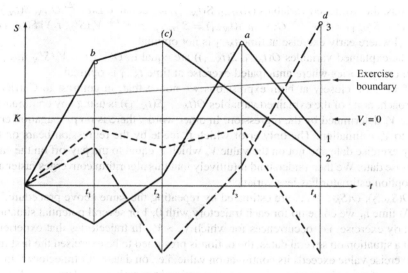

Fig. 26.2 Longstaff-Schwartz approach

Alternative approaches to the regression have been proposed for calculating the conditional expectation $E[O(t_j + 1, S(t_j + 1)) \mid I_j]$. Rogers (2002) for instance relies on the Black–Scholes price of the European equivalent of the American option being valued.

Another generic class of approach is grounded on a parameterization of the ("optimal") exercise frontier. The exercise frontier is a function of time representing the critical value $S_e(t)$ below which a put (resp., above which a call) is exercised on date t. In Fig. 26.2, we have briefly represented this exercise frontier in dotted lines (it is in fact not linear in spite of the stylized representation of our figure). We do not develop this approach here and we limit ourselves to noting that once this efficient frontier is determined, the simulation easily yields the option price. Indeed, it then makes it possible to determine, for each trajectory i, the first date τ_i on which this frontier is reached and to evaluate the option accordingly as explained previously, by using the following formula: $\frac{1}{M}\sum_{i=1}^{M} e^{-r\tau_i} V_e(S_{\tau_i})$.

The procedure may be generalized in the case of an option whose value depends on several underlying processes. Let us consider for example the case of an option whose value depends on two processes $S(t)$ and $X(t)$ (exchange option, underlying asset price and interest rates both stochastic, etc.).

Equations (26.9a and 26.9b) then become:

$$V_c(t_j, S_j, X_j) = e^{-r\Delta t} \, \mathrm{E}\big[O(t_{j+1}, S(t_{j+1}), X(t_{j+1})) \mid S(t_j) = S_j, X(t_j) = X_j\big].$$

The conditional expectation can be calculated with several regressors that are functions of the two variables S and X, for example, X, Y, X^2, Y^2, and XY.

26.5 Monte Carlo and American Options

Example 7

The following numerical example is very similar to the one presented by Longstaff and Schwartz (2001).

The aim is to evaluate an American put option with a three-year maturity, which may be exercised early at the end of years 1, 2, or 3. The current price of the underlying is 1.0, and the strike of the put is set to 1.1.

The procedure is illustrated for 10 trajectories ($M = 10, N = 3$) presented in Table 26.3. This is a purely illustrative example because simulations are typically performed over several tens of thousands of trajectories and hundreds of dates.

- Let us first consider date 3. The put expires *in-the-money at the* end of the five trajectories 3, 4, 6, 7, and 9. It should be noted that, for trajectory 9, it will certainly be exercised on the maturity date, because its intrinsic value is constantly negative on previous dates and it will never be exercised on paths 2, 5, and 10 (because it is *out-of-the-money* on all dates).
- Then consider date 2 on which the put is *in-the-money* on trajectories 1, 3, 4, 6, and 7. We therefore have five pairs (S_2, P_3): (1.08, 0), (1.07, 0.07), (0.97, 0.18), (0.77, 0.20), (0.84, 0.09) to perform the regression. The least squares estimation performed with regressors 1, S and S^2 gives as continuation value at the end of period 2 (conditional on $K > S_2$): $V_c(2) = e^{-0.06}$ $P_3 = -1.07 + 2.983\ S_2 - 1.813(S_2)^2$ which gives the values $V_c(2)$ of the following Table 26.4. Note that, for clarity, in all Tables 26.4 to 26.7, the option values appear in bold in case of early exercise. Comparison of these continuation values with the exercise values $V_e(2)$ indicates that the exercise takes place on trajectories 4, 6, and 7.

Table 26.5 shows the option payoffs assuming that they could only be exercised on dates 2 and 3 and their corresponding present value on date 2, over the 10 trajectories.

- Let us now consider date 1 where the put is *in-the-money* on trajectories 1, 4, 6, 7 and 8; we still have 5 pairs (S_1, P_2) to perform the regression: (1.09, 0), (0.93, 0.13), (0.76, 0.33), (0.92, 0.26), (0.88, 0). P_2 is equal to the present value of the subsequent payoffs (0 or 1 payoff per trajectory) on early exercise dates. Careful readers will notice that, to calculate P_2, option payoffs (Table 26.5) are used rather than Max $(V_c(2), K-S_2)$ (Table 26.4) so as to minimize the estimation bias (see Longstaff and Schwartz). Thus, on the first trajectory $P_2 = 0$, and not 0.0369, is selected.

This regression gives:
$V_c(1) = e^{-0.06}\ P_2 = 2.038 - 3.335\ S_1 + 1.356(S_1)^2$, which results in Table 26.6:

(continued)

> **Example 7** (continued)
>
> Table 26.7 summarizes the payoffs generated on the 10 trajectories and the 3 dates, the option being exercised "optimally" on date 1 for trajectories 4, 6, 7, and 8 and on date 3 for trajectories 3 and 9; there is no exercise on trajectories 1, 2, 5, and 10.[17]
>
> The estimated value of the put is then equal to the empirical expectation of the present value of its payoffs, i.e.:
>
> $$\frac{1}{10}\left[e^{-0.06}(0.17 + 0.34 + 0.18 + 0.22) + e^{-0.18}(0.07 + 0.10)\right] = 0.100$$

Table 26.3 Trajectories of the underlying price and the terminal value of the put

Dates trajectories	0	1	2	3	Terminal value of the put
1	1.00	1.09	1.08	1.34	0
2	1.00	1.16	1.26	1.54	0
3	1.00	1.22	1.07	1.03	0.07
4	1.00	0.93	0.97	0.92	0.18
5	1.00	1.11	1.56	1.52	0
6	1.00	0.76	0.77	0.90	0.20
7	1.00	0.92	0.84	1.01	0.09
8	1.00	0.88	1.22	1.34	0
9	1.00	1.15	1.12	1.00	0.10
10	1.00	1.10	1.12	1.15	0

Table 26.4 Analysis of the early exercise decision at $t = 2$

Path	S_2	$V_e(2)$	$V_c(2)$	$\max(V_c(2), K\text{-}S_2)$
1	1.08	0.02	0.0369	0.0369 (continue)
3	1.07	0.03	0.0461	0.0461 (continue)
4	0.97	0.13	0.1176	**0.13** (early exercise)
6	0.77	0.33	0.1520	**0.33** (early exercise)
7	0.84	0.26	0.1565	**0.26** (early exercise)

---

[17] Note that the absence of exercise is easily explained for trajectories 2, 5 and 10 because the put is constantly *out-of-the-money*; this is not the case on trajectory 1 where it is *in-the-money* on dates 1 and 2. However, on these dates it is optimal not to exercise in the expectation of a higher subsequent gain (the continuation value is higher than the intrinsic value as shown in Tables 26.5 and 26.6).

26.6 Summary

Table 26.5 Payoffs at $t = 2$ and 3 in absence of exercise at $t = 1$

Dates trajectories	Present value in $t = 2$	Payoff 2	Payoff 3
1	0		
2	0		
3	0.0659		**0.07**
4	0.13	**0.13**	
5	0		
6	0.33	**0.33**	
7	0.26	**0.26**	
8	0		
9	0.10		**0.10**
10	0		

Table 26.6 Analysis of the exercise decision at $t = 1$

Path	S_1	$V_e(1)$	$V_c(1)$	Max($V_c(1)$, K-S_1)
1	1.09	0.01	0.0139	0.0139 (continuation)
4	0.93	0.17	0.1092	**0.17** (early exercise)
6	0.76	0.34	0.2866	**0.34** (early exercise)
7	0.92	0.18	0.1175	**0.18** (early exercise)
8	0.88	0.22	0.1533	**0.22** (early exercise)

Table 26.7 Option payoffs on the three dates and the 10 trajectories

Dates trajectories	Payoff 1	Payoff 2	Payoff 3
1			
2			
3			**0.07**
4	**0.17**		
5			
6	**0.34**		
7	**0.18**		
8	**0.22**		
9			**0.10**
10			

26.6 Summary

- Monte Carlo simulation (MCS) requires first drawing a random sample from a given distribution function.
- When A is uniform on $(0, 1)$ the random variable $F^{-1}(A)$ has F as distribution: Proba$[F^{-1}(A) \leq x] = F(x)$. We can thus *build a sample from any given distribution function F from a random sample drawn from a uniform* variable.
- Also, random samples of normal distributions can be built using the Box-Muller technique. Many commercial softwares (Excel, ...) allow generating random samples drawn from standard distributions.

- We often consider a portfolio or a security whose value $V(t, \underline{Y}(t))$, depends on time and on m random factors $\underline{Y}(t)$. MCS is useful when the probability distribution of $V(t, \underline{Y}(t))$ fails to be represented by a closed formula and its empirical distribution must hence be simulated.
- A large sample is drawn from the probability distribution of $\underline{Y}(t)$ and mapped onto the corresponding values $V(t, \underline{Y}(t))$, building hence the empirical distribution $V(t, \underline{Y}(t))$. Depending on the problem, either a large number M of *paths* $[\underline{Y}_t, V(t, \underline{Y}_t)]_{t=t_1,\dots,t_N}$, or only M values of $(\underline{Y}(T), V(T, \underline{Y}(T))$ at a single time T, will be generated.
- Simulating paths of a *univariate* $Y(t)$ is grounded upon the stochastic equation which drives $Y(t)$ over time, for instance, a diffusion process: $dY = \mu(t, Y(t)) dt + \sigma(t, Y(t))dW$, where $\mu(.)$ and $\sigma(.)$ are known functions. A discretized version of this diffusion is obtained by splitting $(0, T)$ into N intervals of equal length Δt: (*) $Y(t_j) - Y(t_{j-1}) = \mu(t_{j-1}, Y(t_{j-1}))\Delta t + \sigma(t_{j-1}, Y(t_{j-1})) \sqrt{\Delta t}\, U(j)$, where $U(j)$ are independent standard normal.
- Any single specific path $Y_t]_{t=t_1,\dots,t_N}$ is generated by drawing a sequence of N independent random U_j from a standard normal distribution and then applying recursively equation (*). This procedure is repeated M times to generate M paths.
- MCS can be used, for instance, for estimating: (i) a VaR or an Expected Shortfall by building a simulated loss (see Chap. 27); (ii) the value of a European option, by simulating an RN distribution of its terminal payoff, or by generating whole paths (for some exotic options); (iii) the Greek parameters of an option.
- When the *m components of* $\underline{Y}$ *are correlated*, simulating the different factors $Y_i(t)$ independently is inappropriate.
- Drawing from a *Gaussian vector* $\underline{Y}(t)$ whose m components are correlated is grounded on the *Choleski decomposition*: $\underline{Y}$ being $N_m\left(\underline{\mu}; \Sigma\right)$ and Σ being the product of a *triangular* matrix Λ and its transpose, $\Sigma = \Lambda\Lambda'$, we have: $\underline{Y} = \underline{\mu} + \Lambda\underline{U}$ where the components of $\underline{U}$ are *independent* standard normal; hence a sample of $N_m\left(\underline{\mu}; \Sigma\right)$ is obtained from *independent* standard normal draws. The λ_{ij} can be determined step by step.
- Drawing from a *non-Gaussian vector* $\underline{Y}(t)$ with m correlated components can be grounded on a *Gaussian copula* by which correlated draws $(y_1, \dots, y_m)$ of $(Y_1, \dots, Y_m)$ are obtained *from correlated Gaussian draws* $(u_1, \dots, u_m)$ by using the formulas: $y_i = F_i^{-1}(N(u_i))$, F_i being the marginal distribution of Y_i.
- Gaussian copulas are *special cases* of a broader family. Student Copulas yield increasing correlation at the extremes, an advantage over the Gaussian copula, especially for assessing risk.
- Different techniques improve the efficiency (trade-off between accuracy and computation time) of MCS:
  The *antithetic variables* method consists in associating to each draw U_j its opposite ("*antithetic*") $-U_j$. This halves the number of draws and ensures that their empirical mean is equal to zero.

The *control variate* method exploits information in the *estimation of a known parameter* to reduce the error of an estimate of an unknown parameter.

Importance sampling relies on the idea that some simulations are more relevant than others and allows increasing the relative number of these « useful » simulations.

Stratified sampling is a method of drawing samples from a population with *heterogeneous* sub-populations (strata). It consists of drawing independent samples from each sub-population whose members are relatively *homogeneous*.

- MCS in its standard form is *not suitable for valuing American options* because the simulated trajectories are "dissociated" (the tree is not recombining), hence they *do not allow a simple determination of the continuation value* at each stage.
- To overcome the previous problem, different *regression methods* rely on an *estimation of the continuation value* at each date t_j considered in the simulation, by a procedure involving *regression and backward induction*. This continuation value may be compared to the exercise value $V_e(S_j)$ to decide whether to early exercise the option or not and thus to price the option in the same vein as in the binomial tree method.

Suggestion for Further Reading

Books

Cherubini, U., Luciano, E., & Vecchiatto, W. (2004). *Copula methods in finance*. Wiley.

Clauss, P. (2011). *Portfolio management: A quantitative approach*. Dunod.

*Fusai, G., & Roncoroni, A. (2008). *Implementing models in quantitative finance: Methods and cases*. Springer.

*Glasserman, P. (2004). *Monte Carlo methods in financial engineering*. Springer.

Hull, J. (2018a). *Options, futures and other derivatives* (10th ed.). Prentice Hall Pearson Education.

Hull, J. (2018b). *Risk management and financial institutions* (5th ed.). Prentice Hall Pearson Education.

*Lamberton, D., & Lapeyre, B. (2007). *Introduction to Stochastic Calculus applied to finance* (2nd ed.). CRC Press.

Rogers, L. C. G. (2002). Monte Carlo valuation of American options. *Mathematical Finance, 17*, 271–286.

Articles

Boyle, P. P. (1977). Options: a Monte Carlo approach. *Journal of Financial Economics, 4*(3), 323–338.

Boyle, P. P., Broadie, M., & Glasserman, P. (1997). Monte Carlo methods for security pricing. *Journal of Economic, Dynamics and Control, 21*, 1267–1322.

Broadie, M., & Glasserman, P. (1996). Estimating security price derivatives using simulation. *Management Science, 42*, 269–285.

Carrière, J. (1996). Valuation of early exercise price of options using simulations and non-parametric regressions. *Insurance, Mathematics and Economics, 19*, 19–30.

Longstaff, F., & Schwartz, E. (2001). Valuing American options by simulation, a Simple Least Squares approach. *Review of Financial Studies, 14*, 113–147.

Value at Risk, Expected Shortfall, and Other Risk Measures

27

The losses in the financial market sustained by numerous financial institutions and some industrial enterprises have drawn attention to the need for all the financial services and the regulatory agencies to understand thoroughly the risks arising from financial activities. On such an understanding depend the rigor and quality of the internal monitoring of risks ensuring their measurement and their management. The crisis of 2007–2009 showed dramatically that the necessary understanding was often lacking. It also emphasized the shortcomings of the traditional measures of risk, which were badly suited to periods of crisis.

The risks inherent in the activities of financial institutions include:

- *Market risk*, that is, the risk of disadvantageous variations of one or more prices, interest rates, indices, exchange rates, volatilities, or other factors affecting the markets.
- *Credit risk* impacting debts and connected with the possibility of client default.
- *Counterparty risk* impacting all transactions and linked to the possibility that a counterparty does not meet its obligations. For example, there is the risk of default by an insurer, or of the issuer of a derivative instrument (to be distinguished from the risk impacting the asset price underlying the derivative).
- *Liquidity risk*, that is, the risk that an institution cannot meet its obligations as a result of bad timing of its cash inflows and outflows.
- *Operational risk*, linked to human error, fraud, or any deficiency in the institution's operational system.
- *Legal risk*, linked to the possibility that a counterparty cannot be forced to meet its obligations as a result of an inadequate contract, or of legal or organizational error.

We are grateful to Philippe Artzner for his incisive remarks.

© The Author(s), under exclusive license to Springer Nature Switzerland AG 2022
P. Poncet, R. Portait, *Capital Market Finance*, Springer Texts in Business and
Economics, https://doi.org/10.1007/978-3-030-84600-8_27

The risk measures at the heart of the problem addressed in this chapter are chiefly concerned with market risks. Credit and counterparty risks are considered in the following chapters. It is obvious that a detailed understanding of all the risks resulting from the use of financial instruments is necessary for:

- The optimal allocation of equity which is a rare and precious resource.
- The effective monitoring and follow-up of risks (in particular of those related to traders' positions, to illiquid assets, to substantially leveraged complex derivatives, etc.) and their management, notably by hedging.
- Performance measurement.
- Conformity to the requirements and recommendations issued by supervising bodies and policy makers.[1]

In regard to this last aspect one must distinguish two levels of regulation:

- At the national level, banking authorities, regulators of the financial markets, as well as the clearing houses organized for the different markets, generally impose obligations intended to improve the monitoring of financial activities.
- At the international level, as a result of the Group of Thirty, the Basel Committee recommended in April 1995 in its *Capital Adequacy Directive* that banks use a flat-rate method of market risk measurement calculated from their own internal models and based on the concept of Value at Risk (henceforward written VaR). The agreements referred to as Basel II (2004) further specified and bolstered this directive. Recently, an extensive study has been undertaken to address the deficiencies found during the 2007–2009 crisis and the presumed weaknesses of Basel II (Basel III and IV).

The notion of VaR was introduced, and operational methods of estimating it were refined, by the American bank JP Morgan at the start of the 1990s. The first developments addressed market risks (the RiskMetrics™ method), and these were followed by credit risks (the CreditMetrics™ method which introduced the notion of Credit-VaR). The idea is to summarize the risk impacting a portfolio or an assets-and-liabilities position in a single measure with a direct interpretation. More precisely, the VaR tries to quantify, within a specified confidence level (typically 95 % or 99 %) the potential loss which could be sustained by a given isolated position, a portfolio, or a bank as a whole, in a short period of time (typically from one to ten trading days for market positions, or a year for credit portfolios) in market conditions that can be considered as normal. At a theoretical level, this notion is inherited from the "probability of ruin" which has long been used by insurance companies.

---

[1] It is important to note that the meaning of risk depends on the point of view adopted, which can also translate into the adoption by the concerned parties of conflicting risk measures. For example, because their liabilities are limited to their initial investments, the shareholders do not necessarily have the same notion of risk as the government concerned with public welfare. In a similar way (concern with moral hazard), the asymmetric scheme of the bonuses given to traders may induce them to take excessive risks seen from the point of view of the shareholders or the government.

27.1 Analytic Study of Value at Risk

Whereas the VaR is conceptually a very simple notion (it is a quantile of the distribution of losses), calculating it may turn out to be complicated if a position including several different instruments, among which are derivatives, is involved.

Furthermore, the VaR is in fact tainted with a number of shortcomings, and other indicators such as Expected Shortfall and risk measuring tools have been developed to overcome these deficiencies.

This chapter is devoted to the different risk measures focusing on market risks, given that credit risks and Credit-VaR are the subjects of the following Chaps. 28, 29, and 30.

Section 27.1 gives the definition and interpretation of the VaR, Section 27.2 deals with methods of calculation and simulation, Section 27.3 analyzes the limits of the VaR, presents the Expected Shortfall (ES) and the conditions for coherence of risk measures and Sect. 27.4 provides tools that allow correction of the large biases affecting the standard measures as a result of the non-normality of distributions, extreme correlations, etc.

27.1 Analytic Study of Value at Risk

We have seen in preceding chapters that a market portfolio comprising different securities, including derivatives, is impacted by numerous sources of risk: (multi-dimensional) interest rate risk for interest-rate instruments, risks of variation in the different factors that impact stock market values, variation in the volatility for optional instruments, etc. We have already given a long list of indicators for these different risks like simple or multiple variations of interest rates, and delta, gamma, and rho for options. This battery of indicators, which is very useful in refined management of market positions and used on a daily basis by traders, is too complex and insufficiently parsimonious at the general management level. In fact, the managers of a trading room and the overseers of a company or financial institution need a synthetic indicator that expresses quantitatively, in as simple and concrete way as possible, global market risk. The VaR does meet this need (imperfectly, as we will see) in that it quantifies the risk in a single number that can be interpreted directly. It thus became a must for the analysis of the risk impacting market positions. It is still much in use, in spite of its different limitations and serious drawbacks which have motivated the development of complementary or alternative methods. Some of these (presented in Sects. 27.3 and 27.4) are variants of the standard VaR while others differ greatly from it.

27.1.1 The Problem of a Synthetic Risk Measure and Introduction to VaR

The problem is to measure the risk impacting the market value of a given position. We use the notation:

V_0 the present value (at date 0) of the position or portfolio being considered.

V_h the value of this portfolio at the time horizon (or holding period) h (e.g., $h = 10$ days).

$L_h \equiv V_0 - V_h$ represents the *loss* between 0 and h; L_h is a random variable which may be positive or negative. $G_h = - L_h$ is the gain (positive or negative). In normal or favorable situations, $E(G_h) > 0$, and therefore $E(L_h) < 0$.

The risk of loss is specified with the help of the probability distribution L_h from which different indicators may be computed.

F_X denotes the distribution function (cumulated probabilities) of a random variable X.

Thus, $F_L(x) = \text{Prob}(L_h \leq x)$.

27.1.1.1 The Variance (or Standard Deviation) of L_h Is a First Measure of Risk

The first synthetic indicator of the risk for a portfolio goes back to Markowitz (1959) and is the cornerstone of classical portfolio theory. This theory brings to light that investors should be interested in the global risk of their portfolios and take into account diversification. It is based on the standard deviation of the portfolio value (or, equivalently, the standard deviation of the variation of this value) as a measure of the risk, in accordance with the mean-variance paradigm (*see* Chap. 21):

1st risk measure of a portfolio = standard deviation of $L_h = \sigma_L$

We have already analyzed the drawbacks of this indicator, notably the negative deviations in the loss L_h from its mean (gains) are taken into account in the same way as its positive deviations (losses) while managers and investors are in general essentially concerned with downside risks ($L_h > 0$). While such a disadvantage disappears when the probability distribution of L_h is symmetrical about its mean (as is the case for the normal distribution), it can be considered crippling in the case of an asymmetric distribution. Figure 27.1, which graphs the loss distributions L_h and L_h' for two portfolios x and x', illustrates this drawback.

The two distributions of L_h and L_h' in Fig. 27.1 have the same means μ and standard deviations σ (they are mutually symmetric about the vertical axis at their common mean $\mu < 0$) but the distribution of L_h excludes any risk of loss (Prob $(L_h > 0) = 0$) contrary to that of L_h'. It is therefore a fallacy to consider as equal the risks induced by holding x or x'.

In a general sense, the standard deviation σ_L has a difficult interpretation since, without clarity on the form of the distribution of L_h, it is not unambiguously linked to the probability of loss. In Fig. 27.1, for example, two different distributions for L_h with the same means and standard deviations, correspond to different loss probabilities (there is no risk of loss in the case of the solid curve).

27.1 Analytic Study of Value at Risk

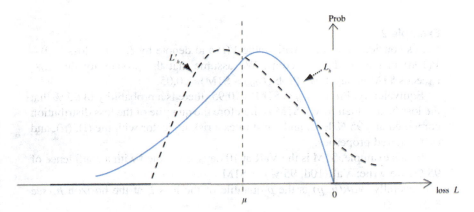

Fig. 27.1 Two probability distributions of possible losses, with the same means and standard deviations

---

Example 1
(i) First let us assume that the distribution of the loss L_h is normal with mean -5 M€ (which corresponds to a positive expectation of gain of $+5$ M€) and standard deviation 6 M€. Denoting a standard Gaussian variable by U, we can write: $L_h = 6U - 5$. From this:
Prob $(L_h > 1) = $ Prob $(U > 1) = 0.1587$: a probability of 15.87 % is therefore inferred that the loss is greater than or equal to 1 M€; but this answer depends crucially on the assumption of normality, which is in general not the case.
(ii) Now let us assume that the distribution of the *gain* $-L_h$ is log-normal: the gain then is always positive whatever the value of σ_h (zero probability of a loss; *a fortiori*, the loss L_h is always less than 1 M€).

27.1.1.2 A Quantile of the Probability Distribution of the Loss L_h as a Second Risk Measure; VaR Defined by Such a Quantile

It follows from the preceding discussion that a risk measuring index ought to have, at a minimum, the following properties: (*i*) it should be simple (preferably a single number); (*ii*) it should be directly interpretable; (*iii*) it should only reflect "downside risk," i.e., the risk of losses. We emphasize that these are minimum conditions that are not enough to ensure the coherence of a risk measure, as we will see in Sect. 27.3.

The standard deviation σ_L only has the first property; as we will see, a quantile of the distribution of L_h, from which VaR is defined, has all three properties.

Example 2

Let us consider a position with value $V(t)$ and denote by L_{10d} the loss $V(0) - V(10d)$ in the 10 days to come. We assume that the probability that L_{10d} exceeds \$1M equals 5 %: Prob $(L_{10d} > \$1M) = 0.05$.

Equivalently: Prob $(L_{10d} \leq \$1M) = 0.95$ (there is a probability of 95 % that the loss be less than \$1M); \$1M is therefore the quantile of the loss distribution computed at a 95 % level and constitutes a risk indicator with the (*i*), (*ii*), and (*iii*) required properties.

In this example, \$1M is the VaR at 10 days, computed with a confidence of 95 %; we write: VaR(10d, 95 %) = \$1M.

Generally, $VaR(h, p)$ is the p-quantile of the loss L at the horizon h (*see* § 27.1.2.2).

This introduction and the example explain intuitively the interest and significance of the VaR, which we will now define more formally.

27.1.2 Definition of the VaR, Interpretations, and Calculation Rules

27.1.2.1 General Definition and Interpretations

The VaR at the horizon h and with probability level p is the smallest number VaR(h,p) such that[2]:

$$Prob \ (L_h \leq VaR(h,p)) = p \tag{27.1a}$$

or, equivalently:

$$Prob \ (L_h > VaR(h,p)) = 1 - p. \tag{27.1b}$$

The VaR is therefore the p-quantile of the loss L_h.

This definition leads directly to the interpretation of the VaR.

Equation (27.1a) shows that *the amount of the loss in the coming period of duration h will be less than the VaR with probability p*; thus, the VaR can be interpreted as a maximum loss with confidence level p.

Equation (27.1b) expresses that *the loss will be greater than the VaR with probability 1-p*.

For example, if the probability level p is 99 % and h equals 1 trading day, the VaR is computed with 99 % confidence and the portfolio will only suffer losses greater than the VaR one trading day in 100.

---

[2] If the distribution of L_h is continuous, the VaR is uniquely defined by Eq. (27.1a) and it is useless to specify that it is the *smallest number* satisfying (27.1a).

27.1 Analytic Study of Value at Risk

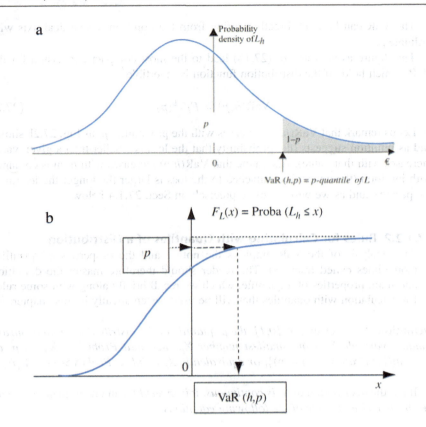

Fig. 27.2 (a) Probability density of the loss L_h and VaR. (b) Distribution function of L_h and VaR

It is important to note that, for reasons of simplicity and since the horizon (holding period) relevant to computing the VaR is in general very short, we will never take into account the *discount factor* even if we are estimating the future value of a portfolio or of the position considered. Let us remark that, in addition, in the case of a *discrete probability distribution*, p % of the observations does not generally give a whole number; thus it is necessary to decide whether the threshold number of observations should be rounded up (the pessimist view) or down (the optimist view).

The graphical interpretation of the VaR is given in Fig. 27.2a showing the probability density of a loss L_h. The grey area, taken to be equal to 1-p, is the probability that this loss will be greater than or equal to VaR(h, p); indeed, VaR(h, p) is such that the grey area is equal to the value 1-p, where p is given a priori.

Figure 27.2b shows the distribution function (cumulative probability) of L_h:

$$F_L(x) \equiv \text{Prob}(L_h \leq x).$$

The VaR can be read directly starting from the point on the vertical axis with ordinate p.

This figure as well as Eq. (27.1b) lead to the most compact expression for the VaR, which holds if the distribution function is invertible[3]:

$$\mathrm{VaR}(h, p) = F_L^{-1}(p). \tag{27.2}$$

Let us remark that $\mathrm{VaR}(h,p)$ increases with the probability p, as Fig. 27.2b shows and as intuition suggests (the probability that the loss is smaller than a given value increases with that value). Also, note that $\mathrm{VaR}(h,p)$ increases with h, in accordance with intuition (the uncertainty attached to the loss is larger the longer the length of the period), and as we will see more precisely in Sect. 27.1.4 below.

27.1.2.2 Rules for Calculating with Quantiles of a Distribution

– Any analysis of the VaR employs the notion and the properties of quantiles (sometimes called fractiles). The reader should therefore master the definition and main properties of a quantile, which we recall briefly along with some rules for calculation with quantiles that will be used systematically in this chapter.

Definition *For a given $p \in [0,1]$, the p-quantile of the distribution of an arbitrary random variable X is the smallest number X_p such that $Prob(X \le X_p) = p$, or $X_p = inf\{x: Prob(X \le x) = p)\}$, or equivalently $X_p = inf\{x: Prob(X > x) = 1-p)\}$.*

If F_X the distribution of X, is continuous, it is invertible, and its p-quantile X_p can be characterized by one of the following equations:

$$F_X(X_p) = p \Leftrightarrow X_p = F_X^{-1}(p).$$

Figure 27.2a, b give graphical representations of the p-quantile of the loss L_h, that is, of $\mathrm{VaR}(h,p)$.

In the following, we will often consider Gaussian variables and denote by α_p the p-quantile of the standard Gaussian distribution $N(0,1)$: $\alpha_p = N^{-1}(p)$; for example $\alpha_{0.95} = 1.645$ (since Prob *(standard Gaussian* $\le 1.645) = 95\ \%$) and $\alpha_{0.99} = 2.33$. Because the standard Gaussian is centered at 0 and symmetric, we have: $\alpha_p = -\alpha_{1-p}$.

The values of α_p at all values of p between 0 and 1 can be obtained with the help of tables (possibly by interpolation) or with a spreadsheet, such as Excel, which includes a statistical module.

– We often have to calculate the quantile of a random variable X that is a function $f(I)$ of a random variable I. The variable I represents, for example, an index or a return and X is often a loss (L) or a profit $(G$ or $V)$. We will encounter in this chapter and in Chaps. 29 and 30 numerous examples of such a situation.

---

[3] As F_L is an increasing function, it suffices it is continuous to be invertible on the interval $(0,1)$.

27.1 Analytic Study of Value at Risk

If f is a monotonous function, the p-quantile of $X = f(I)$ can be expressed very simply as a function of the p-quantile of I, denoted by I_p, in accordance with the following three calculation rules:

Calculation Rule 1

Assume $X = f(I)$ with f monotonous and continuous:

(a) If f is increasing, *p-quantile of $X = f(p$-quantile of I)*: $X_p = f(I_p)$
(b) If f is decreasing, *p-quantile of $X = f((1\text{-}p)$-quantile of I)*: $X_p = f(I_{1-p})$.

Proof

If f is increasing, we need to show that $\text{Prob}[f(I) \leq f(I_p)] = p$; now
$I \leq I_p \Leftrightarrow f(I) \leq f(I_p)$, therefore $\text{Prob}[f(I) \leq f(I_p)] = \text{Prob}[I \leq I_p] = p$.
If f is decreasing, $\text{Prob}[f(I) \leq f(I_{1-p})] = p$ since $I > I_{1-p} \Leftrightarrow f(I) \leq f(I_{1-p})$, therefore
$\text{Prob}[f(I) \leq f(I_{1-p})] = \text{Prob}[I > I_{1-p}] = p$.

In what follows, we will often consider a Gaussian variable X or a function of a Gaussian variable I, in which case we can apply rule 1. Recall that α_p denotes the p-quantile of the standard Gaussian $N(0,1)$ and I_p that of I.

In such a context we will apply the following rules:

Calculation Rule 2

If X has the distribution $N(\mu, \sigma)$, its p-quantile is given by $X_p = \alpha_p \sigma + \mu$.

Proof

We write $X = \mu + \sigma I$ where I is $N(0,1)$ (therefore $I_p \equiv \alpha_p$) and apply rule 1.

Calculation Rule 3

Assume $X = f(I)$, f monotonous and continuous, and I distributed according to $N(\mu, \sigma)$:

(a) If f is increasing: *p-quantile of $X = f(\alpha_p \sigma + \mu)$*
(b) If f is decreasing: *p-quantile of $X = f(\alpha_{1-p} \sigma + \mu) = f(-\alpha_p \sigma + \mu)$*

Proof

This rule is a consequence of rule 1 and the fact that $I_p = \alpha_p \sigma + \mu$ (rule 2).

The calculation rules can be applied to the VaR with $X = L_h$ and $VaR(h,p) = p$-quantile of X.

27.1.2.3 Alternative Expressions for the VaR

- The VaR is often given in terms of the value V_h of the portfolio or of its variation, that is, in terms of the gain $G_h \equiv V_h - V_0 = -L_h$, rather than the loss L_h.

Denoting by F_G the distribution function of the gain G_h, Eqs. (27.1b) and (27.2) can, respectively, be written:

$$\text{Prob } (G_h \leq \text{VaR}(h,p)) = 1 - p \quad \text{and} \quad \text{VaR}(h,p) = F_G^{-1}(1-p).$$

The VaR(h,p) is therefore the ($1-p$)-quantile of the gain.

Rule 1 also allows giving the VaR as a function of a quantile of the return:

- If R_h denotes *the arithmetic return* $R_h = \frac{V_h - V_0}{V_0} = -\frac{L_h}{V_0}$, so that $L_h = -V_0 R_h$, Rule 1 implies:

$$\text{VaR}(h,p) = -V_0 \times (1-p)-\text{quantile of } R_h.$$

- If R'_h denotes *the logarithmic return* $V_h = V_0 e^{R'_h}$, so that $L_h = V_0 - V_h = V_0(1 - e^{R'_h})$, and Rule 1 implies $\text{VaR}(h,p) = V_0 [1- exp((1-p)\text{-quantile of } R'_h)]$.

27.1.3 Analytic Expressions for the VaR in the Gaussian Case

When a hypothesis on the probability distribution of L_h is formulated (i.e., in fact, on the distribution of V_h), explicit calculation of the VaR is possible. This is so notably when V_h (and so L_h) is assumed to be normal or log-normal or, more generally, to be a function of a normal variable (i.e., the Gaussian case). Some results obtained in the Gaussian case are assumed to remain true, approximately, in the general case and are frequently used in different contexts. In this section and the following, we study these results in detail and examine their validity in some more general contexts.

Most of the results are obtained directly with the help of calculation rules 2 and 3 (applicable when the loss L is Gaussian or a function $f(I)$ of a Gaussian variable I; *see* Sect. 27.1.2.3 above).

27.1.3.1 Calculation of the VaR for a Gaussian Loss

An analytic expression for the VaR is easily found if the distribution of the loss L_h is assumed to be Gaussian with known parameters. Let us remark that $L_h = V_0 - V_h$ is Gaussian if and only if V_h is itself normal (since V_0 is known), therefore if and only if the *arithmetic return* $(V_h - V_0)/V_0$ is also Gaussian.

Let us, therefore, assume that the loss L_h is normally distributed with mean μ_h and standard deviation σ_h. Rule 2 immediately implies

$$\text{VaR}(h,p) = \alpha_p \ \sigma_h + \mu_h. \tag{27.3a}$$

The term μ_h is often neglected and the VaR is calculated from the approximate identity.

27.1 Analytic Study of Value at Risk

$$\text{VaR}(h,p) = \alpha_p \, \sigma_h. \tag{27.3b}$$

Indeed, if h is small, the term μ_h, which is proportional to h, is also small with respect to $\alpha_p \sigma_h$ which is proportional to $\sqrt{h}$ (and tends to 0 less quickly than μ_h). We will return to this point in more detail in Sect. 27.1.4.

Sometimes, the VaR is calculated from the *arithmetic return* $R_h = \frac{V_h - V_0}{V_0} = -\frac{L_h}{V_0}$ which is assumed to have a normal distribution $N(\mu'_h, \sigma'_h)$. As a result, we have $L_h = -V_0 R_h$, and Rule 27.3b directly gives the VaR as a function of the parameters μ'_h and σ'_h of the distribution of the *return*:

$$\text{VaR}(h,p) = V_0 \, (\alpha_p \, \sigma'_h - \mu'_h). \tag{27.4a}$$

Equation (27.4a) is equivalent to (27.3a) but shows that the VaR as a function of the parameters of the distribution of the arithmetic return R_h of the portfolio rather than of that of the loss L_h. In accordance with intuition, the VaR is thus larger when the standard deviation of R_h is large and the expectation of R_h is small.

Since, from the symmetry of the normal distribution, we have $\alpha_p = -\alpha_{1-p}$, Eq. (27.4a) can also be written.

$$\text{VaR}(h,p) = -V_0 \, (\alpha_{1-p} \, \sigma'_h + \mu'_h). \tag{27.4b}$$

Example 3

Let us consider a portfolio whose value today is \$ 100M; suppose that, in the coming 10 days, its arithmetic return R_{10d} is normally distributed with mean 0.4 % and standard deviation 5 %. We have $V_{10d} - V_0 = -L_{10d} = V_0 R_{10d}$ and L_{10d} is normally distributed with a mean of $-100 \times 0.004 = -0.4$ and a standard deviation of $100 \times 0.05 = 5$. We can thus write $L_{10d} = 5\,U - 0.4$.

From this, Prob $(L_{10d} \leq x) = $ Prob $(U \leq (0.4 + x)/5)$. Furthermore, Prob $(U \leq 1.645) = 95$ %; the VaR(10d, 95 %) is therefore such that $(0.4 + \text{VaR})/5 = 1.645$; or again VaR(10d, 95 %) = 7.825. Directly applying (27.4a) gives in addition $100 \times (1.645 \times 5\,\% - 0.4\,\%) = 7.825$. Thus, there is a 5 % probability that the loss over the next 10 days exceeds \$ 7.825M, i.e., the value of the portfolio is less than \$ 92.175M.

Equations (27.3a), (27.3b), (27.4a), and (27.4b) are very commonly used to calculate a VaR simply, but they can lead to large errors when the distributions are away from normal. We will see in Sect. 27.4 how to adjust these formulas (with the help of the Cornish-Fisher expansion) to take into account the asymmetry and fat tails which, to various degrees, characterize the actual distributions of financial variables.

Let us note in addition that, assuming normal distributions, the choice of a portfolio under a VaR constraint is in accordance with standard Markowitz portfolio theory (Chap. 21) and the mean-variance criterion. Indeed, looking for a portfolio that minimizes the VaR under a constraint on the expectation leads to

$$\text{Min } \text{VaR}(x) \text{ subject to (the constraint) } \mu_x = k.$$

This optimization problem can be rewritten, taking into account Eq. (27.3a), for example, as Min $\alpha_p \sigma_x + \mu_x$ s.t. $\mu_x = k$, which is equivalent to Min σ_x s.t. $\mu_x = k$, which is nothing but Markowitz's mean-standard-deviation problem. Assuming normality, an efficient portfolio in Markowitz's sense which minimizes the standard deviation but is constrained by the expected return requirement also minimizes the VaR under the same requirement for the expected return.

This remark allows us to employ the results of standard portfolio theory to handle numerous problems related to the VaR. Sect. 27.1.3.3 below constitutes an example of this close connection between the VaR and the standard portfolio risk analysis in the Gaussian framework.

27.1.3.2 VaR Calculation When V_h Is Assumed Log-Normal

For reasons developed in Chap. 8, it is usually preferable to assume for V_h a log-normal distribution than a normal one, as is now standard in financial modeling. Therefore, we assume here that V_h is log-normally distributed, which implies that over the period $(0, h)$, the *log-return* $R_h = \ln(V_h) - \ln(V_0)$ is normal.

We thus can write $V_h = V_0 e^{R_h}$ and $L_h = V_0 - V_h = V_0(1 - e^{R_h})$.

The loss can then be written as a decreasing function of the Gaussian variable R_h, for which we will denote the mean and the standard deviation by μ'_h and σ'_h, respectively. Applying Rule 3-b gives:

$$\text{VaR}(h, p) = V_0\left(1 - e^{\alpha_{1-p}\sigma'_h + \mu_h}\right) = V_0\left(1 - e^{-\alpha_p\sigma'_h + \mu'_h}\right). \tag{27.5}$$

The reader will note that for small values of h, corresponding to the short horizons over which the VaR is usually calculated, a first-order expansion of the exponential in Eq. (27.5) leads to (27.4b) or to (27.4a). Equations (27.3a) and (27.4a) then are usually good approximations to the VaR if the value of the portfolio can reasonably be assumed to be log-normally distributed.

Example 4

Let us again consider the portfolio from the preceding Example 3, with an initial value of $ 100M, but now suppose that the log- return is Gaussian for 10 business days.

We assume the log-return has a mean of 0.4 % and a standard deviation of 5 %.[4]

(continued)

---

[4]This, assuming 250 trading days per year for simplicity, corresponds to an annual mean of 10% ($=$ 0.4% $\times$ (250/10) trading days) and an annual volatility of 25% $\left(= 5\% \times \sqrt{\frac{250}{10}}\right)$.

27.1 Analytic Study of Value at Risk

> **Example 4** (continued)
>
> Equation (27.5) implies, for p equal to 95 %, that $\text{VaR}(10\text{d}, 95\ \%) = 100$ $(1 - e^{-0.07825}) = \$\ 7.527\text{M}$.
>
> The probability is therefore 5 % that the portfolio loses between $\$\ 7.527\text{M}$ and $\$\ 100\text{M}$, within the next 10 days.
>
> Compare this result with that in example 27.3 where the VaR was estimated to be $\$\ 7.825\text{M}$ and note the relatively limited error resulting from assuming normality when log-normality actually holds.

27.1.3.3 Contribution of One Component to the VaR of a Portfolio

In this subsection, we will consider a portfolio x made up of N securities and defined by the weights $(x_1, \ldots, x_N)$ of these securities. We denote by R_i the simple return of security i and by R_x the portfolio's so that

$R_x = \sum_{i=1}^{N} x_i R_i$, and as a result of the decomposition of the variance $\sigma^2(R_x)$ we have already used

$$\sigma^2(R_x) = \sum_{i=1}^{N} \sum_{j=1}^{N} x_i x_j \text{cov}(R_i, R_j) = \sum_{i=1}^{N} x_i \sum_{j=1}^{N} x_j \text{cov}(R_i, R_j)$$

$$= \sum_{i=1}^{N} x_i \text{cov}\left(R_i, \sum_{j=1}^{N} x_j R_j\right),$$

from which $\sigma^2(R_x) = \sum_{i=1}^{N} x_i \text{cov}(R_i, R_x)$.

In addition, we adopt the Gaussian framework and neglect the influence of the mean of R_x on the VaR of x so that the latter is simply proportional to the standard deviation of R_x. It follows, in accordance with (27.4a):

$$\text{VaR}_x = V_0\ \alpha_p\ \sigma(R_x).$$

The combination of the last two equations thus leads to an expression for the VaR of a portfolio as a function of the covariances of the returns of the components with the overall return:

$$\text{VaR}_x{}^2 = (V_0 \alpha_p)^2 \sum_{i=1}^{N} x_i \text{cov}(R_i, R_x).$$

The contribution of security i *to the square* of the global VaR is therefore equal to $x_i(V_0\,\alpha_p)^2\text{cov}(R_i, R_x)$.

This contribution is then not linked to the VaR of i itself (which is proportional to $\sigma(R_i)$) but to the covariance $\text{cov}(R_i, R_x)$ of its return with that of the portfolio. This result is well-known and at the core of standard portfolio theory. But we have remarked that in the Gaussian case the standard (mean-variance) portfolio theory does apply to the VaR since it is then proportional to the standard deviation. In the Gaussian case, the VaR and the standard deviation thus are two equivalent indicators of risk. It is then not surprising that the VaR (or the squares of the VaR) are not simply additive and that a conjunction of risks in a situation of partial diversification leads, as in standard portfolio theory, to an indicator of global risk that involves sums of covariances (and not sums of variances).

In the same vein, often arises the question of the effect of adding an asset y on the VaR of an existing portfolio x: this effect will depend on $\text{cov}(R_y, R_x)$ rather than on the VaR of the added asset y.

We will devote Sect. 27.3.4 to a more general and deeper analysis of portfolio risk and to different meanings for the risk attributable to one of its components (marginal risk, incremental risk, and contribution to global risk).

27.1.4 The Influence of Horizon h on the VaR of a Portfolio in the Absence or Presence of Serial Autocorrelation

There is no consensus as to the choice of the horizon h and the probability p on which the VaR depends. For example, for the analysis of market risk, the Basel Committee prescribes $h = 10$ trading days and $p = 99\ \%$ while the RiskMetrics™ method of JP Morgan works with a horizon of $h = 1$ trading day and a probability $p = 95\ \%$. Indeed, the choice of these two parameters depends on the portfolio management type (notably the *turnover* in securities) and the context in which it occurs. Nonetheless, so as to have sufficient data, most estimates are made, in the first instance, on the basis of daily data and end up with a VaR calculated for 1 day. In a second stage, as needed, the 1-day VaR is converted into a VaR for a longer term. Generally arises the question of converting the VaR(h, p) into a VaR(h', p) calculated for a horizon h' different from h. Let us note that for the calculation of the Credit-VaR, which pertains to analyzing credit risk, the horizon usually set is 1 year (*see* Chap. 29).

The manner in which the VaR depends on h is expressed in different ways according to whether or not the successive values of the portfolio (or its log-returns) are assumed independent.

27.1.4.1 In the Absence of AutoCorrelation

- The loss L_h accumulated between 0 and h equals the sum of the different losses on the different days j making up the time interval $(0, h)$: $L_h = \sum_{j=1}^{h} \Delta_j L$. Let us first assume that the successive daily losses, $\Delta_j L \equiv L_j - L_{j-1} = V_{j-1} - V_j$, are Gaussian, *uncorrelated* and with the same mean μ_1 and standard deviation σ_1 for each day $j = 1, \ldots, h$. This gives

- $\mu_h = h\mu_1$; $(\sigma_h)^2 = h\,(\sigma_1)^2$; $\sigma_h = \sigma_1\sqrt{h}$, and by substitution in (27.3a) we obtain

$$\text{VaR}(h, p) = \alpha_p \sigma_1 \sqrt{h} + \mu_1 h. \tag{27.6}$$

We emphasize the fact that this result depends on the assumption that there is no correlation among the successive losses. We will see later how to modify this result when that no longer holds.

- Assume now that the successive daily log-returns are i.i.d. and normally distributed with mean μ_1 and standard deviation σ_1 on every day $j = 1, \ldots, h$.

The logarithmic return over $(0, h)$ equals the sum of the h daily log-returns on the h days, is normally distributed with a mean $\mu'_1 h$ and a standard deviation $\sigma'_1 \sqrt{h}$ so that (27.5) entails:

$$\text{VaR}(h, p) = V_0\left(1 - e^{-\alpha_p \sigma'_1 \sqrt{h} + \mu'_1 h}\right)$$

or, to the first order:

$$\text{VaR}(h, p) = V_0(\alpha_p \sigma'_1 \sqrt{h} - \mu'_1 h). \tag{27.7}$$

Another approximation, which becomes more useful as h becomes smaller, consists in neglecting $\mu'_1 h$ in comparison with $\alpha_p \sigma'_1 \sqrt{h}$ (see § 27.1.3.1 above). Indeed, let us choose the year as the unit period so that μ'_1 and σ'_1 are the annualized expectation and volatility and let h be a small fraction of a year. When μ'_1 and σ'_1 are of the same magnitude, the term in h can be neglected in comparison with $\sqrt{h}$ and we can write

$$\text{VaR}(h, p) = V_0 \alpha_p \sigma'_1 \sqrt{h}. \tag{27.8}$$

Because of this, the VaR of the same portfolio, relative to two different horizons h and h', are approximately related by

$$\text{VaR}(h', p) = \sqrt{\frac{h'}{h}}\, \text{VaR}(h, p).$$

Summing up, the VaR(h,p) thus is approximately proportional to the standard deviation of the return over $(0, h)$ and *varies as the square root of the length h of the period used in the calculation*; this result also depends on the assumption that the successive log-returns are uncorrelated.

Example 5

Let us consider a portfolio X of \$50M made up of American "blue chips" and such that it can be regarded as equivalent to the Nasdaq index. One estimates at 20 % the annual volatility of the Nasdaq and at 12 % the expectation of its annualized return.

We obtain, for the 1-day VaR of X:

- $\text{VaR}(1d, 95\%) = 50 \times \left(1.654 \times 0.2 \times \sqrt{\frac{1}{250}} - 0.12 \times \frac{1}{250}\right) = \$1.022 \text{ M}$, using Eq. (27.7);
- $\text{VaR}(1d, 95 \ \%) = 50 \times \left(1.654 \times 0.2 \times \sqrt{\frac{1}{250}}\right) = 1.046 \text{ M}$ with approximation (27.8).

We get, for the 10-day VaR:

- $\text{VaR}(10d, 95\%) = 50 \times \left(1.654 \times 0.2 \times \sqrt{\frac{10}{250}} - 0.12 \times \frac{10}{250}\right) = 3.068 \text{ M}$, using (27.7);
- $\text{VaR}(10d, 95\%) = 50 \times \left(1.654 \times 0.2 \times \sqrt{\frac{10}{250}}\right) = 3.308 \text{ M}$, with (27.8).

Therefore, the approximate Eq. (27.8) gives accurate results for the 1-day VaR and is obviously less precise for a 10-day horizon.

Let us remark that the same Eq. (27.8) leads to an approximate relationship between two VaRs for the same portfolio computed with different probabilities p and p':

$$\text{VaR}(h, p') = \frac{\alpha_{p'}}{\alpha_p} \text{VaR}(h, p) = \frac{\alpha_{1-p'}}{\alpha_{1-p}} \text{VaR}(h, p).$$

More generally, Eq. (27.8) gives the following conversion formula:

$$\text{VaR}(h', p') = \sqrt{\frac{h'}{h}} \frac{\alpha_{p'}}{\alpha_p} \text{VaR}(h, p). \qquad (27.9)$$

27.1 Analytic Study of Value at Risk

Example 6

The foregoing considerations allow us to convert a VaR(10d, 99 %), such as prescribed by the Basel Committee, into a VaR(1d, 95 %) (RiskMetrics™):

$$\text{VaR}(10d, 99\%) \# \frac{\alpha_{0.99}}{\alpha_{0.95}} \sqrt{10} \ \text{VaR}(1d, 95\%) = 4.48 \times \text{VaR}(1d, 95\%).$$

Sometimes it is admitted that Eq. (27.8), and approximations such as (27.9) which derive from it, give acceptable results. It should be remembered that this presumption has some foundation only if the horizon h is short, the composition of the portfolio is not changed between 0 and h and the distributions of the variations in prices (or the log-returns) are near to Gaussian and serially uncorrelated. Indeed, the assumption of normality, while possibly acceptable for well-diversified portfolios involving a great number of securities, is often far from reality, notably for portfolios containing options; using Eqs. (27.8) and (27.9) then can lead to large errors.

We examine in the following subsection the case of serial correlation, and further on, in Sect. 27.4, Subsect. 27.4.1 and 27.4.2, the methods of calculating the VaR in non-Gaussian cases.

27.1.4.2 Serial Autocorrelation

Serial autocorrelation in price variations or their logarithms (log-returns), and therefore in successive losses, is in practice observed, in particular in illiquid markets. The simplest way to take account of this is to assume an autocorrelation of order 1. Recall that $\Delta_j L \equiv L_j - L_{j-1}$, $variance(L_j) = \sigma_1^2$ for every j, $L_h = \sum_{j=1}^{h} \Delta_j L$, and that the autocorrelation of order 1 of daily losses is written $cor(\Delta_j L, \Delta_{j-1} L) = \rho$ for all j. More precisely, ρ is calculated by the time-series regression: $\Delta_j L = \rho \Delta_{j-1} L + \varepsilon_j$, where the ε_j are assumed i.i.d.

The result is that $cor(\Delta_j L, \Delta_{j-k} L) = \rho^k$ and we can write:

$$\text{Variance}(L_h) \equiv \sigma_h^2 = \text{Variance} \left(\sum_{j=1}^{h} \Delta_j L \right) =$$

$$\sum_{j=1}^{h} variance(\Delta_j L) + 2 \sum_{i>j} covariance(\Delta_i L, \Delta_j L)$$

$$= h\sigma_1^2 + 2 \sum_{j=1}^{h-1} \sum_{i=j+1}^{h} \rho^{i-j} \sigma_1^2, \text{ or in other words :}$$

$$\sigma_h^2 = \left(h + 2 \sum_{k=1}^{h-1} (h-k)\rho^k \right) \sigma_1^2 > h\sigma_1^2.$$

Autocorrelation thus increases the variance of the sum of variables, and therefore the VaR is calculated for a horizon of several days. To fix an order of magnitude, the reader can check that, for $h = 10$, the variance of the loss L_h, thus of the square of the VaR, is bounded above by about 10 % for an autocorrelation $\rho = 10 \%$.[5]

27.2 Estimating the VaR

Although the concept of VaR is theoretically simple, its practical implementation is not. The methods allowing estimation or calculation of the VaR are numerous and varied, often ad hoc, and employ statistical tools that are more or less sophisticated. Nonetheless, notwithstanding their diversity, the different methods of calculating the VaR rest on the following basic principles:

– All positions must be expressed in terms of *marked-to-market values*;
– The global risk should be reckoned by aggregating the individual risks of the instruments making up the portfolio in a way that correctly takes account of *diversification effects*;
– The horizon h should be short, typically one, five, or ten trading days, and the probability confidence level p high (at least 95 %);
– If the portfolio or the position whose VaR is to be calculated contains a majority of complex instruments and/or gives rise to numerous cash flows, it should be subject to a preliminary analysis that may simplify its representation.

Aside from these principles, which are shared by all the methods, different approaches and a large spectrum of assumptions and techniques are available, concerning for example:

– The general methodology;
– The techniques for parameter estimation (notably the relevant variance-covari-ance matrices);
– The form assumed for the different distributions of the rates of return;
– The stability or instability assumed for the distributions and their parameters.

After a first subsection devoted to a preliminary analysis of the position under scrutiny, and to its representation with a simplified portfolio, and a very general presentation of the different possible approximations to the function expressing its value, we present the principal methods which we divide into three families: simulations based on historical data, treated in Sect. 27.2.2, parametric and analytic methods presented in Sect. 27.2.3, and Monte Carlo simulations examined in Sect. 27.2.4. The different methods are compared in Sect. 27.2.5.

---

[5]The upper bound coefficient is $\dfrac{h+2\sum_{j=1}^{h-1}(h-j)\rho^j}{h} = \dfrac{10+2\left(9\times 0.1+8\times 0.1^2+...+0.1^9\right)}{10} = 1.0988.$

27.2.1 Preliminary Analysis and Modeling of a Complex Position

27.2.1.1 Standard Analysis

Let us recall briefly standard modeling, which is applicable in some cases to calculate a VaR.

Consider a portfolio x of assets and liabilities comprising M securities.

The return on security i, between 0 and h, is denoted by R_i, for $i = 1, \ldots, M$.

The return on the portfolio writes $R_x = \sum_{i=1}^{M} x_i R_i$ where x_i is the weight of i in the portfolio ($x_i < 0$ if i is a short position).

The distribution law of R_x (which determines the VaR) depends on the co-variations of the different R_i; it will, in particular, be necessary, for the methods described subsequently, to know the variance-covariance matrix of these M returns ($\text{Cov}(R_i, R_j)$ for $i, j = 1, \ldots, M$).

In fact, this method is above all suitable for stock portfolios. Complex portfolios (including interest rate products or derivatives) require a preliminary analysis intended to model them, i.e., to represent them in a simplified way. This modeling may take the form of a decomposition of the portfolio into a number of standard securities and/or be based on identifying the risk factors that determine the value of the portfolio and its variations.

27.2.1.2 Representation of a Portfolio as a Combination of Elementary Standard Securities

Very often the securities making up a portfolio are too numerous to be exactly and individually taken into account. It is then necessary to determine the number of standard securities (or "primitive" securities) from which any asset can be synthesized. For example:

- Zero-coupon bonds with the main maturities (1 month, 3 months, 6 months, 1 year, 2 years, ..., 20 years) allow an approximate replication of any fixed-income instrument and of any forward contract involving interest rates or any interest rate *swap*;
- An option is equivalent to a portfolio containing delta (δ) units of the underlying asset and a zero-coupon bond;
- A well-diversified portfolio x of stocks held in a currency i is equivalent to $\beta_x \times \text{Index}(i)$, where β_x is the sensitivity of the portfolio to variations of Index (i).

In this way, we try to make any position, however complex, correspond to a portfolio made up of standard securities to which it is approximately equivalent. The procedure is sometimes referred to as *mapping*.

Let us handle in more detail the example of a portfolio of fixed-income instruments.

Let us consider a portfolio yielding the cash flow sequence $\phi \equiv \phi_1, \ldots, \phi_N$ at some dates $t_1, \ldots, t_N$. The proposed simplification by RiskMetrics™ consists in replacing the cash flows ϕ by "equivalent" flows $\underline{F}$, paid or received at standard maturity dates (e.g., 1, 2, 3, 6, 12, 24, ... months).

Thus suppose that the expected cash flow ϕ is due at $t = 4$ months; the nearest standard maturities are 3 months and 6 months. We decompose ϕ into two cash flows: 2/3 ϕ due in 3 months and 1/3 ϕ in 6 months. Such a procedure allows transforming any sequence ϕ into a sequence $\underline{F}$ with standard maturities.

Generally, *mapping* makes it possible to represent a portfolio of fixed-income securities, yielding many cash flows on numerous arbitrary dates, by a portfolio of standard zero-coupon bonds whose variance-covariance matrix may be estimated by a statistical method (e.g., using RiskMetrics™).

27.2.1.3 Determining Risk Factors on Which the Value of the Portfolio Depends

A second approach, which incidentally does not rule out the first, consists in finding factors denoted $\underline{Y} = Y_1, \ldots, Y_M$ which determine the portfolio's value so that the latter can be written as a function $V(Y_1, \ldots, Y_M)$ of these M factors.

We note that this formulation is more general than the one which underlies the classic portfolio analysis in the previous subsection and that depends on the equation $V = V_0 \left(1 + \sum_{i=1}^{M} x_i R_i \right)$ (it is enough to interpret the R_i as the risk factors in the last equation, which is a particular case of the equation $V = V(Y_1, \ldots, Y_M)$). Indeed the vector $\underline{Y}$ may contain, in addition to the prices or returns of the assets or of standard securities, different interest rates, the slope and convexity of the yield curve, interest rate spreads associated with counterparty risk or liquidity risk, implicit volatilities, exchange rates, stock indices, etc.

Let us remark that, in the case where the portfolio has been previously decomposed into standard securities, most of the M factors are nothing but the values or returns of these securities.

Once the preliminary analysis and modeling of the portfolio are done, the VaR can be estimated using either a global valuation method (*full valuation*) or a partial one (*partial valuation*).

27.2.1.4 Full Valuation and Partial Valuation

Some methods of estimating the VaR use the valuation function $V(Y_1, \ldots, Y_M)$ without relying on any approximation of it. They amount to a *full valuation* in contrast to methods of *partial valuation* which are based on an approximation of V (linear or quadratic).

- Actually, an initial simplification, necessary for the implementation of some methods discussed in the sequel, consists in linearizing the valuation function $V(Y_1, \ldots, Y_M)$. This method is called *delta valuation*. It estimates the first-order effect on $V(Y_1, \ldots, Y_M)$ of a variation $\Delta \underline{Y}$, in terms of different sensitivities $\frac{\partial V}{\partial Y_i}$ through the formula: $\Delta V(\mathbf{Y}) = \sum_{i=1}^{M} \frac{\partial V}{\partial Y_i} \Delta Y_i$.

Linearization of the valuation function V leads to some approximations that may be inconvenient, notably when the presence of options introduces strong nonlinearities.

– A second, less inexact approximation, mitigating some problems of the linear approximation at the cost of some implementation difficulties that may be prohibitive, is based on an expansion to the second order. It leads to approximating $\Delta V(\underline{Y})$ by a quadratic expression and is called the *delta-gamma* method, referring to the two greek letter parameters used in managing option positions.

Some methods which we will describe in what follows can be implemented in both a *full valuation* or a *partial valuation framework*; other methods depend on a simplification (a linear or quadratic approximation of the loss) and thus are part of *partial valuation methods*.

27.2.2 Estimating the VaR Through Simulations Based on Historical Data

This is the simplest method. It is based on the assumption of stationarity of the joint distribution of assets making up the portfolio under consideration and/or of the risk factors that affect its value. It is as such compatible with both full and partial valuations.

Generally, the method is based on *estimation* of the theoretical ("real") p-quantile of the distribution of L which is in fact unobservable; the estimation is performed using past daily values (risk factors, returns, prices, interest rates, etc.), which allow simulating a sample of size n of daily losses. The estimator of the VaR is just the empirical p-quantile of this sample. The empirical p-quantile is simply obtained as follows:

– Start with a sample of N elements and denote by N_p the integer between pN and $pN + 1$ (if pN is an integer $N_p = pN$); for example, if $N = 1000$ and $p = 0.95$ (95 %), $pN = 950$. Then order the elements of the sample in increasing order: $L^{(1)}$, ..., $L^{(N_p)}$, ..., $L^{(N)}$.
– The empirical p-quantile, $\widehat{L}_p$, is the result of a linear interpolation between $L^{(N_p)}$ and $L^{(N_p+1)}$ (if pN is an integer $\widehat{L}_p = L^{(N_p)}$). For example, for a sample of 1000 elements ordered by increasing value, the empirical quantile at a threshold of 99 % will be the 990th element. We will present numerous examples of this calculation and study statistically the degree of confidence that we may grant it.

27.2.2.1 Calculating the VaR of an Individual Asset
First, let us consider the very simple case of a portfolio made up of a single asset. We have $N + 1$ observations of values v_j taken on by this asset on days $j = 0, -1, \ldots, -N$

(the date $-N$ is the most distant past date and 0 today's date). These observations allow us to calculate N returns or relative variations in the price: $R(j) = \frac{v_j - v_{j-1}}{v_{j-1}}$, $j = 0, \ldots, -(N-1)$.

These N returns then allow us to construct an empirical distribution for the price v_1 which will be quoted tomorrow (from the N values $v_1^{(j)} = v_0(1 + R(j))$), as well as the empirical histogram for the loss (from the N values $L_1^{(j)} = v_0 - v_1^{(j)}$). We can thus simply estimate the VaR(1d, p) from the empirical p-quantile of the histogram of losses.

It is sufficient, for example, to rank the N values $L_1^{(j)}$ in increasing order and the $(p \times N)^{th}$ term in this list will be the estimate for the VaR we are looking for.

Example 7

Consider a portfolio whose value is $50 M today, with one single asset, which has shown over the most recent 100 trading days the 100 returns *arranged in decreasing order* in column 2 of the table below (which implies a ranking of the losses in increasing order).

Observation	Return on the security	Simulated value of the portfolio tomorrow	Simulated value of the loss (gain if < 0)		
1	11 %	$50 \times 1.11 = 55.5$	-5.5		
2	8.2 %	$50 \times 1.082 = 54.1$	-4.1		
3	7.5 %	$50 \times 1.075 = 53.75$	-3.75		
⋮	⋮	⋮	⋮		
93	-3.8 %	$50 \times 0.962 = 48.1$	1.9		
94	-4.2 %	$50 \times 0.958 = 47.9$	2.1		
95	-4.6 %	$50 \times 0.954 = 47.7$	2.3	←	VaR (95 %)
96	-5 %	$50 \times 0.95 = 47.5$	2.5		
97	-5.6 %	$50 \times 0.944 = 47.2$	2.8		
98	-6 %	$50 \times 0.94 = 47$	3		
99	-6.8 %	$50 \times 0.932 = 46.6$	3.4		
100	-10 %	$50 \times 0.9 = 45$	5		

On the basis of past data, and in the absence of variation over time in the distribution of the security return, the VaR(1d, 95 %) can be estimated to be $2.3 M since the return turns out to be less than the return leading to a corresponding loss (-4.6 %) in 5 % of the cases.

27.2.2.2 The Case of a Portfolio of M Securities

The method is based on the principle demonstrated in the case of a single security, i.e., on a simulation of the variation in value of the portfolio considered between today and tomorrow using the M variations in value of the component securities over the previous N days; to each past day correspond M joint returns on the individual securities and a portfolio value simulated for the morrow.

More precisely, we consider a portfolio x composed of M different securities denoted $i = 1, \ldots, M$. The value *today* of this portfolio is $V_x(0)$ and the weight given *today* to security i is denoted x_i. We also have a history of the M prices over $N + 1$ days. From this history, for each day j ($j = 0, \ldots, -(N-1)$):

1. We calculate the past returns of the M securities. The return observed on day j for security i is $R_i(j)$; a daily observation or scenario corresponds to an M-tuple $(R_1(j), \ldots, R_M(j))$.

2. We simulate N returns $R_x(j)$ on portfolio x *calculated on the basis of its composition and value today and of the M-uple $R_1(j), \ldots, R_M(j))$: $R_x(j) =$

   $\sum_{i=1}^{M} x_i R_i(j))$; in this way we get N values of R_x simulated for the next day to come.

3. We deduce the N *corresponding* variations in the value of the portfolio from today until tomorrow; expressed as losses (variations with a minus sign), these variations may be written:

$$L_x^{(j)} = -V_x(0)\, R_x(j), \text{ or :}$$

$$L_x^{(j)} = -V_x(0) \sum_{i=1}^{M} x_i R_i(j) \text{ for } j = 0, \cdots, -(N-1). \tag{27.10}$$

4. We construct, from the N simulated losses, an empirical loss distribution for the present portfolio, with a horizon of one day. We obtain, using this empirical distribution, an empirical VaR (which is an estimate for the real or theoretical VaR(1d, p)), as in the simple case of a portfolio consisting in single security analyzed in Sect. 27.2.2.1).

Example 8

Let us consider a portfolio, whose value today is \$100 M, made up of three assets in the respective proportions 40 %, 30 %, and 30 %. These three assets have exhibited in the past 250 trading days the returns shown in columns 2, 3, and 4 of the table below. Column 5 simulates what the return of the portfolio will be tomorrow if the returns on the three assets continue to conform to what

(continued)

Example 8 (continued)

they did on day j for $j = 1, \ldots, 250$ and column 6 simulates the corresponding losses.

Jour j	R_1	R_2	R_3	$R_x = 0.4R_1 + 0.3R_2 + 0.3R_3$	$L_{1j} = -100R_x$ (M€)
1	1.5 %	1.2 %	0.8 %	1.20 %	−1.20
2	3 %	2.2 %	1.7 %	2.37 %	−2.37
3	2.5 %	1.9 %	2.0 %	2.17 %	−2.17
⋮	⋮		...	...	...
245	−3.8 %	−2.1 %	−1 %	−2.45 %	2.45
246	0 %	−1 %	0 %	−0.30 %	0.30
247	−4.6 %	−3.1 %	−2.4 %	−3.49 %	3.49
248	−5 %	−3 %	−2.2 %	−3.56 %	3.56
249	0.5 %	0 %	0.2 %	0.26 %	−0.26
250	−2 %	0.2 %	−1.1 %	−1.07 %	1.07

Ordering the simulated losses in increasing order (differing from the chronological order of the preceding table) gives the following table:

Simulation:	1	2		242	243	244	245	246	247	248	249	250
L_x:	−4.4	−3.9	...	3.49	3.53	3.56	4.0	4.12	4.52	4.82	5.0	5.9

Assuming the joint distribution of the three securities' returns behaves today as it was in the past year, the VaR(1d, 97 %) can be estimated as $351 M (between 3.49 and 3.53) since the loss is greater than or equal to this number in 8 simulations out of the 250 (3.2 % of the cases).

In addition, Eq. (27.9) permits us to estimate the VaR over 10 days with a probability of 99 % (in accordance with the Basel directive); indeed, since $\alpha_{0.99} = 2.32$ and $\alpha_{0.97} = 1.88$, one may write:

$$\mathrm{VaR}(10d, 99\%) \# \frac{\alpha_{0.99}}{\alpha_{0.97}} \sqrt{10}\, \mathrm{VaR}(1d, 97\%) = \$13.70 \text{ M}.$$

In other words, the probability that this portfolio loses more than 13.7 % of its value in the coming 10 days is estimated to be of the order of 1 %. In fact, this example is purely pedagogical and we will see later that such an estimate is based on too small a number of observations to be trusted.

27.2.2.3 VaR of a Portfolio Whose Value Depends on Different Risk Factors

In the method just discussed, the returns of the portfolio components are chosen to be independent variables that fix the portfolio's value at a horizon h, and thus the loss from 0 to h. This can then be written $L_h = -V_0 \sum_{i=1}^{M} x_i R_i$.

Such a method sometimes turns out to be inadequate, especially if the portfolio contains derivatives and interest rate products.

In fact, the amount lost L_h can be more generally considered as a function of M risk factors denoted $Y_1, \ldots, Y_M$, so that:

$$L_h = f (Y_1, \ldots, Y_M). \qquad (27.11)$$

We have already remarked the formulation is more general than that which underlies the method exposited in the preceding subsection ($L_h = -V_0 \sum_{i=1}^{M} x_i R_i$) in which it is enough to interpret as R_i as the risk factors to recover a *special* form of (27.11). The technique of simulating the losses L_h staring from historical data for the risk factors Y_i is analogous to that discussed previously: for each date j on which an M-tuple $\underline{Y} = Y_1^{(j)}, \ldots, Y_M^{(j)}$ is observed, there corresponds a simulated loss $L_h^{(j)} = f (Y_1^{(j)}, \ldots, Y_M^{(j)})$. Using the different $L_h(j)$ calculated in this way, we construct the empirical histogram for the loss L_h from which we estimate the VaR with the chosen probability level.

Example 9

Let us consider a position made up of forward contracts on the \$ (against the €) whose VaR we wish to compute. Such contracts are identical, with 3-month maturity, and contracted on the basis of a forward rate €/\$ of 0.8, which entails that one will pay 8 M€ and will receive 10 M\$ in 3 months. Such a portfolio is therefore equivalent to the combination of a loan of 10 M\$ and a debt of 8 M€. Let S denote the spot price of the \$ in terms of the €; i_S and $i_{€}$ denote the 3-month rates on the \$ and € (proportional rates), respectively.

The value V of this portfolio can be expressed as the difference between the present value V_1 of the 10 M\$ dollars to be received and the present value of the 8 M€ to be paid. Expressing these values in euros, we get

$$V_1 = S \times 10/(1 + i \times 0.25) ; V_2 = 8/(1 + i \times 0.25) ; V = V_1 - V_2 \text{ (euros)}. \qquad (27.12)$$

V thus depends on three risk factors: S, i_S and $i_{€}$. At inception, the contract has a market value V of zero; immediately after, V fluctuates along with

(continued)

Example 9 (continued)

variations in the three risk factors. In particular, on the first day V *will undergo variations* ΔV determined by ΔS, Δi_S and Δi_e. We seek a confidence interval for ΔV at 95 % probability, which amounts to finding the VaR(1 day, 95 %) of this position.

On the 260 most recent trading days $j = 1, \ldots, 260$, we have noted the relative variations $(\Delta_j S)/S$, $(\Delta_j i_S)/i_S$ and $(\Delta_j i_e)/i_e$. They would have caused for the position considered today the variations $\Delta_j V \ (= -L_j)$, for $j = 1, \ldots, 260$, calculable using (27.12).

The (empirical) distribution of these 260 values $\Delta_j V$ can serve as a basis for calculating the VaR. Arranging the 260 $\Delta_j V$ in increasing order (from the most unfavorable to the most favorable), the 13th value is the empirical VaR that yields an estimate for VaR(1 day, 95 %).

Example 10

Now consider a more complex portfolio made up of x shares of a stock with a unit price of S, y bonds priced at B and z options on the stock valued P.

The bond price is at each moment a known function $B(r)$ of the bond rate r. The option price, which can be computed with the Black-Scholes formula, is a known function $P(S, \sigma, i)$ of the price S of the underlying, the short-term rate i and the (implicit) volatility σ. We can therefore give the global value V of the position as a function of the four risk factors S, r, σ, and $i : V(S, r, \sigma, i) = x S + y B(r) + z P(S, \sigma, i)$.

Because of this, the variation ΔV between today (date 0) and tomorrow (date 1) can be written

$$\Delta V = -L_{1j} = V(1) - V(0) = x\Delta S + y\Delta B + z\Delta P$$
$$= x S(0)R_S + y(B(r(1)) - B(r(0)))$$
$$+ z[P(S(0)(1 + x R_S), \sigma(1), i(1)) - P(S(0), \sigma(0), i(0))].$$

R_S represents the return on the share over the next day and may substitute for S to express the risk affecting the stock price. The last equation and the past observations on the joint variations of the four factors R, r, σ, and i thus allow the simulation of the empirical distribution of L_{1j} from which we deduce an estimate of the VaR.

27.2.2.4 Reliability and Precision of the Empirical VaR

The p-quantile or empirical VaR, here denoted $\widehat{L}_p$, is an estimator of the "true VaR", called the theoretical VaR, which is unobservable. It is a random variable calculated from a sample of losses (observed or simulated) as in the preceding examples. If this

27.2 Estimating the VaR 1129

estimator is constructed from a sample containing n independent and identically distributed elements, one can show that its standard deviation equals

$$\sigma\left(\widehat{L}_p\right) = \frac{1}{f(L_p)} \sqrt{\frac{p(1-p)}{n}}, \tag{27.13}$$

where f is the density for the loss L evaluated at the point L_p, the p-quantile or theoretical VaR.

Note that, for high probability thresholds (near 1), $f(L_p)$ is usually very small (since it is placed at one extremity of the distribution where the density is very small). From this, it follows that the standard deviation of the estimator is large. Intuitively, since the empirical p-quantile is a local estimator, only the observations around L_p are really relevant and these are few in number. This is why the standard deviation of such a p-quantile is larger than that of estimators of the mean or variance of X.

Also, note that since the function f and the theoretical p-quantile L_p are unobservable, we must estimate $\sigma\left(\widehat{L}_p\right)$ and calculate confidence intervals from estimates of f and the VaR.

Example 11

As an example, and to provide some orders of magnitude, let us take the case of a Gaussian loss L with zero mean and standard deviation of 2 M€ (one could think of the daily loss affecting a portfolio of 200 M€ whose annual return has a standard deviation of 16 %).[6] The one-day VaR at a probability level of 95 % is $\mathrm{VaR}(1d, 0.95) = L_{0.95} = \alpha_{0.95} \times 2 = 3.29$ M€.

Let us think of a risk manager who does not know this VaR and who uses an empirical quantile estimated from a sample with only 256 values (1 year of observations). By (27.13), the standard deviation of this estimator is 0.265. There is therefore a probability of 5 % that the estimator is outside the interval $[3.29 - \alpha_{0.975} \times 0.265; 3.29 + \alpha_{0.975} \times 0.265]$, i.e., $[2.77; 3.81]$. The confidence interval is thus very large (and the estimator very unreliable). If the VaR were calculated with a confidence level of 1 %, the confidence interval for the empirical quantile would be even larger ($[2.37; 4.20]$).

Furthermore, the risk manager does not know L is Gaussian[7] (undoubtedly it is not in reality), and will not center the confidence interval about 3.29 but about the value obtained for the empirical quantile (for instance, 3). An additional problem is that successive losses are truly not i.i.d. That is the

(continued)

---

[6]Using 256 trading days per year, the daily standard deviation of the return equals $16\%/\sqrt{256} = 1\%$, that is a standard deviation for the loss $= 200 \times 1\% = 2$ M€.

[7]If she knew it, she would be advised to use an estimator for the variance and to apply Eqs. (27.3a) or (27.3b).

> **Example 11** (continued)
> reason why empirical VaR with high probability levels p and/or those estimated with a small number of data are questionable.
>
> A more fundamental, and probably more challenging, issue stems from the non-stationary nature of the rate, price, or return processes on which the VaR depends. In particular, we observe that, during financial crises, not only asset price volatilities increase but correlations also do, and sometimes dramatically so, which may considerably reduce the benefits of portfolio diversification. The VaR during a crisis thus is substantially larger than the VaR estimated from past, non-crisis data. This is one of the reasons why the losses reported during the 2007–2009 crisis have been unexpectedly large. The simplest way to account for the extra risk inherent to crisis periods consists in supplementing the "normal" VaR calculation by that of a "stressed" VaR, i.e., a VaR estimated from a larger sample that includes a period of crisis.

27.2.3 Partial Valuation: Linear and Quadratic Approximations (the Delta-Normal and Delta-Gamma Methods)

Simulations based on historical data as presented in Sect. 27.2.2 above require very little modeling and no parameter estimation (except possibly when a *bootstrap* method is employed). The methods discussed in this subsection require more complicated modeling, are based on a (linear or quadratic) approximation of the valuation function, and necessarily involve an estimation procedure. This concerns the parameters controlling the joint variations in the risk factors that determine the gain or the loss of the portfolio under consideration.

We present in Sect. 27.2.3.1 and Sect. 27.2.3.2 below the analytic method called *delta-normal* which is based on linearizing the valuation function V (first order expansion of ΔV); usually this method requires an assumed (joint) normality for the factors affecting V, so as to obtain normality for ΔV. In Sect. 27.2.3.3 we describe the quadratic (or *delta-gamma*) model, which is more precise but more complicated to implement and is based on a second-order approximation to ΔV.

27.2.3.1 General Sketch of the Linear Model (Delta-Normal Method)

The starting point is the same as for the simulation from historical data discussed in § 27.1.2.3: we need to determine the *risk factors* (prices, stock indices, segments of the yield curve(s), commodity prices and exchange rates) that impact the different components of the portfolio.

Denote by $\underline{Y}' \equiv (Y_1, \ldots, Y_M)$ these M factors; $\underline{Y}$ is a column vector and its transpose $\underline{Y}'$ is a row vector. $V(t, \underline{Y}(t))$ denotes the value of the portfolio at any date t, and the loss from date 0 to date h can be written

$$L_h = -[V(h, \underline{Y}(h)) - V(0, \underline{Y}(0))]. \tag{27.14a}$$

27.2 Estimating the VaR

A series expansion leads to the linearized form

$$L_h = -\left(\frac{\partial V}{\partial t}(0, \underline{Y}(0))h + \sum_{i=1}^{M} \frac{\partial V}{\partial Y_i}(0, \underline{Y}(0))\Delta Y_i \right) \qquad (27.14b)$$

where $\Delta Y_i \equiv Y_i(h) - Y_i(0)$ represents the change in the ith factor from 0 to h.

As already explained, Eq. (27.14a) expresses a *full valuation* while (27.14b) is the approximation called *delta valuation*. It is this last that is often used when one wants an analytic formula for the VaR.

Furthermore, while a simulation based on historical data does not require a formula for the probability distribution of L_h, such a formula is necessary for the analytic method (which is also called, for this reason, the parametric method); it will be obtained from an econometric estimate of the parameters characterizing the distribution of $\underline{Y}(h)$. More precisely, the *analytic method* is implemented in three stages:

- An a priori assumption on the distribution of the ΔY_i, the standard assumption being that they are described by a multivariate Gaussian;
- An estimate of the parameters of the joint distribution of the ΔY_i (notably their variance-covariance matrix):[8]
- An application of Eq. (27.14b) (the linearized form) which makes L_h Gaussian (if the ΔY_i are themselves Gaussian) thus allowing the unique characterization of the distribution by just its mean and variance and the calculation of a confidence interval, i.e., a VaR.

The calculation in the third stage can be carried out as follows:
Rewrite Eq. (27.14b) in vector form:

$$-L_h = \frac{\partial V}{\partial t}(0)h + \Delta \underline{Y}' \frac{\partial V}{\partial \underline{Y}}(0) \qquad (27.14c)$$

(the second term on the right is a scalar product).

Denote by Γ the variance-covariance matrix of $\Delta \underline{Y}$ and by μ_Y its mean, estimated in stage 2, and express the mean μ_h and variance σ_h^2 of L_h using (27.14c):

$$-\mu_h = \frac{\partial V}{\partial t}(0)h + \underline{\mu}_Y' \frac{\partial V}{\partial \underline{Y}}(0) \qquad (27.15)$$

---

[8] There are different ways of estimating the variance-covariance matrix of the ΔY_i, in particular, the *simple historical correlation* where each past observation has equal weight in the calculation, the *moving average with exponential weights* where each observation has a weight that becomes weaker the further it is in the past, and *GARCH* models (*Generalized AutoRegressive Conditional Heteroskedasticity*, see Chap. 8, Sect. 8.2).

$$\sigma_h^2 = \frac{\partial V'}{\partial \underline{Y}}(0)\Gamma\frac{\partial V}{\partial \underline{Y}}(0). \tag{27.16}$$

That L_h is Gaussian and we know its two moments from (27.15) and (27.16) implies we can calculate a VaR (e.g., by using Eq. (27.3a)).

The normality assumption, the linearization of the value function and the necessity of estimating the parameters of distributions (mainly the variance-covariance matrix of the ΔY_i) have earned this analytic method the names *delta-normal*, *parametric* or *variance-covariance method*.

27.2.3.2 Illustration of the *Delta-Normal* Method: RiskMetrics™

As an illustration of a parametric method of *delta valuation*, we briefly present the RiskMetrics™ procedure which can be sketched as follows:

- The constituents of the portfolio are decomposed into elementary cash flows (*mapping*), each of which is sensitive to a *unique* risk factor Y_i and defined by an amount in a certain currency and a payment date. Therefore, there is a bijective correspondence between these elementary cash flows and the risk factors $\underline{Y}$.
- The *risk factors* may be:
  - Zero-coupon rates (by country);
  - Forward interest rates (by country);
  - Exchange rates (spot and forward);
  - Market indices (by country);
  - The global prices of some raw materials.
- One computes the **DEaR** (Daily Earnings at Risk) for an *isolated position i*:

In the RiskMetrics terminology, the DEaR is the one-day VaR. For a position i considered *unique and isolated,* this is calculated from the equation:

$$DEaR_i = V_i \times S_i \times X_i \tag{27.17}$$

where:

V_i is the market value of the position,

S_i is the sensitivity of position i to a unit change in the price or interest rate of factor Y_i

$$S_i = \frac{1}{V_i}\frac{\partial V_i}{\partial Y_i},$$

and X_i is the size of an adverse shock to Y_i (a price or an interest rate). This unfavorable change is defined, under the assumption of normality, as the quantile $\alpha_{1-p}\sigma(Y_i)$ of the variation in the factor Y_i *per day*; for example: $X_i = 1.65 \times \sigma(Y_i)$ for a probability level of $p = 95~\%$ or $2.33 \times \sigma(Y_i)$ for $p = 99~\%$.

Note that Eq. (27.17) is in agreement with (27.8) (with $h = 1$).

– *Calculating the* **DEaR** *for a portfolio with N elements*

27.2 Estimating the VaR

By Eq. (27.16), the DEaR (or VaR1d) of a portfolio can be written

$$DEaR_p = \left(\underline{D} \times \Gamma \times \underline{D}'\right)^{0.5} \tag{27.18}$$

where:

$\underline{D}$ is the $(1 \times N)$ vector of the $DEaR_i$, $i = 1, \ldots, N$, where' denotes a transpose, and

Γ is the $(N \times N)$ *correlation matrix of the risk factors.*

– *Calculating the* **VaR** for a horizon h:

$$VaR = DEaR_p \times (\text{number of holding days})^{0.5} \tag{27.19}$$

where the number of holding days (h) is typically 1, 5, or 10 trading days.

Example 12

In this hypothetical example, the exposed position is a portfolio of European fixed-rate bonds with a value of 43 million euros including several issues whose maturities extend from 1 month to 5 years. We would like to calculate the DEaR or *one-day VaR* of this portfolio with a confidence interval of 95 % using Eq. (27.3b) that is here written

$$VaR(1d, 95\%) = \alpha_{0.95} \ \sigma = \ 1.65 \times \sigma.$$

In a first step (*mapping*), all the bonds are decomposed into the cash flows of which they are composed (amounts in K€ and dates) *regrouped into the 8 maturities* 1, 3, and 6 months and 1 to 5 years (1st and 2nd columns of the table below) considered relevant.

Rates	Position (K€)	1.65 × σ	DeaR (K€)	Correlations							
				1 month	3 month	6 month	1 year	2 year	3 year	4 year	5 year
1 month	1463	0.92 %	13.5	1.00	0.92	0.87	0.79	0.67	0.65	0.63	0.61
3 month	595	1.04 %	6.2	0.92	1.00	0.88	0.81	0.79	0.76	0.73	0.70
6 month	2146	1.20 %	25.8	0.87	0.88	1.00	0.87	0.82	0.79	0.77	0.73
1 year	6565	1.83 %	120.1	0.79	0.81	0.87	1.00	0.86	0.85	0.84	0.82
2 year	10,208	2.81 %	286.8	0.67	0.79	0.82	0.86	1.00	0.99	0.99	0.98
3 year	5900	3.04 %	179.4	0.65	0.76	0.79	0.85	0.99	1.00	0.99	0.99
4 year	8863	2.92 %	258.8	0.63	0.73	0.77	0.84	0.99	0.99	1.00	0.99
5 year	6999	2.76 %	193.2	0.61	0.70	0.73	0.82	0.98	0.99	0.99	1.00
			Total	% of the position							
Total	42,739	Non diversified	1083.8	2.54							
		Diversified	974.5	2.28							

Then, the volatilities σ for the 8 different rates and their correlations are estimated and reported respectively in column 3 and columns 5 to 12. Column

(continued)

Example 12 (continued)

3 displays 1.65 times the standard deviation of the price variations of the 8 relevant zero-coupon bonds and columns 5 through 12 display their correlation coefficients (lying therefore between -1 and $+1$).

Column 4 (column 2 × column 3) reports the $DEaR_i$ for each cash flow row, and the sum of the 8 rows gives the total *non-diversified* $DEaR_p$ calculated by simple addition of the different $DEaR_i$, that is without taking into account the imperfect correlations between the changes in zero-coupon prices (1,083,800 €).

The last row of the table displays the total *diversified* $DEaR_p$ or VaR (974,500 €) taking into account the correlation matrix and calculated with formula (27.18) above. Note that there is little difference in this example between the two $DEaR_p$ because the correlations between the different European rates are strong. Also remark that the VaR and the $DEaR_p$ are here equal since the holding period is $h = 1$.

27.2.3.3 The Quadratic or Delta-Gamma Model

The *delta-normal* method just described depends on a linearization of $V(h, \underline{Y}(h))$ around $(0, \underline{Y}(0))$ which is insufficient in the presence of nonlinear positions such as options.

Much better precision can be obtained by expanding V to the second order, that is, approximating ΔV with a quadratic form. The quid pro quo for this enhanced precision is the impossibility of finding a formula for VaR (except at the cost of approximations and/or in some special cases), which means much greater difficulty in implementing this method.

We introduce the method by considering the case of a position made up of a call whose value $C(t,S)$ is calculated using a Black-Scholes (BS) type model. Here, the only risk factor Y is the underlying's price, called S in current practice. Such a position is sufficiently simple to allow for different methods of computing the VaR (global methods, partial, analytic). We will present and compare them to provide intuition.

Let us consider the effect $C(t + \Delta t, S + \Delta S) - C(t, S)$ on the call of a change ΔS of the underlying in the time interval Δt.

Let us first calculate the VaR using the two models of partial valuation.

The linear model is based on: $C(t + \Delta t, S + \Delta S) - C(t, S) \equiv \Delta C \# \frac{\partial C}{\partial t} \Delta t + \frac{\partial C}{\partial S} \Delta S$.

The quadratic model based on: $\Delta C \# \frac{\partial C}{\partial t} \Delta t + \frac{\partial C}{\partial S} \Delta S + \frac{1}{2} \frac{\partial^2 C}{\partial S^2} (\Delta S)^2$.

These two expressions for ΔC can be written using greek parameters (*see* Chap. 12):

27.2 Estimating the VaR

$$\theta = \frac{\partial C}{\partial t}; \quad \delta = \frac{\partial C}{\partial S}; \quad \Gamma = \frac{\partial^2 C}{\partial S^2}$$

$$\Delta C = \theta \, \Delta t + \delta \, \Delta S, \text{ for the linear model;} \qquad (27.20a)$$

$$\Delta C = \theta \, \Delta t + \delta \, \Delta S + 0.5\Gamma \, (\Delta S)^2, \text{ for the quadratic model.} \qquad (27.20b)$$

Recall that an option (bought) is characterized by a negative theta and a positive gamma, delta being positive (negative) for a call (put).

Note that, in the quadratic model, if ΔS is Gaussian ΔC is not so because of the term in $(\Delta S)^2$.

Denote by s_{1-p} the $(1-p)$-quantile of the distribution of ΔS such that $\text{Prob}(\Delta S \leq s_{1-p}) = 1-p$ (in general $s_{1-p} < 0$ when $p \gg 50\ \%$). Remark that:

$$\text{VaR}(\Delta t, p) = C(t, S) - C(t + \Delta t, S + s_{1-p}).$$

Denote by $\text{VaR}_l(p)$ and $\text{VaR}_q(p)$ the $\text{VaR}(\Delta t, p)$ respectively calculated with the linear and with the quadratic model.

Equations (27.20a and 27.20b) mean we can write:

$$\text{VaR}_l(p) = -\theta \, \Delta t - \delta \, s_{1-p}; \text{VaR}_q(p) = -\theta \, \Delta t - \delta \, s_{1-p} - 0.5\Gamma \, s^2_{1-p}$$

with $\delta > 0$ and s_{1-p} likely < 0.

We remark that the VaR is *smaller* in the quadratic model than in the linear model (since $0.5\Gamma \, s^2_{1-p} > 0$): in fact, convexity mitigates the negative impact of unfavorable variations in S.

The reader will verify for a put:

$$\text{VaR}_l(p) = \theta \, \Delta t + \delta \, s_p; \text{VaR}_q(p) = \theta \, \Delta t + \delta \, s_p + 0.5\Gamma \, s^2_p \; (\delta > 0 \text{ and } s_p > 0).$$

The same remark about convexity applies here.

Remark that, in the simple case of an option valued with a BS model, "exact" valuing of the VaR (*full valuation*) is also possible with the help of a quasi-analytic expression. Indeed, in the BS world, the value of a call is influenced by a single random factor, S, and increases with its price. As a result, in the case of a call, the bad states of nature occur at $t + h$, with a confidence level p, when $S(t + h) = S(t) + s_{1-p}$ and we have, as already observed:

$$\text{VaR}(h, p) = C(t, S(t)) - C(t + h, S(t) + s_{1-p}).$$

In this expression for the VaR, the premiums are calculated using the quasi-analytic BS formula.

Remark that for a put: $\text{VaR}(h, p) = P(t, S(t)) - P(t + h, S(t) + s_p)$.

Example 13

Let us consider a call on one share of stock X with strike $33 and maturity 0.5 year. X quotes $30 and its annualized volatility is 30 %; the interest rate is 3.9 %. The price of the call, according to the Black-Scholes (BS) model, is $C = \$1.614$, and the greek parameters are: $\delta = 0.40$; $\Gamma = 0.060$; daily theta $= -0.008$.

Using the assumed annual volatility of 0.3, one approximates the daily variation ΔS (expressed in $) by a Gaussian centered at 0 with standard deviation $= (0.3 \times 30)/\sqrt{250} = \$0.569$. The quantile of ΔS at 95 % probability is therefore $s_{95\%} = 1.65 \times 0.569 = + \$0.94$: a probability of 5 % thus prevails that the stock price X gains $0.94 or more in the next 24 h.

Now consider a position made up of 100,000 *sold* calls whose VaR will be calculated from $s_{95\%}$. The daily theta of the position equals $0.008 \times 100,000 = + 800$, its delta est $-0.4 \times 100,000 = -40,000$ and its gamma equals $-0.06 \times 100,000 = -6000$. In this very simple example, the VaR can be calculated using the linear and quadratic approximations, and by using the BS formula (*full valuation*).

– The linear model (VaR$_p = -\theta \Delta t - \delta s_{1-p}$) gives:

VaR$_l$(1 day, 95 %) $= -800 + 40,000 \times 0.94 = \$36,800$ (the theta could possibly be ignored).

– The quadratic model (VaR$_p = -\theta \Delta t - \delta s_{1-p} - 0.5\Gamma s_{1-p}^2$) leads to:

VaR$_q$(1 day, 95 %) $= 36,800 + 0.5 \times 6000 \times (s_{95\%})^2 = 36,800 + 2650 = \$39,450$.

The VaR of a *short* option position is, therefore, larger (by $2650) if it is estimated using the quadratic model rather than the linear model (convexity is unfavorable to the seller of options).

Generally, the linear model overvalues the VaR of positions with positive gamma and undervalues those of portfolios with negative gamma. In this example, the effect of gamma contributes $2650 to the global VaR, which is 6.7 % of it; this effect would be enhanced if the VaR were calculated for 5 days (with s_{1-p} even higher, the relative weight of the term $0.5\Gamma s_{1-p}^2$ would be larger, both in magnitude and in relatively), or if the fact that the Gaussian model underestimates the tails of distributions (here the right tail) were taken into account.

Finally remark that the assumption of a Gaussian variation ΔS is theoretically incompatible with normality of ΔC (because of the term $0.5\Gamma (\Delta S)^2$), in contrast to the case of the linear model.

(continued)

27.2 Estimating the VaR

Example 13 (continued)
- The *full valuation* method uses the BS formula, which gives[9]: $C-(S + s_{95\%}) = C(30.94) = \$2.009$, from which

$$\begin{aligned} \text{VaR}(1\text{ day}, 95\%) &= 100,000\ (C(30.94) - C(30)) \\ &= 100,000\ (2.009 - 1.614) = 39,500, \end{aligned}$$

a result almost identical to that obtained with the quadratic approximation.

Even more generally, when the position's value depends on several random factors $\underline{Y}$, the second-order expansion on which the quadratic approximation depends, and which generalizes (27.14b), can be written:

$$\Delta V = -L_h = \frac{\partial V}{\partial t}(0)h + \Delta \underline{Y}' \frac{\partial V}{\partial \underline{Y}}(0) + \frac{1}{2}\Delta \underline{Y}' \frac{\partial^2 V}{\partial \underline{Y}^2}(0)\Delta \underline{Y} \qquad (27.21)$$

where $\frac{\partial^2 V}{\partial \underline{Y}^2}$ is the $M \times M$ matrix with typical entry $\frac{\partial^2 V}{\partial Y_i \partial Y_j}$ and the second-order terms involving $\frac{\partial V}{\partial t}$ ($\frac{\partial^2 V}{\partial t^2}h^2$ and the $\frac{\partial^2 V}{\partial t \partial Y_i}h\Delta Y_i$) are neglected.

Equation (27.21) is difficult to use since the distribution law for ΔV is difficult, or impossible, to determine analytically, and contrary to the one-factor case as in Example 13, the VaR is not simply and directly connected with the p-quantile of a single factor but depends on the joint law of M factors $\underline{Y}$. Furthermore, the method requires, as we will see, an estimate for the variance-covariance matrix of $\Delta \underline{Y}$.

Several approaches, however, are possible that lead to different approximations of the VaR.

- A first approach relies on the approximation to ΔV's law by a normal law for which, by (27.21), the variance and mean are respectively equal to:

$$\sigma_{\Delta V}^2 = \sum_{i=1}^{M} \sum_{j=1}^{M} \frac{\partial^2 V}{\partial Y_i \partial Y_j}(0)\ \text{cov}(\Delta Y_i, \Delta Y_j);$$

$$E(\Delta V) = \frac{\partial V}{\partial t}(0)h + \underline{\mu}' \frac{\partial V}{\partial \underline{Y}}(0) + \frac{1}{2}\sigma_{\Delta V}^2,$$

where $\underline{\mu}'$ is the row vector of means of the ΔY_i.

---

[9] Using an option maturity of 0.5 year minus 1 day, i.e., 0.4973 year.

1138   27 Value at Risk, Expected Shortfall, and Other Risk Measures

- It is also possible to find a *non-Gaussian* law that approximates the distribution of ΔV. For example, one may find in a certain class of distributions those whose first four moments best correspond to those of ΔV.[10] Or one can use the Cornish-Fisher expansion (explained in Sect. 27.4, § 27.4.1) to adjust the VaR in order to take into account moments of order 3 and 4. Finally one can, as we explain in Sect. 27.4.2, by employing extreme value theory, determine the generalized Pareto distribution whose parameters best fit the tail of the distribution of ΔV.[11]
- Finally, the quadratic model can be combined with a simulation (Monte Carlo or historical) involving the construction of a sample with a great number of $\Delta \underline{Y}$; using this representative sample, an empirical distribution for ΔV is built using Eq. (27.21). Simulations from historical data have already been described; calculating the VaR by using Monte Carlo simulations is the subject of the next subsection.

27.2.4 Calculating the VaR Using Monte Carlo Simulations

Monte Carlo simulations were the subject of Chap. 26. Different examples of calculating the VaR are presented there. This is a probabilistic method intended to simulate empirically the distribution of $V(t, \underline{Y}(t))$, either at different instants t along a trajectory between the dates 0 and T, or at the single instant T (the terminal date). Monte Carlo simulations are above all useful when it is difficult or impossible to solve a problem in closed form, which is true in numerous complicated situations that present themselves in practice. The method consists in synthesizing a sample of large size M drawn according to the law of $\underline{Y}(t)$ and associating to each element $\underline{Y}_j$ in this sample the corresponding value $V(t, \underline{Y}_j)$. According to the type of problem at hand, Monte Carlo simulation involves the creation of a great number M of possible *paths* $\underline{Y}_{j,t}]_t = t_1, t_2, \ldots, t_N$, from which one calculates the paths of the corresponding values $V(t, \underline{Y}_{j,t})]_t = t_1, t_2, \ldots, t_N$, or one simply calculates M values $V(h, \underline{Y}_j)$ on a *single date h*.

In order to reduce the overlap with Chap. 26, we will describe the simulation of the values $Y(h)$ and $V(h, Y(h))$ on the single date h near 0 (from 1 to 5 days) and in the presence of a single risk factor. The simulation of several factors $\underline{Y}$ (in general correlated) is analyzed in Chap. 26. In the following formulas, Y_j denotes a particular realization of the random variable $Y(h)$.

We start by simulating $Y(h)$ from the stochastic equation supposed to govern its evolution, for example, the one that a diffusion process obeys:

---

[10] Closed-form expressions for the mean, variance, skewness and kurtosis of ΔV are first calculated from (27.21). Next one seeks in a sufficiently large class of distributions, for example that of Johnson distributions (which notably includes both normal and log-normal laws) those whose first four moments best conform to the results. The VaR is finally calculated from the p-quantile of this distribution.

[11] See, for example, F. Longin: "The asymptotic distribution of stock market returns", *Journal of Business* n° 69, 1996.

27.2 Estimating the VaR

$$dY = \mu(t, Y(t))dt + \sigma(t, Y(t))dW, \tag{27.22}$$

where dW is a standard Brownian (Wiener) increment and $\mu(.)$ and $\sigma(.)$ are known functions.

Next, we proceed to discretize this equation between 0 and h, which leads to

$$Y(h) - Y(0) = \mu(0, Y(0))h + \sigma(0, Y(0))\sqrt{h}U \tag{27.23}$$

where U is a standard normal variate and $Y(0)$ is a given initial value.

The sample $Y_j]_{j=1, \ldots, M}$ is computed from Eq. (27.23) and M independent drawings $U_j]_{j=1, \ldots, M}$ from a standard Gaussian variate. These drawings U_j can be carried out with a suitable program (e.g., the spreadsheet Excel, useful for small applications) or programmed directly employing some method such as the Box-Muller method described in Chap. 26.

The sample $Y_j]_{j=1, \ldots, M}$ thus allows constructing a sample $V(h, Y_j)]_{j=1, \ldots, M}$ (the function $V(.)$ is assumed given) from which the VaR can be estimated.

To sum up, simulation consists in, successively: (*i*) drawing, using a standard normal distribution, M values $U_j]_{j=1, \ldots, M}$; (*ii*) calculating from the M drawn values the M corresponding values of $Y(h)$ denoted $Y_j]_{j=1, \ldots, M}$; and (*iii*) calculating the M corresponding $V(h, Y_j)]_{j=1, \ldots, M}$.

Example 14

Consider a share of stock s whose price is S and a position made up of the share s as well as some products derived from s such as options. Let us operate in a Black and Scholes (BS) universe in which the value $V(t, S(t))$ of this position depends on time t and a single random factor $Y(t) \equiv S(t)$; the function $V(t, S)$ is assumed known (the option prices obey, e.g., the BS formula).

According to BS, we suppose that $S(t)$ obeys a Geometric Brownian motion:

$$dS/S = \mu dt + \sigma \, dW \Leftrightarrow \tag{27.24a}$$

$$S(t) = S(0) \, e^{\left(\mu - 0.5\sigma^2\right)t + \sigma W(t)} \tag{27.24b}$$

where $W(t)$ is a standard Brownian motion, μ and σ are given annualized constants, and time is measured in years.

We will simulate the value $S(h)$ with $h = 1$ (open) day ($=1/252$ year) to estimate the VaR for 1 day. Thus, we write (27.24a) in the discrete form:

$$S(h) = S(0)\left(1 + \mu\, h + \sigma\sqrt{h}\, U\right) \tag{27.25a}$$

where U is a standard normal variate.

(continued)

Example 14 (continued)

Of course, a VaR can be directly computed here from the corresponding p-quantile of U, but for the sake of the example we proceed with a Monte Carlo simulation. Assume the share s quotes $S(0) = 100$ € on date 0 and that its expected rate of return and its volatility, annualized, respectively equal $\mu = 12\%$ and $\sigma = 35\%$. With these parameters and $h = 1/252$, Eq. (27.25a) can be rewritten (neglecting the term $\mu h = 0.00047$) as

$$S(h) = 100 + 2.2 \ U. \qquad (27.25b)$$

Equation (27.25b) allows simulating $S(h)$ and $V(h, S(h))$ using random drawings for U.

We might make, for example, 2000 drawings $U_j]_{j=1, \ldots, 2000}$, to generate discretely, from (27.25b), 2000 simulations $S_j = 100 + 2.2 \ U_j$. Notice that one could also have used Eq. (27.24b) which, discretizing h, can be written

$$S(t) = S(0) \ e^{\left(\mu - 0.5\sigma^2\right)h + \sigma\sqrt{h} \ U}.$$

Remark in addition that one may associate to every drawing U_j its *antithetic* $- U_j$ (*see* Chap. 26) to obtain 4000 values for U and so 4000 simulations $S_j]_{j = 1, \ldots, 4000}$. Using the *antithetic* variables implies one can be sure the sample of U is well centered about 0 and in this way improves the simulation's quality.

Finally, we calculate the 4000 values $V(h, S_j)$, and consider, for example, the 40th value V^* in increasing order to obtain the VaR(1d, 99 %) (VaR $= V(0, - S(0)) - V^*$).

The preceding description of the method and Example 14 assume that the position's value depends only on a single random factor. In general, the value $V(t, \underline{Y}(t))$ of the portfolio considered depends on M risk factors $\underline{Y} \equiv (Y_1, \ldots, Y_M)$, which are usually correlated; thus one cannot independently simulate realizations of the different factors $Y_i(h)$. The simulations must take account of the correlation structure as it is, for example, expressed through the variance-covariance matrix of the variations of the M factors. The methodology of such simulations is explained in Chap. 26.

Monte Carlo simulations involving several sources of risk $\underline{Y}$ such as prices, indices, interest rates, and volatilities constitute a powerful and sensitive tool for analyzing risk and calculating the VaR. They are also employed for other purposes, such as valuing options or derivatives, or studying portfolio strategies, credit risks (*see* Chaps. 29 and 30), etc.

27.2.5 Comparison Between the Different Methods

As already mentioned, the various parametric methods with or without simulation (Monte Carlo, historical simulation) can be associated with different levels of approximation (linear, quadratic, exact valuation).

The following table puts together the different possibilities.

		Determining the sensitivity of the portfolio	
		Partial valuation (delta or delta-gamma)	Global valuation (full valuation)
Determination of the variations in risk factors	Parametric method (e.g., normal distributions)	– RiskMetrics™ – approximate closed formula – Monte Carlo	– Monte Carlo
	Non-parametric method (arbitrary distributions)	– Simulation from historical data	– Simulation from historical data

Generally, simulations from historical data are limited by the number of observations, which is often insufficient to give an accurate and reliable estimator. Parametric methods stumble over the insufficiencies of the models used, which are often based on an assumption of normality. However, we will see that adjustments are possible to correct deficiencies stemming from this assumption (Sect. 27.4).

Depending on the context, one method will be preferred to another:

– For a position with a large number of components of which few involve non-linearities (e.g., with few options), one can be satisfied with the delta-normal method.
– For a position subject to a limited number of risk factors and large non-linearities (e.g., many options) the quadratic model or exact valuation are appropriate; if the complexity of the position means that the closed formula for the VaR (exact or approximate) is difficult or impossible to derive, these models can be associated with simulations to yield an empirical histogram of the loss.
– For a complicated position with strong non-linearities, performing simulations is unavoidable.

27.3 Limitations and Drawbacks of the VaR, Expected Shortfall, Coherent Measures of Risk, and Portfolio Risks

VaR measures suffer from numerous limitations, both in theory and practice, that lead either to major modifications or to the adoption of other risk measures. After analyzing these limitations in Sect. 27.3.1, we discuss Expected Shortfall (or conditional VaR) in Sect. 27.3.2, the conditions required of a coherent risk measure in Sect. 27.3.3, and portfolio risk and the risk induced by its components in Sect. 27.3.4.

27.3.1 The Drawbacks of VaR Measures

The limitations and drawbacks of VaR measures may be classified into two categories, technical issues on the one hand and conceptual difficulties, which are more fundamental, on the other.

27.3.1.1 Technical Issues
We may draw up the following, non-exhaustive, list of technical issues met in practice and which can be particularly serious in periods of financial crisis:

- The distributions of short-term returns on investment are in general not normal and are leptokurtic, which means that extreme events occur much more frequently than for the normal distribution. This is especially annoying for analytic methods which mostly assume that the log-returns are normal. In addition, the returns are required to display no *serial autocorrelation*, which is often not the case.
- Financial instruments are assumed *liquid* and valued in *normal market conditions*, which is rather contradictory to the notion of VaR itself.
- When we use *delta valuation*, we only consider *local risk* (because of linearization), which may turn out to be very imprecise and thus dangerous if market prices change much and rapidly. Even taking gamma (second-order variations) into account can be insufficient if there are jumps in the trajectories of the market prices of assets and liabilities (crisis periods).
- If the portfolio includes several classes of assets, aggregation is not simple: if the VaR measures for the different classes are simply added, which requires perfect correlations between classes, diversification is not accounted for, therefore the risk is overestimated and the resulting *over-hedging* eats up too much equity, which makes it expensive. If the imperfection in correlations is, as it rightly should be, taken into account, then the danger is *under-hedging* if the correlations, that are usually unstable, have been underestimated. This danger occurs mainly during crisis periods when correlations increase (since all market prices tend to crash simultaneously), which reduces the beneficial effects of diversification exactly when it is most needed.
- The VaR is *insufficient* to calculate the adequate amount of equity required: it is not enough to know that there will be a loss greater than or equal to the VaR, for example, on one day in a hundred, since *on average* over a long period the equity indicated by this VaR will effectively be lost if the market conditions are in agreement with the assumptions made in the calculation.[12]

All these problems proved to be serious during the 2007–2009 crisis which was characterized by: extreme events, serial autocorrelation, price, and rate jumps, increased correlation between different instruments that sharply reduced the effects of diversification, etc. These ingredients are at the root of the failure of the models on

---

[12] The profits made on average, however, will increase the amount of equity.

which the control of banking risks are based and justify using a "stressed" VaR (see Sect. 27.2.4) as well as the alternative methods explained in Sect. 27.4.

Remark that the problems created by the VaR measure are made worse for alternative investments and *hedge funds*. Indeed, the opportunism of the strategies and the massive use of optional instruments which characterize them worsen, in particular, the leptokurtosis, asymmetry, and non-stationarity of their return distributions (in particular, their volatility). Some partial solutions to these problems can be found in alternative risk measures.

27.3.1.2 Conceptual Difficulties

VaR measures face the following additional conceptual difficulties, of which some are prohibitive both in theory and in practice:

- A unique chosen quantile (VaR) of a distribution function cannot replace knowing the complete distribution of the value of the portfolio. In particular, the method gives a probability (e.g., 1/100) for the occurrence of a loss in excess of the VaR calculated, but does not say anything definite about the *actual size of the loss* should it occur. We will return to this important flaw.
- The VaR is calculated on the assumption of an unchanged portfolio during the period it is held: consequently, if it is not calculated with a very short-term horizon h, the VaR is all the less relevant if the portfolio experiences a large turnover and strong growth.
- The recommendation of the Basel Committee to adopt a holding period of 10 (trading) days in calculating the VaR is surprising since the latter has to be recalculated every day. Such a procedure leads to inconsistencies to the extent that, on the one hand, the volatilities and correlations change with time, and on the other, the size and composition of the portfolio evolve.
- Intuitively, an adequate risk measure should have (at least) the property of *sub-additivity*. A risk measure m is called sub-additive if, and only if, for two arbitrary assets or portfolios X_1 et X_2, one has:

$$m(X_1 + X_2) \leq m(X_1) + m(X_2). \tag{27.26}$$

This property, intuitively linked to the effect of diversification on the $(X_1 + X_2)$ portfolio, has some decisive advantages for management and regulation:

- It encourages consolidation of the risk evaluation system to the firm's global level since in this way it uses less equity than do decentralized systems;
- It avoids the temptation to circumvent regulations by artificial creation of ad hoc subsidiaries, which can prove dangerous.

1144   27 Value at Risk, Expected Shortfall, and Other Risk Measures

But, unfortunately, the VaR is not in general sub-additive.[13] The following simple example will show this.

Example 15

Consider a portfolio made up of two positions A and B. The following table gives the losses for each position in each of seven possible scenarios with the probabilities assigned to them.

Losses	Very large decline	Significant decline	Small decline	Stability	Small increase	Significant increase	Very large increase
Position A	−0.5	−0.75	−1.5	−2	−3.25	0	6.5
Position B	6.5	0	−1.65	−2.7	−2.8	−0.75	−0.5
Probability	2.5 %	2.5 %	25 %	40 %	25 %	2.5 %	2.5 %

The VaR with 95 % confidence in *each* of these positions equals 0, since in 97.5 % of the cases, the loss is zero or negative (gain). The sum of the two VaR is therefore zero. Nonetheless, the VaR with 95 % confidence of the *global* portfolio equals 6 since there are 5 % of the chances to lose 6 (= 6.5–0.5, for a very large decline or increase). The risk measure here is therefore *super-additive*, and not sub-additive. This might encourage the manager of the two positions to split the portfolio and to compute the VaR of each component separately, which is clearly dangerous.

It is nonetheless interesting to note that *above the threshold of* 97.5 %, this VaR becomes sub-additive, the portfolio's VaR remaining 6 and each individual VaR being 6.5. One would then have: 6 < (6.5 + 6.5), in accordance with property (27.26). This underlines both the importance of the choice of the threshold and the limitations of the method.

A fundamental criticism of the VaR is formulated by detractors of the use of probabilities in economic calculations. We will return to this criticism in Chap. 29.

---

[13] Roughly speaking, VaR is only a sub-additive measure when the distribution of the losses experienced by the portfolio is elliptic, that is Gaussian or "almost Gaussian," which is rarely the case in practice for positions in financial instruments. This property in itself invalidates the VaR as a relevant risk measure. Meanwhile, at an even more basic level, as of now no coherent set of reasonable axioms for a measure of financial risk, and thus no unambiguous definition of such a risk, has been associated to the VaR statistic (see Sect. 27.3 below).

27.3.2 An Improvement on the VaR: *Expected Shortfall* (or Tail-VaR, or C-VaR)

Although the method gives the probability of a loss in excess of the calculated VaR, it says nothing, as we have seen, about *the size of the loss* when it occurs, or, for that matter, about *the distribution of losses exceeding the VaR* (except if a theoretical distribution, such as the Gaussian, is assumed from the start). This is particularly inconvenient for portfolio return distributions exhibiting fat tails (leptokurtosis). This is the reason why it is preferable to use (at least as a complement) a risk measure called *Expected Shortfall*, henceforth denoted ES, or else *Expected Tail Loss, Tail-VaR,* or *Conditional-VaR* (C-VaR), which does take into account the amount of losses exceeding the VaR. More formally, we calculate the conditional expectation of the loss knowing that it is larger than (or equal to) the VaR, using the Bayes formula:

$$ES(h,p) \equiv E[L|L \geq \text{VaR}(h,p)] = \frac{1}{1-p} \int_{VaR(h,p)}^{+\infty} x f_L(x) dx \qquad (27.27a)$$

where $f_L(x)$ is the density function of the loss and $F_L(x)$ denotes its distribution function.

The Expected Shortfall *ES* thus expresses the *mean size of the losses above the VaR*.

Making the change of variable $u = F_L(x)$ in the integral in (27.27a), we obtain an alternative expression for the ES[14]:

$$ES(h,p) = \frac{1}{1-p} \int_p^1 VaR(h,u) du. \qquad (27.27b)$$

This equation amounts to an alternative definition of the ES and expresses it *as a mean of the VaRs* calculated for thresholds u stricter (more demanding) than p.

The significant advantage of the ES is that, in contrast to the VaR, this measure allows one to *distinguish two distributions* for the loss which *have the same p-quantile but are otherwise different* (in particular not having the same tails, or the same asymmetry). For example, for a standard centered Gaussian distribution, the VaR with a 99 % confidence for a portfolio worth 100 equals 2.33 and its ES is 2.66. For a stable Pareto-Lévy distribution, centered at 0 and symmetric, with exponent 1.5 and scale parameter 0.372, the VaR at 99 % confidence for a portfolio worth 100 also equals 2.3 but the ES is 3.22 because the tail of the distribution is considerably fatter than that of the normal distribution (leptokurtosis).

---

[14]Since $u = F_L(x)$, $du = f_L(x)\, dx$, and $F_L(\text{VaR}(h,p)) = p$, we have: $\frac{1}{1-p} \int_{VaR(h,p)}^{+\infty} x f_L(x) dx = \frac{1}{1-p} \int_p^1 F_L^{-1}(u) du = \frac{1}{1-p} \int_p^1 VaR(h,u) du$.

A second significant advantage of the ES is that it is a *sub-additive risk measure*. We show this with an example.

Example 16

Let us pursue the preceding Example 15. First, let us consider each position in isolation. If the threshold for the VaR (at 95 %) is attained or exceeded, each of the positions loses either 0 or 6.5 with equal probability (2.5 %). The mean loss under this assumption, and so the ES, is equal to 3.25 for each position [the computation using (27.27a) is the following: $(0 \times 2.5 \% + 6.5 \times 2.5 \%)/5 \% = 3.25]$.

Now let us consider the portfolio as a whole. When the threshold of the VaR (at 95 %) is attained or exceeded for the entire portfolio, the loss is always equal to 6. The ES of the portfolio is thus 6 [$= (6 \times 5 \%)/5 \%$]. This ES is less than the sum of the two individual ES ($= 6.5$), which conforms to property (27.26).

A closed form for the ES can be obtained for some special loss distributions. Assume, for example, that the loss with horizon h is a Gaussian $N(\mu_h, \sigma_h)$ which leads to Eq. (27.3a): $VaR(h,p) = \alpha_p \sigma_h + \mu_h$. From this, by (27.27b):

$$ES(h,p) = \frac{1}{1-p} \int_p^1 VaR(h,u)du = \frac{1}{1-p}\sigma_h \int_p^1 \alpha_u du + \mu_h.$$

Defining $\kappa_p \equiv \frac{1}{1-p}\int_p^1 \alpha_u du \equiv \frac{1}{1-p}\int_p^1 N^{-1}(u)du =$ mean of the u -quantiles α_u ($u > p$) of the standard Gaussian, we get:

$$ES(h,p) = \kappa_p \sigma_h + \mu_h. \tag{27.28}$$

For short horizons h, the μ_h term is negligible and $ES(h, p) \# \kappa_p \sigma_h$.

Consequently, in the Gaussian case, like the VaR the ES is proportional to the standard deviation of the loss; however, the proportionality coefficient κ_p is a mean of Gaussian quantiles and not the simple p-quantile α_p.

The preceding considerations suggest that ES is not appreciably more difficult to compute than VaR. This impression is confirmed by the following analysis that goes beyond the assumption of normality (*see*, e.g., the theory of extreme values in Sect. 27.4.2). In addition, ES is no more difficult to estimate than VaR using a historical method, or simulations: it is enough to compute the mean of the losses exceeding the VaR obtained from the simulation (historical or theoretical). The following example illustrates such a calculation.

27.3 Limitations and Drawbacks of the VaR, Expected Shortfall,...

Example 17

Let us again take up the simplified Example 8 which provides in the following table, in increasing order, 250 simulated losses (horizon 1 day, in $ M).

Simulation:	1	2		242	243	244	245	246	247	248	249	250
Losses	−4.4	−3.9	...	3.49	3.53	3.56	4.0	4.12	4.52	4.82	5.0	5.9

The VaR(1d, 97 %) can be estimated as \$3.51M (between 3.49 and 3.53) since the loss turns out to be greater than or equal to this number in 8 simulations out of 250 (3.2 % of the cases).

The ES(1d, 97 %) can be estimated at

$$\frac{1}{8}(3.53 + 3.56 + 4.0 + 4.12 + 4.52 + 4.82 + 5.0 + 5.9) = 4.43M.$$

In this example, the ES exceeds the VaR by \$0.92M, or in other words by 26 %. This difference would have been even larger if the simulation had produced larger losses in the tail of the distribution (from the place #244 on the table).

It is obvious that this pedagogical example is simplified in the extreme: a more realistic example would have included a much larger number of simulated losses (but would have been much more "cumbersome" to present).

Remark that ES is all the more different from (and preferable to) VaR as the loss distribution is further from normal and exhibits fat tails and/or a stronger asymmetry due to the presence of options in the portfolio. This is the main reason why regulators, who are most concerned with tail events, are progressively substituting ES for VaR, as we will see in Chap. 29.

To compute ES, one can use the empirical distribution (as in Example 17) or a theoretical law (Student's t, extreme value laws, etc.) for the left *tail* of the distribution of the *value* of the portfolio (*right tail* of the loss distribution; see Sect. 27.4). For estimating the empirical ES one can use a data series covering normal periods or/and crisis period(s) (stressed ES). Similarly, the parameters used for theoretical laws can be estimated from either kind of series (then leading to either ES or stressed ES). Estimation of ES has been extensively studied in financial econometrics.[15]

In fact, the only advantage of VaR compared to ES is the ease of back-testing when using it. Back-testing checks how well a procedure or a model would have worked in the past. In the case of a VaR, it suffices to check how often the daily loss exceeds the VaR estimated with the current procedure. For instance, if the proportion of days for which the actual loss exceeds the presumed VaR(1d, 99 %) is

---

[15] See, in particular, Scaillet (2004), Fermanian and Scaillet (2005) for the sensitivity of ES in the presence of netting and collateral, and more recently Patton et al. (2019).

"consistently and substantially" higher than 1 %, it is likely that the methodology underestimates the VaR. Back-testing is important for the risk managers as well as for the regulators who both want to check the validity of the bank's risk assessment and the capital requirements.

27.3.3 Coherent Risk Measures

27.3.3.1 Conditions for the Coherence of a Risk Measure

Sub-additivity is not the only important property required of a coherent risk measure. In fact, coherence depends as much on how the measure conforms to a number of other properties that are considered reasonable on a priori grounds. This aspect has not only a theoretical nature but bears important practical consequences. These concern as much the *shareholders* for the calculation of the *desirable amount* of equity and the *optimal allocation* of it amongst different possible activities, as the *regulators* who must supervise the risks to which the different participants in the financial market are exposed. The first advances in this area were due to Artzner et al. (1997, 1999).

We consider the framework of a one-period investment horizon $(h = 1)$[16] and we examine the *value* V of a position or a portfolio over the period, and not the *loss* L relative to this position (we have $L = V_0 - V$).

A coherent risk measure m assigns a number $m(V)$ to each position whose net future value is V such that the four following requirements are satisfied: for each pair of positions V and W (independent or not), for any positive number a and any number b (denoting by r the riskless interest rate):

(i) $m(V + W) \leq m(V) + m(W)$ (sub-additivity)
(ii) $m(aV) = a\, m(V)$ (homogeneity)
(iii) $m(V) \geq m(W)$ *if* $V \leq W$ *a.s.*   (monotonicity)
(iv) $m(V + (1 + r)b) = m(V) - b$ (translation invariance)

It is useful in interpreting these equations to identify the degree of risk with the capital necessary to hedge it.[17]

Property (*i*) has already been discussed and means that the risk measure of two merged portfolios is smaller or equal than the sum of their own risk measures; it is related to the diversification effect that can reduce (but not increase) risk. Property (ii) means that changing the scale of the portfolio by a factor a, everything else being kept equal, should multiply a risk measure by a. It should be noted that homogeneity (*ii*) can be a problem for large positions that are *not very liquid*. Moreover, (*ii*)

---

[16] See Artzner, Delbaen, Eber and Heath, "Thinking Coherently", *Risk*, vol. 10, n° 11, November 1997, for a simple and intuitive presentation and "Coherent Measures of Risk," *Mathematical Finance*, 9, 1999, pp. 203–228, for a mathematically rigorous exposition. The extension of their analysis to two (or more) periods poses difficult problems.

[17] We will see in Chap. 29 that, in the context of Basel agreements, regulatory risk measures are employed to assess capital requirements.

implies that $m(0) = 0$: a position which will certainly be worth nothing tomorrow does not need hedging. Property (*iii*) signifies that if a position W is worth at least as much as another position V in all states of the world, its risk measure cannot exceed that of V. It can be deduced from (*ii*) and (*iii*) that $m(V)$ is a negative number (the shareholders of the firm concerned, or the owners of the portfolio under scrutiny, can withdraw capital without danger) if the future value of V is positive whatever the state of the world. Property (*iv*) implies that if an amount b, invested (algebraically) at the riskless rate, is added to the position whose risk exposure is being calculated the capital required to hedge it is decreased (algebraically) by the amount b. In particular, (*iv*) implies the following intuitively satisfying result: $m(V + (1 + r) m(V)) = 0$: once the necessary capital ($m(V)$) is added to the initial risky position and invested in the risk-free asset, the risk measure becomes zero and no supplementary capital is required.

27.3.3.2 Construction of Coherent Risk Measures

VaR has the last three properties but not the first, as we have seen; because of this it is not a coherent risk measure. Expected Shortfall has them all and consequently is a coherent risk measure. It is to be remembered that both the VaR and the ES enjoy the important property of homogeneity whose interpretation and utilization we will return to later.

The ES is not the only risk measure coherent in the sense of the 4 axioms of Artzner et al. (1997). Indeed it is possible to construct infinitely many coherent risk measures and different ways of doing so are imaginable.

– One approach consists in defining a risk measure as a weighted sum of quantiles of the loss L. For the ES, the weighting coefficients of the u-quantiles are all equal for $u > p$ and 0 for $u < p$ (*see* Eq. (27.27b)): $ES_p = \frac{1}{1-p} \int_p^1 VaR(u)du$. For the VaR the weights all vanish except that for the p-quantile. More generally, one may assign different weights ω_u to the u-quantiles, which gives a value for the risk proportional to $\int_0^1 \omega_u VaR(u)du$. One can show that a condition for coherence of such a risk measure is that the weights ω_u do not decrease with u. In addition, one verifies immediately that this condition is satisfied by ES while it is violated by VaR.

As an example of this approach, let us mention the *exponential spectral risk measure* which uses the weights $\omega_u = e^{-\gamma(1-u)}$, where γ is a constant. This gives a risk value proportional to $\int_0^1 e^{-\gamma(1-u)} VaR(u)du$ and therefore is a coherent measure in the sense of Artzner et al. (1997, 1999).

– It is also possible to start from a different (more or less stringent) set of a priori reasonable requirements to build a different axiomatic theory of risk measuring, yielding more or less narrow classes of "valid" risk measures[18] (we keep the qualification "coherent" for those risk measures that satisfy the axioms of Artzner

---

[18] See for instance Föllmer and Schied (2002), Fritelli and Rosazza Gianin (2002) for the class of convex risk measures, Chen et al. (2013) for systemic risk measures, and Acerbi and Tasche (2002) for spectral risk measures. As to VAR, see Kou and Peng (2016), He and Peng (2018) and Liu and Wang (2020).

et al.). For example, Ruodu Wang and Ričardas Zitikis (2020) ground their theory on four conditions: three of them reflect "natural requirements" (Monotonicity, Law Invariance,s and Prudence),[19] while the last express an important regulator's criterion (No Reward for Concentration).[20] They show that ES is the only risk measure satisfying these four requirements (while it is one among others in the wider class of coherent risk measures).

– Another approach relies on the construction of *generalized scenarios*[21] which one can show lead to the most general coherent risk measures. Specifically, for an arbitrary set of probability distributions Π, the following calculation yields a coherent risk measure:

–Calculate the mean of the loss if it exceeds a certain threshold, and do this for every probability distribution P belonging to Π.

–Then compute the largest of the numbers found: $m(L) = \sup_{P \in \Pi} [E_P(L \mid L \geq threshold)]$.

It is interesting to note that the ES_P is a *special case* of the generalized scenario method (when Π is reduced to the set of measures with densities bounded by $1/(1\text{-}p)$, and *threshold* $= VaR_p$).

27.3.4 Portfolio Risk Measures: Global, Marginal, and Incremental Risk

The critical question concerning a portfolio risk measure and the risks induced by each of its components has already been raised and explored several times in the Chaps. 21–25 devoted to portfolios and in Sect. 27.1.3.3 of this chapter; here we will take it up again from a different angle, in a more general and deeper way and by introducing alternative risk measures.

27.3.4.1 Portfolio Risk Measures

We consider a portfolio made up of n assets here defined by the vector $\underline{x} \equiv (x_1, \ldots, x_n)$ of the *amounts* invested (or *exposures*) in each asset; x_i dollars are thus invested today (date 0) in security i for $i = 1, \ldots, n$ (the values are actual amounts and not weights, contrary to our usual modeling).

---

[19] Monotonicity means: $V \geq W \Rightarrow m(V) \geq m(W)$. Law invariance implies: if V and W have the same probability law, then $m(V) = m(W)$. Prudence means: $\xi_k \to V$ (point wise) $\Rightarrow m(\xi_k) \to m(V)$, so that a statistical estimate converges to the "right" value.

[20] Two components V and W of a portfolio are concentrated (or diversified without benefit) if they share the same tail event (the worst case scenario is the same for both). No reward for concentration means that, in this case, $m(V + W) = m(V) + m(W)$, i.e., there is no required capital reduction.

[21] See the references in footnote 16.

27.3 Limitations and Drawbacks of the VaR, Expected Shortfall,... 1151

Today (at date 0), the total value of the portfolio is: $V_0 = \sum_{i=1}^{n} x_i$.

If there are no transactions between 0 and h, its value at h is $V_h(x_1, \ldots x_n) = \sum_{i=1}^{n} x_i(1 + R_i)$ where R_i denotes the return on security i between 0 and h.

The return on portfolio $\underline{x}$ from 0 to h is $R_x = \frac{\sum_{i=1}^{n} x_i R_i}{V_0}$; the loss is $L_h = V_0 - V_h = -\sum_{i=1}^{n} x_i R_i$.

Note that the loss from component i is $L_i = -x_i R_i$ and that $L_h = -V_0 R_x$. Thus remark that generally, we have:

loss = −Exposure × Return, or in other words : *Loss per invested = −Return*.

Indeed, we may equally well use returns or losses in equations for the dynamics of portfolios or analysis of their risks. For market positions, returns are generally preferred, while for portfolios of debt instruments whose risk of default is the concern (*see* Chaps. 28–30) losses are preferred.

In the preceding Sect. 27.3.3, a risk measure was defined as a mapping m which assigns a real number $m(V_h)$ to a random variable V_h here representing the future value (at h) of a portfolio. In the sequel, we characterize the risk of a portfolio as a function of its composition $\underline{x}$ rather than in terms of its final value or its loss. Such a characterization is derived from any measure m since $m(V_h)$ allows defining a function $Risk_m$ of $\mathbb{R}^n$ onto $\mathbb{R}$ by

$$Risk_m(x_1, \ldots, x_n) = m(V_h(x_1, \ldots, x_n)) = m\left(\sum_{i=1}^{n} x_i(1 + R_i)\right).$$

Thus, $Risk_m(\underline{x})$ is the risk of portfolio $\underline{x}$ if one uses the risk measure m.

Let us consider the examples of the VaR and the ES of a Gaussian portfolio to make this formalization more palatable and to make explicit the transition from m to $Risk_m$.

Denote by $\underline{\mu}$ the vector of expectations and by Γ the variance-covariance matrix of returns R_i on the individual securities. By using $L_h = -\sum_{i=1}^{n} x_i R_i$, we obtain

$$\text{Standard−deviation}(V_h) = \text{Standard−deviation } (L_h) \equiv \sigma_h$$
$$= \sqrt{\underline{x}'\Gamma\underline{x}}; \quad \text{and} \quad \text{E}(L_h) \equiv \mu_h = -\underline{\mu}'\underline{x}.$$

If the vector of returns is Gaussian, we have by Eq. (27.3a):

$$VaR(p, h) = \alpha_p \, \sigma_h + \mu_h,$$

whence

$$Risk_{VaR}(\underline{x}) = \alpha_p \sqrt{\underline{x}' \Gamma \underline{x}} - \underline{\mu}' \underline{x},$$

which we will denote by $VaR_p(\underline{x})$ in this context.
From this, since portfolio $\underline{x}$ is Gaussian:

$$VaR_p(\underline{x}) = \alpha_p \sqrt{\underline{x}' \Gamma \underline{x}} - \underline{\mu}' \underline{x}. \tag{27.29}$$

Similarly, if the portfolio is Gaussian, by (27.28)

$$ES_p(\underline{x}) = \kappa_p \sqrt{\underline{x}' \Gamma \underline{x}} - \underline{\mu}' \underline{x}, \tag{27.30}$$

where $\kappa_p \equiv \frac{1}{1-p} \int_p^1 \alpha_u du$ denotes the mean of the u-quantiles α_u for $u > p$.

The coherence conditions on the measure m have their counterparts on $Risk_m$. Among the properties required for a risk measure m to be coherent, recall that of homogeneity: $\frac{m(aV_h)}{m(V_h)} = a$ (equation (ii) in Sect. 27.3.3.1). Now the homogeneity of m implies that of the $Risk_m$ (since $\frac{Risk_m(a\underline{x})}{Risk_m(\underline{x})} = \frac{m(aV_h)}{m(V_h)}$). Thus we can write:

$$Risk_m(a \, \underline{x}) = a \, Risk_m(\underline{x}) \text{ for all } a \geq 0.$$

Mathematically, this condition expresses that the function of n variables $Risk_m(.)$ is homogeneous of degree 1. Financially it means that if two portfolios only differ by a scale factor a (their structures $\underline{x}$ and $a\underline{x}$ and their values V_h and aV_h are homothetic, i.e., simple multiples of each other), the ratio of their risks equals a; therefore the equity amounts required to hedge them are also related by the ratio a. We emphasize that homogeneity is a necessary condition for coherence but not a sufficient one, and that it does hold for ES, VaR, and even the simple standard deviation of the loss or of the value ($\sigma(L_h)$ or $\sigma(V_h)$).

The homogeneity of $Risk_m$ allows us to use Euler's theorem[22] and to write

---

[22] Euler's theorem asserts that if f is homogeneous of degree 1 and differentiable, then: $f(\underline{x}) = \left(\frac{\partial f}{\partial \underline{x}}\right)' \underline{x} \equiv \sum_i \frac{\partial f}{\partial x_i} x_i.$

Proof: If f is homogeneous: $f((1 + d\lambda)\underline{x}) = (1 + d\lambda) f(\underline{x}) = f(\underline{x}) + f(\underline{x}) \, d\lambda$;

In addition, if f is differentiable: $f((1 + d\lambda)\underline{x}) = f(\underline{x}) + \left(\frac{\partial f}{\partial \underline{x}}\right)' \underline{x} d\lambda.$

Putting these two equations together yields: $f(\underline{x}) \, d\lambda = \left(\frac{\partial f}{\partial \underline{x}}\right)' \underline{x} d\lambda$, and therefore $f(\underline{x}) = \left(\frac{\partial f}{\partial \underline{x}}\right)' \underline{x}.$

$$Risk_m(x_1, \ldots, x_n) = \sum_{i=1}^{n} x_i \frac{\partial Risk_m}{\partial x_i}(x_1, \ldots, x_n). \tag{27.31}$$

In particular, since the VaR and the ES are homogeneous,

$$VaR(x_1, \ldots x_n) = \sum_{i=1}^{n} x_i \frac{VaR(x_1, \ldots x_n)}{\partial x_i}; \quad ES(x_1, \ldots, x_n)$$

$$= \sum_{i=1}^{n} x_i \frac{ES(x_1, \ldots, x_n)}{\partial x_i}. \tag{27.32}$$

These equations which follow from only homogeneity of the risk measure have significant consequences both theoretically and in practice for managing portfolio risk, as we explain in what follows.

27.3.4.2 Risk Induced by a Component of a Portfolio: Marginal Risk, Contribution to Risk and Incremental Risk

Let us consider a portfolio $\underline{x}$ whose overall yield is R_x and whose risk, evaluated using a measure m (such as VaR, ES, ...), is $Risk_m(\underline{x})$. We will analyze the fraction of the risk attributable to one of its components, for example, the ith; to this end, we will distinguish three notions of the risk induced by i: the marginal risk, the contribution to total risk and the incremental risk.

(i) *The marginal risk* induced by i is defined as $\frac{\partial Risk_m}{\partial x_i}(x_1, \ldots, x_n)$; it can be interpreted as the impact on the risk of portfolio $\underline{x}$ of a small variation (\$1) in its exposure to i.

We may obtain a closed expression for the marginal risk if the return distributions are specified: for instance, if the returns are Gaussian:

– The marginal VaR is $\dfrac{\partial VaR_p(\underline{x})}{\partial x_i} = \alpha_p \dfrac{\partial \sqrt{\underline{x}'\Gamma\underline{x}}}{\partial x_i} - \mu_i = \alpha_p \dfrac{\sum_{j=1}^{n} \sigma_{ij} x_j}{\sqrt{\underline{x}'\Gamma\underline{x}}} - \mu_i,$

or

$$\text{marginal VaR}_p \text{ of } i = \alpha_p \frac{\text{cov}(R_i, R_x)}{\sigma(R_x)} - \mu_i; \tag{27.33a}$$

– Analogously,

$$\text{marginal ES}_p \text{ of } i = k_p \frac{\text{cov}(R_i, R_x)}{\sigma(R_x)} - \mu_i. \tag{27.33b}$$

Equations (27.32), (27.33a) and (27.33b) only hold if the returns R_i are Gaussian but others may be obtained for other sorts of distribution.[23]

(ii) *The contribution to the total risk of the component i*, also called *component risk* and denoted $C_i(\underline{x})$, is, by definition, the product of the exposure x_i by the marginal risk of i: $C_i(\underline{x}) = x_i \frac{\partial Risk_m(\underline{x})}{\partial x_i}$.

This definition has the merit of leading, with Eq. (27.31), to an important property of coherence: *the total risk of a portfolio is equal to the sum of the contributions of the different components*:

$$Risk_m(\underline{x}) = \sum_{i=1}^{n} C_i(\underline{x}).$$

This property is valid for all the risk measures m that satisfy the condition of homogeneity (standard deviation, VaR, ES, etc.) as shown by Eq. (27.32). The coherence that it expresses has not only a conceptual appeal: it facilitates the practical management of risk. Indeed, due to this additive property of contributions, one can measure risks, assign equity amount and manage separately each line of risk. Then adding the individual risks and assigned equity amounts leads smoothly to the global risk and global amount of equity required. The *bottom-up* and *top-down* procedures for managing a complicated portfolio of securities or activities, or a balance sheet, are in this way made easier.

(iii) *Incremental risk*[24] is defined as the difference between the risk of the portfolio estimated with the exposure x_i and without it:

---

[23] An expression for the marginal ES, valid for all types of distribution, can be obtained from the following decomposition:

$$ES_p(\underline{x}) = E\{L(\underline{x})|L(\underline{x}) > VaR_p(x)\} = E\left\{ \sum_{i=1}^{n} x_i \; l_i|L(\underline{x}) > VaR_p(x) \right\}$$

$$= \sum_{i=1}^{n} x_i E\{l_i|L(\underline{x}) > VaR_p(x)\}$$

where l_i is the loss on i per \$ invested ($l_i = - R_i$) and $x_i \, l_i =$ the total loss on component i. From this we have:

$\partial ES_p(\underline{x})/ \partial x_i =$ marginal ES_p of $i = E\{l_i \mid L(\underline{x}) > VaR_p(x)\}$.

Care must be taken not to confuse the *marginal* ES_p of i with the ES_p of $i = E\{l_i \mid l_i > VaR_p(l_i)\}$, since the marginal ES_p is derived by conditioning on a *global* loss $L(\underline{x}) > VaR_p(\underline{x})$ and not on the level of the loss l_i.

[24] Also called the *discrete marginal contribution*.

$$\Delta_i(\underline{x}) = Risk_m(x_1, \ldots, x_i, \ldots, x_n) - Risk_m(x_1, \ldots, x_{i-1}, 0, x_{i+1}, \ldots, x_n). \quad (27.34)$$

The incremental risk thus allows estimating the reduction in portfolio risk that would be achieved by completely eliminating i; alternatively, it measures the variation of the risk brought about by an investment of $\$x_i$ in i starting from no exposure at all to this component.

We can see that $\Delta_i(\underline{x})$ differs from $C_i(\underline{x})$ and that $\sum_{i=1}^{n} \Delta_i(\underline{x})$ differs from $Risk_m(\underline{x})$.[25]

27.4 Consequences of Non-normality and Analysis of Extreme Conditions

The assumption of normality or multi-normality, on which most parametric or semi-parametric methods of computing the VaR are based, does not fit with the observed changes in financial variables. Notably in periods of crisis, on the one hand the extreme values observed are more frequent than would result from Gaussian variations, and on the other hand the correlations between variables are greatly increased. This section presents methods for evaluating risk which are independent on the assumption of normality and appropriate for evaluating the consequences of crises.

Tools for analyzing non-Gaussian distributions and correlations at the extremes are presented in Sect. 27.4.1, the theory of extreme values and the calculation of the VaR and ES based on this theory are examined in Sect. 27.4.2 and *stress tests and the analysis of scenarios are discussed* in Sect. 27.4.3.

27.4.1 Non-normal Distributions with Fat Tails and Correlation at the Extremes

The assumption of normality and of multi-normality for risk factors underlying the standard parametric methods is at odds with the behavior of most financial variables in two respects: (i) extreme values are more frequent than those produced by Gaussian distributions (presence of fat tails); (ii) correlations increase at the extremes (in a crisis) while the correlations between the components of a Gaussian vector are constant over any range of variation. These are phenomena whose implications for financial risks are significant, and which the analytical and empirical tools must account for or else be found useless, as became painfully apparent in 2007–2009.

---

[25] Because the $\frac{\partial Risk_m}{\partial x_j}$ are different for $x_i = 0$ and $x_i \neq 0$. Nevertheless, they approximately coincide for well diversified portfolios.

27.4.1.1 Skewness, Kurtosis, and the Cornish-Fisher Method of Computing a Quantile

– The probability densities of relative variations in prices (log-returns) are often characterized by asymmetry and by leptokurtosis (*see* Chap. 8, Sect. 8.2.1 and the Appendix to Chap. 8).

The asymmetry of the probability density of a random variable X can be measured by the *skewness* Sk defined as the third moment of the standardized variable X:

$$Sk = E\left(\frac{X - E(X)}{\sigma(X)}\right)^3.$$

This coefficient vanishes for a symmetric distribution (such as the Gaussian) and is negative if the density of very small values is larger than the density of very large values (asymmetry to the left), which is the case for log-returns of numerous financial assets (principally those of securities subject to credit risk). As a telling example, the *skewness* of the Dow Jones index has been estimated to be -0.9 over the period 1985–2003.

– The existence of fat and thin tails can be determined, either using the kurtosis K, or from the asymptotic behavior of the distribution's tails. The latter method is the subject of Sect. 27.4.2 below. Recall that the kurtosis K of a random variable X can be defined as the fourth moment of the standardized variable:

$$K = E\left(\frac{X - E(X)}{\sigma(X)}\right)^4.$$

Standard normal distributions have a kurtosis equal to 3. A distribution whose kurtosis is larger than 3, is called leptokurtic, and is characterized by a density of extreme values (that is very large or very small ones) exceeding that for a normal variable with the same mean and the same standard deviation (fat tails on the left and/or on the right) and a smaller proportion of average values. Most financial risk sources are assigned a kurtosis larger than 3. As an example, the kurtosis of the Dow Jones has been estimated to be 7.5 over the period 1985–2003.

– Kurtosis and skewness can be used to estimate the VaR of a non-Gaussian distribution using the Cornish-Fisher expansion. This allows us to obtain an approximate closed form for the p-quantile of a distribution as a function of its moments. Just using the first terms of the Cornish-Fisher expansion, we get a formula for the VaR involving the mean μ, the standard deviation σ, the skewness Sk and the kurtosis K of the loss L:

$$VaR(p) = \mu + z_p\,\sigma, \tag{27.35}$$

27.4 Consequences of Non-normality and Analysis of Extreme Conditions

with $z_p = \alpha_p + \frac{1}{6}\left(\alpha_p^2 - 1\right)Sk + \frac{1}{24}\left(\alpha_p^3 - 3\alpha_p\right)(K-3) - \frac{1}{36}\left(2\alpha_p^3 - 5\alpha_p\right)Sk^2$
(recall that α_p is the p-quantile of the standard Gaussian).

This expression for the VaR, sometimes called the *modified VaR*, has the same form as that used in the Gaussian case ($VaR(p) = \mu + \alpha_p\,\sigma$), with the factor z_p simply replacing α_p. It is the factor z_p that is computed using the Cornish-Fisher expansion. We note that one recovers the formula for the Gaussian, $z_p = \alpha_p$, if the loss is normally distributed (since $Sk = 0$ and $K = 3$).

We can either improve the precision of the formula by using a Cornish-Fisher expansion with more terms or content ourselves with less precision using a shorter expansion. For example, in a situation where the kurtosis is not assumed important but asymmetry cannot be ignored, we could use the simplified

$$z_p = \alpha_p + \frac{1}{6}\left(\alpha_p^2 - 1\right)Sk.$$

Modification of the formula for the VaR using the Cornish-Fisher expansion is a simple and effective means of handling non-normal distributions by taking into account the skewness and kurtosis of the loss and the global yield.

Example 18

Let us consider a portfolio of \$40M whose weekly return has zero mean, standard deviation $= 0.025$, skewness $= -0.7$ and kurtosis $= 5$.

The weekly loss ($L = -V \times$ Return) is therefore of zero mean, standard deviation $= 1$, skewness $= +0.7$ and kurtosis $= 5$.[26]

The VaR(1 week, 99 %) calculated with the simple formula (loss assumed normal, wrongly) equals $VaR_1 = \alpha_{0.99} = 2.33$.

The VaR(1 week, 99 %) modified using the Cornish-Fisher expansion (27.35) is

$$VaR_2 = \alpha_{0.99} + \frac{1}{6}(\alpha_{0.99}^2 - 1)Sk + \frac{1}{24}(\alpha_{0.99}^3 - 3\alpha_{0.99})(K-3) - \frac{1}{36}$$
$$\times (2\alpha_{0.99}^3 - 5\alpha_{0.99})Sk^2$$
$$= 3.12.$$

The VaR(1 week, 99 %) modified with the simplified one-term expansion is

(continued)

---

[26] A change of sign in the variable changes the sign of moments of odd order, the even order moments remaining unchanged. A change of scale modifies the standard deviation proportionately and leaves the kurtosis and the skewness unchanged (because they are defined for centered variables).

> **Example 18** (continued)
> $$VaR_3 = \alpha_{0.99} + \frac{1}{6}\left(\alpha_{0.99}^2 - 1\right)Sk = 2.84.$$
>
> The first adjustment provides an increase for the VaR of 34 % while the second correction (less exact) amounts to 22 % only.

27.4.1.2 Correlation of Financial Variables Over the Extreme Ranges of Their Variation

Another important concept is related to the way that random variables co-vary over the extreme ranges of their variation (*tail correlation*). If we consider two variables X_1 and X_2, representing returns, for example, whose marginal distributions are F_1 and F_2, this correlation of the extremes can be defined as follows. Fix a small probability p and consider the two p-quantiles of X_1 and X_2, respectively, equal to $F_1^{-1}(p)$ and $F_2^{-1}(p)$ situated on the left of the tails of the two distributions. Denote by $\lambda(p)$ the conditional probability that X_1 is smaller than $F_1^{-1}(p)$ knowing that X_2 is smaller than $F_2^{-1}(p)$:

$$\lambda(p) = \text{Prob}\left\{X_1 \le F_1^{-1}(p) \mid X_2 \le F_2^{-1}(p)\right\}.$$

$\lambda(p)$ expresses the correlation of the two left tails of the distributions of X_1 and X_2, and is called the *lower tail correlation*[27]: For a very small p, $\lambda(p)$ is the probability that X_1 is extreme knowing that X_2 is extreme.

If $\lim_{p \to 0} \lambda(p) = 0$, the lower extremes of X_1 and X_2 are uncorrelated; *on the contrary, if* $\lim_{p \to 0} \lambda(p) = 1$, the two extremes are perfectly correlated; in addition, if $\lambda(p)$ increases as p decreases, the variables are the more correlated as their values are smaller. It is this last characteristic that is often observed in the markets, as the correlations of price variations, interest rates, and returns increase during crises. This phenomenon mitigates the benefit of diversification and translates as an increase in the VaR and the Expected Shortfall.

In the same way as for the lower tail correlation $\lambda(p)$, we also define the *upper tail correlation*: $\lambda'(p) = \text{Prob}\{X_1 > F_1^{-1}(p) \mid X_2 > F_2^{-1}(p)\}$ for high values of p. According to whether one is working in terms of returns or losses, it is the lower or upper correlations that are relevant during periods of crisis.

One may show that the correlation coefficients of the components of Gaussian vectors are *constant* over all ranges of variation ($\lambda(p)$ and $\lambda'(p)$ are independent of p for a Gaussian pair). As a result of this, the increasing correlation at the extremes on top of the asymmetry and the kurtosis observed in return distributions and in sources of risk question again the assumptions of normality and multi-normality

---

[27] It is also called a quantile-*quantile dependence measure*.

27.4 Consequences of Non-normality and Analysis of Extreme Conditions

(Gaussian vectors) on which most of the standard parametric methods presented above for computing the VaR are grounded. The use of certain types of copulas and the characterization of distributions' tails with the help of extreme value theory allow us to go beyond the normality and obtain more realistic mathematical representations.

27.4.1.3 Use of Copulas to Represent Non-Gaussian Multivariate Laws

The treatment of multivariate variables $(X_1,\ldots, X_m)$ with non-Gaussian marginal laws $(F_1(.),\ldots, F_m(.))$ faces the difficulties of assessing what is the correct representation of their joint law $F(x_1,\ldots, x_m)$ and of making proper drawings from such a law. As was explained in Chap. 26 (Sect. 26.3.2), a practical method for solving these problems is to be found in using a copula C which preserves the marginal laws and ensures correlations between its components.

$C(F_1(.), \ldots, F_m(.))$ thus represents $F(x_1,\ldots, x_m)$. To obtain the desired characteristics of skewness and kurtosis, it is necessary and sufficient that the marginal laws $F_i(.)$ (theoretical or empirical) to which the copula is applied preserves these characteristics. Different examples of applying the copula method are given in Chaps. 26, 29, and 30, in the context of credit risk.

Further, recall that certain copulas (Student copulas) yield a stronger correlation of the components over the distribution tails than over the intermediate ranges of the distributions.[28] This property is interesting since, as we have already observed, one finds that correlations increase strongly during crises (tails of distributions), thereby reducing the benefits of diversification when they are the most needed.

27.4.2 Distributions of Extreme Values

Schematically, the theory of extreme values embraces two types of models: those that represent the distribution of the maximum (or of the minimum) of a sample of identically distributed elements; those that characterize the tails of distributions (left or right) above some threshold. The latter, called *peaks over threshold* (POT), are the most recent and the most useful for financial applications; they are the only ones we explain (briefly) in this book. POT models are based on generalized Pareto distributions.

27.4.2.1 Generalized Pareto Distributions

First, we present the generalized Pareto distributions that allow characterizing, asymptotically, most of the tails of probability distributions used in practice.

A generalized Pareto distribution $G_{\xi,\beta}(x)$ depends on two parameters β and ξ and is written

---

[28] This correlation is constant for a Gaussian copula.

$$G_{\xi,\beta}(x) = 1 - \left(1 + \xi\frac{x}{\beta}\right)^{-1/\xi} \quad \text{for } \xi \neq 0 \quad \text{and} \quad G_\beta(x)$$

$$= 1 - e^{-x/\beta} \quad \text{for} \quad \xi = 0. \tag{27.36}$$

Its domain of definition is $\{x : 1 + \xi\frac{x}{\beta} > 0\}$, β is a positive scale parameter (analogous to the standard deviation) and ξ is a shape parameter that determines the fatness of the distribution's tails (they are fatter the larger ξ is); in general, ξ can be positive or negative but the case $\xi \geq 0$ is the most useful in financial applications.

Moreover, the existence of moments depends on the value of ξ, the kth moment only being defined for $k < 1/\xi$.

27.4.2.2 The Asymptotic Approximation of Distribution Tails

Let us now formulate an important and very general asymptotic result about the right tail of random variates. Although, in the context of calculating the VaR, we consider the extreme right of the distribution of a loss L, the results that follow can be transposed to the left extreme of the distributions and they apply to a vast class of random variables.

Denote by $F(.)$ the cumulative distribution of the random variable L and consider a value s on the right of this distribution (a threshold) and a number $x > 0$.[29] Denote by $F_s(x)$ the probability that L is less than $s + x$ *knowing that it is greater than* s:

$$F_s(x) = \text{Prob}\{L \leq x + s | L > s\} = \frac{\text{Proba}(s < L \leq s + x)}{\text{Proba}(L > s)}, \quad \text{or}$$

$$F_s(x) = \frac{F(s + x) - F(s)}{1 - F(s)}. \tag{27.37}$$

$F_s(x)$ is therefore the probability that L exceeds the threshold s by an amount smaller than x knowing that it does exceed the threshold. This is a conditional probability that allows characterizing the unconditional distribution of L to the right of the threshold s (*right tail distribution*), as we will shortly explain.

We formulate, without proof, the following important proposition which characterizes, for a very large class of distributions[30] F, the asymptotic behavior of $F_s(x)$ for extreme values of s:

Proposition

There exist a parameter ξ and a function $\beta(s)$ depending on F such that $F_s(x)$ converges to a generalized Pareto distribution $G_{\xi,\,\beta(s)}(.)$ as s increases, so that

---

[29] We assume in what follows that the support of the distribution is unbounded ($F(l) < 1$ for all l); in the contrary case, ($F(l)$ attains the value 1 for $l = L_{max}$), we consider $0 < x < L_{max} - s$.

[30] This class includes notably the uniform, normal, log-normal, exponential, χ^2, t, gamma, and beta distributions, etc.

27.4 Consequences of Non-normality and Analysis of Extreme Conditions 1161

$$\underset{\substack{s\to\infty \\ x>0}}{Lim}\ \sup\left[F_s(x) - G_{\xi,\beta(s)}(x)\right] = 0.$$

$\beta(s)$ and ξ depend on the distribution F but the asymptotic distribution F_s is generalized Pareto with no particular presumption as to the form of F. The proposition means that, above a fixed threshold s sufficiently large, and for a very large class of distributions, $F_s(x)$ can be brought into a form $G_{\xi,\beta}(x)$ with two parameters ξ and β. [31] These two parameters may possibly be estimated, as we explain below. For a normal distribution, $\xi = 0$, which implies that, on the tails, a Gaussian is similar to an exponential distribution ($\underset{s\to\infty}{Lim}F_s(x) = 1 - e^{-x/\beta}$ if F is Gaussian). The larger ξ is the fatter the tail is, and for most financial variables $0 < \xi < 0.5$. Moreover, the kth moment, $E(L^k)$, is only defined when $k < 1/\xi$: the moments are thus all defined for a Gaussian and, for most financial variables, the existence of the mean and the standard deviation may be presumed.

The interest of this asymptotic result is that it holds for a very large class of distributions.[32] The disadvantage is that exploiting it requires estimating the two parameters β and ξ and this estimation can only be based on the observation of extreme values, which are rare by nature.

27.4.2.3 Estimation of the Parameters β and ξ

Estimation of β and ξ can be based on the maximum likelihood method, in which case it can be conducted as follows.

We first write the probability density $dG_{\xi,\beta}(x)/dx$ to which $dF_s(x)/dx$ may be equated for large values of s. By virtue of (27.36), we have

$$\frac{dG_{\xi,\beta}(x)}{dx} = \frac{d\left(1 + \xi\frac{x}{\beta}\right)^{-1/\xi}}{dx} = \frac{1}{\beta}\left(1 + \xi\frac{x}{\beta}\right)^{-(1+\xi)/\xi}. \tag{27.38}$$

Starting from a sample of n observations of the variable L we fix the threshold s at the level of the desired p_s-quantile: $s = F^{-1}(p_s) = \text{VaR}_{p_s}$. and we consider the values $L > s$, that is $x = L - s > 0$.

The choice of the threshold results from a tradeoff between two contradictory requirements: s should be sufficiently high (p_s sufficiently large) that the tail of the distribution of L to the right of s is close enough to the asymptotic distribution; *on the contrary*, s must not be chosen too high so that there are enough observations of L larger than s, since these are the only ones useful in assessing the tail of the distribution to the right of s. We select for example the quantile exceeding the

---

[31] β will thus be considered more as a parameter than as a function $\beta(s)$.

[32] In this respect, this asymptotic result has been compared to that of the Central Limit Theorem.

threshold with 95 % confidence ($p_s = 95$ % or $s = \text{VaR } 95$ %). We only retain the $n_s = (1 - p_s) n$ observations exceeding the threshold s (the 5 % highest values if p_s is fixed at the 95 % level); these n_s extreme observations constitute a subsample $L_1, \ldots,$ L_{n_s}. We construct, from the subsample, the likelihood function $\prod_{i=1}^{n_s} \times$ $\frac{1}{\beta} \left(1 + \xi \frac{L_i - s}{\beta} \right)^{-(1-\xi)/\xi}$, and the log-likelihood $-\sum_{i=1}^{n_s} \frac{1+\xi}{\xi} \ln \left(1 + \xi \frac{L_i - s}{\beta} \right) - n \ln (\beta)$.

The estimators we seek are the parameters β and ξ that maximize the log-likelihood.

The estimated β and ξ give an approximation to $F_s(x)$ for large values of s, therefore of the conditional distribution $\text{Prob}(L \leq l | L > s) \# \widehat{G}_{\xi, \beta}(l - s) = 1 - \left(1 + \widehat{\xi} \frac{l-s}{\widehat{\beta}} \right)^{-1/\xi}$.

27.4.2.4 The Right-Hand Tail of the Loss Distribution L

The *conditional probability* $\text{Prob}(L \leq l \mid L > s) \# G_{\xi, \beta} (l - s)$ allows appreciating the tail of the distribution of F to the right of s defined simply as the *unconditional* probability $F_L(l) = \text{Prob}(L \leq l)$ for values of $l > s$. Indeed:

$\text{Prob}(L > l) \equiv \text{Prob}(L > l \text{ and } L > s)$ (since $l > s$), thus, by Bayes formula:

$$\text{Prob}(L > l) = (1 - p_s) \ \text{Prob}(L > l \mid L > s)$$
$$= (1 - p_s) \ [1 - \text{Prob}(L \leq l \mid L > s)],$$

$$\text{Prob}(L > l) = (1 - p_s) \left(1 + \xi \frac{l - s}{\beta} \right)^{-1/\xi}.$$

Since $\text{Prob}(L > l) = 1 - F_L(l)$, we obtain, by approximation, the right tail of the cumulative distribution of L:

$$F_L(l) = 1 - (1 - p_s) \left(1 + \xi \frac{l - s}{\beta} \right)^{-1/\xi}, \qquad (27.39)$$

for large values of s and p_s and $l > s$.

Empirically, from the sample of n values of L of which n_s are *bigger than* the threshold s, we estimate $(1 - p_s)$ as equal to $\frac{n_s}{n}$ (s is thus the empirical VaR_{ps}); using estimates $\widehat{\xi}$ and $\widehat{\beta}$ (obtained from maximum likelihood), we obtain the following estimator for the right-hand tail of L:

$$\widehat{F}_L(l) = 1 - \frac{n_s}{n} \left(1 + \widehat{\xi} \frac{l - s}{\widehat{\beta}} \right)^{-1/\widehat{\xi}}. \qquad (27.40)$$

27.4.2.5 Calculating the VaR and the Expected Shortfall (ES) from Extreme Distributions

The preceding formulas provide theoretical expressions for and empirical estimates of a VaR_p at a confidence level $p > p_s$ ($VaR_p > s$) as well as the ES_p ($ES_p = E\{L \mid L > VaR_p\}$).

– Starting from the equation defining VaR_p, $\text{Prob}(L \leq VaR_p) = p$, and applying (27.39) with the threshold $s < VaR_p$ and $l = VaR_p$, we have:

$$F_L(VaR_p) = p = 1 - (1 - p_s)\left(1 + \xi \frac{VaR_p - s}{\beta}\right)^{-1/\xi}.$$

Solving this last equation and recognizing that $s = VaR_{p_s}$ we obtain the theoretical formula for VaR_p as a function of VaR_{p_s} with $p > p_s$:

$$VaR_p = VaR_{p_s} + \frac{\beta}{\xi}\left(\left(\frac{1 - p}{1 - p_s}\right)^{-\xi} - 1\right). \tag{27.41a}$$

Empirically, starting from the sample of n values of L of which n_s exceed the cutoff s (which is the empirical VaR_{p_s}) and the estimates for β and ξ, we obtain:

$$VaR_p \text{ estimated} = VaR_{p_s} \text{ empirical} + \frac{\widehat{\beta}}{\widehat{\xi}}\left(\left(\frac{n}{n_s}(1 - p)\right)^{-\xi} - 1\right). \tag{27.41b}$$

The advantage of using this equation, rather than directly estimating the *empirical* VaR_p based on few observations, is that it is grounded on an empirical VaR_{p_s} based on a larger number of observations (since $p_s < p$).

– The Expected Shortfall can be computed either from the conditional distribution

$$\text{Prob}(L \leq l | L > VaR_p) \# G_{\xi,\beta}(l - VaR_p) = 1 - \left(1 + \xi\frac{l - VaR_p}{\beta}\right)^{-1/\xi}$$

which one integrates according to Eq. (27.27a), or alternatively starting from (27.41a) (written with $p = u$) which one integrates according to (27.27b)). This computation shows that

$$ES_p = \frac{1}{1 - \xi}\left[VaR_p + \beta - \xi s\right]. \tag{27.42}$$

The ES estimate results from (27.42) upon replacing the theoretical parameters β and ξ by their respective estimates.

Example 19

Suppose we have 2000 observations (over 8 years) of the daily returns of the securities that make up a portfolio x. They allow us to simulate 2000 daily returns on portfolio x.

The smallest 100 returns extend from -1.15 % to -6.67 %. Portfolio x is worth $100M, thus the loss in $M is equal to minus the return in %. The table below is an extract from that listing the losses corresponding to the 100 smallest returns. These 100 most significant losses are numbered 1 to 100, extend from 1.15 M$ to 6.67 M$ and are given in increasing order (losses numbered 3 to 78 are not shown for clarity).

100 largest losses of 2000 displayed in increasing order

1	2	...	79	80	81	82	83	84	85	86	87	88
1.15	1.16		2.54	2.61	2.66	2.73	2.76	2.86	2.89	3.04	3.12	3.19
89	90	91	92	93	94	95	96	97	98	99	100	
3.24	3.32	3.53	3.54	4.7	4.26	4.72	4.83	5.01	5.85	5.90	6.67	

The 1-day empirical VaR (denoted VaR_e in this example), calculated from these historical data, are $VaR_e(95\ \%) = \$1.15M$; $VaR_e(99\ \%) = \$2.61M$.

From the fact that the 20 observations relevant to the estimate of $VaR(99\ \%)$ constitute a much more restricted sample, one might be tempted to use an equation that holds for the Gaussian case and to write $VaR'(99\ \%) = (\alpha_{0.99}/\alpha_{0.95})VaR_e(95\ \%) = \$1.62M$; in this way one obtains a second estimate for $VaR(99\ \%)$ equal to 62 % of its empirical VaR. The significant difference can be explained by the non-Gaussian character of the return on x.

Relying on extreme value theory and estimating from the maximum likelihood (which uses 100 data) yields $\widehat{\beta} = 0.0070$ and $\widehat{\xi} = 0.20$.

Equation (27.41b) (with $p_s = 0.95$, $p = 0.99$, $s = VaR_e(p_s) = 1.15$ and $n/n_s = 2000/1900$) thus provides a third estimate $VaR''(99\ \%) = \$2.48M$.

This last result is slightly smaller than the empirical VaR ($\$2.61M$) and markedly larger than the VaR obtained from a normality assumption ($\$1.62M$). We may take it that it is both more reliable than the first (since it is based on 1000 observations and not on 20), and than the second (since it does not assume the return to be Gaussian).

27.4.3 Stress Tests and Scenario Analysis

In accordance with the Basel recommendations and to account for the problems raised in the preceding analyses, the VaR or ES measures are often supplemented with other procedures, stress tests, and the scenarios method. The crisis of 2007–2009 sharpened interest in these procedures.

27.4 Consequences of Non-normality and Analysis of Extreme Conditions

Stress tests, for which there is no standard implementation protocol, aimed at examining the effects on the value of a portfolio of *extreme market conditions* and/or the *violation of certain fundamental assumptions* underlying the risk models used and thus at supplementing measures of the VaR sort. The notions of probability and quantile tend to disappear, since *stress testing* concerns catastrophic scenarios that are possible but to which one cannot necessarily assign a probability.

27.4.3.1 Developing Hypotheses and Scenarios

One can distinguish two large families of *stress testing* methods and scenario development: those which emphasize historical data and relate to a more "objective" approach; those which are largely based on hypotheses about the future state of the economy and relate to a more "subjective" approach.

– The approaches termed objective are based on historical distributions of returns and interest rates over very long periods.

Typically, one chooses periods of crisis to develop scenarios (the 2007–2009 financial and 2020–2021 health crises are already used as references). These stress scenarios concern either several decisive financial variables (rates, market prices, historical or implicit volatilities and correlations, etc.), or the global economic and financial situation. When assumptions are made on a limited number of variables only, the other economic and financial variables are often held constant; however, an alternative (called conditional stress tests) consists in varying the unstressed variables as a function of the stressed variables by using, for example, a regression model. With a holistic scenario, one avoids having to postulate definite theoretical forms, which has the advantage of preserving the possible nonlinearities present in relations between factors.

The first difficulty with this method is that there are often too few data, which can be dangerous for nonlinear portfolios (notably those that include options). For example, there could be present in the history a drop of 15 % in share prices but no drop of 10 %, so that this last event might not be taken into consideration although it represents a catastrophic loss. A second limitation is that one is interested in extreme variations of the observations, which requires numerous data over a very long period.

– An alternative to the scenario development based essentially on facts observed in the past consists in constructing crisis scenarios, *starting from plausible* ad hoc *assumptions* about the future economic circumstances. This approach is sometimes termed subjective for its assumptions are as much based on the experience and intuition of the strategists and top management as on historical observations.
– Finally, *stress testing* often uses simulations based on models, for example, *structured Monte Carlo simulations*.

27.4.3.2 Analysis of the Consequences of Scenarios

Typically, the value of a portfolio is estimated in each of the postulated or simulated scenarios. This valuation employs *internal evaluation models*. The difference between the current value of a portfolio and its lowest possible value is the retained measure of risk.

Scenario analysis aims at raising and answering the following questions: What will happen if ...? What is the worst scenario? What are the losses if this scenario occurs? Can the institution survive such a crisis? What measures should the institution undertake to ensure survival?

An example where this method is applied is that of clearinghouses and margin call systems put in place by many officially organized derivative markets (futures and options), based on a *worst-case scenario* method (e.g., the SPAN system). Their goal is to avoid that some market participants let other participants bear the risk of their bankruptcy. A second example, to which we will return at the end of Chap. 29, concerns the stress tests imposed since 2008 on American and European banks by their respective Central Bank.

As a conclusion to this chapter, the ES should be preferred to the VaR, calculated from realistic distributions and supplemented with stress tests and scenario analysis. The complexity of risk management processes should obviously depend on the size, nature of activities, and objectives of the financial institution concerned. In any event, the crisis of 2007–2009 made clear the danger of just sticking to a measure of the standard VaR type. The chapter indicates several ways and means of improving on VaR (i.e., ES), of increasing its relevance (e.g., the use of historical data covering a crisis period in order to estimate a stressed ES, use of the Cornish-Fisher formula, of extreme value theory, and the like) and supplementing it (e.g., by stress tests).

27.5 Summary

- The Value at Risk (VaR) measures the risk affecting the market value V of a given position.
  $L_h \equiv V_0 - V_h$ is the *loss* between 0 and h, and F_L its distribution function ($F_L(x) =$ Prob($L_h \leq x$)). The VaR at the horizon h and with probability level p, $VaR(h,p)$, is the p - quantile of L_h:
  Prob $(L_h \leq VaR(h,p)) = p$, or equivalently : Prob $(L_h > VaR(h,p)) = 1$-p.
- If L_h is normal $N(\mu, \sigma)$, its p-quantile is $L_p = \alpha_p \sigma + \mu$, where α_p is the p-quantile of $N(0, 1)$. Closed-form expressions are also obtained if L_h is assumed log-normal.
- When a portfolio x contains several assets, its VaR, depends on the covariances of the returns and the contribution of a component i to the VaR of x depends on cov (R_i, R_x).
- If the successive *daily* losses, $\Delta_j L \equiv L_j - L_{j-1} = V_{j-1} - V_j$, are Gaussian $N(\mu_1, \sigma_1)$ and *uncorrelated* we obtain, for an *horizon* of h days: $\mu_h = h\mu_1$; $\sigma_h = \sigma_1\sqrt{h}$, and $VaR(h,p) = \alpha_p \sigma_1 \sqrt{h} + \mu_1 h$.
- These formulas can be generalized in case of autocorrelation.

27.5 Summary

- In calculating VaR, positions are *marked-to-market,* global risk should account for *diversification,* the horizon h is short (1, 5, or 10 trading days), and the confidence level p is at least 95 %.
- If the portfolio contains complex instruments or numerous cash flows, a simplified representation of it must be constructed.
- This simplification can be obtained, for instance, by (i) representing a multi-cash-flow portfolio of fixed-income securities by a portfolio of zero-coupon bonds whose variance-covariance matrix may be estimated, and (ii) representing the portfolio value $V(\underline{Y})$ as a function of some factors $\underline{Y}$, with or without relying on approximations.
- *Partial valuation* in contrast with *full valuation* is based on an approximation of V. The linear approximation of $V(\underline{Y})$ (*delta valuation*) estimates the first-order effect of a variation $\Delta\underline{Y}$ and may yield closed-form expressions of the VaR in the Gaussian case.
- Estimating VaR through *simulations* based on *historical data* relies on the assumption of *stationarity* of the joint distribution of asset returns whose past values allow constructing a simulated distribution of the loss. VaR(h, p) is the p-quantile of this simulated loss.
- Data series may cover normal periods and/or crisis periods (*stressed* VaR). Similarly, the parameters used for theoretical laws can be estimated from either kind of series.
- *Monte Carlo simulations* (see Chap. 26), useful in complex situations, create a sample of large size M drawn from the presumed law of $\underline{Y}(t)$ and associating to each element $\underline{Y}_j$ the corresponding value $V(t,\underline{Y}_j)$. According to the problem at hand, Monte Carlo simulation involves the simulation, either of a great number M of possible *paths,* or of M values $V(h, \underline{Y}_j)$ on *a single date h* only, yielding M simulated losses. VaR(h, p) is the p-quantile of such a simulated loss distribution.
- VaR suffers from numerous limitations, both technical (such as non-normality and instability of (log) returns, jumps, unstable correlations between assets), and conceptual (it does *not indicate the size of the losses exceeding the VaR* and is *not a coherent measure of risk*).
- The *Expected Shortfall* (ES), is the *conditional expectation of the loss knowing that it exceeds the VaR*: $ES(h,p) \equiv E[L|L \geq \text{VaR}(h,p)] = \frac{1}{1-p}\int_{VaR(h,p)}^{+\infty}xf_L(x)dx.$
- ES is *a coherent measure of risk* that overcomes most of VaR conceptual and theoretical drawbacks. Estimating an ES is not substantially more difficult and relies on the same methods as VaR; it may be based on data covering normal periods and/or crisis periods (stressed ES).
- Actual returns are characterized by *skewness, kurtosis (fat tails) and higher correlations at the extremes* (observed in times of crisis), which are *inconsistent* with the *Gaussian* assumption. They weaken the relevance of analytical VaR and ES assessments but can be improved by:
  A *Cornish-Fisher expansion* (taking skewness and excess kurtosis into account), the representation of *non-Gaussian multivariate distributions by copulas,* and

1168   27 Value at Risk, Expected Shortfall, and Other Risk Measures

resorting *to extreme value models* of the type *peaks over threshold* grounded on generalized *Pareto distributions* which allow *asymptotic representations of fat tails* that yield more realistic assessments of VaR and ES.

- VaR or ES are often supplemented with stress tests and scenarios examining the effects on a portfolio of *extreme market conditions*, raising and answering "what if" type of questions.

Suggestions for Further Reading

Books

*Cherubini, U., Luciano, E., & Vecchiatto, W. (2004). *Copula methods in finance*. Wiley.
Dowd, K. (2005). *Measuring market risk*. Wiley.
*Embrechts, P., Klüppelberg, C., & Mikosch, T. (1999). *Modelling extremal events for insurance and finance*. Springer.
*Fusai, G., & Roncoroni, A. (2008). *Implementing models in quantitative finance: Methods and cases*. Springer.
Gumbel, E. J. (1958). *Statistics of extremes*. Columbia University Press.
Hull, J. (2009). *Risk management and financial institutions* (2nd ed.). Prentice Hall Pearson Education.
Hull, J. (2018). *Options, futures and other derivatives* (10th ed.). Prentice Hall Pearson Education.
Jorion, P. (2006). *Value at risk* (3rd ed.). McGraw-Hill.
Markowitz, H. (1959). *Portfolio selection: Efficient diversification of investment*. John Wiley.
McNeil, A. J., Frey, R., & Embrechts, P. (2015). *Quantitative risk management: Concepts, techniques and tools* (revised ed.). Princeton University Press.
Rüschendorf, L. (2013). *Mathematical risk analysis: Dependence, risk bounds, optimal allocations and portfolios*. Springer.

Articles

Acerbi, C., & Tasche, D. (2002). On the coherence of expected shortfall. *Journal of Banking and Finance, 26*(7), 1487–1503.
Artzner, P., Delbaen, F., Eber, J. M., & Heath, D. (1997). Thinking coherently. *Risk, 10*(11), 68–71.
Artzner, P., Delbaen, F., Eber, J. M., & Heath, D. (1999). Coherent measures of risk. *Mathematical Finance, 9*, 203–228.
Bollerslev, T. (1986). Generalized autoregressive conditional heteroscedasticity. *Journal of Econometrics, 31*, 307–327.
Boudhouk, J., Richardson, M., & Whitelaw, R. (1998). The best of both worlds. *Risk, 1998*(11), 64–67.
Chen, C., Iyengar, G., & Moallemi, C. C. (2013). An axiomatic approach to systemic risk. *Management Science, 59*(6), 1373–1388.
Duffie, D., & Pan, J. (1997). An overview of value at risk. *Journal of Derivatives, 4*(3), 7–49.
Embrechts, P., Puccetti, G., Rüschendorf, L., Wang, R., & Beleraj, A. (2014). An academic response to Basel 3.5. *Risks, 2*(1), 25–48.
Emmer, S., Kratz, M., & Tasche, D. (2015). What is the best risk measure in practice? A comparison of standard measures. *Journal of Risk, 18*(2), 31–60.
Engle, R. F. (1982). Autoregressive conditional heteroscedasticity with estimates of the variance of UK inflation. *Econometrica, 50*, 987–1008.

Suggestions for Further Reading

Fama, E. (1965). Behavior of stock market prices. *Journal of Business, 38*, 34–105.

Fermanian, J. D., & Scaillet, O. (2005). Sensitivity analysis of VaR and expected shortfall for portfolios under netting agreements. *Journal of Banking and Finance, 29*, 927–958.

Föllmer, H., & Schied, A. (2002). Convex measures of risk and trading constraints. *Finance and Stochastics, 6*(4), 429–447.

Frittelli, M., & Rosazza Gianin, E. (2002). Putting order in risk measures. *Journal of Banking and Finance, 26*(7), 1473–1486.

He, X., & Peng, X. (2018). Surplus-invariant, law-invariant, and conic acceptance sets must be the sets induced by value-at-risk. *Operation Research, 66*(5), 1268–1276.

Hull, J., & White, A. (1998a). Incorporating volatility updating into the historical simulation method for value at risk. *Journal of Risk, 1*, 5–19.

Hull, J., & White, A. (1998b). Value at risk when daily changes are not normally distributed. *Journal of Derivatives, 5*(3), 9–19.

Jamshidian, F., & Zhu, Y. (1997). Scenario simulation model: Theory and methodology. *Finance and Stochastics, 1*, 43–67.

Kou, S., & Peng, X. (2016). On the measurement of economic tail risk. *Operations Research, 64*(5), 1056–1072.

Kupiec, P. (1995). Techniques for verifying the accuracy of risk management models. *Journal of Derivatives, 3*, 73–84.

Kupiec, P. (1999). Stress testing in a value at risk framework. *Journal of Derivatives, 6*, 7–24.

Liu, F., & Wang, R. (2020). A theory for measures of tail risk. *Mathematics of Operations Research, 90*, 66–79.

Longin, F. (1996). The asymptotic distribution of extreme stock market returns. *Journal of Business, 69*, 383–408.

Longin, F. (2000). From value at risk to stress testing: the extreme value approach. *Journal of Banking and Finance, 24*, 1097–1130.

Longin, F. (2001). Beyond the VaR. *Journal of Derivatives, 8*(4), 36–48.

McNeil, A. J. (1999). Extreme value theory for risk managers. In *Internal models and CAD II, risk books*. Accessed from www.math.eth.ch/mcneil

Mina, J., & Yi Xiao, J. (2001). *Return to RiskMetrics: The evolution of a standard*. RiskMetrics.

Morgan, J. P. *RiskMetrics™, Technical Documents*. Morgan Guarantee Trust Company.

Neftci, S. N. (2000). Value at risk calculations, extreme events and tail estimation. *Journal of Derivatives, 7*(3), 23–38.

Patton, A. J., Ziegel, J. F., & Chen, R. (2019). Dynamic semi-parametric models for expected shortfall (and value-at-risk). *Journal of Econometrics, 211*(2), 388–413.

Scaillet, O. (2004). Nonparametric estimation and sensitivity analysis of expected shortfall. *Mathematical Finance, 14*(1), 115–129.

Wang, R., & Zitikis, R. (2020). An axiomatic foundation for the expected shortfall. *Management Science, 2020*, 1–17.

Modeling Credit Risk (1): Credit Risk Assessment and Empirical Analysis

28

Until the end of the 1990s, theoretical as well as analytical developments concerning financial risks focused mainly on market risks: interest rate risk, exchange rate risk, and risk affecting the price of shares and portfolios. The few theoretical developments concerning credit risk, initiated by Merton and based on option theory, had not gone beyond the academic sphere. In banking and nonbanking companies, the issuer default risk was the prerogative of *credit risk teams in charge of implementing* qualitative methods based on business experience and quantitative *credit scoring* methods. Those methods are based on statistical analysis[1] and are useful to "rate" potential issuers/borrowers and to assess the appropriateness of granting them a credit or not. In particular, Edward Altman introduced 50 years ago the Z-score, based on discriminant analysis and involving several accounting ratios.[2] The original Altman's method has been improved and used to compute probabilities of default. However, these methods are insufficient to effectively manage credit risk throughout the life of the credit, which includes both default risk and the risk of a deterioration of creditworthiness (e.g., rating), especially from a *marked-to-market* perspective. Most of all, they do not provide any answer to the problems posed by the management of credit risk in a debt portfolio, the more or less diversified nature of which plays an important role. Moreover, traditional approaches to portfolio theory (Markowitz) are insufficient, or require significant adaptations, to apply to the treatment of a portfolio credit risk. The available analytical methods were basic and this explains why, until the late 1990s, banks and regulators applied

---

We thank Riadh Belhaj and Andras Fulop for their helpful comments on this chapter.

[1] For example, data analysis, such as Main Component Analysis, Discriminant Analysis or Factor Analysis, or processing historical data relating to the issuer/borrower behavior according to age, revenues, various solvency ratios, and so on.

[2] The original Z ratio was a linear combination of 5 ratios: Working capital/Total assets; Retained Earnings/Total assets; EBIT/Total assets; Sales/Total assets; and Market value of equity/Book value of total liabilities.

© The Author(s), under exclusive license to Springer Nature Switzerland AG 2022
P. Poncet, R. Portait, *Capital Market Finance*, Springer Texts in Business and Economics, https://doi.org/10.1007/978-3-030-84600-8_28

1171

very simple criteria for measuring and hedging credit risk. The lack of sound theoretical and technical framework in credit risk management also explains why, until the late 1990s, credit risk was barely addressed in graduate finance education. However, credit risk has become the most important risk borne by credit institutions.[3] This is why, in recent years, credit risk analysis and management methods have generated considerable interest in the academic and professional worlds as well as among regulators. In addition, awareness of the importance of counterparty risk (distinct from credit risk) has also fallen under the scrutiny of those worlds (academic and professional), particularly in the aftermath of recent financial crises.

This book devotes two chapters to the assessment and management of credit risk, in addition to the third chapter on credit derivatives. This chapter, which addresses the analysis and valuation of securities subject to credit and/or counterparty risk, is split into two sections. The first section focuses on empirical analysis and evaluation tools, while the second section presents the main valuation models and some applications.

28.1 Empirical Tools for Credit Risk Analysis

In the first paragraph, we recall the basics of credit risk analysis (presented in particular in Chaps. 5 and 6) and clarify the credit rating framework. These basics are a prerequisite for the study of modern methods of credit risk management, which are grounded on different approaches. The one we present in this section, which can be described as "empirical and actuarial," is based on the issuer/borrower rating (developed according to more or less traditional methods), on the historical default rates and rating changes observed in the past and on the yield curves applied to each rating (a one-to-one mapping between a yield curve and a rating level).

28.1.1 Reminder of Basic Concepts, Empirical Observations, and Notations

28.1.1.1 Basic Concepts and Notations

Let us recall from Chap. 5 that debt securities are affected, to various extent, by credit risk that justifies the existence of a (credit) spread. Letting r^{max} be the rate of return on risky but default-less assets ("promised" rate), r the risk-free rate, and s the spread, we obtain:

---

[3] According to surveys performed by the IMF or the EBA (European Banking authority) for not only significant credit institutions but also less significant players, credit risk represents more than two thirds of the overall risk; the remaining third corresponds to operational risk and market risks (interest rates and FX rates).

28.1 Empirical Tools for Credit Risk Analysis

$$r^{max} = r + s.$$

r^{max} is an ex ante promised rate, i.e., the maximum yield to maturity[4] that must be distinguished from the rate at maturity actually observed, denoted by R in this chapter. R is a random variable (due to the possible event of default affecting the debt security) so that we have:

$$R = r^{max} \text{ in absence of default}; R < r^{max} \text{ in case of default}; E(R) < r^{max}.$$

Finally, the risk-free rate r is often defined as the one borne by government securities displaying the same maturity and profile (zero-coupon or money market yield); however, in some circumstances, it is preferable to use a different reference for the risk-free rate. For instance, the Libor (or Euribor) curve will be used for the analysis of short-term or floating-rate securities. The fixed-rate (zero-coupon) yield curve from vanilla *swaps* may be used to value credit derivatives (*see* Chap. 30). Let us recall (from Chap. 7) that since the aftermath of the subprime crisis and the evolution of the regulation affecting the interest swap markets, those swaps are traded through Central Counterparty Clearing Houses ("CCP") and subject to collateral deposits and margin calls. Those swaps are therefore little subject to credit risk (more precisely called *counterparty risk* when dealing with derivatives), which make them good candidates to build a risk-free rate curve.

- Credit risk includes default risk[5] and the risk of creditworthiness deterioration (spread risk). The former concerns the likelihood of a borrower failing to repay any interest or nominal on a due date (*default*). The latter addresses the possible deterioration of the borrower creditworthiness perceived by the market (without the borrower necessarily being in default)[6] leading to an increase of the credit spread and a decrease of the market value of the security. Since the credit spread is expected to be an inverse function of the issuer's rating, spread risk is linked to rating risk. However, there are other aspects affecting this relationship.

In the remainder of this chapter, we will use in many instances the term default. We stress again that, barring exception, the term *default* is a legal concept, which encompasses various notions. The most intuitive notion for the reader not familiar with this concept is that the default automatically triggers a file for bankruptcy and the liquidation of the company. However, it is possible that an event of default does

---

[4]This rate is obtained by holding the security until maturity in absence of issuer default. One may come up with a higher rate through the sale of this security above the par (following a drop the interest rate and/or credit spread) before it matures.

[5]For market players (trading CDS) and rating agencies, the "default" on a corporate debt issuer is triggered by one of the following events: bankruptcy, failure to pay (one of more cash flows), repudiation (e.g., the issuer does not recognize the debt anymore), debt (modified) restructuring. In Europe, current developments regarding "Resolution" may also affect this definition.

[6]This risk is called *Credit Deterioration* and must be distinguished from *Default*.

not lead to liquidation but to a debt restructuring process in which the issuer will for instance reduce the coupon or the nominal to be paid. In some instance, the event of default also corresponds to a case where the issuer announces it does not acknowledge its debt anymore ("debt repudiation" usually met in practice for governmental debt). In any of those cases, the debt security holders will lose part (or all) of their investment. Therefore, we will have to model this loss of value due to an event of default. This event is called "credit event" for credit derivatives, as we will see in Chap. 30.

Noting that the expected return must be equal to the risk-free rate plus a risk premium π and a possible liquidity premium l, so that $E(R) = r + \pi + l$, we obtain a breakdown of the credit spread $r^{max} - r$ into three components: the expectation of the drop in profitability linked to the possibility of a default which is equal to $r^{max} - E(R)$, a risk premium (in the strict sense) and a liquidity premium, i.e.

$$r^{max} - r \equiv s = \left((r^{max} - E(R)) + \pi + l \right).$$

It should be noted that only the first two components of the spread ($r^{max} - E(R)$ and π) are attributable to credit risk.[7]

One should also bear in mind that the credit spread may vary, even if the issuer's rating and the liquidity of the security remain unchanged: Any change affecting the market price of credit or liquidity risk may impact the premiums π or l which, in this sense, have a systemic component.

In addition, other components may be added to the required rate when the credit instrument contains an option held by the issuer (such as the prepayment option) or subtracted when the option is held by the investor (such as the early resale option) or when the security has a tax benefit. These possible additional components will not be considered in the remainder of this chapter and the spread will be written as the sum of three components only.

Finally, recall that the term *counterparty risk* is reserved for the risk that affects the credit risk of the parties engaged in a derivative deal and that it has been greatly limited by widespread recourse to margin procedures (initial margin and margin calls) and to central clearing counterparties (CCPs) or clearinghouses for vanilla OTC derivatives.

– We have shown in Chap. 5 that the spread s is related to the probability of default p and the expected recovery rate α (fraction of the debt value recovered by the creditor in the event of default). *If we neglect the default risk and liquidity premiums,* we come up with the credit triangle formula valid for the yield to maturity of a security over a single period but also, under certain assumptions, for a coupon-bearing security:

---

[7]The $r^{max} - E(R)$ component is sometimes referred to as the *default premium*, which must again be distinguished from the *risk premium* π.

28.1 Empirical Tools for Credit Risk Analysis

$$r^{max} - r \equiv s = r^{max} - E(R) = p(1 - \alpha),$$

where the term $p(1 - \alpha)$ corresponds to the expected loss attributable to a default on an exposure of \$1, $r^{max} - E(R)$, and the credit triangle formula means that the (credit) spread must compensate this expected loss. We stress that premiums π and l are here disregarded. We will see later that, if the premium l may be considered as nil for a liquid security, the premium π is always positive and materializes the market's risk aversion.

For an exposure of \$ M, the expected loss is $M p(1 - \alpha)$. In the terminology used among practitioners, M is the *Exposure at Default* (*EAD*), and $(1 - \alpha)$ the *loss given default* for \$1 exposed (*Loss Given Default* (LGD)). The *Expected Loss* (*EL*) hence writes: $EL = EAD \times LGD \times p$. For the sake of simplicity, we do not use these notations albeit they should be familiar to the interested reader as those notations are common in professional documents (regulatory texts, internal procedures, and so on).

These preliminary analyses and observations highlight the importance of measuring default probabilities on the basis of historical data and trying to map these historical probabilities onto observed spreads.

28.1.1.2 Empirical Observations on Yield Curves

Due to differences in credit spreads, the yield curve as a function of maturity is not unique because it depends on the issuer's credit risk. To avoid any ambiguity, we ground our reasoning upon zero-coupon rate curves. The range of zero-coupon rates free of credit risk (usually Treasury-bonds, -notes, and -bills from AAA OECD countries) may stand slightly below the curve for AAA-rated securities, which in turn is below the AA range, etc.

By definition, the credit spread s is the difference between the zero-coupon rate r_θ^{max}, with maturity θ, "promised" by an issuer with a given credit risk and rating, and the corresponding rate r_θ of a Treasury security, presumed free of credit risk:

$$s_\theta \equiv r_\theta^{max} - r_\theta.$$

Recall that the "promised" rate r_θ^{max} is the rate of return at maturity obtained in the absence of default. Figure 28.1 shows the typical profile of the range s of these *spreads*.

Empirically, we can see that these spreads are higher when the rating is low and the maturity is remote.[8] Moreover, for a given rating and a given maturity, the spread is not constant over time; in particular, it increases in periods of recession and decreases in bullish market conditions. It is also important to remember that the various Treasury bonds obtain different ratings and that Governments/States from

---

[8] The spread curve for credit derivatives corresponding to a given rating, which should theoretically be close to the above cash spread curve for the same rating, differs from it for reasons explained in Chap. 30; in particular, the spread curve for credit derivatives does not always increase with maturity.

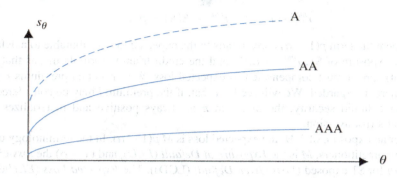

Fig. 28.1 Spreads over Treasury securities displaying the best creditworthiness, by rating and maturity

Table 28.1 Example of cumulative probabilities of default in %

	1 year	2 years	3 years	4 years	
AAA	0.00	0.00	0.02	0.05	...
AA	0.01	0.03	0.07	0.14	...
A	0.03	0.07	0.12	0.25	...
BBB	0.29	0.60	1.00	1.80	...
BB	2.30	3.80	6.50	9.00	...
B	5.29	12.00	16.10	19.40	...
CCC	45.56	56.80	62.10	68.50	...
D	100.00				

different countries do not borrow at the same rate and, therefore, there exist so-called "sovereign" spreads between the borrowing rates of those different States.

We will now go more in depth and complement these preliminary observations with an empirical approach at first, then introduce a more theoretical framework.

28.1.2 Historical (Empirical) Default Probabilities and Transition Matrix

28.1.2.1 Historical Probabilities of Default

The probabilities of default sorted by rating and at different time horizons are estimated from historical data by major rating agencies such as Standard and Poor's or Moody's. They periodically publish tables that typically present *cumulative* probabilities of default (see Table 28.1).

For example, the probability that a company currently rated A will default in the coming year is 0.03% and reaches 0.07% over the next 2 years (implying an *unconditional* probability of 0.04% in the second year). The probability of default in the second year *conditional on the absence of default in the first year*, called the *hazard rate*, is equal to 0.04% /0.9997. Although in the case of low default

28.1 Empirical Tools for Credit Risk Analysis

probabilities, as in this example, these two probabilities have almost equal numerical values, they are conceptually different and must be carefully distinguished, as we see later on.

We observe that for the companies displaying high credit ratings, the annual probability of default is higher when considering remote maturities. This phenomenon is not as significant when the company's credit rating is lower, and it is even reverse for companies initially in financial distress, such as those rated CCC. These intuitive results are easy to explain.[9]

Finally, it should be noted that D (default) is an absorbent state: Once this state is reached, the company is presumed to have been liquidated and in some instances the claims partially repaid.

28.1.2.2 Transition Probabilities from One Rating to Another: The Transition Matrix

We have already emphasized that credit risk does not only lie in the possibility of a default during the holding period of the debt but also in the possibility of a creditworthiness deterioration in the same period, which results in a rating downgrade and a drop in the debt market value.

The transition probabilities from one category to another, estimated using empirical frequencies, are shown in a double entry matrix called a transition or migration matrix. As an example, Table 28.2 shows a transition matrix established by Standard and Poor's.

For example, there is a probability of 77.07% that a debt rated AA will keep the same rating in the coming year but may be downgraded to AA− with a probability of 13.10% or to A+ with a probability of 4.26% in the same timeframe and so on.

It should be noted that consistent relationships exist between Tables 28.1 and 28.2.

The transition probabilities to D, i.e., the default probabilities displayed in the last column of Table 28.2, must coincide with the 1-year default probabilities in the first column of Table 28.1.

The iterative multiplication of the annual transition matrix by itself gives the multi-year rating transition matrix. Let us represent the different ratings by an integer i, with $i = 1, \ldots, d$, where 1 and d respectively denote the best (AAA) and the worst rating, the latter corresponding to the default state (D). p_{ij} is the probability that a company rated i at the beginning of the year will be rated j at the beginning of the following year. p_{id} represents the probability of default over the coming year, $p^{(2)}_{ij}$ the same probability over 2 years and $p^{(n)}_{ij}$ over n years.

Denote by $\mathbf{M}$ the annual transition matrix of general term p_{ij} and by $\mathbf{M^{(n)}}$ the transition matrix over a period of n years with general term $p^{(n)}_{ij}$. If the annual

---

[9] A near-perfect situation can only get worse; in a very serious situation, the "imminent" default is likely, but if the crisis is overcome, the chances of recovery increase.

Table 28.2 Example of a transition matrix over 1 year (probabilities expressed in %)

(%)	AAA	AA+	AA	AA−	A+	A	A−	BBB+	BBB	BBB−	BB+	BB	BB−	B+	B	B−	CCC+	CCC	CCC−	D
AAA	83.32	8.92	5.20	1.01	0.53	0.25	0.38	0.09	0.17	0.02	0.05	0.05	0.00	0.00	0.00	0.00	0.00	0.00	0.00	0.00
AA+	1.06	78.41	13.10	4.72	1.06	0.84	0.25	0.14	0.20	0.10	0.09	0.01	0.01	0.00	0.00	0.00	0.00	0.00	0.00	0.00
AA	0.58	0.56	77.07	13.10	4.26	2.68	0.68	0.54	0.31	0.03	0.01	0.04	0.04	0.01	0.01	0.03	0.00	0.03	0.00	0.01
AA−	0.09	0.22	1.28	75.53	13.57	6.78	1.31	0.44	0.26	0.15	0.09	0.06	0.02	0.05	0.12	0.01	0.00	0.00	0.00	0.01
A+	0.02	0.10	0.57	2.46	76.39	13.05	4.74	1.25	0.69	0.20	0.07	0.14	0.08	0.08	0.12	0.01	0.01	0.00	0.00	0.02
A	0.07	0.09	0.49	0.87	2.13	79.20	11.75	3.17	1.20	0.35	0.19	0.20	0.13	0.08	0.04	0.01	0.00	0.00	0.01	0.02
A−	0.14	0.05	0.21	0.39	1.17	2.81	77.11	13.52	3.21	0.75	0.20	0.15	0.06	0.11	0.03	0.03	0.00	0.00	0.00	0.03
BBB+	0.09	0.08	0.12	0.26	1.01	3.37	2.64	75.69	14.09	1.56	0.36	0.28	0.08	0.08	0.05	0.02	0.00	0.01	0.00	0.20
BBB	0.04	0.03	0.19	0.22	0.62	1.50	2.92	8.87	71.49	11.42	1.08	0.72	0.22	0.17	0.15	0.03	0.01	0.02	0.00	0.29
BBB−	0.12	0.01	0.12	0.28	0.38	1.01	1.30	3.87	12.81	64.39	4.87	5.67	2.26	1.15	0.45	0.37	0.09	0.05	0.02	0.81
BB+	0.19	0.01	0.02	0.20	0.17	0.47	1.00	1.16	4.83	11.79	68.09	8.24	1.24	0.64	0.27	0.13	0.03	0.03	0.01	1.48
BB	0.01	0.04	0.09	0.06	0.03	0.32	0.24	0.51	1.41	3.96	13.35	69.32	6.02	1.48	0.47	0.21	0.09	0.08	0.02	2.30
BB−	0.00	0.01	0.01	0.06	0.12	0.20	0.35	0.49	1.01	1.60	4.85	12.37	66.63	7.08	1.08	0.38	0.16	0.10	0.04	3.46
B+	0.00	0.07	0.02	0.13	0.06	0.11	0.31	0.41	0.24	0.42	0.93	3.40	14.42	68.09	5.66	0.92	0.35	0.24	0.10	4.10
B	0.01	0.00	0.13	0.02	0.06	0.36	0.37	0.30	0.28	0.16	0.71	1.49	3.58	16.42	63.13	5.08	1.44	0.79	0.36	5.29
B−	0.02	0.00	0.01	0.01	0.19	0.18	0.19	0.36	0.22	0.17	0.40	0.59	1.32	6.81	15.08	58.99	4.60	2.00	0.72	8.14
CCC+	0.02	0.00	0.01	0.02	0.86	0.09	0.08	0.62	0.35	0.05	0.17	0.19	1.20	3.64	7.00	15.58	40.65	4.20	1.70	23.58
CCC	0.36	0.01	0.01	0.01	0.04	0.38	0.39	0.73	0.43	0.12	1.06	1.09	0.68	2.09	8.08	6.87	12.57	17.02	2.51	45.56
CCC−	0.01	0.00	0.00	0.00	0.01	0.02	0.02	0.04	0.04	0.06	0.71	0.20	2.08	2.36	1.79	2.71	1.72	12.48	9.34	66.41
D	0.00	0.00	0.00	0.00	0.00	0.00	0.00	0.00	0.00	0.00	0.00	0.00	0.00	0.00	0.00	0.00	0.00	0.00	0.00	100.00

28.1 Empirical Tools for Credit Risk Analysis

transition probabilities, p_{ij}, remain stable from 1 year to the next,[10] we then have: $p^{(2)}_{ij} = \sum_{k=1}^{d} p_{ik}p_{kj}$, i.e. $\mathbf{M}^{(2)} = \mathbf{M}^2$; and more generally: $\mathbf{M}^{(n)} = \mathbf{M}^n$.

The last column of the $\mathbf{M}^n$ matrix should therefore theoretically coincide, in this case, with the nth column of Table 28.1 displaying the cumulative default probabilities over n years.[11]

28.1.3 Risk-Neutral Default Probabilities Implicit in the Spread Curve and Discounting Methods in the Presence of Credit Risk

While empirical or historical probabilities are relevant when assessing the credit risk borne by an investor, modern valuation methods use risk-neutral (RN) probabilities and, more generally, martingale probabilities. Historical probabilities can be inferred from an historical sequence of quoted prices, migrations, or defaults, as explained in the previous paragraph, whereas RN probabilities can be deduced from current observed prices, rates, or spreads.

28.1.3.1 Risk-Neutral or Forward-Neutral Default Probabilities Implied in Credit Spreads

We resume the analysis from the above relationship between the credit spread and the default probability of a zero-coupon bond with duration θ and use continuous rates for ease of exposition.

The zero-coupon bond which is worth €1 on date 0 delivers a contractual ("promised") cash flow $X_{max} = \exp(\theta\, r_\theta^{max})$ on date θ. The payoff X is not certain as this security bears a default risk. There is a positive probability that the issuer defaults on date θ and, should the case occur, only repays a fraction α of its debt.

Let $\phi(\theta)$ be the *cumulative probability* of default between 0 and θ. Let p_t be the *unconditional* probability of default during period t alone, so that we can write

$$\phi(\theta) = \sum_{t=1}^{\theta} p_t.$$

Let $\gamma(\theta)$, with $\gamma(\theta) = 1 - \phi(\theta)$, be the probability of absence of default until time θ and called in the sequel the *survival probability* at time θ.

Note again that the *unconditional* probability p_t must be distinguished from the probability of default in period t, *given that the issuer had not defaulted* before that period, i.e., from the *conditional* probability h_t of default in period t, called the *hazard rate*.

---

[10] It follows that the rating is a Markovian process, because, at any given date, the probabilities of transition p_{ij} from one rating to another do not depend on past events. The matrix $\mathbf{M}$ then represents a Markov chain. Unfortunately, the stability over time of transition probabilities is not empirically verified.

[11] This is not the case in our examples, columns 2, 3, and 4 of Table 28.1 being ad hoc.

The conditional probabilities h^θ and unconditional probabilities p^θ are linked by the recurrence relationship:

$$p_1 = h_1, \text{and } p_\theta = h_\theta \prod_{t=1}^{\theta-1} (1 - h_t) = h_\theta \gamma(\theta - 1), \text{ for } \theta = 2, \ldots, T \text{ and } p_\theta < h_\theta.$$

First, we will assume the recovery rate α to be known with certainty.

The expectation of the random cash flow X actually paid by the security at time θ can be written as follows:

$$E(X) = (1 - \phi(\theta)) \exp(\theta\, r_\theta^{\max}) + \phi(\theta)\alpha \exp(\theta\, r_\theta^{\max}), \text{i.e.} :$$
$$E(X) = \exp(\theta\, r_\theta^{\max}) [1 - \phi(\theta)(1 - \alpha)].$$

It should be noted that $1 - \phi(\theta)(1 - \alpha)$ is an adjustment coefficient (<1) by which the contractual ("promised") cash-flow must be multiplied to obtain the expectation of the cash-flow actually paid. Besides, $\phi(\theta)(1 - \alpha)$ is the expected loss for \$1 invested. These two terms will be used frequently in the following analyses.

Let us consider the *risk-neutral* (RN) world where probabilities are different from those in the real world. We will hence note p^*, h^*, and ϕ^* the RN probabilities. Similarly, the RN expectation (calculated with the corresponding RN probabilities) will be also indicated by a star in order to distinguish it from the real-world expectation.

The RN expectation of the random cash flow X actually paid by the security which is worth \$1 on date 0 and which contractually delivers a cash flow equal to exp $(\theta\, r_\theta^{\max})$ at time θ then writes:

$$E^*(X) = \exp(\theta\, r_\theta^{\max}) [1 - \phi^*(\theta)(1 - \alpha)]. \tag{i}$$

In a risk-neutral world, the risk premium π is equal to zero. If we disregard the liquidity premium l, an investment of \$1 that gives $E^*(X)$ should yield the risk-free rate r:

$$E^*(X) = e^{\theta r}. \tag{ii}$$

By bringing together (*i*) and (*ii*), we obtain: $\exp(\theta\, r_\theta^{\max}) [1 - \phi^*(\theta)(1 - \alpha)] = e^{\theta r}$, or:

$$1 - \phi^*(\theta)(1 - \alpha) = e^{\theta(r - r\theta\max)}, \text{i.e.}$$

$$1 - \phi^*(\theta)(1 - \alpha) = e^{-\theta\, s\theta} \iff s_\theta = -\frac{1}{\theta} \ln(1 - \phi^*(\theta)(1 - \alpha)) \tag{28.1a}$$

which, for small values of $\phi^*(\theta)$ and/or high values of α, maybe approximated at the first order by:

28.1 Empirical Tools for Credit Risk Analysis

$$s_\theta = \frac{1}{\theta} \, \phi^*(\theta) \, (1 - \alpha).$$ (28.1b)

Recall that $\phi^*(\theta) \, (1 - \alpha)$ represents the RN expected loss attributable to an exposure of \$1. The relationship (28.1b) means that this loss must be compensated by a spread s_θ perceived *on an accruals basis (pro rata temporis)* over θ periods. The relationship (28.1a) means that this loss is equivalent to a continuous interest rate loss $\frac{1}{\theta} \ln(1 - \phi^*(\theta) \, (1 - \alpha))$ which must be compensated by a spread s_θ of the same value.

We stress that the relationships (28.1a) or (28.1b) only hold for liquid assets and in the absence of a risk premium, i.e., *in a risk-neutral world*. That is why the probabilities $\phi^*(\theta)$ associated with the spreads s^θ of zero-coupon bonds as shown above in expression (28.1a) or (28.1b) are RN probabilities and not historical probabilities $\phi(\theta)$.

At this stage, we invite the reader to carefully distinguish the different notions of probability used which we summarize as follows:

p denotes an *unconditional annual historical* probability and p^* its RN equivalent,
h is an *annual conditional* historical probability of the absence of default in previous years (*hazard rate*) and h^* its RN equivalent,
ϕ and ϕ^* denote respectively *cumulative* probabilities of default in the historical and RN world,
γ and γ^* denote respectively *survival* probabilities in the historical and RN world.

When using historical probabilities, the zero-coupon bond credit spread formula equivalent to (28.1b) is as follows:

$$s_\theta = \frac{1}{\theta} \, \phi(\theta) \, (1 - \alpha) \; + \; \pi$$ (28.2)

where π refers to the risk premium, and where the liquidity premium l is assumed to be zero.

In the paragraph above, the recovery rate α is assumed to be known with certainty. Recall that this rate depends in particular on whether the debt is subordinated or not and on the existence of collateral (e.g., housing mortgages, guarantees, and pledges) that can serve as a guarantee. For example, an average recovery rate for senior (unsubordinated) debt with a guarantee can be expected to be 75%,[12] 50% for a senior unsecured debt (i.e., without guarantee) and 25% for a subordinated unsecured debt. However, even if averages can be established by category, the percentage of the value of a particular claim actually recovered by

---

[12] These averages vary over time and across sectors. In addition, these are historical probabilities. For valuation, for which the RN probabilities are relevant, and for the stripping operations (extraction of successive default probabilities from the CDS spread curve) described below, practitioners often use a 40% recovery rate as a market convention.

the creditor in the event of default is uncertain. Taking this risk factor into account adds an additional layer to credit risk and complicates the analysis and calculations. For the sake of simplicity, this recovery rate is often assumed to be statistically independent from the other risk factors coming into play. This makes it possible to write $E(X) = (1 - \phi(\theta)) \exp(\theta\, r_\theta^{max}) + \phi(\theta)\, \alpha \exp(\theta\, r_\theta^{max})$, where α is an *average* recovery rate (specific to the issuer category), to carry out the previous calculation in a similar manner and to come up with Eqs. (28.1a) and (28.2) under this definition of the parameter α. Such a simplifying assumption will be adopted throughout this chapter.

Since the recovery rate α is thus defined as an average and, given the relationships (28.1a) or (28.1b) can be used in two ways: (*i*) by calculating theoretical spreads based on separately estimated RN probabilities of default and (*ii*) by calculating RN probabilities of default based on observed spreads (*stripping*).

(i) (28.1a) or (28.1b) allows computing theoretical spreads s_θ from presumed known default probabilities; in fact, the probabilities that can be claimed to be known (such as those shown in Table 28.1) are the historical probabilities. The use of (28.1a)–(28.1b) relationships without premiums with historical probabilities $\phi(\theta)$ ($s'_\theta = \frac{1}{\theta}\phi(\theta)(1 - \alpha)$) instead of Eq. (28.2), leads to theoretical spreads s' that *are significantly lower* than the observed spreads s: it is then possible to attribute the difference between observed and computed spreads to the risk and liquidity premiums (π and l).

(ii) For an observed spread curve corresponding to a given rating, we can calculate the cumulative RN probability $\phi^*(\theta)$ *implicit* in each spread s_θ and deduce from it the sequence p^*_θ of RN (*unconditional*) default probabilities; indeed, the observation of s_θ as well as Eqs. (28.1a) or (28.1b) allows to compute $\phi^*(\theta)$.

For example (28.1b) leads to:

$$\phi^*(\theta) = \theta\frac{1}{1 - \alpha} s_\theta \text{ for } \theta = 1, \ldots, T, \text{ then} : p^*_\theta = \phi^*(\theta - 1) \text{ for } \theta = 1, \ldots, T.$$

Example 1

From the zero-coupon yield curve of A-rated bonds, the following spreads are calculated:

Maturity:	1 year	2 years	3 years
ZC spread in comparison to AAA government bonds:	0.60%	0.70%	0.80%

In addition, the average recovery rate α on securities belonging in this category is estimated to be 43%.

(continued)

28.1 Empirical Tools for Credit Risk Analysis

Example 1 (continued)
We first derive the cumulative RN probabilities of default by using Eq. (28.1b):

$$\phi^*(1) = 1.75 \times 0.6\% = 1.05\%; \phi^*(2) = 2 \times 1.75 \times 0.70\% = 2.45\%;$$
$$\phi^*(3) = 3 \times 1.75 \times 0.80\% = 4.2\%.$$

We then deduce the annual unconditional RN probabilities p^*_t.
The RN probability of default during the first year: $p^*_1 = 1.05\%$.
The RN probability of default in the second year: $p^*_2 = 2.45\% - 1.05\% = 1.40\%$.
Similarly, $p^*_3 = 4.20\% - 2.45\% = 1.75\%$.

It should be noted that the derivation of the RN default probabilities from the observed spread curve is called *stripping* and is not grounded on any particular modeling assumption.

In practice, we observe that RN default probabilities $p^{*\theta}$ implicit in the range of observed spreads are generally much *higher* than the historical probabilities p^θ such as seen in Table 28.1. Indeed, the RN probabilities somehow compensate for the absence of risk premiums by overweighting the probability of adverse events.

The RN probabilities deserve our interest because they allow the valuation of a risky asset (i.e., subject to credit risk) by simply calculating the expectation of its discounted payoff at the risk-free rate.

In fact, since the yield curve is in practice never flat, the value V_0, on date 0, of a risky payoff X paid at time θ is calculated by discounting the expectation of this payoff at the risk-free zero-coupon rate (with the same maturity θ) denoted by r_θ: $V_0 = E^*(X)e^{-\theta\, r\theta}$.

More precisely, the expectation to be used in this case is the forward-neutral expectation (FN).[13] Therefore, the probabilities $\phi^*(\theta)$ implicit in the range of observed spreads are forward-neutral and not risk-neutral although they are often mistakenly confounded in practice.

Remark also that the formula (28.1b) is similar to the credit triangle. In the particular case of constant probabilities p^*_t (for each period t) and equal to p^*, we have $\phi^*(\theta) = \theta p^*$ from which we derive: $s_\theta = (1 - \alpha)\, p^*$.

---

[13] Recall that forward-neutral probabilities are those that make forward asset prices martingales, i.e., spot prices expressed using the zero-coupon bond as the numeraire (*see* Chap. 11 (Sect. 11.2.6), 16, and 19). Recall also that the probabilities FN and RN are identical if interest rates are assumed to be constant or deterministic. More generally, if the instantaneous rate $r(t)$ is not correlated with the risky payoff X, the FN expectation E^* involved in the discounting relationship $(V_0 = E^*(X)e^{-\theta r_\theta})$ is also equal to the RN expectation. In this case, expectations RN and FN are identical because

$$V_0 = E^{RN}[X\, e^{\int_0^\theta -r(u)du}] = E^{RN}[X]E^{RN}[e^{\int_0^\theta -r(u)du}] = E^{RN}(X)e^{-\theta r_\theta} = E^*(X)e^{-\theta r_\theta}.$$

Note that this last relation, although similar, is somewhat different from that of the credit triangle, $\widehat{s}_\theta = (1 - \alpha)\, h^*$ derived in the appendix of Chap. 5, under certain assumptions, for a multi-period *coupon-bearing* security. In that formula, $\widehat{s}_\theta$ represents the spread added to the *yield to maturity* of a coupon-bearing security (to be distinguished from the spread s_θ added to the *zero-coupon rate*), and h^* denotes the *conditional* probability of default assumed to be constant from one period to the next (to be distinguished from p^* which represents an *unconditional* default probability[14]). Therefore, for the same issuer, $\widehat{s}_\theta$ and s_θ generally differ.

Finally, recall that, for a given rating, the spread usually increases with time t, which seems to indicate that the market generally perceives the unconditional RN-FN probability of default p^*_t during the future period t as increasing with t (and not constant or decreasing).

28.1.3.2 Cash-Flow Discounting of a Fixed-Income Security Affected by Credit Risk

Let us first consider a fixed-income security which should contractually pay the following cash-flow sequence $X_\theta^{\max}]_{\theta=t_1,\dots,T}$. The contractual ("promised") cash-flows are not random but, since the security is subject to default risk, the effective sequence will be $\underline{X} \equiv X_\theta]_{\theta=t_1,\dots,T}$ with $X_\theta \leq X_\theta^{\max}$ and X_θ random.

The discounting of such cash-flows, and therefore the valuation of the security, can be carried out using three different approaches:

(i) The traditional and simplest approach consists in *discounting the contractual cash-flows* X_θ^{max} and may be performed in two ways:
- The basic way is to use a *single discount rate*, which corresponds to the yield required by the market for securities with a comparable profile and risk and equal to $y_T + \widehat{s}_T$, where, for example, y_T is a risk-free *yield* and $\widehat{s}_T$ is the credit spread applied to this yield:

$$VP_{\underline{X}}(0) = \sum_{\theta=t_1}^{T} X_\theta^{\max} e^{-\theta(y_T + \widehat{s}_T)}.$$

- But it is theoretically preferable to discount each X_θ^{max} flow with a specific rate, namely the zero-coupon rate required by the market for cash-flows of comparable maturity and risk (equal to $r_\theta + s_\theta$):

$$VP_{\underline{X}}(0) = \sum_{\theta=t_1}^{T} X_\theta^{\max} e^{-\theta(r_\theta + s_\theta)}. \tag{28.3}$$

---

[14]Note that a *constant* conditional probability implies a *decreasing* unconditional probability.

28.1 Empirical Tools for Credit Risk Analysis

Here, r_θ and s_θ denote respectively a zero-coupon interest rate and a zero-coupon credit spread.

(ii) The second approach grounded on a probabilistic method, consists in discounting the *expected cash-flows under the historical probability* with the risk-free zero-coupon interest rate curve to which the risk premium π_θ is added:

$$VP_{\underline{X}}(0) = \sum_{\theta=t_1}^{T} E(X_\theta) \ e^{-\theta(r_\theta + \pi_\theta)}.$$

(iii) The third approach, which is also probabilistic, consists in discounting the *expected cash-flows under the RN or FN probability* with the risk-free zero-coupon interest rate curve (i.e., without adding any risk premium π_θ):

$$VP_{\underline{X}}(0) = \sum_{\theta=t_1}^{T} E^*(X_\theta) \ e^{-\theta r_\theta}. \tag{28.4}$$

This last relationship is generic and applies to any type of security generating a sequence of random cash-flows X_θ (options, etc.). As for fixed-income securities subject to default risk, the cash-flow expectation of X_θ is often calculated by expressing that this (random) cash flow can take on three values. First, $X_\theta^{\max}$, which corresponds to the survival event at time θ (with a RN probability equal to $\gamma^*(\theta)$). Second, αV, which corresponds to the event of default *during* the period θ (probability p^*_θ with $p^*_{t_i} = \gamma^*(t_{i-1}) - \gamma^*(t_i)$), where V is often the nominal value of the security. Third, 0 in the event of default during a period preceding θ (with a probability $= 1 - \gamma^*(\theta) - p^*_\theta$).

The relationship (28.4) (third method) then gives the value of the fixed-income security:

$$VP_{\underline{X}}(0) = \sum_{\theta=t_1}^{T} e^{-\theta r_\theta} \left(X_\theta^{\max} \gamma^*(\theta) + \alpha V p^*_\theta \right). \tag{28.5}$$

In the empirical methods that have just been presented, the RN-FN ($\gamma^*(\theta)$ and p^*_θ) probabilities are inferred from the spread curve obtained by the stripping method explained above. In the analytical methods presented in Sect. 28.2, these probabilities result from modeling default events.

28.1.3.3 Discounting of a Random Cash-Flow Bearing Default or Counterparty Risk: Valuation of Derivatives Affected by Counterparty Risk

In the previous paragraph, we examined the credit risk of a fixed-income security. By definition, *fixed-income* securities are claims on *certain cash flows* X^{max}, even in the presence of a default risk affecting the cash flow X actually paid.

We now consider the case of a security, such as a derivative product (option, forward contract, etc.), whose *contractual ("promised") cash-flows are themselves random.* When the contractual payoff is affected by default or counterparty risk, the cash flow actually paid is $X \leq X^{max}$.

To ease our explanation, we first examine the example of a call with a contractual payoff $X^{max} = (S_T - K)^+$. When the issuer of the call is likely to forfeit his commitment (to deliver the contractual payoff), *a second risk* affects the *actual payoff X ($X \leq X^{max}$)*. This second risk is referred to as *counterparty risk* and must be carefully distinguished from the risk affecting the underlying asset price S_T or the contractual payoff. This distinction between credit risk and counterparty risk is important, in particular in the case of a credit derivative (see Chap. 30).

Suppose that the underlying of the aforementioned call is a receivable whose value S_T is random due to the credit risk borne by this receivable. Credit risk then affects the underlying receivable whilst counterparty risk is related to the possible default of the option seller.

Such a security may be valued by discounting at the risk-free rate the RN-FN expectations of the cash flows X actually paid, in accordance with relationship (28.4). However, in this context, these expectations are more complex to calculate because two combined risks affect each cash flow: the one that affects the payoff in the absence of default by the counterparty, X^{max}, and the one that is linked to the counterparty risk itself. We will now explain how to value a security bearing these two types of risk.

Let X_θ^{max} be a random cash-flow due on date θ. For the sake of simplicity, we consider here a single cash flow. The case of a sequence of several cash flows is addressed by a simple aggregation. Assume that a counterparty risk affects this payoff. We may think of an option or a forward contract with a risky counterparty. Due to the counterparty risk, the actual payoff is $X_\theta \leq X_\theta^{max}$.

Three cases must be distinguished:

(i) First case: X_θ^{max} and X_θ can only take on positive or zero values (e.g., long position on an option). Due to the counterparty risk, the actual payoff is:

$$X_\theta = (1 - (1 - \alpha)\mathbf{1}_D) X_\theta^{max},$$

where $\mathbf{1}_D$ is the indicator function which is equal to 1 if the counterparty defaults on date θ and 0 otherwise. α denotes the recovery rate in the event of default. The value of this contract is obtained by discounting the RN-FN expectation of the above payoff, namely (see Eq. (28.4)):

$$VP(X_\theta) = e^{-\theta\,r_\theta}\, E^*[X_\theta] = e^{-\theta\,r_\theta}\, E^*\big((1 - (1 - \alpha)\mathbf{1}_D)\, X_\theta^{max}\big).$$

If the contractual payoff and the counterparty default risk are not correlated[15] and the recovery rate is deemed as non-random, we simply get:

---

[15] Such an assumption is reasonable if the instrument in question does not represent a significant part of the issuer's balance sheet.

28.1 Empirical Tools for Credit Risk Analysis

$$VP(X_\theta) = e^{-\theta\, r_\theta}\, E^*\left[(1-(1-\alpha)\mathbf{1}_D)\, X_\theta^{max}\right]$$
$$= e^{-\theta\, r_\theta}\, E^*\left[(1-(1-\alpha)\mathbf{1}_D]\, E^*\left[X_\theta^{max}\right];\right.$$

Noting that $E^*(\mathbf{1}_D) = \phi^*(\theta) =$ RN default probability of the counterparty at date θ, it comes:

$$VP(X_\theta) = \underbrace{[1-(1-\alpha)\phi^*(\theta)]}_{\text{Adjustment coefficient}} \times \underbrace{e^{-\theta\, r_\theta} E^*[X_\theta^{max}]}_{VP(X_\theta^{max})} \tag{28.6}$$

(value of the payoff without counterparty risk)

We then come up with a simple computation rule: *the value of a random cash-flow affected with counterparty risk is calculated as its value in absence of counterparty risk multiplied by an adjustment coefficient* that takes into account the expected loss in the event of the counterparty default $((1-\alpha)\phi^*(\theta))$.

Moreover, since according to (28.1a), $1-(1-\alpha)\phi^*(\theta) = e^{-\theta s_\theta}$, we have:

$$VP(X_\theta) = e^{-(r_\theta + s_\theta)T}\, E^*\left[X_\theta^{max}\right] = e^{-\theta s_\theta}\, VP\left(X_\theta^{max}\right). \tag{28.7}$$

This translates into a second computation rule: *the random cash flow affected with counterparty risk is valued from the contractual cash flow (as if it were free of counterparty risk) by increasing the discount rate with the counterparty's own credit spread.*

These two equivalent calculation rules, therefore, generalize the result obtained for "deterministic" contractual cash flows affected with default risk (Eqs. (28.3), (28.4), and (28.5)) to random contractual cash flows with (additional) counterparty risk. However, we emphasize that the two computation rules (28.6) and (28.7) only hold under the assumption that the risk of the underlying and the counterparty risk on the contractual payoff are not correlated. It should also be noted that the recovery rate α was considered fixed. Recall that this restriction can be disregarded if the random recovery rate is independent from both the payoff and counterparty risk (α is then an expected recovery rate).

(ii) Second case: In the absence of counterparty risk, the payoff X_θ^{max} can only take on negative or zero values (e.g., short position on an option), in which case the counterparty risk is obviously zero for the counterparty A (e.g., the option seller) but not for counterparty B. The market value of the contract is calculated as in the previous case, but the default probabilities and credit spread taken into account are obviously those of counterparty A.

(iii) Third case: In the absence of default risk for both counterparties A and B, the payoff X_θ^{max} can take on positive or negative values (e.g., a swap or a forward position involving counterparties A and B).

If counterparty risk affects B and/or A, we simply consider X_θ^{max} as the sum of its positive and negative components and apply the above computation rules to each component:

$$X_\theta^{max} \equiv X^+ + X^-, \quad \text{where } X^+ = \text{Max}\left(X_\theta^{max}, 0\right) \text{ and } X^- = \text{Min}\left(X_\theta^{max}, 0\right).$$

Let D_A and D_B be the default events of respectively A and B and $\mathbf{1}_{D_A}$, $\mathbf{1}_{D_B}$ the indicators functions for D_A and D_B events. If B defaults, A is only affected if X_θ^{max} is positive; in a symmetric way if A defaults, B is affected only if X_θ^{max} is negative. The payoff X_θ is hence as follows:

$$X_\theta = \left[1 - (1 - \alpha_B) \mathbf{1}_{D_B}\right] X^+ + \left[1 - (1 - \alpha_A) \mathbf{1}_{D_A}\right] X^-.$$

In this relationship, $(1 - \alpha_B) \mathbf{1}_{D_B} X^+$ and $(1 - \alpha_A) \mathbf{1}_{D_A} X^-$ represent, respectively, the losses attributable to the defaults of A and B.

The value of the contract is hence written as the sum of the values of its two components:

$$VP(X_\theta) = e^{-\theta r_\theta} E^*\{\left[1 - (1 - \alpha_B) \mathbf{1}_{D_B}\right] X^+\} + e^{-\theta r_\theta} E^*\{\left[1 - (1 - \alpha_A) \mathbf{1}_{D_A}\right] X^-\}.$$

If we assume the independence between: (i) the payoff underlying value, (ii) the counterparty default risk, and (iii) the recovery rates, by applying formulas (28.6) and (28.7) separately to the two positive and negative components of the payoff, we hence come up with:

$$VP(X_\theta) = \left[1 - (1 - \alpha)\varphi_B^*(\theta)\right] e^{-\theta r_\theta} E^*[X^+] + \left[1 - (1 - \alpha)\varphi_A^*(\theta)\right] e^{-\theta r_\theta} E^*[X^+]; \text{or :}$$

$$VP(X_\theta) = e^{-\theta(r_\theta + s_B)} E^*[X^+] + e^{-\theta(r_\theta + s_A)} E^*[X^-].$$

Let us note that $VP(X_\theta)$ is calculated as the sum of two options whose respective payoff are: $[1 - (1 - \alpha_B) \text{Proba}(D_B)^*] X^+$ and $[1 - (1 - \alpha_A) \text{Proba}(D_A)^*] X^-$.

Finally, it should be noted that counterparty A, when analyzing the risk it is faced with, only considers the loss due to counterparty B default, i.e., $(1 - \alpha_B) \mathbf{1}_{D_B} X^+$.

Example 2

Let us consider a European put option on a stock, with a one-year maturity valued at 100 by the Black-Scholes model. This model gives the option price in the absence of counterparty risk of the issuer. However, this put option is issued by an A-rated counterparty that borrows at 1 year with a spread of 0.60%. The Black-Scholes price will simply be adjusted by a multiplying factor equal to $e^{-0.006}$, which leads to a value of 99.40 for the option affected by counterparty risk.

(continued)

> **Example 2** (continued)
>
> It should be noted that this 0.60% spread corresponds to a default probability of 1.05% in the coming year for an expected recovery rate of 43% (see Example 1) and that the same price for this option could have been determined by the alternative calculation:
>
> $$(1 - 0.57 \times 1.05\%) \times 100 = 99.40.$$

28.2 Modeling Default Events and Valuation of Securities

The risk analysis and valuation methods presented in the previous section are essentially empirical in nature because the historical probabilities are derived from the historical default occurrences or rating migration observed in the past. In contrast, RN-FN probabilities are inferred from the observed spread curve (stripping), without the default event being modeled. Theoretical models make it possible to specify and support this empirical approach by a mathematical representation of the default based either on a stochastic process or on option theory. Two types of credit risk models can then be distinguished:

- Those that fall under the so-called *reduced* approach, which starts directly from the stochastic process that is supposed to drive the price of the risky security and which represent default events as *jumps*.
- Those that are qualified as *structural* models, which start from the evolution of the value of the issuer's assets and attribute the default to an insufficient value for these assets and are based on an option pricing model.

The *reduced* approach is presented in § 28.2.1 and the *structural* approach in § 28.2.2.

28.2.1 Reduced-Form Approach (Intensity Models)

The models that fall under the reduced-form approach represent the evolution of the price of the risky asset by a stochastic process with jumps, which reflect default events. In many models, a jump process comes in addition to a continuous process. Thus, the variation in the risky asset price $B'(t)$ often depends on a standard Brownian $W(t)$ and a Poisson process $N(t)$.

We first introduce[16] the mathematical tool applicable to this type of process and then describe a model that falls under this approach.

---

[16]Poisson processes were briefly introduced in Chap. 18 dedicated to stochastic calculus (Sect. 18.6).

28.2.1.1 Mathematical Tool: Generalized Poisson Process, and Default and Survival Probabilities

Any event seriously affecting the issuer's creditworthiness, such as a rating downgrade or a failure to pay a coupon, translates into a drop of the value $B^r(t)$ of the issuer debt, i.e. a negative jump in that value.

The Poisson process is the simplest model to represent jumps.

Definition *A Poisson process with intensity $\lambda > 0$ is an increasing process N (t) having integer values, with initial value $N(0) = 0$ and independent increments, and such that:*

$$Proba\{N(t_2) - N(t_1) = n\} = \frac{\lambda^n (t_2 - t_1)^n}{n!} \exp\left(\lambda(t_2 - t_1)\right) \quad for\, 0 \leq t_1 \leq t_2.$$

In particular:

- By setting $t_1 = 0$ and $t_2 = t$, we obtain:

$$Proba\{N(t) = n\} = \frac{(\lambda t)^n}{n!} e^{-\lambda t}.$$

- Choosing $t_1 = t$ and $t_2 = t + \Delta t$ and defining $\Delta N(t) = N(t + \Delta t) - N(t)$ yields:

$$Proba\{\Delta N(t) = n\} = \frac{(\lambda \Delta t)^n}{n!} e^{-\lambda \Delta t}.$$

And by making Δt tend to 0 we obtain, according to the rules of differential calculus:

$$Proba\left\{dN(t) = 1\right\} = \lambda\, dt \text{ for } n = 1 \text{ and } Proba\left\{dN(t) > 1\right\} = 0.$$

This last relationship means that, over the infinitesimal interval of length dt, the process does not jump with a probability $1 - \lambda\, dt$, jumps once with a probability $\lambda\, dt$, and jumps more than once with probability zero. This property characterizes the Poisson process and is an alternative way to define it.[17]

Such a process is used to represent credit events, a jump representing any possible credit event met in practice (failure to pay a coupon, bankruptcy, debt restructuring, etc.).

Let us emphasize the main difference between the Poisson and the Itô process (based on a Brownian motion): when the time interval dt tends to 0, the probability of the jump of the Poisson process tends to zero as well. However, the size of the jump

---

[17] The process is then characterized as meeting the conditions: $N(0) = 0$, independent increments, Proba $\{dN = 1\} = \lambda\, dt$ and Proba $\{dN > 1\} = 0$.

28.2 Modeling Default Events and Valuation of Securities

is finite (therefore the trajectory of the Poisson process is discontinuous), whereas for the Itô process, the magnitude of the variation tends to 0 (the trajectory is continuous) but the probability of a non-zero variation is equal to 1, whatever the considered time interval may be.

The standard Poisson process that has just been defined is quite restrictive because its intensity is assumed to be constant. This is why, to represent the number $N(t)$ of jumps between 0 and t, generalized Poisson processes, also known as *marked point processes*, are also used, such as:

$dN = N(t + dt) - N(t)$ is equal to 1 if a jump occurs in the interval $(t, t + dt)$ and to 0 otherwise.

The intensity parameter (also called "compensator"), however, is not necessarily constant and is denoted by $\lambda(t)$, which means that the probability of a jump in the interval $(t, t + dt)$ is equal to $\lambda(t)dt$. In general models, the intensity $\lambda(t)$ is itself stochastic but in the remainder of this chapter, it is assumed to be a deterministic function of time.

dN is hence equal to 1 with a probability $\lambda(t)dt$ and 0 with a probability $1 - \lambda(t) dt$. As a result, $E(dN) = \lambda(t)dt$ and the process $N(t) - \int_0^t \lambda(u)du$ is a martingale. The dynamics of this process can be written under the historical, RN or FN probability measure but the intensity parameter $\lambda(t)$ depends on the chosen measure. In the following paragraph, to ease the exposition we do not distinguish by different notations the historical RN and FN intensities.

In what follows, we denote by Λ_t the average of $\lambda(u)$ in the interval $(0, t)$: $\Lambda_t \equiv \frac{1}{t} \int_0^t \lambda(u)du$.

In the simplest models, default (i.e., bankruptcy) is triggered *at the first* credit event, i.e., at the first jump, and survival is identified as the prior absence of a jump; survival at t is then equivalent to $N(t) = 0$ and the probability of survival at time t is: $\gamma(t) = \text{Proba} \{N(t) = 0\}$.

The probability of survival is related to the probability of default $\phi(t) = \text{Proba} \{N(t) \geq 1\}$, i.e., the probability that at least one jump occurred between 0 and t, by the equation: $\gamma(t) = 1 - \phi(t)$.

In addition, $-d\gamma = \gamma(t) - \gamma(t+dt)$ corresponds to the probability that a jump will occur for the first time in the interval $(t, t+dt)$, hence:

$$-d\gamma = \text{Proba} \{N(t) = 0\} \times \text{Proba} \{dN = 1\}, \text{i.e.} : -d\gamma = \gamma(t) \lambda(t) dt,$$

which in turn implies:

$$\text{Survival probability} \equiv \gamma(t) = e^{\int_0^t -\lambda(u)du} = e^{-t\Lambda_t}. \tag{28.8}$$

In the particular case of a constant intensity λ, we find (see above):

$$\gamma(t) = \text{Proba}\{N(t) = 0\} = e^{-\lambda t}. \tag{28.8'}$$

An alternative and equivalent point of view consists in considering that the date τ of default (first jump) is a stopping time $\tau = inf\{t \in (0,T) \mid N(t) = 1\}$ which has $\phi(t) = 1 - \gamma(t)$ as distribution law:

$$\text{Proba}\{\tau \leq t\} = \phi(t) = 1 - \gamma(t) = 1 - e^{\int_0^t -\lambda(u)du}.$$

In models for which default occurs on the first jump, the intensity $\lambda(t)$ is interpreted as a *conditional* probability of default (jump of order 1), or "*hazard rate*", as well as an *unconditional* probability of a jump of any order. To avoid any confusion, it is necessary to clearly distinguish the probabilities relative to the first jump from those relative to a jump of any order. This point is to be clarified in the paragraph below.

Let us note at first that, since the increments in the Poisson process are independent, we have:

Proba (*jump* between t and $t + dt$ | *no jump* before t) = Proba (*jump* between t and $t + dt$) = $\lambda(t)$ dt.

Therefore, $\lambda(t)dt$ is interpreted both as the conditional probability of default between t and $t + dt$ knowing that default did not occur before (*hazard rate*, the continuous time analog of the discrete time probability denoted by h) and as the unconditional probability of *any jump* (not necessarily the first).

In addition, the *unconditional* probability of default between t and $t + dt$ = Proba (*first jump between t and $t + dt$*) writes:

Proba (*default* between t and $t + dt$) = Proba (*jump* between t and $t + dt$ $\cap$ *no jump* before t).

Since the event {*jump* between t and $t + dt$} and the event {*no jump* before t} are independent, the probability of their intersection (joint occurrence) is the product of their respective probabilities and we can write:

Proba (*default* between t and $t + dt$) = Proba (*jump* between t and $t + dt$) $\times$ Proba (*no jump* before t),

and, since Proba (*no jump* before t) = Proba (*survival* at t) = $\gamma(t)$, we have:

Proba (*default* between t and $t + dt$) = $\gamma(t)$ $\lambda(t)dt$.

In a Nutshell
- When considering *any* jump, there is no need to distinguish unconditional and conditional probabilities and Proba (*jump* between t and $t + dt$) = $\lambda(t)dt$;
- When considering the *first* jump or a default occurring on the first jump:

Proba (*default* between t and $t + dt$) = $\gamma(t)$ $\lambda(t)dt$ (unconditional probability of default).

28.2 Modeling Default Events and Valuation of Securities

Proba (*default* between t and $t + dt$ | *survival* at t) $= \lambda(t)dt$ (conditional probability or *hazard rate*).

The calculation rules and notations for jumps, default (first jump), the default dates, and the related probabilities result from what has been described previously. A brief summary of these notations and rules is presented in the appendix and may be useful to the reader.

In the analysis above, credit events are represented as jumps without the amplitude of these jumps being specified. This representation is sufficient in the simplest models where the first jump triggers the default and the recovery rate of the claim is an exogenously given parameter. Jarrow and Turnbull (1995) model, presented in § 28.2.1.2 below, falls into this category.

These models lead to simple formulas for discounting flows affected by default risk.

Let us consider, for example, a contractual cash flow of \$1 "promised" at date t and subordinated to the survival of the issuer at that date; assume also that the RN-FN survival probability is $\gamma(t) = e^{\int_0^t -\lambda(u)du} = e^{-t\Lambda_t}$, where $\lambda(u)$ represents the RN-FN intensity of the marked process that represents the default. If the recovery rate is assumed to be zero in the event of default, the value on date 0 of this cash flow is $e^{-tr_t}e^{-t\Lambda_t} = e^{-t(r_t + \Lambda_t)}$, where r_t refers to the zero-coupon rate of maturity t prevailing at time 0. The discount factor to be applied to the "promised" contractual cash flow thus is $e^{-t(r_t + \Lambda_t)}$, which simplifies into $e^{-(r + \lambda)t}$ in a single interest rate and constant intensity model.

In the more general case of security generating a contractual ("promised") continuous cash flow $F(t)$ and giving rise to a recovery amount $\alpha M(t)$ in the event of default (the claim being extinguished once this amount is recovered), the present value writes for example:

$$V = \lambda\alpha \int_0^T M(t)e^{-(r+\lambda)t}dt + \int_0^T e^{-(r+\lambda)t}F(t)dt.$$

The first term represents the present value of the expectation of recovered amount in the event of a default (Probability of defaulting in the interval $(t, t + dt) = e^{-\lambda t}\lambda dt$). The second term is the present value of the expectation of the contractual cash flows which, in fact, are paid solely in the event of survival (Probability $= e^{-\lambda t}$).

In more complex models, the default (i.e., bankruptcy) may be preceded by several jumps and the consequences of a jump must be specified. A jump may reflect a credit event (change of rating, failure to pay a coupon, debt restructuring, etc.). Moreover, jumps do not necessarily have the same magnitude. If $J(t)$ denotes the relative magnitude of the possible jump on date t (which translates into a decrease in market value), the process governing the price $V(t)$ of any asset affected by such jumps can be written:

$$\frac{dV}{V} = [\mu(t) + \lambda(t)J(t)]dt + \sigma(t)\,dW - J(t)dN.$$

It should be noted that the expected instantaneous return on the asset is $= E\left(\frac{1}{V}\frac{dV}{dt}\right)$ $= \mu(t)$, because $E[\sigma(t)dW] = 0$ and $E[J(t)dN] = \lambda(t)\,J(t)\,dt$.[18]

If this dynamic is written in the RN world, in which both the intensity and the Brownian motion differ from their respective counterparts in the historical world, we have: $\mu(t) = r(t)$.

28.2.1.2 The Jarrow and Turnbull Model (1995)

This model allows translating the empirical approach presented in Sect. 28.1 into a precise mathematical framework. The default occurs on the first jump modeled with a generalized Poisson process of intensity $\lambda(t)$, which implies a survival probability at t equal to $\gamma(t) = e^{\int_0^t -\lambda(u)du}$ and a default probability between 0 and t equal to $\phi(t) = 1 - \gamma(t) = 1 - e^{\int_0^t -\lambda(u)du}$, as explained in the preceding paragraph. In the case of default, the creditor recovery rate is equal to a constant parameter α.

In the following development, intensities $\lambda(t)$, $\phi(t)$, and $\gamma(t)$ are *forward-neutral* (FN) probabilities.

The relationship (28.1a) that we recall below, states that the spread s_θ under the FN probability for a zero-coupon bond maturing at time θ equals:

$$s_\theta = -\tfrac{1}{\theta}\ln(1 - \phi(\theta)\,(1-\alpha))\,,\text{ hence: } s_\theta = -\tfrac{1}{\theta}\ln[1 - (1 - e^{\int_0^t -\lambda(u)du})\,(1-\alpha)]\text{, and}$$

finally:

$$s_\theta = -\frac{1}{\theta}\left[e^{-\theta\Lambda_\theta} + \left(1 - e^{-\theta\Lambda_\theta}\right)\alpha\right],$$

where $\Lambda_\theta \equiv \tfrac{1}{\theta}\int_0^\theta \lambda(t)dt$ represents, as previously defined, the average FN intensity between 0 and θ.

We can thus calculate the different values of Λ_θ (average FN intensity) from the various observed spreads s_θ, as seen in Example 1 of Sect. 28.1.3.1.

More generally, let us consider a coupon bearing security x whose coupons k_j are paid on dates θ_j for $j = 1, \ldots, n$ and with nominal value of \$1 to be contractually paid on date $\theta_n \equiv T$. Denote by $B^r_x(0)$ its value on date 0, where r indicates the presence of a credit risk, while $B_\theta(0) = e^{-\theta\,r_\theta(0)}$ the price of a zero-coupon of maturity θ without credit risk. The value $B^r_x(0)$ of the security x can be calculated, in accordance with (28.5), by discounting the FN expectations of all cash flows (coupons and nominal)[19]:

---

[18] When $J(t)$ and $dN(t)$ are correlated, this statement is true only for E_t (conditional expectation on date t).

[19] That is, calculated under the probabilities FN Qj.

28.2 Modeling Default Events and Valuation of Securities 1195

$$B^r{}_x(0) = \sum_{j=1}^{n} k_j B_{\theta j}(0)\gamma(\theta_j) + B_T(0)[\gamma(T) + (1 - \gamma(T))\alpha], \qquad (28.9a)$$

or else, since $\gamma(\theta) = \exp(-\theta \Lambda_\theta)$:

$$B^r{}_x(0) = \sum_{j=1}^{n} k_j B_{\theta j}(0) + B_T(0)\left[\alpha + e^{-T\Lambda_T}(1 - \alpha)\right]. \qquad (28.9b)$$

In this model, the default at any date triggers the recovery of a fraction α of the nominal value at date T. The total recovery expectation is equal to $\alpha\,[1 - \gamma\,(T)]$ and its present value to $B_T(0)\,\alpha\,[1 - \gamma\,(T)]$, in accordance with (28.9a).

28.2.1.3 Default Model with Nonconstant Recovery Rate: Duffie and Singleton model (1999)

Like the Jarrow and Turnbull model, other models represent credit events using a Poisson process but use different assumptions about the recovery date in the event of default.

In particular, it is convenient to assume, as in Duffie and Singleton (1999), that the recovered amount is paid *immediately* after the default event (and not on date T as in the previous model). This amount can also be conveniently expressed as a proportion α of the market value of the security in the absence of default (rather than as a proportion of its nominal value). Under these assumptions, the credit triangle formula is obtained in its simplest form.

More precisely, let us consider a security with a payoff $A(T)$ at maturity T in absence of default: it is, therefore, the contractual ("promised") payoff, possibly random. The Duffie-Singleton model simply indicates that the value $A(t)$ of the security on date t, writes *in absence of default*:

$$A(t) = E_t\left[e^{-\int_t^T [r(u)+(1-\alpha(u))\lambda(u)]du}A(T)\right].$$

This equation states that, in the RN world, the discount rate applicable to values A (possible values of the risky security in absence of default on date T), to which the "promised" RN rate must be equal, is the instantaneous rate $r + (1 - \alpha)\lambda$.

The term $(1 - \alpha)\lambda$ is the *instantaneous spread* that is added to the risk-free rate to compensate for the expected loss: it is again the formula of the credit triangle already encountered above, which involves here a RN-FN conditional probability (*hazard rate*).

The proof may be conducted as follows. It is assumed that, under the RN probability, the first jump in a Poisson process of intensity $\lambda(t)$ (possibly random[20]) triggers the default of the considered security. Note that $A(t)$ is the market value of

---

[20]Duffie and Singleton (1999) proposed several models for the evolution of the intensity $\lambda(t)$ inspired by interest rate models (mean-reverting process, etc.).

this security on date t *conditional on the absence of default before that date*. We must distinguish the value $A(t)$ from $V(t)$ which represents the *unconditional* value of the security in the absence of a default ($V(t)$ and $A(t)$ only coincide in the absence of a default between 0 and t). In the event of a default between t *and* $t + dt$ the issuer pays to the creditor a fraction $\alpha(t)$ (possibly random) of the value $A(t)$ at time $t + dt$ and the security ceases to exist (i.e., the issuer liability is extinguished).

Let us consider any date t on which the security is not in default, therefore its value $V(t) = A(t)$. On date $t + dt$ the price $V(t + dt)$ of the security can take on two values:

$V(t + dt) = A(t + dt)$ in absence of default (with probability $1 - \lambda(t)dt$),
$V(t + dt) = \alpha(t)A(t)$ in case of a default occurring between t and $t + dt$ (with probability $\lambda(t)dt$).

Therefore, $E_t V(t + dt) = [1 - \lambda(t)dt]\, A(t + dt) + \lambda(t)\alpha(t)A(t)dt$. Hence by subtracting $V(t)$ from the left-hand term and $A(t)$ ($=V(t)$) from the right-hand term, we have:

$$E_t V(t + dt) - V(t) = E_t\, dV = [1 - \lambda(t)dt]\,(A(t) + dA) + \lambda(t)\alpha(t)A(t)\, dt - A(t).$$

By dividing the left-hand side of this equation by $V(t)$ and the right-hand side by $A(t)$, by eliminating the term $(dAdt)$ of order greater than 1 and by simplifying the expression, we have:

$$E_t \frac{dV}{V} = E_t \frac{dA}{A} - (1 - \alpha(t))\lambda(t)dt.$$

This means that the possibility of a default reduces the expected growth rate of the security value by the amount $(1 - \alpha(t))\lambda(t)$.

Moreover, under the RN probability: $E_t \frac{dV}{V} = r(t)dt$; therefore, under this probability:

$$E_t \frac{dA}{A} = [r(t) + (1 - \alpha(t))\lambda(t)]dt.$$

If $A(t)$ follows an Itô process, its RN dynamics becomes:

$$\frac{dA}{A} = [r(t) + (1 - \alpha(t))\lambda(t)]dt + \sigma_A(t)dW, \tag{28.10}$$

or else: $A(t) = A(0)\, \exp[\int_0^t [r(s) + (1 - \alpha(s))\lambda(s) - 0.5\, \sigma_A^2(s)]\, ds + \int_0^t \sigma_A(s)dW_s]$.

In the particular case of $r(t)$, $\lambda(t)$ and $\alpha(t)$ being constant and respectively equal to r, λ, and α, (28.10) implies:

$$E\,[A(T)] = A(0)\, e^{[r+(1-\alpha)\lambda]T}, \text{i.e. } A(0) = E\,[A(T)]\, e^{-[r+(1-\alpha)\lambda]T}.$$

28.2 Modeling Default Events and Valuation of Securities

In the general case of a non-constant and even stochastic interest rate $r(t)$, intensity $\lambda(t)$, and recovery rate $\alpha(t)$, (28.10) implies that $A(t)$ $\exp[-\int_0^t [r(s) + (1 - \alpha(s))\lambda(s)]\,ds$ is a martingale. It follows that on any date t, in the absence of a default between 0 and t:

$$A(t) = V(t) = E_t\left[e^{-\int_t^T [r(u) + (1 - \alpha(u))\lambda(u)]du} A(T)\right].$$

This discounting rule of the contractual ("promised") cash flows at the rate $r(t)$ plus the spread $(1 - \alpha(t))\lambda(t)$ applies to the simple case of a fixed cash flow $A(T)$ (fixed income security) as well as to the complex case of a *random payoff* $A(T)$ affected by counterparty risk (see Sect. 28.1 § 28.1.3.3 of this chapter). Recall again that $\lambda(t)$ denotes here the *RN-FN* intensity (hazard rate).

In particular, the last equation gives the value $B_T^r(t)$ of a zero-coupon security with default risk for which the contractual payoff is $A(T) = 1$:

$$B_{T^r}(t) = E_t\left[e^{-\int_t^T [r(u) + (1 - \alpha(u))\lambda(u)]du}\right], \text{ where } B_{T^r}(t)$$

$$= e^{-[r + (1 - \alpha)\lambda](T - t)} \text{ if } r, \alpha \text{ and } \lambda \text{ are constant.}$$

This approach, and in particular the dynamics (28.10), may ground numerical methods (trees, simulations), as we show in § 28.2.3.

28.2.2 Structural Approach: Merton's Model and Barrier Models

In the case of limited liability companies (the most frequent case for large corporations), the shareholders have the option to decide to file for bankruptcy (following the suspension of payments declaration).

Here, we note $V(t)$ the total value of the company, i.e., the market value of all its assets. Due to the asset-liability identity characterizing any balance sheet, this total value is equal to the sum of the market values of the shares $S(t)$ and the debts $B^r(t)$ (where the exponent r indicates the presence of a credit risk). The suspension of payments option is exercised by the shareholders as soon as $V(t)$ becomes equal to the value of the debts, to prevent the value of the shares from becoming negative. Thus, the shares can be considered as calls written on the assets of the company. Merton developed in 1974 an equity and debt valuation model grounded on the above principle. This model was used to analyze and quantify credit risk and has undergone various extensions.

28.2.2.1 The "Seminal" Model (Merton's Model (1974))
This model is consistent with the Black and Scholes model (1973).

- The company does not distribute dividends or coupons (on the issued debt) and does not repay any debt between 0 and T.

- The interest rate r is constant.
- The total value of the company $V(t)$ (equal to the value of its assets) follows a diffusion process with constant volatility σ_V governed by the SDE:

$$\frac{dV}{V} = \mu(t)dt + \sigma_V dW_t, \text{ in the real world;}$$

$$\frac{dV}{V} = rdt + \sigma_V dW_t^*, \text{ in the RN world.}$$

- The company debt consists of a single zero-coupon security with a maturity T that contractually pays an amount D (principal and interest) and whose market value at date t is denoted by $B^r_T(t)$ (as previously, the exponent r indicates that the debt bears credit risk).

The value of the shares obviously depends on the value of the assets. At maturity T, one of the following two situations will prevail:

- If $V(T) < D$, shareholders file for bankruptcy, i.e., they abandon the company's assets to the creditors: $B^r_T(T) = V(T)$ and $S(T) = 0$;
- If $V(T) \geq D$, the shareholders reimburse the creditors and have the ownership over the company's remaining assets whose value is equal to $V(T) - D$: $B^r_T(T) = D$ and $S(T) = V(T) - D$.

The shares are hence equivalent to the claim over the payoff $S(T) = Max(V(T) - D, 0)$ and can therefore be considered and evaluated as a call written on the company's assets of strike D and maturity T.

Similarly, the debt issued by the company represents a right (claim) over the following payoff, which can be written in three equivalent ways:

$$B^r_{T'} = Min(V(T), D)$$
$$= V(T) - Max(V(T) - D, 0)$$
$$= D - Max(D - V(T), 0)$$

According to the last expression, a risky debt is equivalent to a long position on a risk-free zero-coupon and a short position on a put of strike D written on the company's assets. The value of this put expresses the reduction in the debt value attributable to credit risk.

Equity and debt can thus be both evaluated using a standard option pricing model, as pioneered by Black, Scholes, and Merton.

On date 0, the value $S(0)$ of equity and the value $B^r_T(0)$ of debt are given by the standard BS formula:

28.2 Modeling Default Events and Valuation of Securities

$$S(0) = V(0)N(d_1) - D\,e^{-rT}\,N(d_2); \tag{28.11a}$$

$$B^r{}_T(0) = V(0) - S(0) = V(0)[1 - N(d_1)] + D\,e^{-rT}\,N(d_2),\text{ where}: \tag{28.11b}$$

$$d_1 = \frac{\ln(V(0)/D) + (r + \sigma_V^2/2)T}{\sigma_V\sqrt{T}}; d_2 = \frac{\ln(V(0)/D) + (r - \sigma_V^2/2)T}{\sigma_V\sqrt{T}}. \tag{28.11c}$$

The risky zero-coupon rate of maturity T, $r_T^{max}(0)$, results directly from the equation $\left[e^{Tr_T^{max}(0)} = \frac{D}{B_T^r(0)}\right]$, i. e.:

$$r_T^{max}(0) = \frac{1}{T}\ln\left(\frac{D}{B_T^r(0)}\right),$$

as well as the credit spread $s = \frac{1}{T}\ln\left(\frac{D}{B_T^r(0)}\right) - r.$

As noted repeatedly in the chapters dedicated to option valuation, the RN probability to exercise the call is $N(d2)$, so that the RN probability of default (bankruptcy) is equal to $1 - N(d_2) = N(-d_2)$ (probability of not exercising the call). This probability, denoted by $p_d{}^*$ in the following, must be carefully distinguished from the historical probability of default, denoted by p_d, whose value will be calculated later on.

It should also be noted that the model allows mapping credit spreads onto default probabilities.

For listed companies, $S(0)$ is observable and the volatility σ_S of the (listed) shares can be estimated from historical prices (historical volatility) and/or from option prices (implied volatility). Most often, σ_V and $V(0)$ are not observable.[21] However, these last two unknowns can be determined from a system of two equations composed of (28.11a) and the relationship between σ_S and σ_V. As for the latter relationship, by applying Itô's lemma to the function $S(0,V)$, we obtain the diffusion parameter σ_S of the dS/S process (the drift is not of interest here):

$$\sigma_S = \frac{1}{S}\frac{\partial S}{\partial V}V\sigma_V.$$

In addition, under (28.11a), we have $\frac{\partial S}{\partial V} = N(d_1)$, from which we derive:

$$\sigma_V = \frac{S(0)}{N(d_1)V(0)}\,\sigma_s. \tag{28.12}$$

Merton's model is therefore sometimes used to deduce from the parameters of the observable share price dynamics the parameters of the unobservable value of the

---

[21] This is because, in general, some debts and most of the company's assets are not *marked to market*.

assets dynamics: the system composed of (28.11a) and (28.12) allows to determine V (0) and σ_V. As there is no analytical solution to this system, an iterative procedure is necessary to obtain the numerical values of the two unknowns $V(0)$ and σ_V. In a second step, (28.11b) allows to calculate the debt market value, i.e., the corresponding rate r_T^{max} and the credit spread as well as the RN probability of default $1 - N(d_2)$. We will come back to these questions in the development of the MKMV method (see Chap. 29).

Example 3

The market capitalization of company X shares amounts to $S(0) = 180$ million and its debt consists of a zero-coupon contractually paying $D = 280$ million in 3 years' time. The volatility σ_S of its shares is estimated at 38.30% and the risk-free interest rate r is 5% for all maturities.

The system of two Eqs. (28.11a) and (28.12) allows to determine by successive approximations the two unknowns: $\sigma_V = 16.76\%$ and $V(0) = 420$.

The value of the debt is then $B'_T(0) = 420{-}180 = 240$, and the corresponding annualized rate of return $r_T^{max} = \frac{1}{3} \ln \left(\frac{280}{240}\right) = 5.14\%$, implying a credit spread of 0.14%.

In addition, the RN probability of default within 3 years is equal to $N(-d_2) = 3.85\%$.

The recovery rate α depends on the value of the company in 3 years: for a value $V(T) = 279.9$ the recovery rate almost equals 100% but for $V(T) = 140$, it amounts to 50%. The *average* recovery rate $\overline{\alpha}$ in the event of default can be assessed using Eq. (28.1b):

$0.0014 = \frac{1}{3} \cdot 0.0385 \, (1 - \overline{\alpha})$, . i.e. $\overline{\alpha} = 89\%$. This rate is higher than that observed in practice, because, in reality, liquidation costs (in case of bankruptcy) reduce significantly the company's residual value in the event of default (see § 28.2.2.2 below).

28.2.2.2 Merton's Model with Bankruptcy Costs

In the previous model, creditors are assumed to recover the full value of the company in the event of default (bankruptcy). In practice, bankruptcy costs (also known as liquidation costs) affect the value of the company in the event of liquidation. These costs are both direct, such as lawyers' and liquidators' fees, legal proceedings, and publicity costs and employees' layoff indemnities, as well as indirect, such as deterioration of brand image, loss of customers, worsening of suppliers' credit terms, loss of synergy due to the dismantling of all or part of the assets, conflicts between shareholders, creditors and managers, etc. They empirically represent up to

28.2 Modeling Default Events and Valuation of Securities

20%, and sometimes even more, of the company's value, the percentage depending directly on the size and nature of the company's activity.

Suppose that in the event of a suspension of payments, the liquidation costs $G(T)$ represent a fraction κ of the company's value $V(T)$, so that:

$$G(T) = \kappa \, V(T) \text{ if } V(T) \leq D \text{ and 0 otherwise}$$

$$S(T) = Max(V(T) - D, 0), \text{ is not affected by bankruptcy costs}$$

$$B'_T(T) = V(T) - S(T) - G(T).$$

The value of bankruptcy costs on date 0 is $G(0) = e^{-rT} E^* \{G(T)\}$. This expectation is obtained by a calculation similar to the one used to derive the BS formula (see Chap. 11), which is briefly recalled here.

We start from: $V(T) = V(0) \exp\left(\left(r - \frac{\sigma_V^2}{2}\right)T + \sigma_V \sqrt{T}Z\right)$ where Z follows a N $(0,1)$ Gaussian distribution; we note that: $V(T) \leq D$ if and only if $Z \leq -d_2$; we then have:

$$G(0) = e^{-rT} E^* \{G(T)\}$$

$$= e^{-rT} \kappa V(0) \frac{1}{\sqrt{2\pi}} \int\limits_{z < -d_2} \exp\left(\left(r - \frac{\sigma_V^2}{2}\right)T + \sigma_V \sqrt{T}z\right) \exp\left(-\frac{z^2}{2}\right) dz,$$

$$G(0) = \kappa V(0) \frac{1}{\sqrt{2\pi}} \int\limits_{z < -d_2} \exp\left(-\frac{\left(z - \sigma_V \sqrt{T}\right)^2}{2}\right) dz$$

$$= \kappa V(0) \frac{1}{\sqrt{2\pi}} \int\limits_{u < -d_1} \exp\left(-\frac{u^2}{2}\right) du,$$

or else:

$$G(0) = \kappa \, V(0)(1 - N(d_1)).$$

$G(0)$ is lower than $\kappa V(0)$ because bankruptcy costs are incurred only if $V(T)$ is lower than D.

Therefore, the debt value on date 0 is:

$$B'_T(0) = V(0) - S(0) - G(0)$$

$$= V(0)(1 - \kappa)(1 - N(d_1)) + D \, e^{-rT} N(d_2). \tag{28.13}$$

> **Example 4**
>
> Consider Example 3 again, assuming that bankruptcy costs represent 40% of the value of the liquidated company ($\kappa = 0.4$), $S(0) = 180$ million, $D = 280$ million contractually promised in 3 years, $r = 5\%$, $\sigma_V = 16.76\%$ and $V(0) = 420$. The calculation gives $N(d_1) = 0.98025$, $N(d_2) = 0.96152$, $B'_T(0) = 236.70$, which corresponds to a rate $r_T^{max} = 5.6\%$, i.e., a credit spread of 60 bps (basis points) and an average recovery rate in the event of default around 53%. The latter is much more in line with the observed rates than the one resulting from the model in the absence of bankruptcy costs (89%).

28.2.2.3 Barrier Models (Dynamic Models)

In the simple model described in Sect. 28.2.1, the debt security is assumed to be a zero-coupon bond delivering a single cash flow composed of a principal and interest payment on date T. The default can therefore only occur on date T, as shareholders have no incentive to file for bankruptcy before the payment date T.[22] However, in practice, debt is composed of various securities (bonds) with different maturities (one part is short-term and the other medium- to long-term), capital repayments and interest payments give rise to cash flows payments throughout the time interval $(0, T)$ and suspension of payments (and hence file for bankruptcy) can occur at any time up to T. We cannot then consider the solvency of the company solely on date T as in static Merton models, but we must look at its entire trajectory between 0 and T. This is why the models in which default can occur at any time are called "dynamic models."

* We present below an overview of a dynamic model in which the default is triggered when the value $V(t)$ of the company decreases until it hits the default threshold. The valuation of the company shares (equity) and debt is then based on a (*path dependent*[23]) *down-and-out option* (see Chap. 14, barrier options, Sect. 28.2.1).

More precisely, a default barrier model has the following general features:

- In the time interval $(t, t + dt)$ the company distributes to its shareholders and creditors, in the form of coupons, principal repayments, and dividends, a total amount δdt of which cdt is allocated to the creditors. Consequently, the RN dynamics of the overall value $V(t)$ (debt and equity), on the one hand, and the

---

[22] In terms of options, limited liability can be analyzed as an American call that shareholders have no interest in exercising early if the underlying asset does not pay a coupon.

[23] That is, whose value depends not only on the terminal value of the underlying but also on its trajectory.

28.2 Modeling Default Events and Valuation of Securities 1203

RN expectation of a change in the value of the debt, on the other hand, write respectively:

$$\frac{dV}{V} = (r - \delta)dt + \sigma_V dW^*; \tag{28.14}$$

$$\frac{E^*(dB^r)}{dt} = (r - c)B^r. \tag{28.15}$$

- The market value of the bond $B^r(t, V)$ is a function of time and $V(t)$. We need to determine this function. From Itô's lemma and relations (28.14) and (28.15), $B^r(t, V)$ is governed by the partial differential equation (PDE):

$$\frac{E^*(dB^r)}{dt} = \frac{\partial B^r}{\partial t} + (r - \delta) V \frac{\partial B^r}{\partial V} + 0, 5\, V^2 \sigma_V^2 \frac{\partial^2 B^r}{\partial V^2} = (r - c)B^r. \tag{28.16}$$

- This PDE determines the desired $B^r(t, V)$ function if boundaries conditions are set such that they express either the value of the debt in the event of bankruptcy (which may occur at any time $\tau \leq T$), or its value at time T in the absence of default.
- The default threshold, i.e., the value at which the suspension of payments occurs on date t, is denoted by $V_d(t)$. In the models presented below, $V_d(t)$ is assumed to be exogenous; for example, the MKMV model, discussed in the next chapter, sets this bankruptcy threshold at the nominal value of the short-term debt due on date t plus half of the long-term (or medium-term) debt nominal value. The date of the possible default is the first occurrence when the value $V(t)$ hits the barrier $V_d(t)$, and we define the stopping time $\tau = inf\{t \in Z (0, T) \mid V(t) = V_d(t)\}$ accordingly: if $\tau < T$, a default (bankruptcy) occurs, whereas $\tau = T$ means that the company avoids bankruptcy in the interval $(0, T)$.
- By denoting $(1 - \kappa)$ the fraction of the company value recovered by the creditors in the event of default (κ representing the fraction attributable to the liquidation costs and hence incurred by the creditors), the value of the debt, in this case, is equal to $B^r(\tau) = (1 - \kappa) V_d(\tau)$. If D denotes the nominal value of the debt including the last coupon on terminal date T, the boundaries conditions set on PDE (28.16) become:

$$B^r(\tau, V_d(\tau)) = (1 - \kappa) V_d(\tau) \text{ if } \tau < T;$$

$$B^r(T, V(T)) = D \text{ if } \tau \geq T \text{ and } V(T) \geq D;$$

$$B^r(T, V(T)) = (1 - \kappa) V(T) \text{ if } \tau \geq T \text{ and } V_d(T) \leq V(T) < D.$$

PDE (28.16) along with the above boundary conditions may be used to determine $B^r(t, V)$ either analytically in some simple cases (such as that of a constant default threshold V_d) as for a barrier option, or numerically.

In the case of a constant default threshold, $V_d < D$, a constant proportion κ for bankruptcy costs, and a continuous coupon c, the value $B^r(t, V, V_d)$ of the risky debt may in fact be obtained in a more convenient manner through a probabilistic approach than by solving a PDE such as seen in (28.16).

We consider the current date $t = 0$ and the stopping time τ defined previously: $\tau = \inf \{t \in [0, T] \mid V(t) = V_d\}$. The density f of this stopping time and its distribution function F, i.e., $F(t) = \int_0^t f(s) ds$, are known. This results directly from Lemma 4 of Appendix 2 of Chap. 14, related to the probability of reaching a barrier by an arithmetic Brownian motion $X(t)$ of drift μ and volatility σ. Indeed, Eq. L4 of that Appendix characterizing the probability distribution of the minimum m_t reached by $X(t)$ between 0 and t (starting from $X(0) = 0$), leads directly to the distribution law of the first instance when such a Brownian motion hits a barrier y from above:

$$\text{Proba}(m_t \geq y) = \text{Proba}(\tau \leq t)) = N\left(\frac{-y + \mu t}{\sigma \sqrt{t}}\right) - e^{\frac{2\mu y}{\sigma^2}} N\left(\frac{y + \mu t}{\sigma \sqrt{t}}\right).$$

In the context of this chapter, $y = \ln(V_d/V(0))$, ($y < 0$), and $\mu = r - \delta - \sigma^2/2$.

The value $B^r(0, V, V_d)$ of the risky bond on date 0 is equal to the RN expectation of the discounted contractual cash flows (final payoff plus coupons), i.e.:

$$B^r(0, V, V_d) = De^{-rT} E_Q\left[1_{\tau > T} 1_{V(T) > D}\right] + (1 - \kappa) e^{-rT} E_Q\left[1_{\tau > T} 1_{V(T) < D} V(T)\right]$$

$$+ (1 - \kappa) V_d E_Q[e^{-r\tau} 1_{\tau < T}] + \int_0^T c e^{-rs}(1 - F(s)) ds.$$

The first term on the right-hand side of the equation is the debt value in absence of default. The second term is the debt value in the event of default at time T; in this case the value is equal to $(1 - \kappa) V(T)$ corresponding to the assets' value less the incurred liquidation costs. The third term is the debt value if a default occurs before T. The last term is the value of all coupons, if any, received up to the date of default.

Note that this last term is rewritten, after integration by parts:

$$\frac{c}{r} - \frac{c}{r} e^{-rT}[1 - F(T)] - \frac{c}{r} \int_0^T e^{-rs} f(s) ds.$$

Using the same technique as developed in Chap. 14 for barrier options with a *rebate at hit* (a fixed lump sum payment given to the option holder when the knockout barrier is hit), we obtain, by using the notation $V(0) = V$:

28.2 Modeling Default Events and Valuation of Securities

$$B'(0, V, V_d) = De^{-rT}\left[N(d_2) - \left(\frac{V_d}{V}\right)^{2(\varepsilon-1)} N(d_3)\right] + (1-\kappa)Ve^{-\delta T}[N(d_4) - N(d_1)]$$

$$+ (1-\kappa)Ve^{-\delta T}\left(\frac{V_d}{V}\right)^{2\varepsilon}[N(d_5) - N(d_6)] + (1-\kappa)V\left[\left(\frac{V_d}{V}\right)^{\varepsilon+a} N(d_7) + \left(\frac{V_d}{V}\right)^{\varepsilon-a} N(d_8)\right]$$

$$+ \frac{c}{r}\left[1 - \left(\frac{V_d}{V}\right)^{\varepsilon-1+a} N(d_7) - \left(\frac{V_d}{V}\right)^{\varepsilon-1-a} N(d_8)\right] - \frac{c}{r}e^{-rT}\left[N(d_9) - \left(\frac{V_d}{V}\right)^{2(\varepsilon-1)} N(d_{10})\right]$$

$$(28.17)$$

with:

$$\mu = r - \delta - \sigma^2/2, \qquad \mu' = r - \delta + \sigma^2/2, \quad a = \frac{\sqrt{\mu^2 + 2r\sigma^2}}{\sigma^2},$$

$$\varepsilon = \frac{\mu'}{\sigma^2} = \frac{r-\delta}{\sigma^2} + \frac{1}{2}, \qquad (\varepsilon - 1) = \frac{\mu}{\sigma^2} = \frac{r-\delta}{\sigma^2} - \frac{1}{2},$$

$$d_1 = \frac{1}{\sigma\sqrt{T}}[\ln(V/D) + \mu'T], \qquad d_2 = \frac{1}{\sigma\sqrt{T}}[\ln(V/D) + \mu T],$$

$$d_3 = \frac{1}{\sigma\sqrt{T}}[\ln(V_d^2/DV) + \mu T], \qquad d_4 = \frac{1}{\sigma\sqrt{T}}\left[\ln\left(\frac{V}{V_d}\right) + \mu'T\right],$$

$$d_5 = d_3 + \sigma\sqrt{T}, \qquad d_6 = \frac{1}{\sigma\sqrt{T}}\left[\ln\left(\frac{V_d}{V}\right) + \mu'T\right],$$

$$d_7 = \frac{1}{\sigma\sqrt{T}}\left[\ln\left(\frac{V_d}{V}\right) + a\sigma^2 T\right], \qquad d_8 = \frac{1}{\sigma\sqrt{T}}\left[\ln\left(\frac{V_d}{V}\right) - a\sigma^2 T\right],$$

$$d_9 = d_4 - \sigma\sqrt{T}, \qquad d_{10} = d_6 - \sigma\sqrt{T}.$$

Remarks

- In the particular case of a debt without default risk, all terms in $N(.)$ are equal to 1 or 0 (because $V_d = 0$ and $\sigma = 0$) and we come up with the classic result in the certain future:

$$B(0, V) = D\,e^{-rT} + \frac{c}{r}\left(1 - e^{-rT}\right).$$

- In the particular case of zero-coupon bond ($c = 0$), no liquidation cost, and a barrier $V_d = D$ (the nominal value of the debt), which implies $d_1 = d_4$ et $d_5 = d_6$, formula (28.17) reduces to:

$$B'(0, V, D) = De^{-rT}\left[N(d_2) - \left(\frac{D}{V}\right)^{2(\varepsilon-1)} N(d_3)\right]$$

$$+ V\left[\left(\frac{D}{V}\right)^{\varepsilon+a} N(d_7) + \left(\frac{D}{V}\right)^{\varepsilon-a} N(d_8)\right].$$

- In the particular case of a perpetual coupon-bearing bond (a perpetuity) $[e^{-rT} = e^{-\delta T} = 0]$, formula (28.17) simplifies to that of Leland (1994):

$$B^r(0, V, V_d, \infty) = (1 - \kappa)V\left(\frac{V_d}{V}\right)^{\varepsilon+a} + \frac{c}{r}\left[1 - \left(\frac{V_d}{V}\right)^{\varepsilon-1+a}\right].$$

- In the particular case of a barrier $V_d = D$, we find the formula derived by Leland and Toft (1996), i.e., formula (28.17) without its second and third terms, because $d_4 = d_1$ and $d_5 = d_6$.

There are other possible applications of this type of dynamics, in particular, to analyze the optimal capital structure of a company with different various debt structures (see, e.g., Attaoui and Poncet (2013, 2015)).

It should be noted in conclusion that in some dynamic models, the default threshold is itself endogenous (not fixed in advance): the interruption of coupon or principal payments occurs *at the time optimally* chosen by the shareholders. For example, they may be incentivized to pay coupons, even if the value of the assets is less than the nominal value of the debt, if they strongly believe that the situation will get better before the upcoming repayment dates. The problem then becomes similar to that of American options (see Chap. 13).

28.2.2.4 Comparison, Merits, and Limitations of Default Models

Structural and intensity models are based on different logics. While structural models are based on a representation of the cause of default (the company's assets value are not sufficient to meet its liabilities), intensity models represent the default event using an ad hoc Poisson process. From a theoretical point of view, the structural approach is, therefore, more satisfactory. However, empirically, structural models have a major weakness that is not present in intensity models: they *strongly underestimate the probabilities of default over the first periods*, and therefore the credit spreads applied to short-term products. This drawback is inherent in the continuous path process supposed to govern the value $V(t)$ of the assets. Because of this continuous process, the probability is very low over a short interval $(0, T)$ (even relative to the duration T of the interval) that $V(t)$ will fall by an amount $V(0) - V_d$ and thus reach the default threshold V_d.

This continuous process even implies that a very short-term *junk bond* (bond with a very high default probability) must be traded at the same spread as an AAA security, which is in contradiction with all empirical observations.[24] This is why market practitioners do not really use these models. In *contrast*, in an intensity

---

[24] By introducing uncertainty into the assessment of the initial value of assets (linked in particular to imprecision or arbitrariness in accounting rules), we can recover realistic credit spreads on short-term securities. This is because, due to this initial uncertainty, the probability to be on the eve of a default date is positive (see Duffie and Lando (2001)).

model, jumps are possible at any time, with a probability $\lambda(t)$ per unit of time that can be calibrated to credit spreads actually observed, even for short-term securities. Yet, in spite of its limitations, the Merton model constitutes a conceptual framework that has proved very useful and has led to different extensions and applications presented in the following chapter (notion of distance to default, MKMV model, Gaussian single-factor model, etc.).

– Among the structural models, it is important to distinguish between static and dynamic models. For example, let us compare a static Merton model to a dynamic barrier model, both models being characterized by the same asset dynamics and the same default threshold. Not only does the timing of default differ in the two models (in the static model it may only occur on date T), but the probability of default obtained with the static model is lower than that of the dynamic model. Indeed, unlike the latter model, the static model allows one or more hits of the default threshold on dates $t < T$ $(V(t) < V_d)$ followed by a recovery situation that allows the company to end up on date T with an asset value above this threshold $(V(T) > V_d)$.

Depending on the problem at hand, a static model is sufficient or a dynamic one is required. For example, in the simple case of a company with highly concentrated maturities the static model is relevant, while it may be unrealistic in the case of a complex debt structure involving short and long-term debts and numerous coupon and principal repayments over time. In practice, Merton's model and its extensions yield an acceptable *ranking* of default probabilities (risk-neutral and historical). The way financial engineers implement this model is by converting through an appropriate monotonic transformation the (theoretical RN) probability of default given by Merton's model into an estimate of both real-world and RN default probabilities that are rather accurate or relevant.[25] Moody's KMV and Kamakura provide such estimations as well as an interesting indication of "distance to default," as explained in the Chap. 29.

28.2.3 A Practical Application: the Valuation of Convertible Bonds

The structural or reduced approach may be used to value options written on debt securities subject to default risk, using an analytical or a numerical method. As an example, we study the valuation of convertible bonds, which are a rather complex case of a hybrid security embedding two optional clauses.

We have already described the main feature of bonds convertible into shares in Chap. 16, Sect. 16.5.1. Recall that a convertible bond (CB) is a bond that the holder

---

[25] This monotonic transformation yields acceptable results because the *rankings* of the theoretical default probabilities produced by Merton's model, the RN default probabilities and the historical default probabilities are the same.

has the right to convert into q shares (the conversion rate q is defined contractually) and that it most often includes an early redemption option at the issuer's hand. It is therefore a hybrid security embedding an optional clause, namely an American option, for the benefit of the investor (bondholder), and often a second one for the benefit of the issuer. Remember that the simplest evaluation method consists in breaking down the CB into two components (bond + option(S)) which are valued separately. The value of the so-called *actuarial floor* bond component can be measured using the discounted value (at the prevailing rate on the bond market, taking into account the issuer's risk of default) of the remaining bond cash flows (coupons + principal repayment).

The optional component(s) can be calculated using an appropriate option pricing model. However, there are many issues when applying the BSM model. First, the CB embeds an American option whose strike is not fixed since it is equal to the market price of the bond, which therefore fluctuates with interest rates and the credit spread (in fact, it is an American exchange option). Besides, the CB is subject to default risk. Finally, the CB contract most often embeds an early redemption option for the benefit of the issuer and, as explained below, the exercise strategy of both embedded options can be optimized (in particular, the issuer may redeem early to prevent the CB holder from converting). We cannot therefore simply value the two options independently.

Numerical methods can overcome some of these difficulties. According to the problem at hand and the degree of precision required, different methods are used to value a CB. These are usually based on either a structural or an intensity (reduced) model.

We first adopt a structural approach, then present a simple intensity model, and finally perform an evaluation using Monte Carlo simulations.

28.2.3.1 Structural Approach

Consider a company whose total asset value is $V(t)$ and the value of its CB is $B(t)$. The nominal value of this bond is D and its maturity T. N denotes the number of shares issued by the company before conversion and hence its market capitalization is $N\,S(t)$.

In the absence of debts other than the CB, at any given moment we have: $V(t) = N\,S(t) + B(t)$.

In the presence of other debts, $V(t)$ may be considered as the value of the assets *net* of these other debts and some elements of the following analysis have to be modified accordingly.

Upon conversion, the company creates n shares. We define $\eta \equiv \frac{n}{N+n}$ (inverse of the dilution factor).[26] The conversion will therefore give the ownership of a fraction η of the company to the bondholders (these ownership rights therefore have a value $\eta\,V(t)$).

---

[26] If m denotes the number of convertible bonds, the *conversion rate* is defined by $q \equiv n/m$.

28.2 Modeling Default Events and Valuation of Securities

It is also assumed that the shareholders have the right to redeem the bond at a price $R(t)$. Generally, $R(t) > D$ for t $< T$ ($R(T) = D$). In the event of a redemption triggered by the shareholders (early or at maturity), bondholders choose either to accept the redemption (if $\eta\ V(t) \leq R(t)$) or go for conversion (if $\eta\ V(t) > R(t)$).

At maturity (on date T), in the absence of prior redemption or conversion and depending on the company's $V(T)$ value, three situations may occur: default, repayment, or conversion. These three situations are presented in the following table:

	Value of the company $V(T)$	Value of the debt $B(T)$	Value of the shares
Default	$V(T) < D$	$V(T)$	0
Repayment	$\eta V(T) \leq D < V(T)$	D	$V(T) - D$
Conversion	$\eta V(T) > D$	$\eta V(T)$	$V(T)(1 - \eta)$

Compactly, we can therefore write, for date T:

$$B(T) = \max\left[\min\left(V(T), D\right), \eta V(T)\right].\qquad (28.18)$$

On an intermediate date $t < T$, in the absence of earlier redemption or conversion, four events may occur:

1. *Continuation*: Neither early redemption by shareholders nor conversion by bondholders;
2. *Early repayment*: Shareholders (in practice, the managers of the company representing the rights and interests of the shareholders) call for repayment and bondholders consent;
3. *Forced conversion*: Shareholders call for redemption but bondholders proceed to the conversion just prior to redemption;
4. *Spontaneous conversion*: Bondholders voluntarily trigger conversion without shareholders calling for early redemption.

These four situations correspond to the different combinations of the decisions of the two groups of stakeholders and are summarized in the following table:

	No conversion	Conversion
No early redemption call	*Continuation:* $B(t) = Bc(t)$	*Spontaneous conversion:* $B(t) = \eta V(t)$
Early redemption call	*Redemption:* $B(t) = R(t)$	*Forced conversion:* $B(t) = \eta V(t)$

$B_c(t)$ represents the continuation value at time t, i.e., the value of the CB in the event of no conversion and no repayment on that date.

In the case of a call for redemption, $B(t) = \max[R(t), \eta V(t)]$.

In absence of a redemption call, $B(t) = \max[B_c(t), \eta V(t)]$.

Since the shareholders aim to minimize the value of the CBs (to maximize the value of their shares), they call for early redemption if and only if: $\max[R(t), \eta V(t)] < \max[B_c(t), \eta V(t)]$.

The value of the convertible bonds on date t then is:

$$B(t) = \min\{\max[R(t), \eta V(t)], \max[B_c(t), \eta V(t)]\}. \qquad (28.19)$$

Relationships (28.18) and (28.19) hence reflect the conditions that enable the CB to be evaluated, either from an evaluation PDE using for instance a finite difference method or from a recombining tree representing the dynamics of $V(t)$.

- As to the PDE approach, it should be noted that, in a world composed of two state variables, $r(t)$ (interest rate) and $V(t)$, the price of the CB writes $B(t, r, V)$ and the PDE is: $D_t B = rB$ where D_t is the Dynkin operator (see Chap. 20).[27]
- As to the recombining tree method, the evaluation starts from the future date T (by applying (28.18)) and goes on by backward induction down to date 0 through iterative use of Eq. (28.19).

This structural approach is not always easy to implement in practice, particularly because there are always other existing debts with various maturities to be evaluated. This makes the dynamics of $V(t)$ and the expression of default conditions more complex. Therefore we present in the next paragraph a recombining tree model based on the dynamics of $S(t)$ rather than further elaborating on the structural approach.

28.2.3.2 Intensity Model, with a Trinomial Tree Representing the Dynamics of S(t)

The following model is based on a representation of the dynamics of the share value $S(t)$ by a trinomial tree for which, at each node, one of the branches represents an event of default denoted by f. As in all numerical methods, the period $(0, T)$ is evenly divided into N time intervals (t_{i-1}, t_i) of equal duration $\Delta t = T/N$ with $t_i = i \Delta t$.

On each date t_i three values of $S(t_{i+1})$ are therefore possible for the following period:

$S(t_{i+1}) = u\,S(t_i)$ (up), or $d\,S(t_i)$ (down), or 0 (default) with RN probabilities respectively equal to $q_u, q_d, q_f = 1 - e^{-\lambda \Delta t}$.

To this tree corresponds a dynamics for the CB price $V(t)$, with $V(t) = \alpha D$ in the event of default. The parameter α accounts for possible bankruptcy costs: if the company is in default the CB holders only recover a fraction $\alpha < 1$ of the redemption value. Besides, default is characterized by a Poisson process of parameter (intensity) λ. In the event of default, the dynamic stops as default branches f are not continued, $S = 0$ and the CB has a value $B(t_i) = \alpha D$ (or $B(t_i) = \alpha B(t_{i-1})$). The tree $S(t)$ recombines on the other branches.

---

[27] Namely: $\frac{\partial B}{\partial t} + \frac{\partial B}{\partial r}\mu_r(.) + \frac{\partial B}{\partial S}rS + \frac{1}{2}\frac{\partial^2 B}{\partial S^2}\sigma_S^2 S^2 + \frac{1}{2}\frac{\partial^2 B}{\partial r^2}\sigma_r(.)^2 + \frac{\partial^2 B}{\partial S \partial r}\sigma_r S(.)S = rB$; this expression depends on $\mu_r(.)$ and $\sigma_r(.)$, hence on the interest rate model adopted.

28.2 Modeling Default Events and Valuation of Securities

As for the model calibration, we rely on the RN dynamics governing $A(t) = S(t)$ in *absence* of default (see Eq. (28.10) with $\alpha = 0$ and a continuous dividend c paid by the share):

$$dA = A(t)[r(t) - c(t) + \lambda]dt + A\sigma dW.$$

This relationship means that the spread λ is added to the risk-free rate to compensate for the expected loss due to the possibility of default (see § 28.2.1.3 above).

Based on one of the proposed settings for the binomial model (calibration 2, Eq. (10.32), Chap. 10), the parameters can be selected as follows[28]:

$$u = e^{\left(r - c + \lambda(1-\alpha) - \frac{1}{2}\sigma^2\right)\frac{T}{N} + \sigma\sqrt{\frac{T}{N}}}$$

$$d = e^{\left(r - c + \lambda(1-\alpha) - \frac{1}{2}\sigma^2\right)\frac{T}{N} - \sigma\sqrt{\frac{T}{N}}}$$

$$q_d = q_u = 0.5e^{-\lambda\frac{T}{N}}; \quad q_f = 1 - e^{-\lambda\frac{T}{N}}.$$

For this calibration, the conditional (on no default) process $A(t)$, which follows a binomial process, is treated separately to adjust the value of the parameters to its first two moments.

The $B(t_i)$ values of the CB are derived from the tree representing the dynamics of $S(t)$ by backward induction. This induction is based on the behavior of both shareholders and bondholders at each node of the tree. As in the case of the structural model studied in § 28.2.3.1 above, the rational behavior of bondholders (who seek to maximize the value of their CBs) and that of shareholders (who seek to minimize this value) make it possible to derive two rules (similar to the rules expressed in Eqs. (28.18) and (28.19)):

Rule #1 (value at maturity): $B(t_N) = \max[D, qS(t_N)]$ on branches u and d, and $B(t_N) = \alpha D$ on branch f, where q is the conversion rate (see footnote 25) not to be confused with the probabilities q_u, q_d, q_f.

Rule #2 (value at any intermediate date before maturity): $B(t_i) = \min\{\max[R(t_i), qS(t_i)], \max[B_c(t_i), qS(t_i)]\}$.

As explained in § 28.2.3.1, this rule reflects the fact that, in both situations (early redemption and absence of early redemption), bondholders maximize the value of their assets (max(.), max(.)), but the decision to call or not early redemption is decided by the shareholders in order to minimize the bondholders' wealth (min(.)).

The computation is performed by backward induction. We start at $t_N = T$, the date on which, in the absence of a default until t_{N-1}, $B(t_N)$ is given by rule #1. In case of i share price increases and $N - i$ share price decreases (node (i, N)), we have: $S(t_N) = u^i d^{(N-i)} S_0$.

---

[28] This is not the only possible calibration, nor necessarily the most efficient.

On date t_{N-1}, the continuation value is $B_c(t_{N-1}) = e^{-r\Delta t} E^{RN}[B(t_N) | S(t_{N-1}) = S_{N-1}]$. This conditional expectation is calculated, at each node $(i, N-1)$ where $S(t_{N-1}) = u^i d^{(N-1-i)} S_0$, as the average of the following three possible share price outcomes weighted by their corresponding probabilities, namely:

$$B_c(i, N-1) = e^{-r\Delta t} E^{RN}\left[B(t_N) | S(t_{N-1}) = u^i d^{(N-1-i)} S_0\right]$$

$$= e^{-r\Delta t}\left\{q_u \max\left[D, q_u u^{i+1} d^{(N-i-1)} S_0\right] + q_d \max\left[D, u^i d^{(N-i)} S_0\right] + q_f \alpha D\right\}.$$

This continuation value is used to apply rule #2 that will give $B(t_{N-1})$. Once $B(t_{N-1})$ has been determined on each node $(i, N-1)$, this calculation is repeated on date t_{N-2} to obtain $B(t_{N-2})$ then, similarly, on dates $t_{N-3}, \ldots, t_0$ to finally come up with $B(0)$.

Additional Remarks
- The rational behavior of shareholders (i.e., CB issuer) should prevent situations where the conversion value $qS(t)$ exceeds the redemption value $R(t)$. Indeed, according to rule #2, if the CB is not converted on date t: $B_c(t) > qS(t)$; therefore, if $R(t) < qS(t)$: $\max[R(t), qS(t)] < \max[B_c(t), qS(t)]$ the CB issuer is incentivized to call an early redemption. However, we often observe in practice cases where the CB issuer misses the optimal date to exercise its early redemption right.
- A generalization of this model is to assume that the default intensity λ is a function of the underlying price. We hence write: $\lambda(t) = \Phi(S(t))$ where Φ is a decreasing function.[29] This assumption is realistic and does not pose any implementation problems because it does not jeopardize the recombining nature of the tree[30] nor the backward induction evaluation procedure introduced above.
- In most CB contracts, early redemption is only permitted from a certain date. The existence of such a contractual restriction does not raise any difficulty in the application of the backward induction method.
- A more complex version of this model takes into account the stochastic nature of interest rates. The tree is then three-dimensional and describes the dynamics of the two processes $r(t)$ and $S(t)$. The construction of such trees is explained in Chap. 13.

28.2.3.3 Evaluation with Monte Carlo simulations
The last, alternative method is based on simulations. It involves the development of a large number of share price and interest rate scenarios using Monte Carlo simulations. The following explanations are rather brief on purpose, as they rely

---

[29] It was suggested, particularly by Duffie and Singleton, to choose $\Phi(S) = constant/S$.

[30] The reader may verify that with the proposed calibration $u_i d_{i+1} = d_i u_{i+1}$, even when l or r depends on time, an *up* followed by a *down* and a *down* followed by an *up* lead to the same value of $S(t_{i+1})$ when starting from a given $S(t_{i-1})$. This ensures the recombining character of the tree governing the value of $S(t)$ ($A(t)$).

28.2 Modeling Default Events and Valuation of Securities

on techniques exposed in particular in Chap. 26 (Monte Carlo Simulations). Example 5 of Chap. 26 presents a simulation of correlated trajectories of the share price, its stochastic volatility, and the interest rate. The adaptation of this simulation framework to the problem of CB valuation requires modeling the correlated evolutions of the interest rate r, the share price S as well as the credit spread s. The $B(t, r, S, s)$ price of the CB derives from this. The three factors r, S, s are assumed to follow the following three-dimensional diffusion process governed by three independent standard Brownian motions W_1, W_2, W_3:

$$dr = a(b - r(t))\, dt + \lambda_{11}\, dW_1, \qquad (r)$$

$$dS/S = (r(t) + \theta)\, dt + \lambda_{21}(.)\, dW_1 + \lambda_{22}(.)dW_2, \qquad (S)$$

$$ds = c(\phi(S(t)) - s(t))\, dt + \lambda_{31}(.)\, dW_1 + \lambda_{32}(.)\, dW_2 + \lambda_{33}(.)dW_3. \qquad (s)$$

By virtue of the last equation, the spread mean-reverts towards $\phi(S(t))$, presumably a decreasing function, because we expect to see s decreasing with S. In addition (or in replacement) to the credit spread process, a Poisson process $Q(t)$ can also be introduced, which generates a *stopping time* representing a payment default or a rating downgrade.

The valuation method of CBs based on these simulations can be briefly described as follows. A scenario is associated with three trajectories r, S, s, or/and Q. In the presence of a contractual clause permitting an early redemption, four types of scenarios arise that we present from the most unfavorable (from the company shareholder perspective) to the most favorable.

In very adverse scenarios, the company is in default at a date $\tau < T$ and the CB holders collect the coupons until τ and recover, on that date, a fraction α of the redemption value (with $\alpha < 1$, 0.4 for instance). In this scenario, the CB cash flow sequence is: $\{c, c, \ldots, c, \alpha V_r\}$ where c refers to the annual CB coupon and V_r to the redemption price of the CB (the above array includes τ cash flows).

In fairly adverse scenarios, there is no early redemption or conversion and the $qS(T)$ conversion value on date T is lower than V_r. Therefore, the CB is not converted but repaid at maturity. The CB then behaves like a conventional bond, resulting in the following sequence of cash flows (coupon and principal redemption at maturity): $\{c, c, \ldots, c, c + V_r\}$ (the above array includes T cash flows).

In fairly favorable scenarios, the conversion value $qS(T)$ on date T, is larger than V_r, but the CB has not been converted or redeemed in advance. The CB delivers the coupons and is converted at maturity, thus implying the sequence: $\{c, c, \ldots, c + qS(T)\}$.

In the most favorable scenarios, the CB is converted early on date $t < T$ into q shares representing a total value of $qS(t)$, thus implying the sequence: $\{c, c, \ldots, c + qS(t)\}$. These conversions are either *spontaneous* (voluntary) or *forced* by a call for early redemption.

Voluntary exercise situations are the most difficult to detect because the optimal exercise date t on a trajectory is not easy to determine in a Monte Carlo simulation (see Chap. 26, Sect. 26.4, § 26.5 for an analysis of this problem as well as for the presentation of a method to address it).

The CB value is calculated by discounting an "average" cash flow sequence, which is calculated over a very large number of simulated cash flows sequences.

The advantage of using simulations over the recombining tree technique lies in their ability to take into account several random factors (share price, credit spread, interest rate, default, etc.) whereas in practice the tree is only tractable with a maximum of two factors. The advantage of the recombining tree over simulations is its ability to deal easily with the American feature of the options. However, the regression method (set out in Chap. 26, Sect. 26.4, § 26.5) makes it possible to address the conversion or the early repayment right at the expense of a significant complication in the approach.

Example 5

This example is extremely simplified in order to meet a double pedagogical objective: to illustrate the evaluation methods and to analyze the economic and financial mechanisms governing the CB.

This involves analyzing and evaluating the issue of convertible bonds by Company X.

These CBs, with a nominal value of $100, *bullet* style and repayable at par in 12 years, are issued at a price of $97 and distribute an annual coupon of $4 (the nominal rate is 4%). The yield on issuance date is 4.33% (it would be exactly 4% if the bond were issued at par). For the sake of simplicity, it is assumed that the CB does not have a contractual prepayment clause benefiting the issuer. The equivalent bond yield for the same issuer is 6%. By buying the CB rather than a traditional bond, the investor, therefore, waives 1.67% interest in exchange for the conversion option (i.e., one bond for one share X) at any time. Share X is now trading at $60 and has just distributed a dividend of $1.5 from earnings per share of $3.

We discuss two methods to evaluate the CB.

The first method consists in valuing the actuarial floor and the conversion option separately, the value of the CB being the sum of these two values.

- The actuarial floor is the value of the traditional bond (free of conversion rights) obtained by discounting the sequence {4, 4, 4, ..., 4, 104} at a rate of 6%, i.e., $83.2. This implies that the conversion option is sold by the issuer at a price of: 97–83.2 = $13.8.
- The value C of the conversion option is measured using an appropriate valuation model. For example, the Merton dividend model (assumed to represent 2.5% of the share value: $1.5 = 0.025 \times 60$) can be used. With a volatility of 30%, this model gives $C = \$16.1$ ($> \$13.8$) and implies that the CB is a good investment. More generally, the theoretical value of the CB is

(continued)

28.2 Modeling Default Events and Valuation of Securities

Example 5 (continued)

equal to $83.2 + C$ and the acquisition of the security is favorable if $C > \$13.8$.

The second method is based on the simulations of a large number of scenarios, each of which involves a cash flow sequence for the CB.

The simulations of the $S(t)$ share prices are performed by taking a drift equal to the interest rate net of the dividend distribution rate (e.g., 6%–2.5% = 3.5%) and a volatility consistent with the annualized standard deviation of historical share price returns.

We detail three different share price, earnings, and dividend situations. Each situation corresponds to a behavior of the CB that is summarized in the form of a cash flow sequence. In each one of them, the PER (equal to $60/3 = 20$) and the percentage of earnings distributed as dividends (1.5/3 = 50%) are assumed to remain constant over the 12-year life of the CB. The adjusted dividend per share will therefore consistently represent 2.5% of the adjusted share price.

In a moderately favorable situation (called S1), the price and dividend per share evolve as follows, starting in year 7:

S1/years	7	8	9	10	11	12
Share price	97.4	107.2	117.9	129.7	140	156
Dividend	2.43	2.68	2.95	3.24	3.50	3.90

In this situation S1:

– At maturity, the CB will be converted (for a share price equal to $156) rather than repaid at the price of $100.
– The CB will never be converted early; indeed, the holders benefit from waiting, as they receive a coupon of $4 always *higher* than the dividend and retain in addition the option's time value.

The cash flow sequence $\underline{S}_1$ generated by the CB in situation S1 is therefore constituted by the coupon payments to which is added the share price at maturity:

S1 / years	1 to 11	12
$\underline{S}_1$	4	$156 + 4 = 160$

In the fairly unfavorable situation S2 (share price in 12 years <100 but no default from the issuer), the CB behaves like a traditional bond and the corresponding sequence $\underline{S}_2$ writes:

S2/years	1 to 11	12
$\underline{S}_2$	4	$100 + 4 = 104$

Let us now consider the following very favorable situation S3:

S3/years	7	8	9	10	11	12
Share price	100	120	140	180	190	180
Dividend	2.5	3.0	3.5	4.5	4.75	4.5

In such a situation it is reasonable to expect that the CB will be converted early around the years 10–11 because the dividend exceeds the coupon.[31] We, therefore, assume that conversion occurs in year 10, on the eve of the dividend distribution, when the adjusted share price is worth \$180. We also assume that the coupon is paid before the dividend. Therefore, the sequence implied by the investment in a CB is:

S3/years	1 to 9	10
S3	4	$180 + 4 = 184$

In fact, the implementation of this method for valuation purposes requires the simulation of a very large number N of scenarios (several tens of thousands) such as those just described, leading to $\underline{S}_1, \underline{S}_2, \ldots, \underline{S}_N$ cash flow sequences, from which the "average" sequence: $\underline{S}_a = (\underline{S}_1 + \underline{S}_2 \ldots + \underline{S}_N)/N$ is calculated. The CB is then evaluated by discounting $\underline{S}_a$.

28.3 Summary

- *Credit risk* justifies the existence of a spread s which is the difference between the zero-coupon rate $r_\theta{}^{max}$ with maturity θ "promised" by an issuer with a given credit risk and rating and the rate r_θ of a security free of credit risk: $s_\theta \equiv r_\theta^{max} - r_\theta$. The *ex ante* "promised" r^{max}, which is the rate of return at maturity obtained in the absence of default, must be distinguished from the ex post realized rate R which is random (due to a possible default): $R = r^{max}$ in absence of default; $R < r^{max}$ in case of default; $E(R) < r^{max}$.

---

[31] This is in fact a questionable assumption, as the time value of the option may be higher than the value of the difference between the dividend and coupon rates. One way to incorporate a rational decision to early exercise an American option into a Monte Carlo simulation is to estimate the option continuation value by regression and compare it to its intrinsic value, as explained in Chap. 26, Sect. 26.4.5.

28.3 Summary

In general: $E(R) = r + \pi + l$, where π is a risk premium and l a liquidity premium, leading to a breakdown of the credit spread $r^{max} - r \equiv s$ into three components: $s = (r^{max} - E(R)) + \pi + l$.

Counterparty risk is a term reserved for the risk of the parties dealing with derivative assets.

- **Empirical approach of credit risk**

  Probabilities of default, for each rating and at different time horizons, are estimated from historical data. The *unconditional* probability of default in period θ, p_θ, must be distinguished from the *hazard rate* h_θ which is the probability of default *conditional on the absence of a previous default*. The cumulative probability of default at time θ is denoted $\phi(\theta)$. Different expressions for the credit triangle can be obtained; for instance, when α is the recovery rate, in absence of a liquidity premium, the zero-coupon credit spread formula may be written either (i) $s_\theta = \frac{1}{\theta} \phi(\theta) (1 - \alpha) + \pi$ when using historical probabilities, or (ii) $s_\theta = \frac{1}{\theta} \phi^*(\theta)$ $(1 - \alpha)$ using RN probabilities.

  Relation (i) is used for calculating theoretical spreads based on estimated probabilities of default and (ii) for inferring (implicit) RN probabilities of default from observed spreads (*stripping*).

  RN default probabilities p^*_θ implicit in the range of observed spreads are generally much *higher* than the historical probabilities p_θ: the RN probabilities compensate for the absence of risk premia by "overweighting" the probabilities of adverse events.

  Three alternative methods may be implemented for discounting a "promised" contractual cash flow X_θ^{max}:

  (a) discount X_θ^{max} with a specific rate $r_\theta + s_\theta$; (b) discount the *historical expected cash-flow* $E(X_\theta)$ with $r_\theta + \pi_\theta$; (c) discount the *RN expected flow* $E^*(X_\theta)$ with the riskless rate r_θ (no risk premium).

  In presence of counterparty risk, a security may also be valued by discounting at the riskless rate the RN expectations of the cash flows. However, these expectations involve two combined risks affecting each cash flow: the risk on the payoff in the absence of default by the counterparty, and the counterparty risk per se.

- **Reduced-form approach of credit risk**

  The evolution of the risky asset price is represented by a stochastic process with jumps reflecting default events, usually a generalized Poisson process $N(t)$ defined as having integer values, with independent increments, and such that: $N(0) = 0$; Proba $\{dN(t) = 1\} = \lambda(t) \, dt$; Proba$\{dN(t) > 1\} = 0$ and Proba $\{dN(t) = 0\} = 1 - \lambda(t) \, dt$. This definition implies that $N(t) - \int_0^t \lambda(u) du$ is a martingale.

  In the Jarrow-Turnbull and the Duffie-Singleton models, default occurs on the first jump of a generalized Poisson process of intensity $\lambda(t)$. In the former model, the default at any date triggers the recovery of a fraction α of the nominal value at terminal date T. In the latter model, the recovery is paid immediately upon default, is proportional to the value of the asset in absence of default, and the *instantaneous spread* added to the riskless rate to compensate for the RN-expected loss is equal to $s(t) = (1 - \alpha)\lambda^*(t)$ (credit triangle).

- **Structural approach of credit risk**

Structural models consider the evolution of the market value $V(t)$ of the firm's assets. $V(t)$ is also equal to the sum of the market values of shares $S(t)$ and debts $B^r(t)$ (the exponent r indicates credit risk). Too low a value $V(t)$ triggers default. In most structural models, the bankruptcy option is exercised by the shareholders as soon as $V(t)$ becomes equal to the value of the debts. Debt and equity are thus valued with an option pricing model.

Merton's model is consistent with the Black-Scholes model: no dividends; $V(t)$ follows a geometric Brownian motion; the debt is a zero-coupon with maturity T and nominal D whose market value is $B^r_T(t)$. Shareholders file for bankruptcy *iff* $V(T) < D$, hence $S(T) = Max(V(T) - D, 0)$, a standard call payoff. The model can easily be adjusted to account for bankruptcy costs.

Contrary to Merton's over-simplified assumption, debt is usually composed of different securities generating cash flows throughout the time interval $(0, T)$. Therefore, suspension of payments can occur at any time up to T. *Dynamic models* consider the whole trajectory over $(0, T)$ and default is triggered at any time when $V(t)$ hits a threshold. The valuation of equity and debt thus involves (*path dependent*) *down-and-out options*.

- **Valuation of convertible bonds**

Structural and reduced form approaches may be used to value debt securities with default risk and embedded options.

A Convertible Bond (CB) often embeds two optional clauses: the holder always has the right to convert it into q shares, and often the issuer has an early redemption option. CB valuation can be performed using a structural, reduced-form, or Monte Carlo approach.

Upon conversion, the company, whose capital is previously composed of N shares, issues n new additional shares. Converting thus gives bondholders the ownership of a fraction $\eta \equiv n/(n + N)$ of the company.

At maturity T, in the absence of prior redemption or conversion, and depending on the company's value $V(T)$, three situations may occur: default $(B(T) = V(T))$, repayment $(B(T) = D)$ or conversion $(B(T) = \eta V(T))$. This implies a first equation: (*) $B(T) = \max[\min(V(T), D), \eta V(T)]$.

On date $t < T$, in the absence of earlier redemption or conversion, four events may occur:

- *Continuation*: Neither early call for redemption nor conversion by bondholders $(B(t) = B_c(t))$;
- *Early repayment*: Shareholders call for repayment for a global value $R(t)$ and bondholders agree $(B(t) = R(t))$;
- *Forced conversion*: Shareholders call for redemption but bondholders proceed to the conversion $(B(t) = \eta V(t))$;
- *Spontaneous conversion*: Bondholders voluntarily trigger conversion $(B(t) = \eta V(t))$.

This implies a second equation: (**) $B(t) = \min\{\max[R(t), \eta V(t)], \max[B_c(t), \eta V(t)]\}$.

Equations (*) and (**) allow computing a CB value $B(t, V)$ either from a valuation PDE ($D_t B = rB$) using, e.g., a finite difference method, or from a recombining tree representing the dynamics of $V(t)$.

Monte Carlo simulations of the RN trajectories of $(r(t), S(t), s(t))$, where s is the credit spread, lead to CB values for each trajectory. The desired CB value is simply the average of these values discounted at the risk-free rate.

Appendix

Notation and calculation rules for jumps and default events in intensity models.

Basic Definitions

$N(t) = $ *number of jumps between* 0 *and* t; $\tau = $ *date of the first jump or default date.*

Events

$$\{jump\ between\ t\ and\ t + dt\} \equiv \{dN = 1\};$$

$$\{default\ between\ t\ and\ t + dt\} \equiv \{\tau \in [t, t + dt]\} \equiv \{dN = 1\ and\ N(t) = 0\}$$
$$\equiv \{dN = 1\ and\ \tau > t\};$$

$$\{default\ between\ t\ and\ t + dt | survival\ until\ t\} \equiv \{\tau \in [t, t + dt]\ | \tau > t\}.$$

Probabilities

$\text{Proba}\{dN = 1\} = \lambda(t)\,dt\ (= \text{Proba}\{dN = 1 \mid N(t) = n\}$ for all n).

$\text{Proba}\{N(t) = 0\} = \text{Proba}\{\tau > t\} = \gamma(t) = $ probability of survival at time t;

$\gamma(t) = e^{\int_0^t -\lambda(u)du} = e^{-t\,\Lambda t}$; if $\lambda(t) = \lambda = $ constant: $\gamma(t) = e^{-\lambda t}$.

$\text{Proba}\{\tau \le t\} = 1 - \gamma(t)$ (distribution law of τ) $= \phi_t$;

$\text{Proba}\{\tau \in [t, t + dt]\} = \gamma(t)\,\lambda(t)\,dt = -\,d\gamma.$

$\text{Proba}\ \{\tau \in [t, t + dt] \mid \tau > t\} = \lambda(t)\,dt$ (hence $\lambda(t) = $ *hazard rate*).

Suggestions for Further Reading

Books

**Brigo, D., & Mercurio, F. (2007). *Interest models - Theory and practice, with smile, inflation and credit* (2nd ed.). Springer Finance.

Duffie, D. (2011). *Measuring corporate default risk*. Oxford University Press.

Duffie, D., & Singleton, K. (2003). *Credit risk: Pricing, measurement and management*. Princeton University Press.

Hull, J. (2018a). *Options, futures and other derivatives* (10th ed.). Prentice Hall Pearson Education.

Hull, J. (2018b). *Risk management and financial institutions* (5th ed.). Prentice Hall Pearson Education.

**Schönbucher, P. (2003). *Credit derivative pricing models: Models, pricing and implementation*. Wiley Finance.

Articles

Altman, E. I. (1989). Measuring corporate bond mortality and performance. *Journal of Finance, 44*, 902–922.

Ammann, A., Kind, A., & Wilde, C. (2003). Are convertible bonds underpriced? An analysis of the French market. *Journal of Banking and Finance, 27*(4), 635–753.

Attaoui, S., & Poncet, P. (2013). Capital structure and debt priority. *Financial Management, 2013*, 737–775.

Attaoui, S., & Poncet, P. (2015). Write-down bonds and capital and debt structures. *Journal of Corporate Finance, 35*, 97–119.

Black, F., & Cox, J. (1976). Valuing corporate securities: Some effects of bonds indentures provisions. *Journal of Finance, 31*, 351–367.

Black, F., & Scholes, M. (1973). The pricing of options and corporate liabilities. *Journal of Political Economy, 81*, 637–659.

Brennan, M. J., & Schwartz, E. S. (1980). Analyzing convertible bonds. *Journal of Financial and Quantitative Analysis, 15*(4), 907–929.

Collin-Dufresne, P., Goldstein, R., & Martin, S. (2001). The determinants of credit spread changes. *Journal of Finance, 56*, 2177–2207.

Duffie, D., & Lando, D. (2001). Term structure of credit spreads with incomplete accounting information. *Econometrica, 69*, 633–644.

Duffie, D., & Singleton, K. (1999). Modelling term structure of defaultable bonds. *Review of Financial Studies, 12*(4), 687–720.

Elton, E., Gruber, M., Agarwal, D., & Man, C. (2001). Explaining the rate spread on corporate bonds. *Journal of Finance, 56*(1), 247–277.

Jarrow, R., & Turnbull, S. (1995). Pricing derivatives on financial securities subject to credit risk. *Journal of Finance, 50*(1), 53–86.

Kealhofer, S. (2003a). Quantifying default risk I: Default prediction. *Financial Analysts Journal, 59* (1), 30–44.

Kealhofer, S. (2003b). Quantifying default risk II: Debt valuation. *Financial Analysts Journal, 3*, 78–92.

Leland, H. (1994). Corporate debt value, bond covenants, and optimal capital structure. *Journal of Finance, 49*(4), 1213–1252.

Leland, H., & Toft, K. (1996). Optimal capital structure, endogenous bankruptcy and the term structure of credit spreads. *Journal of Finance, 51*, 987–1019.

Litterman, R., & Iben, T. (1991). Corporate bond valuation and the term structure of credit spreads. *Journal of Portfolio Management, 1991*, 52–64.

Madan, D., & Unal, D. (2000). A two factor model for pricing risky debt and the term structure of credit spreads. *Journal of Financial and Quantitative Analysis, 35*, 43–65.

Merton, R. (1974). On the pricing of corporate debt: The risk structure of interest rates. *Journal of Finance, 29*, 449–470.

Toft, K., & Prucyk, B. (1997). Options on leveraged equity: Theory and empirical tests. *Journal of Finance, 52*, 1151–1180.

Zhou, R. (1997). A jump-diffusion approach for modelling credit risk and valuing defaultable securities,. *Working Paper*.

Website

www.defaultrisk.com

Modeling Credit Risk (2): Credit-VaR and Operational Methods for Credit Risk Management

29

Until the late 1990s, banks and regulators lacked adequate tools to measure, analyze, manage, and control credit risk. For example, regulators applied a very simple criterion to assess bank credit risk, the Cooke ratio, which determined the required capital to cover the bank's default risk.

Under the impetus of a few pioneers such as JP Morgan (CreditMetrics™) or KMV, operational methods for measuring and managing credit risk have been developed in recent years and applied broadly, in particular in banking management and for regulation purposes. Yet these methods, which are based on the empirical analyses and models presented in the previous chapter, pose numerous and sensitive limitations and implementation problems that restrict their validity, especially during periods of severe financial crisis such as 2007–2012. Those who mastered the core assumptions of these models should have been aware of these limitations but most decision makers largely ignored them. Their use actually requires great care and acute judgment, as explained in this chapter. Moreover, in order to make the methods tractable, they have been implemented from an engineering rather than mathematical viewpoint, without fear of technical complexity but at the cost of compromises with the required theoretical rigor, and it is in this spirit that they are presented here.

Finally, among the numerous methodologies available, we present only those that we consider the most significant. Readers wishing a deeper analysis and a more thorough comparison will refer to specialized books.[1]

This chapter introduces analytical tools (such as Credit-VaR and Expected Shortfall) and operational methods for credit risk measurement and management (CreditMetrics™, MKMV, default correlation models). Banking regulation (Basel

---

We thank Riadh Belhaj for his comments on this chapter and Eric Boutitie for an interesting discussion on banking risk management. The usual disclaimer of course applies.

---

[1] The following books cover the content of this and the following chapters: Hull J., *Risk Management and Financial Institutions*, Wiley (fifth ed. 2018), and Schönbucher P., *Credit Derivative Pricing Models*, Wiley (2003).

© The Author(s), under exclusive license to Springer Nature Switzerland AG 2022
P. Poncet, R. Portait, *Capital Market Finance*, Springer Texts in Business and Economics, https://doi.org/10.1007/978-3-030-84600-8_29

1221

accords) which not only encompasses credit risk but extends to other types of risk is addressed in Chap. 30.

The empirical approaches and models described in the previous chapter can be used to quantify the credit risk affecting a security, a loan, a receivable, or a portfolio of these assets, for example, using VaR or Expected Shortfall (*ES*) (see Chaps. 26 and 27). In fact, calculating the VaR or ES of an asset or a portfolio subject to credit risk is a major issue and raises specific and complex theoretical, empirical and practical problems. Various methods have been proposed, in particular by JPMorgan-CreditMetrics™ and Moody's-KMV (MKMV hereafter).[2] They have been adopted by the banking regulatory authorities and other market players. In addition, under the Basel agreements, the regulator invites banks and financial institutions to develop and implement internal tools for measuring and managing risks, particularly credit risk, based on the principles we are about to set out, but which leave considerable leeway to build in-house assessments. Consequently, various methods are implemented by practitioners.

As far as terminology is concerned, it is sometimes useful to make a distinction between various debt assets subject to default risk, namely: loans, receivables, and debt securities. Recall that a security is a tradable asset which in consequence has a market value, whilst a loan or a receivable is not traded, hence is not marked-to-market (but may be marked-to-model). Besides, we use the general term "obligor" for the debtor in case of loans, receivables, or debt securities, but restrict the term "issuer" to the case of securities. We use "exposure" or "credit exposure" when it comes to the default risk of the asset itself.[3] Finally, for ease of exposition, the term "*asset*" is meant to be "*debt asset*" in the whole chapter.

The chapter is organized as follows. Section 29.1 presents the generic principle for determining the Credit-VaR and Sect. 29.2 describes the empirical methods based on observed migration frequencies and credit spreads (which inform on credit worthiness). Given the theoretical and practical limitations of these empirical methods, Sects. 29.3 and 29.4 introduce analytical methods, based in particular on structural models (Merton 1974; Vasicek 2002; Kealhofer 2003a, b), which are used both for determining the probabilities of default and migration of the securities and for their valuation. Section 29.5 presents the calculation of the capital required to cover the credit risk of a portfolio or balance sheet using the concept of *Unexpected Loss* (UL). The methods above also apply to the management and control of banking risks. Section 29.6 then shows how they fit into the prudential framework of banking activities, describes the rules applied to the determination of a bank's minimum

---

[2] KMV is named after the three founders (Stephen Kealhofer, John McQuown and Oldrich Vasicek) of KMV. A few years after its foundation, this company was taken over by Moody's under the name MKMV, which later became Moody's Analytics. In this chapter, we continue to refer to the MKMV models.

[3] In fact, from a legal perspective, it is improper to speak of the default of an exposure (e.g., the default risk of a loan) as solely the obligor (issuer) is subject to default.

29.1 Determining the Credit-VaR of an Asset: Overview and General Principles

capital and liquidity, and the current Basel 3 dispositions aiming at remedying Basel 2 deficiencies.

29.1 Determining the Credit-VaR of an Asset: Overview and General Principles

Regardless of the method used, the aim is to estimate the risk affecting the creditworthiness of a financial asset or a portfolio of assets at a horizon H, generally one year. We make a distinction between the MTM (*Marked to Market*) approach and the DM (*Default Mode*) approach. In the (MTM) approach, securities are evaluated at their market value at time H. What is measured is the total credit risk, which concerns both losses due to the actual default of the issuer and the depreciation attributable to deterioration in its creditworthiness (possibly measured by its rating). The DM approach,[4] on the other hand, focuses only on losses due to the obligor's default. Because, in contrast to securities, receivables, and loans are generally not traded assets and as such do not have explicit market prices, the MTM approach applies mainly to debt securities[5] whilst the DM approach is applicable to loans (e.g., mortgages) and receivables as well as securities.

As in some of the developments of the previous chapter, the rating is represented here by an integer i, with $i = 1,..., d$, where 1 represents the highest rating (AAA) and d the worst, i.e., the one corresponding to the default state (D).

We first consider a *single security* x whose rating today (date 0) is i and whose value depends on its rating. Let $B^x_i(0)$ be its value on date 0 (observed) and $B^x_j(H)$ its value on date H with a j rating (random).

In general, the methods used to calculate at horizon H the Credit-VaR of the single security x, be it empirical or analytical, consist of two steps:

- *Step* 1: Determining the p_{ij} rating migration probabilities between 0 and H from rating i to j, in the case of an MTM model (for a DM model, the determination of the default probabilities p_{id} are sufficient, those probabilities being relative to the migration from rating i to the default state d).
- *Step* 2: Calculating the $B^x_j(H)$ market values of the security at time H for each possible rating j, in the case of an MTM model (for the DM model, the losses incurred in the event of default are sufficient).

In the case of a *portfolio*, the calculation of the Credit-VaR includes a *third step*, as shown in Fig. 29.1; this extra step addresses the determination of the distribution of the total value $B^p(H)$ of portfolio p composed of several securities x. This

---

[4]The interest rates are assumed to be given, thus separating interest rate risk (which is studied elsewhere, in Chaps. 5 and 27) from credit risk.

[5]Loans and receivables could, however, be *marked to model*.

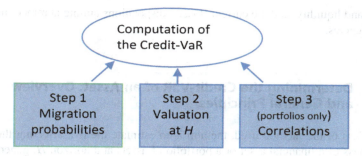

Fig. 29.1 3 steps related to Credit-VaR computation

distribution is constructed using the joint probability laws (usually entailing correlations) of the different $B^x_j(H)$ related to the rating migration of the different securities x included in the portfolio. As exposed in Sect. 29.4.2, this step is simple when considering a well-diversified portfolio whose systematic risk is represented by a single risk factor I. $B^p(H)$ therefore depends only on the value of $I(H)$ and the overall risk results from the simple aggregation of individual systematic risks.

Moreover, while all operational methods have both theoretical and empirical components, they differ in the relative importance given to each one of them.

Broadly speaking, the empirical approach is based on migration probabilities such as those provided by rating agencies (step 1) and on the valuation of an asset by discounting the contractual cash-flows using the appropriate yield curve observed for assets with the same rating (step 2). The theoretical approach aims at determining both migration probabilities and asset values using a model, for example, a Merton-style structural model.

29.2 Empirical Credit-VaR of an Asset Based on the Migration Matrix

The empirical analysis relies on two ingredients, i.e., a *presumably known* rating migration matrix and *observed* credit spreads. The latter makes it possible to value risky assets by *discounting contractual cash-flows with a rate curve embedding credit spreads*. We start with the case of an individual asset and set the limitations of the method prior to examining portfolios in Sect. 29.4.

29.2.1 Computation of the Credit-VaR of an Individual Asset

The simplest conceptual method for analyzing the credit risk of an asset with current rating i is to calculate its value at time H by discounting its contractual cash-flows occurring after time H for each rating j that it can be given at time H. Discounting is

29.2 Empirical Credit-VaR of an Asset Based on the Migration Matrix

generally carried out using the forward rate curves prevailing today (one curve attached to each rating j). A distribution of this value is then constructed from the credit-rating migration probabilities. This two-step approach is broadly the one adopted by CreditMetrics™ to calculate the Credit-VaR of an individual asset. It can be described as follows:

- Step 1: The transition probabilities p_{ij} from a rating i to j, between 0 and H, are provided in a given rating migration matrix. We assume that all assets having the same rating have identical transition probabilities.
  Various methods can be used to build a rating migration matrix. The simplest but not the best way is to estimate each individual transition probabilities from the historically observed rating migration frequencies. We address the problems raised by this empirical approach in Sect. 29.2.2 and present in Sect. 29.3 alternative methods to evaluate transition probabilities, which are theoretically and empirically sounder.
- Step 2: The valuation of j-rated assets at horizon H is carried out by discounting the contractual cash flows with the appropriate forward rate curve that currently applies to j-rated assets. Let $f^{j}_{H,n}$ be the n-period zero-coupon forward rate starting at future time H prevailing today for a j-rated asset. The value $B^x_{j}(H)$ on date H of a j-rated asset x will be valued by discounting the cash-flows (using the rate $curve\, f^{j}_{H,n}$ for all relevant n) remaining at time H.

Thus, for each asset x currently rated i, its different possible values $B^x_{j}(H)$ ($j = 1,..., d$) in one year are calculated by discounting the remaining contractual cash-flows and each of these possible discounted values has a probability equal to the corresponding $[p_{ij}]$ available in the migration matrix. The whole probability distribution of $B^x(H)$ then obtains. This method is called the *empirical approach*. It solely considers credit risk (taken from the migration matrix) disconnected from interest rate risk (which is nil here as interest rates are assumed to be constant).

Once the distribution of the asset value prevailing in one year (H) is established, one can estimate its moments and compute its quantiles (such as the VaR). Let us consider an asset x whose current rating and value are respectively denoted by i and $B^x_{i}(0)$. Also, let $B^x(H)$ be the random value of this asset on date H.

We have $B^x(H) = B^x_{j}(H)$ with a probability p_{ij} ($j = 1, \ldots, d$) and we can write:

$$E\{B^x(H)\} = \sum\nolimits_{j=1}^{d} p_{ij} B^x_{j}(H); \text{Variance}\{B^x(H)\}$$

$$= \sum\nolimits_{j=1}^{d} p_{ij} \left[B^x_{j}(H)\right]^2 - [E(B^x(H))]^2;$$

$$\text{VaR}_x(p, H) = p - \text{quantile of the distribution of } B^x(H).$$

Example 1 (Simplified)

Let us consider an A-rated asset with a nominal value of $100 and a maturity of 3 years, bearing a 6% fixed annual coupon (the next coupon being distributed in one year), and let us assume that, as functions of the issuers' rating, the zero-coupon forward rate curves are as follows:

	1 year in one year (%)	2 years in one year (%)
AAA (1)	5.00	5.01
AA (2)	5.02	5.06
A (3)	5.06	5.15
BBB (4)	5.60	6.20
BB (5)	6.60	7.50
B (6)	8.30	10.00
CCC (7)	15.50	18.50
D (8)	Recovery of 40% of the nominal amount	

For example, in one year just after the distribution of the next coupon, in the absence of a rating change and based on current forward rates, this asset is worth:

$$B_A(H) = \frac{6}{1.0506} + \frac{106}{(1.0515)^2} = \$101.58.$$

If its rating improves to AA, it is worth:

$$B_{AA}(H) = \frac{6}{1.0502} + \frac{106}{(1.0506)^2} = \$101.75.$$

In the event of default, the creditor only recovers $40 a time H. The asset will therefore be worth: $B_D(H) = \$40$.

This allows us to calculate the different possible values of the asset in one year.

Assuming that the transition probabilities are:

$p_{31} = 0.5\%, p_{32} = 3.5\%, p_{33} = 75\%, p_{34} = 14\%, p_{35} = 5\%, p_{36} = 1.6\%,$ $p_{37} = 0.3\%, p_{38} = 0.1\%$, the value of the asset in one year and the relative gain/loss over one year are as follows:

	Value	Relative gain/loss[a]	Probability (%)
AAA (1)	101.84	0.26	0.5
AA (2)	101.75	0.17	3.50
A (3)	**101.58**	**0**	**75**
BBB (4)	99.67	−1.91	14
BB (5)	97.35	−4.23	5
B (6)	93.14	−8.44	1.6
CCC (7)	86.61	−14.97	0.3
D (8)	40	−61.58	0.1

[a]Calculated with respect to the value of the asset in one year in the absence of a rating change (101.58)

The 99.6% Credit-VaR over one year is therefore equal to $14.97 and the 98% Credit-VaR is equal to $8.44. The 98% Expected Shortfall is equal to: (8.44 × 1.6% +14.97 × 0.3% + 61.58 × 0.1%)/2% = $12.08.

29.2 Empirical Credit-VaR of an Asset Based on the Migration Matrix

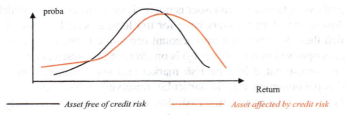

Fig. 29.2 Impact of credit risk on the probability distribution of an asset return before its maturity

It should be emphasized that the above analysis disregards other risks that affect asset values. In particular, the interest rate risk between today and time H is not taken into account, since the value of the security at time H is calculated using the *current* forward rate curve. We will not present here the complex models that combine these risks.[6] It should simply be noted that the combination of credit and interest rate risks changes the probability distributions of asset values and returns. Indeed, these distributions differ from that of a similar asset free from credit risk as they are less concentrated around their mean and more skewed and have left fat tails, as shown in Fig. 29.2.

These differences are explained by default occurrences which result in significant losses with low probabilities.

29.2.2 Limitations of the Empirical Approach

The empirical approach based on historical rating migration probabilities (step 1) and present value calculation (step 2) has a number of drawbacks, the most important of which are the following:

- Rating transition and default probabilities are unstable over time, making it problematic to use migration probabilities observed in the past to represent present or future probabilities. In particular, in bearish economic conditions the probability of a rating downgrade increases sharply whilst the probability of a rating upgrade decreases. This bad surprise was observed in 2007–2009. The opposite is true in bullish economic conditions.
- There is no one-to-one relationship between rating migration or default probabilities and rating because the latter depends on various other factors, such as the legal conditions defining a default (which differ for instance between Europe and the USA). Thus, a senior debt may display both migration probabilities and probability of default lower than a subordinated debt bearing a similar rating.

---

[6]Complex models combine both interest rate uncertainty and credit risk, e.g., Vasicek (2002).

- The rating of assets from a given issuer is not reviewed continuously, which leads to the following problems: issuers have for too long a period of time the same rating which then does not take into account economic turns; the probability of migrating (as opposed to not migrating) is on average underestimated; ratings are on average overestimated in a bearish market and underestimated in a bullish market. This phenomenon was, in particular, observed in 2007–2009.
- Simply discounting the contractual cash flows does not account properly for the (possible) optional components embedded in an asset and is irrelevant for credit derivatives.

Finally, it should be remembered that, as a result of diversification, the credit risk of a portfolio composed of many assets from different issuers is less than the sum of the credit risk of each individual security. It is therefore mandatory, when measuring the overall credit risk of a portfolio to take into account the dependence between its individual components without, however, disregarding the increase of this dependence during crises. The methods for assessing the credit risk of a portfolio and the calculation of its Credit-VaR are presented in Sect. 29.4.

29.3 Credit-VaR of an Individual Asset: Analytical Approaches Based on Asset Price Dynamics (MKMV...) and on Structural Models

The drawbacks of the empirical approach described above, in particular the lack of accuracy and instability of the empirical migration frequencies, may be circumvented by using an analytical model that deduces the migration probabilities from more stable economic variables.

Different methodologies (including MKMV and, to a lesser extent, CreditMetrics™)[7] are based on the *dynamics of the value $V(t)$ of the firm's (issuer's) assets.*

The probabilities of default and migration are deduced from this value, for instance by using a Merton-style structural model (as in the MKMV method) which, as we have seen, makes it possible to infer the parameters of the process governing the assets' value $V(t)$ from those which govern the dynamics of the equity value. Determining these migration probabilities constitutes the first step in calculating the VaR under an MTM model. The second step requires modeling the value of the assets on date H for each rating j. This evaluation can be carried out by using a structural model or discounting the RN expectation of future cash flows. It uses the concepts of *standardized return* and *distance to default* that we now define.

---

[7]CreditMetrics uses asset profitability to assess the correlation parameters needed to calculate the Credit-VaR of an entire portfolio.

29.3.1 Asset Dynamics, Standardized Return, Default Probabilities, and Distance to Default

The simplest assumption regarding the process governing the assets' value $V(t)$ is that it follows a geometric Brownian motion, as in Merton's model, which implies future log-normal values $V(t)$. The log-normality assumption is consistent with empirical observations when it comes to the firm's total asset value, while being relatively inconsistent with observed equity and debt prices, as mentioned earlier.

We write, under the historical probability: $\frac{dV}{V} = \mu dt + \sigma_V dW$, or, by posing $Z \equiv \frac{W(H)}{\sqrt{H}}$, standard Gaussian variate $N(0, 1)$:

$$V(H) = V(0)e^{\left(\mu - \frac{\sigma_V^2}{2}\right)H + \sigma_V\sqrt{H}Z}. \tag{29.1}$$

$\ln(V(H)/V(0)) = (\mu - \sigma_V^2/2)H + \sigma_V\sqrt{H}Z$ represents the gross asset return over the period $(0, H)$ and Z is interpreted as a *standardized return*.

We thus posit:

$$Z = \frac{\ln\left(\frac{V(H)}{V(0)}\right) - \left(\mu - \frac{\sigma_V^2}{2}\right)H}{\sigma_V\sqrt{H}}. \tag{29.2}$$

This convenient standardized return is used systematically in the remainder of this chapter. The nonstandardized gross return $\ln\left(\frac{V(H)}{V(0)}\right)$ is obtained from Z by a simple affine transformation with a scale parameter $\sigma_V\sqrt{H}$ and a position parameter $(\mu - \sigma_V^2/2)H$.

The relationship (1) makes it possible to map a default probability to each default threshold. Recall that the default threshold V_d is defined as the asset value on date H below which the issuer is in default. The issuer is in default at H if and only if $V(H) \leq V_d$, i.e., according to (1), if:

$(\mu - \sigma_V^2/2)H + \sigma_V\sqrt{H}Z \leq \ln\left(\frac{V_d}{V(0)}\right)$, i.e. $Z \leq -DD$, with:

$$DD \equiv \frac{\ln\left(\frac{V(0)}{V_d}\right) + \left(\mu - \frac{\sigma_V^2}{2}\right)H}{\sigma_V\sqrt{H}}. \tag{29.3}$$

DD is called the *distance to default*. The issuer is hence in default whenever the standardized return on the assets hits or goes below the threshold $-DD$. $Z \leq -DD$ expresses in a simple and compact way the necessary and sufficient condition of default on date H in a static model.

The *historical* probability of default p_d has a one-to-one relationship with the distance to default:

$$p_d = \text{Proba}\,(Z \le -DD),$$

and since Z is distributed as $N(0,1)$, we have:

$$p_d = N(-DD) = 1 - N(DD) <=> DD = -N^{-1}(p_d), \qquad (29.4)$$

where $N(.)$ represents the cumulative distribution function of the $N(0,1)$ distribution.

Example 2

An issuer X whose probability of default in the current year is 0.2% (rated for example BBB+) has a distance to default equal to $DD_X = -N^{-1}(0.002) = 2.878$.

It should be noted that the distance to default DD:

- Is specific to each issuer (firm issuing debt(s)).
- Depends on the structure of its balance sheet (it increases with the ratio $\frac{V(0)}{V_d}$ where V_d is closely related to the market value of the debt(s)) and on the horizon H.
- Increases with the average return μ on its total assets and decreases with their volatility σ_V.
- And is linked by relation (29.4) to one probability of default (conversely, each probability of default corresponds to one distance to default).

29.3.2 Derivation of the Rating Migration Quantiles Associated with the Standardized Return

From any given rating scale (from AAA to D for instance) mapped onto a one-to-one array of default probabilities, it is possible to derive the distance to default attached to each rating.

Recall that if each rating i *is* associated with a default probability p_{id} at the given horizon H (as in the last column of a transition matrix), the distance to default of all issuers bearing the same rating i should, from relation (29.4), be $DD_i = -N^{-1}(p_{id})$.

It should be noted that the computation procedure of a distance to default from a default probability p_{id} associated to a *known* rating i (given exogenously) is *not* based on relationship (29.3), and therefore has the advantage of not requiring the knowledge of $V(0)$, μ, σ_V or V_d.

This procedure, which consists in mapping the quantile $-DD_i$ of the standardized return on the probability of default p_{id}, can be generalized. Indeed, we can define several quantiles, denoted by z^i_j, corresponding to the distances from rating i to any rating j. Any standardized return variable Z_i can in effect be split into different

29.3 Credit-VaR of an Individual Asset: Analytical Approaches Based on Asset...

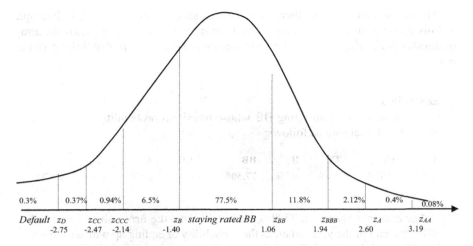

Fig. 29.3 Decomposition of the standardized return of a BB-rated firm

tranches, the jth tranche corresponding to the distance from rating i to rating j (see Fig. 29.3). Thus $z^i_D = -DD_i$ is the *particular* distance from rating i to the default state (i.e., the standardized return below which an issuer with rating i defaults between dates 0 and H). Similarly, z^i_{CCC} is the standardized return threshold at which the rating i becomes CCC between dates 0 and H; z^i_B is the threshold at which it changes to B and so on. To each threshold corresponds a transition probability:

$$N\left(z^i_j\right)$$

= Proba (issuer currently rated i will have a rating equal to or lower than j on date H).

The thresholds z^i_j can be obtained from the migration rating probabilities p_{iD}, $p_{i,CC}, p_{i,CCC}, p_{i,B} \ldots$ recursively (from worst to best j), as follows:

First, the probability that the company currently rated i will be in default at horizon H is $p_{id} = N(z^i_D)$, corresponding to the area delimited by the standard normal distribution on the left of z^i_D. Therefore, inverting the standard normal cumulative distribution function yields: $z^i_D = N^{-1}(p_{id})$.

Second, the threshold z^i_{CC} is derived from the relation: $p_{i,CC} = N(z^i_{CC}) - N(z^i_D)$, which is the area delimited by the standard normal distribution between z^i_{CC} and z^i_D. Thus $N(z^i_{CC}) = p_{i,CC} + N(z^i_D) = p_{i,CC} + p_{i,D} \Rightarrow z^i_{CC} = N^{-1}(p_{i,CC} + p_{i,D})$.

Next, in the same vein, we derive the quantile z^i_{CCC} from the transition probability $p_{i,CCC}$ by the relation $N(z^i_{CCC}) - N(z^i_{CC}) = p_{i,CCC}$, (the area delimited by the standard normal distribution between z^i_{CCC} and z^i_{CC}) implying that $N(z^i_{CCC}) = p_{i,CCC} + p_{i,CC} + p_{i,D} \Leftrightarrow z^i_{CCC} = N^{-1}(p_{i,CCC} + p_{i,CC} + p_{i,D})$.

All the relevant quantiles then are recovered successively as shown in Example 3. This decomposition of the standardized return into tranches thus yields the array of thresholds $z^i_D, z^i_{CC}, \ldots, z^i_{AA}$ derived from a set of transition probabilities given *ex ante*.

Example 3

Consider an issuer with rating BB whose transition probabilities within one year ($H = 1$ year) are as follows:

D	CC	CCC	B	**BB**	BBB	A	AA	AAA
0.3%	0.37%	0.94%	6.5%	**77.50%**	11.8%	2.12%	0.4%	0.07%

For example, there is a 77.50% probability that the firm will still be rated BB at the end of the year, whereas the probability of ending up with a rating A is 2.12% and the probability of default is 0.3%.

In terms of standardized return Z, the rating thresholds derived from the above migration probabilities are:

$z_D = -DD$	z_{CC}	z_{CCC}	z_B	z_{BB}	z_{BBB}	z_A	z_{AA}
−2.75	−2.47	−2.14	−1.40	1.06	1.94	2.60	3.19

For example, $\text{Proba}(Z \leq z_{CC}) = 0.003 + 0.0037$, hence $z_{CC} = N^{-1}(0.0067) = -2.47$.

The decomposition of the standardized return distribution presented in Fig. 29.3 gives both the transition probabilities and the corresponding threshold attached to an issuer currently rated BB.

Recall once more that the decomposition of the standardized return Z as described above and introduced by CreditMetrics™, does not require the computation of the distance to default (using formula (29.3)), and therefore does not require the knowledge of $V(0)$, V_d, μ and σ_V. Thresholds are simply based on rating migration probabilities given exogenously. Again, these probabilities are estimated through an empirical approach and therefore suffer from the shortcomings attached to the empirical estimation and already discussed in Sect. 29.3.2. It is in general not unambiguous to decompose standardized returns Z_i into z^i_j quantiles by simply referring to an external synthetic rating system, such as S&P or Moody's, grounded on rating classes supposedly composed of *homogeneous* (in terms of credit risk) firms.

This weakness has motivated the development of alternative methodologies such as MKMV, to which we turn now.

29.3.3 Computation of the Distance to Default and Expected Default Frequency (MKMV-Moody's Analytics Method)[8]

Following an opposite approach to the one described above, the distances to default DD can be *calculated* using definition (29.3) from which can be *derived* a probability of default $p_d = N(-DD)$. This is the approach proposed by MKMV—Moody's Analytics.

The calculation of the distance to default through definition (29.3) requires the knowledge of the parameters μ and σ_V governing the dynamics of the assets' value, of the overall value of the company $V(0)$ and of the default threshold V_d. However, the total market value $V(0)$ and the volatility are rarely known. In addition, some assets are either not traded (e.g., bank loans) or are simply not admitted for quotation on a listed market or even not traded on an OTC market; their market value is hence not available either, which prevents $V(t)$ from being calculated as the sum of corporate debt value and equity market value. However, as noted in the previous Chap. 28 (Sect. 28.2.2.2, Merton's model), the current share price $S(0)$, as well as the past share prices of a listed company are observable, and the share price volatility can be estimated from historical data. If the default threshold V_d is assumed to be known, $V(0)$ and σ_V can be determined from a structural model, for example, using the system of Eqs. (28.12 and 28.11a, 28.11b, 28.11c) presented in the previous chapter which writes respectively at any time t:

$$S(t) = V(t)N(d_1) - D\,e^{-r(T-t)}\,N(d_2) = f_1(V(t), \sigma_V), \tag{29.5}$$

$$\sigma_V = \frac{S(t)}{N(d_1)V(t)}\,\sigma_S = f_2(V(t), S(t), \sigma_S). \tag{29.6}$$

The use of these two relationships to derive the value of $V(t)$ from $S(t)$ and σ_S has been illustrated in Example 3 of Chap. 28.

In addition, the value of μ can be assessed by applying a risk premium over the risk-free rate r. This premium can be estimated using the CAPM or an APT model (see Sect. 29.3.6 below).

Note that the distance to default calculated under a RN dynamic is obtained with the riskless rate r and not μ.

Also note that the debt's market value derives from the calculated value $V(0)$ and the observed market share price $S(0)$: $B'(0) = V(0)-S(0)$. (See Sect. 28.2.2 and Example 3 in the previous chapter).

Obviously, the greater the distance to default DD is, the lower is the probability of default $N(-DD)$. The values DD can therefore be considered as a score from which one can derive a rating system where each value DD is univocally mapped onto a probability of default (it does not depend on the recovery rate). Such a rating system

---

[8]For a discussion of the model, see http://www.moodysanalytics.com/~/media/Insight/Quantitative-research/Default-and-Recovery/2012/2012-28-06-Public-EDF-Methodology.ashx

is, by construction, not ordinal (i.e., composed of a limited number of CCC, B, BB,..., AAA classes) but cardinal (it is possible to define an "infinite" number of different ratings). Hence it avoids the overlapping issue where a given rating class comprises heterogeneous default probabilities. However, it is possible (but not necessary) to build an ordinal scoring system composed of a finite array of ratings AAA, AA,...,CCC where each rating is defined by a range of default probabilities and overlapping ranges are avoided.

In practice, applying these principles faces numerous tricky technical problems. We present briefly some of them as well as the main choices made by MKMV to address them. The interested reader is referred to the documents of Moody's Analytics (Expected Default Frequency methodology) for a description of the various tweaks used to overcome these issues.

- Issue #1: Estimates of σ_S based on historical share prices turn out to be unstable as being too sensitive to the range of historical share prices considered. This undermines the determination of $V(t)$ and σ_S by solving the system of Eqs. (29.5) and (29.6). To circumvent this issue, MKMV directly estimates σ_V by iterating Eq. (29.5), $S(t) = f_1(V(t), \sigma_V)$, based on an array of historical share prices $S_\theta]_{\theta = 1,...,N}$.

The iteration process can be described as follows:

- We arbitrarily set the first value $\sigma_V^{(1)}$ from which we compute N values $V_\theta^{(1)}$ using the historical set of share prices S_θ mentioned above and the equation $S_\theta = f_1\left(V_\theta^{(1)}, \sigma_V^{(1)}\right)$ for $\theta = 1,..., N$.
- From these N values $V_\theta^{(1)}]_{\theta=1,\cdots,N}$ we derive $\sigma_V^{(2)}$ which generally differs from $\sigma_V^{(1)}$. We then plug $\sigma_V^{(2)}$ into Eq. (29.5) which now writes $S_\theta = f_1\left(V_\theta^{(2)}, \sigma_V^{(2)}\right)$ and, by repeating the previous computation, we derive N values $V_\theta^{(2)}]_{\theta=1,\cdots,N}$.
- This process is repeated p times, where p denotes the number of iterations being sufficient to allow determining an array $V_\theta^{(p)}]_{\theta=1,\cdots,N}$ whose empirical volatility $\sigma_V^{(p+1)}$ is "close enough" to $\sigma_V^{(p)}$. At this stage, $\sigma_V^{(p)}$ is deemed to be an accurate estimate of parameter σ_V and the iteration process stops.[9]
- Issue #2: It is empirically observed that default is generally triggered when the value of assets falls between that of short-term debts and the sum of the values of short and medium to long-term debts. MKMV thereby sets the default threshold V_d at the nominal value of short-term debt plus a weighted sum of the nominal

---

[9]This computation process is analogous to a "solver" seeking a fixed point (namely when $\sigma_V^{(p)}$ is "sufficiently close" to $\sigma_V^{(p+1)}$). However, the fixed point may not be unique and depend from the initial choice made for $\sigma_V^{(1)}$.

29.3 Credit-VaR of an Individual Asset: Analytical Approaches Based on Asset...

values of the various long-term debts. The weights, all between 0 and 1, decrease with the debt maturities and depend on the considered default horizon H.

$V(0)$, V_d, and σ_V being previously determined, we have the necessary and sufficient elements to apply Eq. (29.3) which, for a default horizon set to one year ($H = 1$), writes:

$$DD = \frac{\ln\left(\frac{V(0)}{V_d}\right) + \left(\mu - \frac{\sigma_V^2}{2}\right)}{\sigma_V}.$$

This equation gives the distance to default of any firm.

– Issue #3: The relation $p_d = N(-DD)$ which gives the theoretical default probability as a function of the distance to default leads to poor empirical results. In particular, we have indicated that it is derived from a static model that only considers time T and therefore disregards all occurrences when $V(t)$ $(t < T)$ falls below the default threshold V_d and subsequently recovers. Thus, it underestimates theoretically the probability of default. MKMV, therefore, *does not use* the relation $p_d = N(-DD)$ in practice. Instead, it maps the DD determined for a given company onto the frequency of default historically observed for firms displaying the same DD distance in the past.

This frequency of default, which embeds both theoretical and empirical features, is called *Expected Default Frequency* (EDF). It may substantially *differ* from the "theoretical" value $p_d = N(-DD)$. Moody's Analytics publishes the EDFs for 30,000 firms every year.

The EDF can be used as a cardinal rating system but it is also useful to build, on such bases, an ordinal rating system, as close as possible to a standard system such as that of S&P. This results in a mapping between EDFs and traditional S&P ratings, in approximate accordance with Table 29.1.

For example, for a BBB+ company, the probability of default (EDF) in the coming year is around 0.2%.

It should be emphasized that the mapping reported in Table 29.1 (established before the 2007–2009 crisis) is necessarily approximate because the S&P rating classes do overlap in terms of default probability (EDF). This mapping is also unstable over time because, as EDFs are more sensitive to economic conditions than S&P ratings, the EDFs corresponding to the same S&P rating are smaller during

Table 29.1 Approximate mapping between EDFs (in basis points "bps") and S&P ratings

EDF (bps)	2–4	4–10	10–19	19–40	40–72	72–100	100–140	140–200	200–350
S&P ratings	≥AA	AA/ A	A/ BBB+	BBB+/ BBB–	BBB–/ BB	BB/ BB–	BB–/ B+	B+/B	B/B–

economic expansions than during recessions or stagnations (which is more conservative from a credit risk perspective).

Finally, it should be noted that, as with default probabilities, a distinction is made between historical EDFs and risk-neutral EDFs (EDFs*). The EDFs* corresponding to a given distance to default can be determined either from the average credit spread applied to firms with the same distance to default (a credit triangle relationship is then used to deduce the RN probability from the spread) or from the analytical relationship between the RN and historical default probabilities (see Sects. 29.3.5 and 29.3.6).

29.3.4 Comparing the Two Approaches

To summarize, Credit-Metrics™ uses an external rating scale such as Moody's or S&P to calculate the distance to default. Conversely, MKMV uses the distances to default calculated using relation (29.3) to derive a rating scale. The advantage of the Credit-Metrics™ method lies in its relative simplicity, since it is based on an external rating scale, not on relation (29.3), and does not require the estimation of $V(0)$, V_d, μ, and σ_V. Its main drawbacks lie in the heterogeneity of rating classes (in terms of default and transition probabilities) derived from the use of external ratings which are built on different criteria. Also, Credit-Metrics™ ratings are updated on the same frequency as the external ratings on which they are built and hence may not be updated frequently enough or not be sensitive to economic conditions.

Empirical studies suggest that MKMV distances to default and their associated EDFs are reliable tools to predict a default in the near future, are more sensitive than the other ratings scales presented in this chapter and also possess a "forward-looking" feature.[10] Indeed, we observe that distances to default decrease sharply (and their EDFs increase):

- Within the year preceding the default.
- In periods of economic downturn.
- Within the year preceding rating downgrades.

29.3.5 Estimation of the Credit-VaR of an Asset Using EDF and a Valuation Model Based on RN-FN Probabilities

As discussed above, computing the Credit-VaR involves two steps. *Step 1* consists in calculating the transition (migration) probabilities and *step 2* is devoted to the asset valuation on date H.

---

[10] We draw the attention of the reader to the fact that most studies we have mentioned emanate from MKMV, which might lack perfect impartiality.

29.3 Credit-VaR of an Individual Asset: Analytical Approaches Based on Asset... 1237

- *The migration probabilities* p_{ij} related to a rating framework based on EDF (MKMV method) are estimated from historical data. They are significantly higher than those provided by rating agencies (the probabilities of remaining in the same rating are therefore lower) because EDFs are more sensitive to economic conditions than standard ratings determined with alternative methods. It is therefore not surprising that this method delivers Credit-VaR estimates that differ substantially from those obtained with the migration probabilities provided by the rating agencies.
- *The asset valuation* on date H is based, in analytical approaches, on risk-neutral (RN) or forward-neutral (FN) probabilities.

We have repeatedly emphasized the distinction between historical probabilities and RN-FN probabilities. The former are used for risk measurement and management purposes (in particular, the VaR is a p-quantile of an historical distribution) whereas the RN-FN computation framework is relevant for valuation purposes only. Unlike standard financial methods, which discount expected (using the historical probability) contractual cash flows at the risk-free rate plus a credit spread, analytical VaR calculation methods discount the RN or FN expectations of contractual flows at the risk-free rate, in accordance with option pricing models.

Recall that the distance to default DD defined by relation (29.3) and the resulting probability of default p_d are calculated under the *historical* probability. However, the definition (29.3) may also be written by using the RN probability. Since the RN and historical dynamics of $V(t)$ differ only by their drift parameter, it is sufficient to replace μ by r to find a RN distance to default equal to the parameter of the Black-Scholes-Merton model:

$$d_2 = \frac{\ln\left(\frac{V(0)}{V_d}\right) + \left(r - \frac{\sigma_V^2}{2}\right)H}{\sigma_V\sqrt{H}}.$$

Thus the standard parameter d_2 of the BSM model is interpreted as a RN distance to default. The probability RN of default in H, denoted here p_d^*, is simply:

$$p_d^* = N(-d_2). \tag{29.7}$$

These RN probabilities can be used to value assets bearing a default risk. For example, consider the valuation of a contractual cash flow stream subject to default risk.

- Let us first calculate the value $B^d_t(0)$ on date 0 of a risky zero-coupon contractually delivering a single cash flow f at time t, where the superscript d denotes the default risk feature. The RN probability of default between 0 and t is $p_d^* = N(-d_2)$, α denotes the recovery rate in case of default and r is the risk-free rate with maturity t.

The cash flow F actually delivered is a random variable equal to f with probability $(1-p_d^*)$ and αf with probability p_d^*. Then: $E^*(F) = (1-p_d^*)f + p_d^* \alpha f = [1-p_d^*(1-\alpha)]f$.

The value at 0 of the risky zero-coupon is therefore: $B^d_t(0) = e^{-rt}[1-p_d^*(1-\alpha)]f.$

- This approach may be extended to an asset x generating a contractual a sequence of cash flows $f_\theta]_{\theta=t_1,\cdots,t_N}$ so that we have:

$$B^d_x(0) = \sum_{\theta=t_1}^{t_N} e^{-r_\theta\theta}[1 - \phi^*(\theta)(1 - \alpha)]f_\theta,$$

where $\phi^*(\theta)$ here denotes the cumulative RN-FN distribution function of default between 0 and θ, under the assumption that in case of default the recovery rate α applies to the discounted value of the remaining contractual cash flows.

This method applies to the valuation of options and credit derivatives and consists in calculating the RN-FN expectation of their discounted payoffs.

The calculation of the Credit-VaR of an asset rated i on date 0 requires, as already discussed, determining its value $B_j(H)$ at H for all the possible ratings j given to the issuer on that date. These values associated with the transition probabilities p_{ij} make up a probability distribution from which the VaR of the asset can be determined.

29.3.6 Relationship between Historical and RN Default Probabilities

We derive here the relationship between the RN probability of default (p_d^*) and its historical equivalent (p_d).

Relation (29.3) and the formula for d_2 lead directly to:

$$DD = d_2 + \frac{\mu - r}{\sigma_V}\sqrt{H}.$$

In addition, since $p_d^* = N(-d_2)$, we have:

$$p_d^* = N\left(-DD + \frac{\mu - r}{\sigma_V}\sqrt{H}\right) \text{hence}:$$

$$p_d^* = N\left(N^{-1}(p_d) + \frac{\mu - r}{\sigma_V}\sqrt{H}\right).$$

We note that $\mu > r$ implies $DD > d_2$, hence $p_d < p_d^*$. We thus recover a result presented above: the RN probabilities of default are higher than their historical peers.

In addition, recall that $\frac{\mu-r}{\sigma_V}$ is the Sharpe ratio of the firm's total assets and that the CAPM implies: $\mu - r = \beta_V(\mu_M - r)$ where μ_M is the expected return on the market portfolio. Hence[11]:

---

[11] This relationship is useful to the extent that μ is not observable, while β_V and the Sharpe ratio can be estimated.

$$\frac{\mu - r}{\sigma_V} = \frac{\beta_V(\mu_M - r)}{\sigma_V}.$$

These relationships can be used to deduce the RN probabilities from the historical probabilities (assumed to be exogenously known) or to calculate risk-neutral EDFs (EDFs*). We obtain for example:

$$EDF* = N\left(N^{-1}(EDF) + \frac{\mu - r}{\sigma_V}\sqrt{H}\right).$$

29.4 Credit-VaR of an Entire Portfolio (Step 3) and Factor Models

For the sake of simplicity, we assume that the number of individual exposures in the portfolio is equal to the number of different obligors, which can be done by aggregating all exposures related to obligor i into a single exposure.

In the previous two paragraphs, we have examined the case of an individual debt asset. We now consider a portfolio of several assets (issued by different entities) subject to credit risk. The diversification effect implies the sub-additivity of individual credit risks (Sect. 29.4.1) or default risk (Sect. 29.4.2), as the aggregated (credit or default) risk of the portfolio depends on the *dependence structure of credit/default risks* that links the different obligors. It should be noted that the notion of *credit/default dependence structure* needs to be clarified. First, the *dependence* may apply (i) to default events (defined as the correlation between the indicator functions of these events, (ii) to asset returns, (iii) to losses attributable to the default of one or several obligors, or iv) to the dates of default. Second, the *dependence structure* needs to be determined by an appropriate model such as a Gaussian model where this dependence is assessed by a correlation parameter or a correlation matrix or even more complex models. Appendix 1 develops these different notions and their relationships. In the remainder of this chapter, we characterize *default correlation* by the correlation parameters in *Gaussian* models only, and we model the dependence between *the returns on the assets issued by the obligors*.

We present first, in a sketchy manner, the Marked-to-Market (MTM) approach to VaR for a portfolio based on Monte Carlo simulations, before examining in more detail a semi-analytical Default Mode (DM) model, as well as approaches involving copulas and copula factorial models. Finally, we discuss the issues that arise when the portfolio includes several assets issued by the same counterparty, particularly in the case of *netting* (see Sect. 29.4.6 below).

29.4.1 Marked-to-Market (MTM) Models Involving Simulations

Consider a portfolio p composed of $N = n_1 + n_2 + \cdots + n_K$ assets marked to market (debt securities, credit derivatives, etc.), n_j denoting the number of assets with rating j and K the total amount of different ratings attributed to the N securities composing the portfolio. We denote by R_i the return on the total value of the ith issuer's assets between 0 and H. Thus, we have: $V_i(H) = V_i(0)e^{R_i}$, where $V_i(0)$ and $V_i(H)$, respectively, denote the ith obligor's total asset value at time 0 and H. Let us express the value at time H of the exposure $B_i^d(H)$ associated with this obligor as a function of R_i, i.e., $B_i^d(H) = f_i(R_i)$. This value is determined, for example, from a Merton-like model or a credit derivative valuation model. The value $B_p(H)$ of the portfolio p then writes:

$$B_p^d(H) = \sum\nolimits_{i=1}^{N} f_i(R_i).$$

The distribution of $B_p^d(H)$ can be simulated using a Monte Carlo approach. Many different values of the N-*dimensional* vector $(R_1, \ldots, R_N)$ are simulated, taking into account the dependence structure (here, simply the correlations) between its components.

- First, let us explain how the correlation matrix of the $(R_i, R_j)_{i = 1, \ldots, N; j = 1, \ldots, N}$ can be estimated. Computational procedures are grounded on factor models similar to those discussed in Chap. 23. Both CreditMetrics™ and MKMV use multifactor models relying on indices that reflect industry and country economic conditions.[12] Thus, the return R_i on the assets of a particular obligor i which operates in various locations and industries depend on several industry indices $I_h (h = 1, \ldots, m)$ and several country indices $I_k (k = m + 1, \ldots, m')$:

$$R_i = \sum_{h=1}^{m} a_{ih}I_h + \sum_{k=m+1}^{m'} a_{ik}I_k + \varepsilon_i.$$

The weights a_{ih} attributed to each industry and each country are commensurate with the respective weights of industries and countries in which the obligor operates.

An additional layer of complexity may be contemplated by considering that country and industry indices are endogenous and can be obtained through factor models whose exogenous variables reflect for instance the world economy. These factor models thus determine the covariances $cov(I_h, I_k)$ between the endogenous indices. We then calculate the covariances between the R_i as: $cov(R_i, R_j) = \sum_{h=1}^{m'} \times$

$\sum_{k=1}^{m'} a_{ih}a_{jk}cov(I_h, I_k).$

---

[12] As an indication, the MKMV model (2007 version) involves fourteen factors.

29.4 Credit-VaR of an Entire Portfolio (Step 3) and Factor Models

Recall, however, that the correlations determined by these models are quite unstable over time (in particular, they increase dramatically during economic downturns).

– Second, the correlation matrix of the R_i thus being estimated and the vector $(R_1, \ldots, R_N)$ being assumed to be Gaussian, a Choleski decomposition (i.e., the triangular decomposition explained in Chap. 26) is used to simulate (in accordance with the correlations previously estimated) several tens of thousands (or more) values for this vector. Each simulation yields an N-*tuple* $(f_1(R_1), \ldots, f_N(R_N))$ and therefore a value $B_p^d(H) = \sum_{i=1}^{N} f_i(R_i)$ for the portfolio p. All these numerous simulated values of $B_p^d(H)$ constitute an empirical distribution from which the VaR can be calculated.

However, if both the number N of securities in the portfolio and the number m' of indices used in the factor model are high, such a simulation may become computationally too intensive and time-consuming and simplifications may become necessary, in particular by limiting the number of factors and grouping the different assets into a limited number of homogeneous asset classes.

Finally, it should be noted that the simulation described is static in the sense that asset market values are simulated at time H. As such, any default at $t < H$ followed by a recovery between t and H is simply recorded as a non-default. However, if necessary and at the expense of an increasing computational burden, a dynamic feature can be obtained by dividing the time interval $(0, H)$ into evenly distributed sub-intervals.

A sequence of q simulations of the R_i^j returns ($j = 1, \ldots, q$) is then carried out, each simulation at time step j of N returns being computed in a similar manner as the one described above. At each time step j, we can value the assets whose obligor has defaulted at time j at their recovery value, and consider that they ceased existing at this date.[13]

It is worth noting that this approach focusing on the obligor's asset return may also allow including the obligor's rating migration to determine the asset value dynamics. This inclusion is addressed briefly in the last chapter dedicated to banking regulation. Securities within a trading book are subject to an additional layer in the VaR method, called Incremental Risk Charge (IRC), aiming at capturing the rating migration impact on the security market value.

---

[13] Here we disregard the case where the obligor restructures its debt which hence remains alive and displays a positive market value at the later stages $j, j + 1, \ldots, m$.

29.4.2 A Single-Factor DM Model of the Credit Risk of a Perfectly Diversified Portfolio (The Asymptotic Granular Vasicek-Gordy One-Factor Model)

From now on, we focus on DM (*default mode*) type models which aim at measuring the losses due to default on credit exposures. These models thereby disregard losses on asset market values attributable for instance to rating migration as in the previous paragraph.

We first describe the two seminal models of Vasicek (2002) and Gordy (2003) which belong to the asymptotic one-factor model family (also called indistinctly *granular* or *fine-grained*). The granular model is used to quantify the minimum capital requirements related to default risk that banks must hold to operate (see the IRB-Basel approach in the next chapter). It leads to an analytical and tractable formula for the Credit-VaR of a portfolio by simply aggregating the systematic default risks of its individual exposures when the portfolio is so well diversified that the largest individual exposure can be considered as an infinitely small share of the total portfolio exposure.

Let Z_i denote the standardized return on the assets of the ith obligor. $-DD_i$ and p_{id}, respectively, denote the distance to default and the probability of default (both assumed to be known) for this obligor. For the sake of clarity, we rewrite below Eqs. (29.3) and (29.4) for a specific i which are used to determine $-DD_i$ and p_{id} (with a default horizon $H = 1$ year):

$$DD_i \equiv \frac{ln\left(\frac{V_i(0)}{V_{di}}\right) + \left(\mu_i - \frac{\sigma_{V_i}^2}{2}\right)}{\sigma_{V_i}} ; p_{id} = Proba(Z_i \leq -DD_i) = N(-DD_i).$$

In addition, the various returns Z_i are assumed to depend on a single common factor I representing systematic risk (such as the global economic condition) and on an idiosyncratic, diversifiable risk ε_i. Let us then posit:

$$Z_i = \rho_i I + \sqrt{1 - \rho_i^2} \varepsilon_i, \tag{29.8}$$

where Z_i, I and ε_i are standardized Gaussian variates, ε_i is independent of I and of ε_j for $i \neq j$, ρ_i is the correlation between the standardized return Z_i and the global index I and Corr(Z_i, Z_j) $= \rho_i \rho_j$.

One drawback of this model is to assume ρ_i constant in addition to its one-factorial character. Under stressed market conditions such as in 2007–2009, all the ρ_i correlation parameters soared to unexpectedly high levels.

Let l_i be the incurred loss on an individual exposure due to a default of the ith obligor over a one-year horizon, α_i the recovery rate and M_i the exposure value

29.4 Credit-VaR of an Entire Portfolio (Step 3) and Factor Models 1243

(nominal of the loan/receivable or market value of the debt security at the time of default). We obtain[14]:

$$l_i = M_i \left(1 - \sigma_i\right).$$

We recall that the number of individual exposures is assumed to equal the number of obligors in the portfolio (possible by aggregating all exposures related to obligor i into a single exposure).

Let us denote by $\omega_{i,N} = \dfrac{M_i}{\sum_{j=1}^{N} M_i}$ the relative weight of obligor i within the total

portfolio exposure.

The overall loss per dollar of exposure, L_N, of the portfolio composed of N debt assets, writes, over a one-year horizon:

$$L_N = \sum_{i=1}^{N} \omega_{i,N}\left(1 - \alpha_i\right)1_{Z_i \leq -DD_i}, \tag{29.9}$$

where $1_{Z_i \leq -DD_i}$ represents the indicator function of a default, which takes on two values: 1 if the obligor i defaults within the one-year horizon and 0 if not. By substituting (29.8) in (29.9) we obtain:

$$L_N = \sum_{i=1}^{N} \omega_{i,N}\left(1 - \alpha_i\right)1_{\rho_i I + \sqrt{1-\rho_i^2}\,\varepsilon_i \leq -DD_i}. \tag{29.10}$$

The objective is to determine the probability distribution of L_N from which we can draw the VaR.

Note that $1_{\rho_i I + \sqrt{1-\rho_i^2}\,\varepsilon_i \leq -DD_i}$ is a function of two random variables, I and ε_i. We start by fixing a *given value* y for I which grounds the following result on the *conditional* law of L_N.

Since the ε_i are independent, the N random (conditional) variables $\left(1_{\rho_i I + \sqrt{1-\rho_i^2}\,\varepsilon_i \leq -DD_i} | I = y\right)$ *are independent* and their respective conditional expectations are equal to:

$$E\left[1_{\rho_i I + \sqrt{1-\rho_i^2}\,\varepsilon_i \leq -DD_i} | I = y\right] = \text{Proba}\left(\rho_i I + \sqrt{1 - \rho_i^2}\,\varepsilon_i \leq -DD_i | I = y\right)$$

$$= N\left(-\frac{DD_i + \rho_i I}{\sqrt{1 - \rho_i^2}} | I = y\right).$$

---

[14]Recall that practitioners use the following terminology: M is the *exposure at default* (EAD) and $(1 - \alpha)$ the *loss following a* default for \$1 of exposure (*Loss Given Default* or LGD). The *Expected Loss* (EL) hence writes as a function of the probability of default PD: EL $=$ EAD $\times$ LGD $\times$ PD. The loss attributable to the default of debt i is equal to: $l_i = EAD_i \times LGD_i$.

It should be noted further that the default probability of the obligor i, conditional on a given value y of I, writes: $q_i(I = y) \equiv N\left(-\frac{DD_i + \rho_i I}{\sqrt{1-\rho_i^2}} | I = y\right)$.

Let us also remark that (conditional on $I = y$), L_N is the sum of N independent but not identically distributed random variables which do not have the same (conditional) mean $q_i(I = y)$.

We now define the concept of an *asymptotic* portfolio (also called *infinitely fine-grained*) which will prove fruitful later on. A portfolio is said to be asymptotic when:

$$\sum_{N=1}^{\infty} \left(\frac{M_N}{\sum_{j=1}^{N} M_j}\right)^2 < +\infty.$$

Recall that in the equation above, M_i denotes the exposure value of each of the debt assets (securities, loans, receivables) composing the portfolio.

Intuitively, this formula means that the portfolio is composed of a very large number of assets, none of which being "significant."

For such an asymptotic portfolio and conditionally on $(I = y)$, L_N is the sum of N independent but not identically distributed $\omega_{i,N}(1 - \alpha_i) 1_{\rho_i I + \sqrt{1-\rho_i^2}\varepsilon_i \leq -DD_i}$ random variables, which allows the application of the extended strong law of large numbers[15]:

$$\lim_{N \to +\infty} \left(L_N - \sum_{i=1}^{N} \omega_{i,N}(1 - \alpha_i)q_i(I = y)\right) = 0, \text{ almost surely.}$$

This means, for practical purposes, that for a sufficiently large number N of obligors, with no "significant" ones, the random variable denoting the loss per dollar of exposure L_N maybe *approximated* by its expectation computed conditionally on a given value of the single common risk factor I, namely:

$$L_N = \sum_{i=1}^{N} \omega_{i,N}(1 - \alpha_i)q_i(I = y)$$

$$= \sum_{i=1}^{N} \omega_{i,N}(1 - \alpha_i)N\left(-\frac{DD_i + \rho_i I}{\sqrt{1 - \rho_i^2}} | I = y\right). \tag{29.11}$$

In other words, *idiosyncratic risks ϵ_i vanish by diversification* and an (asymptotic) expression of the loss L_N is obtained, which *depends on the value y of the systematic risk factor I but not on the ϵ_i.* This result will prove useful below.

We are now in a position to compute the VaR of L_N directly from the VaR of its conditional expectation. As I is Gaussian, its (1-p) quantile is equal to: $I_{1-p} = N^{-1}(1 - p)$. The p-quantile of L_N, i.e., the VaR calculated at the probability

---

[15] The proof of this theorem is difficult and beyond the scope of this book. An interested reader will find it for instance in A.N. Shiryaev, "Probability," Springer (1996), pp. 389–391.

29.4 Credit-VaR of an Entire Portfolio (Step 3) and Factor Models 1245

threshold p (e.g., 99%) at a one-year horizon, is obtained for a value of y equal to I_{1-p} (e.g., for the worst economic situation occurring with a 1% probability).[16]

If we now multiply the per dollar loss L_N by the total exposure $\sum_{j=1}^{N} M_i$, we obtain the portfolio VaR:

$$VaR_L(p, 1\ year) = \sum_{i=1}^{N} l_i q_i (I_{1-p}) = \sum_{i=1}^{N} l_i N \left(-\frac{DD_i + \rho_i N^{-1}(1-p)}{\sqrt{1 - \rho_i^2}} \right). \quad (29.12a)$$

VaR is therefore calculated as the expected loss on the portfolio, conditional on the systemic risk (global economic situation) I equal to its $(1-p)$-quantile I_{1-p}. In addition, the contribution of the obligor i to the VaR of the granular portfolio is equal to $q_i(I_{1-p})$, where this latter quantity is the conditional expectation of default for $I = I_{1-p}$.

Alternative formulas, but nonetheless similar, make use of the unconditional probabilities of default $p_{id} = N(-DD_i)$, i.e., $-DD_i = N^{-1}(p_{id})$, or directly EDFs, and write:

$$VaR_L(p, 1\ year) = \sum_{i=1}^{N} l_i q_i (I_{1-p})$$

$$= \sum_{i=1}^{N} l_i N \left(\frac{N^{-1}(p_{id}) - \rho_i N^{-1}(1-p)}{\sqrt{1 - \rho_i^2}} \right) \quad \text{or else}: \quad (29.12b)$$

$$VaR_L(p, 1\ year) = \sum_{i=1}^{N} l_i N \left(\frac{N^{-1}(\text{EDF}) - \rho_i N^{-1}(1-p)}{\sqrt{1 - \rho_i^2}} \right). \quad (29.12c)$$

It should be noted that:

– VaR is an increasing function of ρ_i, the risk attached to the obligor i being less efficiently diversified when ρ_i is high. And, as already noticed, ρ_i soars rapidly in times of crisis. To be conservative, ρ_i is set to a level that corresponds to a downturn economic situation associated with the quantile I_{1-p}. This value ρ_i should be estimated from a set of historical data including a period of stress, which makes it higher than when estimated from average economic conditions.
– For sufficiently negative I_{1-p} values, i.e., for high confidence levels p,[17] which are those chosen in practice to be conservative, the conditional probability

---

[16]Calculation rule n° 2 of Chap. 27 applies (the loss is a monotonous decreasing function of a Gaussian variate I).

[17]Specifically, for $I_{1-p} < -\frac{DD_i\left(1-\sqrt{1-\rho_i^2}\right)}{\rho_i}$, if $\rho_i \neq 0$, which is the case for the very high *p-values* chosen in practice. In the degenerate case $\rho_i = 0$, we have: $q_i(I_{1-p}) = p_{id} = N(-DD_i)$ independent from I_{1-p}: see Example 4 where we examine orders of magnitude for those values.

$q_i(I_{1-p})$ is higher than the unconditional probability of default $p_{id} = N(-DD_i)$. The probability $q_i(I_{1-p})$ is therefore interpreted as an *augmented* default probability corresponding to bearish economic conditions at the I_{1-p} level.

- In the particular case of a portfolio composed of identical exposures (same DD_i, M_i, α_i, ρ_i) we have $\omega_{i,N} = \frac{1}{N}$ and the probability of default writes $q(I) \equiv N\left(-\frac{DD+\rho I}{\sqrt{1-\rho^2}}\right)$. This simplified model, originally developed by Vasicek, is sometimes referred to as the *homogeneous* granular asymptotic model.

- The one-factor model is very simple to use because the marginal contribution of each obligor i to the VaR of the portfolio is equal to $q_i(I_{1-p})$ and depends only on the characteristics attributable to the obligor i and is *independent* from the composition of the portfolio. This makes it easier to deploy a *bottom-up* approach to assess the risk of the portfolio since the overall risk is obtained by simply aggregating the individual exposures (see Eqs. (29.11) or (29.12a, 29.12b, 29.12c)). In fact, this property is of the utmost importance for both credit risk managers and regulators. The latter have adopted this model for assessing the capital requirements related to credit risk for banks using the IRB approach (see next chapter). Therefore, the capital requirement solely depends on the intrinsic features attached to each obligor (DD_i or p_{id}, M_i, α_i, ρ_i) and not the composition of the portfolio if the latter is sufficiently diversified to be considered infinitely fine-grained. In practice, to avoid burdensome computations and restrict the number of data to be stored and maintained, banks regroup the various exposures into homogenous portfolios whose individual components bear the same characteristics.

Finally, it should be noted that diversification is not the only key assumption from which the invariance property (i.e., portfolio Credit VaR solely depends on the characteristics of each individual exposure) is derived but that the unicity of the index reflecting systematic risk is also required.[18]

Example 4

To assess the order of magnitude of the diversification effect as well as the systematic risk, let us consider a portfolio of $100 Million composed of 10,000 debt assets with the same nominal value ($10,000) and whose distance to default are all equal to 2.878, implying a probability of default over one year of 0.2% (these are, e.g., BBB+ rated securities). We assume that the standardized returns of the obligors' assets are represented by the single-factor model

(continued)

---

[18] It can be shown that only *asymptotic* portfolios with a *single* risk *factor* possess such an invariance property (see Gordy 2003).

29.4 Credit-VaR of an Entire Portfolio (Step 3) and Factor Models

(8) and that their correlation coefficients ρ_i are all equal to 0.5 (homogeneous granular portfolio). Let us also assume that in the event of default the recovery rate is equal to 40% for all assets.

The 99.7% VaR is then derived using relationship (29.12a) where:

$$I_{0.3\%} = N^{-1}(0.3\%) = -2.748; q_i(I_{0.3\%}) = N\left(-\frac{2.878 - 0,5 \times 2.748}{\sqrt{1 - 0.25}}\right)$$

$$= 4.1\%.$$

Therefore:

VaR(99.7%,1 year) $= 0.041 \times 0.6 \times 100 = \$2.46$ Million.

In addition, the (unconditional) loss expectation due to defaults over one year is:

$$E(L) = 0.002 \times (1 - 0.4) \times 100 = \$ 0.12 \text{ million}.$$

It should be noted that:

- The probability of default of an obligor i, conditional on a global economic situation among the worst 0.3% occurrences is $q_i(I_{0.3\%}) = 4.1\%$. It is *twenty times higher* than the unconditional probability of default which only amounts to 0.2%.
- In the absence of any correlation between the different asset returns Z_i (i.e., all $\rho_i = 0$), the diversification is asymptotically perfect and the loss L is certainly equal to its expectation, i.e., \$0.12 million. In such a situation, the one-year degenerated VaR is equal to \$0.12 million *for all p-values*.

Notwithstanding its tractability, the single-factor model is considered too simplistic to determine suitable economic capital requirements by sophisticated banks (in contrast with the regulatory capital requirements which are measured under the IRB approach with a single-factor model). The most technically advanced banks have developed over the years more complex models to determine their *economic capital requirements* (see Sect. 29.5 for the difference between economic and regulatory capital) such as the multi-factorial types described in Sects. 29.4.1 above and 29.4.3 and 29.4.4 below.

29.4.3 Extensions of the Asymptotic Single-Factor Granular Model

We briefly examine different extensions of the asymptotic granular single-factor model described above.

- The basic DM model disregards the potential defaults occurring beyond time H and affecting the exposures still alive in H, i.e., those with a maturity greater than H ($H = 1$ year in practice). Adjustments were thereby determined to consider these maturity effects and remedy the above weakness. They simply consist of increasing the Credit-VaR for each exposure by a coefficient that depends on its maturity and probability of default. We examine in the next chapter the adjustments proposed by the regulator under the IRB procedure.
- It is not necessary to assume a Gaussian distribution for I because, in any case, the VaR is calculated as the expected loss on the portfolio, conditional on I being equal to its $(1-p)$-quantile $I_{1-p} = F^{-1}(1-p)$. This quantile can be determined for *any* given distribution $F(.)$, not necessarily the normal distribution $N(.)$.
- The single-factor model can be extended to the multifactor case for which Eq. (29.11) is generalized in a way to ease simulations. With K standardized and orthogonalized factors (all following a Gaussian distribution),

$$\left(Z_i = \sum_{k=1}^{K} \rho_{ik} I_k + \sqrt{1 - \sum_{k=1}^{K} \rho_{ik}^2} \, \varepsilon_i \right), \text{ we obtain:}$$

$$L(I_1, \ldots, I_K) = \sum_{i=1}^{N} l_i N \left(-\frac{DD_i + \sum_{k=1}^{K} \rho_{ik} I_k}{\sqrt{1 - \sum_{k=1}^{K} \rho_{ik}^2}} \right).$$

In contrast to the single-factor model, obtaining an analytical relationship for the VaR such as (29.12a, 29.12b, 29.12c) under the multi-factor approach is much more complex because the loss $L(I_1, \ldots, I_K)$ is no longer a monotonic function of a single variable whose p-quantile is simply a function of the p-quantile of this variable. Simulations are then necessary but are, however, greatly facilitated by the factor structure of the model.

- A factor model can also be used to determine the loss distribution of an *imperfectly diversified* portfolio where at least one exposure is dominant, which violates the condition of the extended strong law of large numbers. We show how to carry out the calculation in the single-factor case. As above, we start by setting a fixed value for I, which makes losses on the different components of the portfolio independent, and thus use conditional probabilities.

The relationship (29.10) $\left(L(I) = \sum_{i=1}^{N} l_i 1_{\rho_i I + \sqrt{1-\rho_i^2} \varepsilon_i \leq -DD_i} \right)$ implies that, conditional on I, the loss $L(I)$ is the sum of N independent variables, each of them taking on two values, 0 or l_i, with respective probabilities $1 - q_i(I)$ and $q_i(I) \equiv$ $N\left(-\frac{DD_i + \rho_i I}{\sqrt{1-\rho_i^2}} \right)$.

Unlike for the asymptotic model, $L(I)$ may not be approximated by its conditional expectation (because the asymptotic condition underpinning the extended strong law

29.4 Credit-VaR of an Entire Portfolio (Step 3) and Factor Models

of large numbers does not hold) but its probability distribution can be determined as follows:

- For $N = 1$, $L(I)$ takes on 2 values, 0 or l_i, with respective probabilities $1 - q_i(I)$ and $q_i(I)$.
- For $N = 2$, $L(I)$ takes on 4 values, 0, l_1, l_2, $l_1 + l_2$,with respective probabilities $(1-q_1(I))(1- q_2(I))$, $q_1(I) (1- q_2(I))$, $q_2(I) (1- q_1(I))$, $q_1(I) q_2(I)$.
- For $N = 3$, $L(I)$ takes on 8 values, 0, l_1, l_2, $l_1 + l_2$, l_3, $l_1 + l_3$, $l_2 + l_3$, $l_1 + l_2 + l_3$, with respective probabilities $(1-q_1(I))(1-q_2(I))(1-q_3(I))$, $q_1(I)(1-q_2(I))(1-q_3(I))$, ..., $q_1(I)q_2(I)q_3(I)$. And so on.

For each value of I, $L(I)$ may take on 2^N values whose probabilities can be calculated. We repeat this procedure for different values of I (following for instance the standard normal distribution $N(0,1)$) and calculate the integral over the integrand I to obtain the loss distribution L.

The complexity of this model increases exponentially with the number N of exposures in the portfolio. To avoid this issue, a simplified version of this model assumes all the exposures are alike and match the characteristics of an "average exposure". We then posit:

- $l_i = l$ for all i, therefore $L(I)$ takes on $N + 1$ values $(0, 1, 2l, \ldots, Nl)$ only.
- The probability of default $q(I) = N\left(-\dfrac{DD+\rho I}{\sqrt{1-\rho^2}}\right)$ is the same for all obligors.

Under these conditions, for a given value of I, the total loss $L_N(I)$ obeys a binomial law:

$\text{Proba}[L(I) = 0] = \text{Proba}[0 \text{ default on all obligors}] = (1 - q(I))^N.$

$\text{Proba}[L(I) = l] = \text{Proba}[1 \text{ default affecting any obligor}] = Nq(I)(1 - q(I))^{N-1}$

....

$\text{Proba}[L(I) = j \, l] = \text{Proba}[\text{default of } j \text{ obligors}] = \binom{N}{j}q(I)^j(1 - q(I))^{N-j}$

(where $\binom{N}{j} = \dfrac{N!}{j!(N-j)!}$ is the number of different combinations of j defaults among N possible).

....

$\text{Proba}[L(I) = N \, l] = \text{Proba}[\text{default of all } N \text{ obligors}] = (q(I))^N$

To obtain the distribution of the total loss, we integrate these conditional probabilities on I, which we recall is assumed to be $N(0,1)$:

$$\text{Proba}[L = j \, l] = \frac{1}{\sqrt{2\pi}} \int_{-\infty}^{+\infty} \binom{N}{j}q(I)^j(1 - q(I))^{N-j} \, exp\left(-\frac{I^2}{2}\right) dI \quad (29.13)$$

with $q(I) = N\left(-\dfrac{DD+\rho I}{\sqrt{1-\rho^2}}\right)$.

The integral is calculated numerically by drawing a (large) Gaussian sample of values from I.

This type of model is used to value multi-asset credit derivatives such as CDS baskets (see Chap. 30).

29.4.4 Alternative Approach: Modeling the Default Dependence Structure with a Copula

Copulas constitute an alternative approach to model a dependence structure for defaults and may under some assumptions lead to results similar, if not identical, to those of the structural modeling presented above. As explained in Chaps. 26 and 27, copulas are used to account for the dependence structure between the different components of a portfolio whose marginal distributions F_i are known and to obtain a sample drawn from a joint distribution of correlated components.

We consider a one-factor Gaussian copula to determine the joint default distribution of a portfolio composed of N exposures, and show that the choice of this copula leads to results similar to those obtained with the models based on standardized returns presented in Sects. 29.4.2 and 29.4.3 above.

Let τ_i be the date of default of obligor i and $\phi_i(t)$ the cumulative distribution function of τ_i: $\phi_i(t) = Proba(\tau_i \leq t)$ for $t \geq 0$ (equal to the probability of default between 0 and t).

We will bear in mind the following: (i) a default occurrence before time H writes: $\tau_i \leq H$; (ii) the default probability between 0 and H is $p_{id} = Proba(\tau_i \leq H) = \phi_i(H)$; and (iii) $\phi_i(0) = 0$ (the obligor is not in default at time 0).

Recall that a Gaussian copula relates a standard Gaussian ($N(0,1)$) variate U_i to the random default date τ_i by the relation $N(U_i) = \phi_i(\tau_i)$. This increasing and monotonous relation has the useful propriety to preserve the quantiles since, if u_i is the p-quantile of U_i, $t_i = \phi_i^{-1}(N(u_i))$ is the p-quantile of τ_i.[19]

In addition, $\tau_i \leq t$ is equivalent to $U_i \leq N^{-1}(\phi_i(t))$ and in particular $\tau_i \leq H$ (default of obligor i between 0 and H) is equivalent to $U_i \leq N^{-1}(\phi_i(H)) = N^{-1}(p_{id})$. Denoting $\nu_i = N^{-1}(p_{id})$, the default of obligor i is therefore characterized by $U_i \leq \nu_i$ and ν_i is interpreted as a default threshold.

When considering the portfolio of N exposures, the dependence structure of defaults is determined by the correlation of the U_i. Let us assume that the latter correlation results from a one-factor Gaussian model governing the U_i, which writes with the usual notations:

$$U_i = \rho_i I + \sqrt{1 - \rho_i^2} \varepsilon_i, \text{ with } I, \varepsilon_i, \text{ and } \varepsilon_j \text{ independent standard Gaussian variates.}$$

---

[19] $N(u_i) = \phi_i(t_i) = p$ means that $Proba(U_i \leq u_i) = Proba(\tau_i \leq t_i) = p$, i.e., u_i and t_i are respectively the p-quantiles of U_i and τ_i.

29.4 Credit-VaR of an Entire Portfolio (Step 3) and Factor Models

We obviously come up with the same results as those obtained with the models developed in Sects. 29.4.2 and 29.4.3, as the assumptions are the same. Namely, U_i plays the role of the standardized returns Z_i and the default thresholds ν_i that of the distances to default $-DD_i$. Indeed, using the same notations as in Sects. 29.4.2 and 29.4.3 yields an equation analogous to (29.10):

$$L = \sum_{i=1}^{N} l_i 1_{\rho_i I + \sqrt{1-\rho_i^2}\varepsilon_i \leq \nu_i}.$$

The one-factor models developed in Sects. 29.4.2 and 29.4.3 derive accordingly. For example, in the case of an asymptotic granular (infinitely fine-grained) portfolio, the same analysis as in Sect. 29.4.2 yields:

$$L(I) = \sum_{i=1}^{N} l_i N\left(\frac{\nu_i - \rho_i I}{\sqrt{1-\rho_i^2}}\right),$$

(analogous to (29.11)), and, since $\nu_i = N^{-1}(p_{id})$, we have: $L(I) = \sum_{i=1}^{N} l_i N\left(\frac{N^{-1}(p_{id})-\rho_i I}{\sqrt{1-\rho_i^2}}\right)$, and we recover (29.12b): $VaR_L(p, 1 \text{ year}) = \sum_{i=1}^{N} l_i N\left(\frac{N^{-1}(p_{id})-\rho_i N^{-1}(1-p)}{\sqrt{1-\rho_i^2}}\right).$

In contrast with the models described in Sects. 29.4.2 and 29.4.3, the copula approach consists in positing an ad hoc copula and considering the marginal default probabilities as exogenously given (disregarding the way they are estimated or computed). This approach relies neither on a structural model nor on asset returns and their correlations. Nevertheless, it yields the same results as structural models or models of asset returns, provided that the correlations ρ_i used in the (Gaussian) copula are identified to the standardized asset return correlations.

29.4.5 Probability Distribution of the Default Dates Affecting a Portfolio

Factor models (structural or with copulas) allow not only to model the distribution of losses due to defaults, as explained in the previous paragraphs, but also that of the obligors' default dates.

Let us consider the one-factor Gaussian copula model presented in Sect. 29.4.4 above and an obligor i whose probabilities of default $\phi_i(t)$ are assumed known for different dates t. These probabilities may be either estimated by empirical methods or based on a model. In the latter case, it may be:

- An intensity model $\left(\phi_i(t) = 1 - e^{-\int_0^t \lambda(u)du}\right)$.
- A dynamic barrier structural dynamic model ($\phi_i(t)$ then results from a formula similar to those presented in Sect. 28.2.2.3).
- Or an adjusted static structural model ($\phi_i(t)$ then can be an Expected Default Frequency, (EDF).

Recall that these default probabilities $\phi_i(t)$ constitute the marginal (unconditional) cumulative distribution function of the first default date τ_i:

$$\phi_i(t) = \text{Proba}\ (\tau_i \leq t).$$

In the case of a portfolio composed of M exposures, the joint distribution of the M default dates$(\tau_1, \ldots, \tau_M)$ maybe characterized by a Gaussian copula involving M standard Gaussian variates $(U_1, \ldots, U_M)$ whose correlation matrix is denoted by ρ, and such that $N(U_i) = \phi_i(\tau_i)$.

The resulting joint distribution of default dates hence writes:

$$\text{Proba}(\tau_1 \leq t_1, \ldots, \tau_M \leq t_M) = \text{Proba}(\phi_1 \tau_1) \leq \phi_1(t_1), \ldots, \phi_M(\tau_M) \leq \phi_M(t_M))$$

$$= \text{Proba}(N(U_1) \leq \phi_1(t_1), \ldots, N(U_M) \leq \phi_M(t_M))$$

$$= \text{Proba}\big(U_1 \leq N^{-1}(\phi_1(t_1)), \ldots, U_M \leq N^{-1}(\phi_M(t_M))\big).$$

Therefore, if $N_M(.; \rho)$ denotes the *M-dimensional* Gaussian distribution whose standardized elements are linked by the copula correlation matrix ρ, we obtain the following expression for the joint distribution of default dates:

$$\text{Proba}(\tau_1 \leq t_1, \ldots, \tau_M \leq t_M) = N_M\big[N^{-1}(\phi_1(t_1)), \ldots, N^{-1}(\phi_M(t_M)); \rho\big]. \quad (29.14)$$

As explained in Chap. 26 dedicated to Monte Carlo simulations, draws $(t_1, \ldots, t_M)$ from such a law are obtained from correlated Gaussian draws $(u_1, \ldots, u_M)$ by using the transformations $t_i = \phi_i^{-1}(N(u_i))$.

Note that copula correlations may also result from a factor model. For example, a one-factor model such as the one presented above $\left(U_i = \rho_i I + \sqrt{1 - \rho_i^2}\varepsilon_i\right)$ leads to a correlation matrix ρ with generic term $\rho_i \rho_j$. Additional assumptions such as those introduced in the previous paragraphs (identical obligors, perfect diversification) allow for further simplifications. Examples of such modeling applied to the valuation of credit derivatives are presented in Chap. 30.

The joint distribution of default dates (relationship (29.14)) allows to determine the probability distribution of the first default date: $\tau_{min} = inf(\tau_1, \ldots, \tau_M)$, namely:

29.4 Credit-VaR of an Entire Portfolio (Step 3) and Factor Models   1253

$$\text{Proba}(\tau_{\min} \le t) = 1 - \text{Proba}(\tau_1 > t, \ldots, \tau_M > t), \text{therefore}:$$

$$\text{Proba}(\tau_{\min} \le t) = 1 - N_M\big(N^{-1}(1 - \phi_1(t)), \ldots, N^{-1}(1 - \phi_M(t)); \rho\big).$$

The distribution of the first default date is useful for the valuation of credit derivatives called *first to default products* encountered in the next chapter.

29.4.6 Portfolio Comprising Several Positions on the Same Obligor: Netting

Consider now a portfolio (or sub-portfolio) held by investor A containing N financial debt instruments issued by the *same obligor* B. The default correlations between these N assets are then all equal to 1 and the diversification effect presented above is nil. However, if some of these assets are likely to take on negative market values (short positions in securities, swaps, etc.), another form of risk reduction prevails if counterparties A and B are *legally or contractually* entitled to *offset* their reciprocal exposures upon default of A or B (e.g., A is liable to repay a debt to B while A has also a claim over B). Such a situation is observed for instance when A and B have entered into two interest rate swaps where at some point (from A perspective) the market value of one swap is positive (A has a claim over B) and the market value of the other swap is negative (A has a debt over B). Whenever there exists a *netting agreement,* A and B are entitled to offset those market values to come up with a single net remaining claim (or debt) to be settled upon the default of A or B.

In practice, counterparties establish *contractual netting* agreements (as the general legal framework requires too stringent conditions to enforce netting), mostly for OTC derivatives and repos, to offset, at least partially, gains, and losses upon one counterparty's default and hence minimize accordingly the residual claim to be recovered and liquidation costs.

Let $\nu_i(T)$ be the value on date T of the ith asset in the *absence* of B's default. We distinguish two cases: (i) all assets display a positive market value (long positions) and $\nu_i(T) \ge 0$ for $i = 1,\ldots, N$; (ii) some of the assets may have a negative market value $\nu_i(T) < 0$ (e.g. short positions, swaps, or forward contracts with a negative market value).

In the case where all $\nu_i(T)$ values are positive, the loss due to default writes:

$$l_1(T) = (1 - \alpha)\sum\nolimits_{i=1}^{N} \nu_i(T).$$

Netting is impossible, the incurred losses attributable to B's default add up and the correlation of defaults is equal to 1. This situation is similar to that of a portfolio composed of a single financial instrument with a market value equal to $\sum\limits_{i=1}^{N} \nu_i(T)$.

Assuming that some positions of A have positive market values and others have (or may take on) negative market values, we must distinguish two cases when B defaults:

- If *netting is enforceable*, the loss due to the default writes:

$$l_2(T) = Max\left(0, (1 - \alpha)\sum_{i=1}^{N} v_i(T)\right);$$

- If *netting is not allowed*, the loss due to the default is higher and writes:

$$l_3(T) = (1 - \alpha)\sum_{i=1}^{N} Max(0, v_i(T)).$$

$l_2(T)$ is the payoff of an option with an underlying made of N financial instruments while $l_3(T)$ is the payoff of N options (a different option for each instrument) and $l_2(T) \leq l_3(T)$. When *netting is allowed*, the inclusion of an additional asset that may take on negative market values may possibly *reduce* the overall risk (negative values being deducted from the loss due to the default).

29.5 Credit-VaR, Unexpected Loss and Economic Capital

29.5.1 Definition of Unexpected Loss (UL)

The aim of this section is to illustrate how the Credit-VaR of a portfolio (composed of various exposures) is used by the regulatory bodies and the banks' managers to determine the equity capital requirements needed to hedge the global credit risk.[20] At a later stage, we will make a distinction between regulatory and economic capital.

Capital requirements are grounded on the following principle: the (potential) losses should never (or almost never) exceed the amount of equity capital to avoid the bank's default. Therefore, to cover the losses due to the default of its credit exposures at an horizon H and a confidence level p, the bank's minimum capital requirement is equal to VaR(H, p).

Let us define the loss L *due to defaults* on the above portfolio as the difference at time H between the value that the portfolio would have in case of no defaults, $V_{H/nd,}$ and its actual value taking into account the defaults of some of its components between 0 and H: $L \equiv V_{H/nd} - V_H$.

In addition, $V_H - V_0$ represents the overall *performance* between 0 and H (a random variable that takes on a positive or a negative value) and we write:

---

[20] In accordance with capital requirement regulations and market practices, we stick to the word "capital" which encompasses not only equity but also additional items such as some subordinated or convertible bonds which display characteristics similar to equity (and are also used to cover unexpected losses).

29.5 Credit-VaR, Unexpected Loss and Economic Capital

$$Gain/Loss \equiv V_H - V_0 = \underbrace{V_{H/nd} - V_0}_{\text{Gain without defaults}} \quad \underbrace{- L}_{\text{Loss due to defaults}}$$

The overall performance thus writes as the sum of the gain calculated in the absence of default (not random if only the default risk is considered, all other risks being disregarded), minus the loss L due to default(s) (random). In practice, we assume that the expected performance is always positive (otherwise it is reasonable to consider that the bank would not do any business), so that the gain in the absence of default remains larger than the *Expected Loss (EL)* due to defaults. From this respect, *EL* is treated as a normal business cost that must be compensated by the credit spread charged to the client. The capital requirement needed to cover the default risk thus concerns only the part exceeding *EL*, called *Unexpected Loss UL* (= L–EL). The VaR is calculated on this component. More precisely, *UL* is computed at the probability threshold p (p-quantile) and horizon H as:

$$UL(p, H) \equiv VaR_L(p, H) - EL. \tag{29.15}$$

The required capital should at least cover the *UL*, i.e.

$$Required\ Capital \geq UL(p, H).$$

This ensures that the amount of capital held by the bank covers the abnormal losses due to defaults with probability p. Therefore, provided the normal operating margin covers the expected losses due to defaults, a capital amount greater or equal to $UL(p, H)$ limits the bank's probability of default to $(1-p)$ between dates 0 and H.

It should be noted that this approach is consistent with the notion that the bank's capital serves as a *buffer* (i.e., a guarantee for the bank's creditors) to cover the risks related to its business that are *not* covered otherwise (by the operating margin nor by specific hedging policies).

Example 5

Let us take again the previous Example 4 of the $100 million portfolio of 10,000 identical but imperfectly correlated exposures, whose probability of default over one year is 0.2%, the recovery rate is equal to 40% and the standardized returns obey the one-factor model (8) with the correlation coefficients ρ_i with the global economic condition I all being equal to 0.5. Under these assumptions, we obtain:

$EL = \$0.12$ M, and relation (29.12a, 29.12b, 29.12c) gives: VaR (99.7%) = \$2.46 M.

As a result, UL (99.7%) = VaR (99.7%)–EL = \$2.34 M. A capital amount of \$2.34 M is required to cover the default risk at the 99.7% confidence level, which corresponds approximately to a BBB rating (see Table 29.1).

(continued)

Recall that the results are highly sensitive to the value of the correlation coefficient. For example, in the extreme case of no correlation between asset returns and global economic conditions (all ρ_i are nil), diversification is maximal and the loss L due to defaults is (quasi) certain and equal to its expectation, i.e., \$0.12 M. In such a situation, $\text{VaR}(p) = \$0.12$ M and UL $(p) = 0$ for all values of p. It is thus logical to allocate zero capital to such a risk-free portfolio, as the expected loss of \$0.12 M due to defaults should be covered by the credit spreads charged to borrowers.

Remark The definition of the loss expressed by (29.11) is clear in the DM approach but should be made more precise in the MTM approach. In the latter case, in addition to losses due to actual defaults, one should include the loss of market value attributable to creditworthiness degradation, such as an adverse rating migration (which translates into a higher credit spread). An algebraic loss (negative if it is a gain) can then be defined by the difference between the values of the portfolio in absence and in presence of rating migrations (upgrades or downgrades of some components). This additional algebraic loss should be accounted for in the computation of the UL.

The approach presented above is used by banks' and financial institutions' top managers and regulatory bodies to set to an appropriate level their capital requirements related to credit risk (see next chapter).

From a credit risk management perspective, the horizon H is most often chosen as 1 year, and this is the horizon chosen by the regulator. One needs to make a clear distinction between regulatory and economic capital requirements.

- Regulatory capital requirement is defined by the regulator and imposes the following mandatory conditions: (i) p threshold is set to $p_{reg} = 99.90\%$ (corresponding to a rating of approximately A-), (ii) the H horizon is set to $H_{reg} = 1$ year, and (iii) the banks adopting the IRB approach must use the formula displayed in Appendix 3 (Credit-VaR) to measure their capital requirement.
- Economic capital requirement is defined by each individual bank's management who: (i) set its own threshold p_{obj} (equal to or higher than the regulatory level) and horizon H_{obj} in conjunction with their profit target and risk appetite including the rating objective, (ii) use its own formula to compute the capital requirements (for the sake of tractability usually banks start from the regulatory formula and include numerous adjustments in particular to address concentration risk which constitutes one the main issues not addressed by the Credit-VaR in DM models. These adjustments are assessed using historical or Monte-Carlo VaR).

In all cases the required capital for a bank satisfies the following relationship:

$$Required\ capital \geq \text{UL}(p_{obj}, H_{obj}) \geq \text{UL}(p_{reg}, H_{reg}).$$

We justify the above relationship as follows:

- If the required capital falls below the regulatory requirement $UL(p_{reg}, H_{reg})$ the bank is in default (or at least is not entitled to operate any longer).
- If the required capital lies between $UL(p_{obj}, H_{obj})$ and $\text{UL}(p_{reg}, H_{reg})$, the capital is not used optimally or the bank's management is too conservative when it comes to setting an appropriate level of economic capital to meet their business objectives.

Note also that having $\text{UL}(p_{obj}, H_{obj}) \leq \text{UL}(p_{reg}, H_{reg})$ is useless from a business perspective as the bank is not entitled to operate with an amount of capital below its regulatory requirements.

29.5.2 Probability Threshold and Rating

We have seen that the rating is linked to the probability of default. This allows to map the rating and the threshold p chosen for the calculation of the UL (p is equal to the theoretical probability of default when the capital is equal to UL. For an horizon H of 1 year, this mapping conforms approximately to Fig. 29.4 representing the probability distribution of L, its expectation EL, different p-quantiles, and the associated $UL(p)$ as well as the corresponding ratings. Therefore, the choice of a level of capital equal to $UL(99.90\%)$ corresponds in theory to a rating of A/A- (see Table 28.1 in previous Chap. 28). For example, a bank targeting an AA−/A+ rating should set its capital to the level $UL(99.97\%)$.

29.6 Control and Regulation of Banking Risks

The above methods also apply to the management and control of banking risks. This section describes how they fit into the prudential framework of banking activities. After a brief and general presentation of the main regulatory bodies and their criteria, we briefly describe the rules applied to the determination of minimum capital and liquidity in the banking sector, and the current Basel 3 framework aiming at improving Basel 2 identified deficiencies. We emphasize that this section does not intend to be comprehensive but solely tries to address the current banking regulation challenges and link them with some other topics addressed in this book.

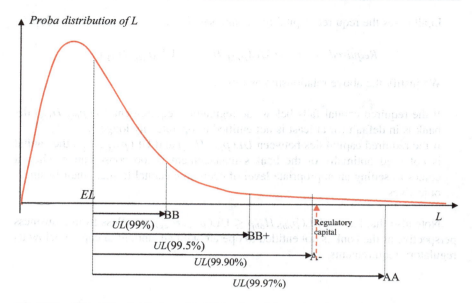

Fig. 29.4 Probability distribution of L, UL (p, 1 year) for various levels of p and corresponding ratings

29.6.1 Regulators and the Basel Committee: General Presentation

Some key definitions are necessary to avoid confusion. First, we distinguish the *regulators* (also called *regulatory bodies*) from the *supervisors* (or *supervisory bodies*). The former are in charge of setting the applicable prudential rules (e.g., Basel Committee, European Commission in the Eurozone, Federal Reserve in the USA) while the latter grant licenses to operate and oversee the enforcement of those rules (e.g., European Central Bank—ECB, US agencies such as the OCC of FDIC).[21]

Second, the word "banking" or "banks" embraces two kinds of undertakings: those which collect deposits (e.g., from retail and corporate clients) and grant loans (e.g., mortgage loans, corporate loans) and those which predominantly run financial market activities (e.g., trade bonds and /or derivatives with their clients, execute client orders on trading venues) but are entitled neither to collect deposits nor to grant credits. Barring exception, *asset managers*, *insurance companies* are outside the scope of these regulations whose objectives aim at ensuring that the risks (market, credit, operational, and liquidity) borne within banks' balance sheets, particularly in their "systemic" aspects, do not jeopardize both the security of depositors and the financial market stability and eventually weaken the economy. Note that in Europe the business model of *universal banks* is predominant. They operate under the legal status of *credit institutions* whereby they are primarily

---

[21] OCC: Office of the Comptroller of the Currency; FDIC: Federal Deposit Insurance Corporation.

29.6 Control and Regulation of Banking Risks

entitled to collect deposits, grant loans, and run financial market activities. Those institutions are the most complex to supervise as they run a large range of activities on a single balance sheet. Undertakings willing to solely run financial market activities (and refinance themselves on the market) operate as *investment firms*.

In OECD countries, banks and financial institutions are supervised by *national regulatory authorities*, which are themselves coordinated at the international level. The main supra-national regulatory body is the *Basel Committee for Banking Supervision (BCBS)*, housed by the Bank for International Settlements (BIS) and is composed of representatives of the central banks and regulatory authorities of the major industrialized countries. The main objectives of the BCBS are to define the role of national regulators in cross-border legal situations, to ensure that no bank escapes the supervision of the regulatory authorities, to promote common rules, in particular with regards to the minimal level of capital needed to operate, in order to avoid distortions of competition, and to improve the practices and security of the financial system. Barring exception, the BCBS recommendations do not have any legal enforcement status. However, the participating countries are bound to translate them into enforceable rules applicable to their relevant jurisdictions with a certain degree of flexibility to allow for some differences between countries regarding the content of the rules and their enforcement dates.

Recent experience also shows that setting minimal levels of capital and liquidity is not sufficient to prevent bank failures and should be accompanied by a strong internal control framework delineated into a set of policies and procedures (e.g., organization, accounting, IT systems). They ensure that the bank's profitability objectives are defined and attained without compromising with the regulatory requirements and that any breach is identified and remedied in a timely manner. In this respect, the BCBS frames the internal control system and the nature and frequency with which the banks should inform, through *regulatory reports*, their respective supervisors who in turn conduct regular *on-site inspections* to assess the seamless functioning of the banks under their supervision.

The set of regulatory rules in (or on the verge of being in) force in most jurisdictions is commonly known as the *Basel 3 provisions* which took over the Basel 1 and then 2 agreements and include the lessons learned from the 2007–2008 financial crisis. In Europe, those provisions have been translated into two *Capital Requirement Regulations*[22] endorsed by the *European Parliament* and directly applicable to all banks within the Eurozone. They have been complemented by two series of *Capital Requirement Directives*[23] which were translated into each EU member's own applicable laws and endorsed by their respective parliaments prior to entering into force. In the USA, the Basel 3 provisions have been gradually translated into the Code of Federal Regulations (CFR). We discuss the rules concerning

---

[22] The provisions of CRR I (2013) and CRR II (2019) were gradually enforced until 2024.

[23] CRD IV (2013) concerns mostly internal control framework, CRD V (2019) mainly addressing additional capital buffer requirements to further strengthen banks' solvency.

banks' required capital and liquidity, the other aspects of the Basel agreements being beyond the scope of this book. Barring exception, we will stick to the wording "Basel 3" and not referred to their respective legal translations previously mentioned.

29.6.2 Capital and liquidity Rules under Basel 3

Basel 3 provisions encapsulate those of Basel 2 and of the so-called Basel 2.5 (addressing specific aspects of market risk) and complete them with some new key aspects regarding liquidity. Basel 3 sticks to the *3 pillars* introduced by Basel 2.

- Pillar 1 sets the *minimum regulatory capital* and *liquidity binding requirements* applicable to all banks (regardless of their activities) to legally operate. The minimum regulatory capital covers their credit, market, and operational risks and newly limits their *leverage*. The 2007–2008 crisis illustrated that in case a bank is *rumored* to reach the point where it will eventually breach its minimum capital, its clients and/or counterparties massively withdrawing their *demand deposits* or stopping lending money in the interbank market. This can provoke a sudden default as the bank does not have sufficient *liquidity* to repay its depositors or is no longer deemed trustworthy enough to get refinancing. Therefore, Basel 3 introduces a new set of *minimum liquidity requirements* to ensure that banks are in a capacity to face a severe liquidity crisis in the short-term and also prevent them from an excessive usage of their short-term liabilities (e.g., demand deposits) to finance their long-term activities (e.g., mortgage loans).
- Pillar 2 deals with prudential supervision. Pillar 1 is comparable to a *one-size-fits-all* suit which by construction cannot perfectly fit each banking model (investment, retail, or universal bank). Pillar 2 remedies this drawback and imposes banks to tailor their *own bespoke* suit, by determining and measuring the risks they do bear and come up with an *additional minimum own level of capital* (also called Pillar 2 *capital add-on* or *economic capital*) needed to cover them. This capital add-on can only be non-negative. Once validated by the supervisory bodies, it is constantly monitored by them as it becomes the *new minimum binding level of capital*. Any enduring breach of this level corresponds to an insolvency situation and a withdrawal of the license to operate. Pillar 2 also applies to liquidity requirements but does not, at least in the Eurozone, entail an additional binding minimum liquidity requirement on top of Pillar 1 (see *solvency ratio* below). The two Pillar 2 exercises are called *Internal Capital Adequacy Assessment Process* (ICAAP) and *Internal Liquidity Adequacy Assessment Process* (ILAAP). Although the methods used to quantify the add-ons are beyond the scope of this book, we have briefly explained (in Sect. 29.5) how to measure the *economic capital* attributable to credit risk.

29.6 Control and Regulation of Banking Risks

- Pillar 3 concerns *market discipline* which is a misnomer. It is simply a series of public and comprehensive disclosures[24] addressing details of Pillar 1 computations, internal control, and risk management practices. The level of comprehensiveness depends on the bank's size and complexity. It informs the public at large about each bank's financial soundness. Basel 3 provisions have streamlined the nature and volume of disclosures to enhance transparency, and regulators and supervisors constantly exert pressure on banks to further improve the quality of the disclosed information.
Barring the exceptions recited in the remainder of this section, Basel 3 provisions have all gradually been enforced since 2013 in the EU and the USA, with a second milestone in 2019, and will be completely enforceable by 2023. Currently, regulators are complementing these provisions with the so-called "Basel 4 package" which is more a refinement of the current Basel 3 framework than a new framework.

29.6.3 Pillar 1 Capital Requirements under Basel 3

Basel 3 sticks to the principles set out by Basel 2 and either refine some of the computations (3.1) or introduces new measures on capital (Sect. 29.6.3.2).

29.6.3.1 From Basel 2 to Basel 3

Pillar 1 is the quantitative part of this reform and the only one we will (briefly) discuss as an illustration of the computation techniques we presented. It determines the amount of regulatory capital, calculated using the *McDonough ratio, a.k.a. solvency ratio, which is inherited from Basel 2.* It is expressed in the following synthetic way:

$$Solvency\ Ratio : \frac{\text{Eligible Regulatory Capital}}{\text{RWAs} = (\text{Credit} + \text{Market} + \text{Operational})\ \text{risks}}$$
$$\geq ((8\% + \text{Pillar 2 add} - \text{on}) + \text{Basel 3 Buffers}).$$

Each bank should compute its own Risk-Weighted Assets (RWAs) attributable to the sum of Credit, Market, and Operational risks and prove that the amount of eligible regulatory capital is equal to or larger than the minimum regulatory capital itself equal to (Minimum Pillar 1: 8% + Pillar 2 capital add-on + Basel 3 newly introduced buffers) × RWAs in a permanent regime and never goes below (8% + Pillar 2 capital add-on) × RWAs under severe stress. We briefly discuss here the concept of eligible regulatory capital and later (Sect. 29.6.3.2) the notion of Basel 3 buffers.

---

[24] The interested reader will search on internet the bank's name with the mention "Pillar 3," which should directly lead to the bank's appropriate document.

The *Eligible Regulatory Capital* is equal to the sum of the following 3 layers:

- *Core Equity Tier 1* (CET1, also called *Common Equity Tier 1*) which corresponds to the capital items also recognized by the international accounting standards[25] (such as book value of the shares, share premiums, retained earnings) from which some elements are disregarded by the regulators (such as fictitious Debit Value Adjustment profits as exposed in Chap. 30, intangible assets). Basel 3 imposes to hold a CET1 amount at least equal to 4.5% of the RWAs.
- *Additional Tier 1* (AT1) is composed of debt instruments which "mimic" shares (i.e., perpetual debt, with discretionary coupon payment and subordination to all other debts in case of liquidation) but do not grant their holders ownership rights. *Tier 1* (T1) is defined as: $T1 = CET1 + AT1$. Basel 3 imposes to hold a T1 amount at least equal to 6% of the RWAs.
- *Tier 2* (T2) is composed of subordinated debts. There is no limit to the minimum T2 to hold as long as the total Eligible Regulatory Capital (CET1 + AT1 + T2) is at least equal to 8% of the RWAs.

Although CET1 constitutes a legal obligation to operate, there is no obligation to issue AT1 and/or T2. Many banks solely operate with CET1, which thereby reduces their realized return on equity (since AT1 and T2 are less costly than sheer equity).

As banks differ in terms of size and complexity, Basel rules allow two sets of methods to compute the RWAs: The standardized approach and the internal model approach. The first is accounting-based (for credit and operational risk) and sensitivity-based for market risk (e.g., interest rate delta) whereas the internal model is model based. The former is designed to be extremely conservative, comes up with large RWAs and therefore high regulatory capital requirements and does not involve mathematical issues. In contrast, the internal model aims at alleviating the RWAs and relies on mathematical formulas but cannot be implemented without prior approval from the supervisory bodies. Its high costs of maintenance (task force and IT systems involved in internal control) deter many banks from adopting it. We briefly illustrate how the internal model relates to some of the mathematical formulas presented in this book.

Operational risk is the risk of losses (or gains) resulting from the failure of systems (cyber-attacks, data leakage, ransom, etc.) and/or willing or unwilling personnel misconduct (errors, fraud, employment and security practices, etc.). The internal model approach, also called *Advanced Measurement Approach*, estimates a distribution for unexpected loss grounded on operational incidents. Regulators, however, may completely discard any such internal model for operational risk in the upcoming Basel 4 framework.

As to market risk, the minimum regulatory capital requirement on the *trading book* (securities and derivatives on all asset classes) is presently calculated from a Market VaR approach (either historical or Monte-Carlo as explained in Chap. 27)

---

[25] European IFRS and American GAAP.

29.6 Control and Regulation of Banking Risks 1263

with a 99% threshold 10-day VaR, derived from a 1-day VaR, calibrated to a historical data set of a least 1 year. To this amount is applied a conservativeness coefficient that increases if the ex post daily loss too frequently exceeds the announced VaR (*back testing; see* Appendix 2). A stressed VaR component (calibrated on a historical stress) is added to come up with the final amount of required capital. *This internal model directly determines the minimum required amount of regulatory capital and the RWAs are derived by multiplying this amount by 12.5 (the inverse of 8%) for regulatory reporting, analysis purposes and to compute the solvency ratio.* The *Fundamental Reform of the Trading Book* (FRTB) will gradually enter into force from January 2022 and redefine the frontier between the *trading* and *non-trading* book with stringent conditions and replace VaR by Expected Shortfall (see Chap. 27).

Regarding *credit risk*, the minimum regulatory capital requirement is computed on each homogenous portfolio identified by the bank with the *Credit-VaR* formula presented in Appendix 3 with a confidence level of $1 - p = 99.9\%$. To assess credit risk, Basel 3 provisions allow banks to choose between two approaches:

- The *Internal Rating Based Foundation* (IRB-F) where the values p_{id} (see Appendix 3) are determined internally and the loss given default l_i is directly provided by the Basel 3 provisions (e.g., 45% for senior exposures).
- The *Advanced IRB* (IRB-A), similar to IRB-F, where the loss given default l_i is estimated by the bank. The IRB-A is designed mainly for medium and large institutions with extensive and technically competent risk analysis and control services.

Both IRB procedures are based on the granular single-factor model (see Sect. 29.4.2) which, as we have seen, has a crucial theoretical property: the overall risk (measured by CreditVaR or Unexpected Loss) results from the simple aggregation of individual risks and not from the portfolio composition (see, in particular, Eq. (29.12a, 29.12b, 29.12c)). Recall that this property depends on the assumption that the specific risks are fully diversified.

When it comes to determining the Pillar 2 capital add-on for credit risk, banks revisit the previous assumptions and re-compute the p_{id} to account for stress conditions and add some concentration factors to deal with the non-homogeneity of portfolios.

The Basel 4 package (still under discussion) contemplates to limit the capacity of the IRB approach to come up with a regulatory capital relief by introducing a *floor* whereby the RWAs could not be less than 72.5% of that computed with the standardized approach. In parallel, the standardized weight applicable to retail mortgage exposures will be lowered to incentivize most retail banks with a large mortgage activity either to go back to the standardized approach or to renounce shifting to the IRB, and hence to improve comparability and level-playing field among large retail banks.

It should be noted that a failure of the financial system can jeopardize the smooth functioning of the global economy. The financial events of 2007–2012 and the present Covid pandemic have underlined the importance of the risks that threaten the banking system and the inadequacies and limitations of the system put in place to control them: concepts, methods and models used, rating companies, control mechanisms, and so on.

Finally, Basel 3 also introduced in July 2021 a new standardized approach to measure the *exposure at default* (EAD, see the definition in Chap. 28) of a derivative portfolio, called the *Standardized Approach for Counterparty Credit Risk* (SA-CCR). The EAD is derived from a simple accounting and sensitivity-based measure of the fluctuations in the derivatives' mark-to-market value.

29.6.3.2 Specific Improvements on Required Capital Achieved by Basel 3: Buffers and Leverage Ratio

Basel 3 imposes a series of *combined buffers,* all to be mandatorily constituted with CET1 capital to further reinforce the stability of the banks as follows:

- The *conservation buffer* (CB) is equal to 2.5% of the RWAs.
- The *countercyclical buffer*[26] (CCyB), bank-specific, imposed at discretion by the supervisors. It aims at reflecting the risk during a bullying cycle of excessive credit growth and/or excessive profit from trading where "mark-to-market" and "mark-to-model" derivatives display significant values which may vanish when entering into a bearish cycle.
- Additional discretionary buffers are specifically dedicated to *Globally/Other Systemically Important Institutions* (G-SII/O-SII buffers) and also in the Eurozone the *systemic risk buffer*. These SIIs defined by specific criteria (such as the size of their balance sheet) generate systemic risk because of their size, their transnational activity, the share of the financial counterparties in their balance sheet or because they run material positions in derivatives.

Supervisory bodies may limit dividend distributions to constitute these buffers at bullish points of the business cycle, which may be used by banks to withstand a severe crisis. The combined buffers entail banks to hold in a permanent regime a minimum of 7% of CET1 (4.5% + 2.5%). For instance, CCyBs were set to 0.25% in France and Germany in 2019 (and lowered to zero during the Covid pandemic) further raising the bar to a minimum of 7.25%% of CET1. Market forces drive banks (especially G-SII and O-SII) to operate far above the minimum CET1 and ratios above 10% are routinely observed. In this respect, note that in the Eurozone the stress-testing exercise for G-SII/O-SSI steered by the ECB and/or the EBA comes up

---

[26]The interested reader may consult (https://www.esrb.europa.eu/national_policy/ccb/html/index. en.html) the current CCyB applicable in the Eurozone.

29.6 Control and Regulation of Banking Risks

with a "Pillar 2 Guidance" which adds to the combined buffers and incentivizes banks to operate above the required level of capital.

Basel 3 provisions also impose an additional minimum capital measure relying upon the leverage ratio applicable since July 2021. This ratio is in essence accounting-based and puts an additional constraint on the size of the balance sheet (and off-balance sheet) with regards to the total T1 capital, namely:

$$\frac{Tier\ 1\ Capital}{Total\ Assets + Off\ Balance\ sheet\ (weighted)\ commitments + Derivatives\ Exposures\ under\ SA - CCR}$$

$\geq 3\%$.

The leverage ratio also fosters interbank comparability as it is an accounting-based, model-free measure, including the exposure to derivatives which should be measured with the standardized approach (SA-CCR) even if the bank uses an internal model to measure its counterparty risk when computing its solvency ratio.

29.6.4 Details on Pillar 1 Liquidity Requirements

Liquidity risk was hardly taken care of in the Basel 2 framework and its control was entrusted to the care of national regulators. To limit this liquidity risk, banks are subject to constraints expressed by two ratios: the *Liquidity Coverage Ratio* or LCR (short-term) and the *Net Stable Funding Ratio* or NSFR (medium-term). Those ratios are *accounting-based* and model-free.

– The LCR is a *ratio* that measures banks' resilience to acute liquidity crises (both systemic and bank-specific) occurring within a period of 30 days:

$$LCR = \frac{HQLA}{Net\ liqudity\ outflows} \geq 100\%.$$

Compliance with this ratio requires banks to hold a portfolio composed of High-Quality Liquid Assets (HQLA) (i.e., cash and cash-equivalents asset whose nature is prescribed by Basel 3 provisions) whose amount is larger than the potential (net) liquidity outflows resulting for instance from a sharp decrease in deposits, and other similar factors (draw-downs on off-balance sheet lines, collaterals, etc.).[27]

---

[27] For instance, in the calculation of the LCR outflows within 30 days, retail deposits are weighted at 5% while financial institutions deposits are weighted at 100% as they are much more volatile.

- The NSFR (*Net Stable Funding Ratio*) (enforced in July 2021) is a ratio that limits the capacity of the bank to transform short-term resources into medium- and long-term assets: Most long-term assets (of over one-year residual maturity) requiring *a stable funding* (RSF) must be financed with sufficient *available stable funding* (ASF)[28]:

$$\text{NSFR} = \frac{\text{Available Stable Funding (ASF)}}{\text{Required Stable Funding (RSF)}} \geq 100\%.$$

where ASF and RSF are *accounting* values, in accordance with the Basel 3 provisions.

29.6.5 Additional Basel 3 Reflections and Reforms

Basel 2 was a definite improvement over Basel 1 and many believe that the 2007–2009 crisis would have been milder had the new framework been fully in place when it was launched. However, several of its aspects have been criticized and in turn been revised or supplemented.

29.6.5.1 Additional Improvements Sought for and Basel 3 Reforms

The Basel 3 package increases the ability of individual banks and the banking system as a whole to withstand severe crises. Without claiming completeness, we present a list of the objectives pursued and the reforms decided upon.

1. To strengthen international cooperation. At the European level, a supranational system of financial supervision ("European Financial Supervision System")[29] articulated around two components has been put in place. A macro-prudential component has been entrusted to the "European Systemic Risk Board," and a micro-prudential component includes a "European Banking Authority" (EBA), a "European Insurance and Occupational Pensions Authority" (in charge of implementing provisions applicable to insurance companies) and a "European Securities and Markets Authority." The Europe Agreement has established a banking union and a single supervisory mechanism transferring the ultimate supervisory power to the ECB responsible for supervising the 130 largest banks in the Euro-zone. In the meantime, the national States will have to solve any problems their banks may have on their own territory, seeking the help of the European Stability Mechanism if necessary. The EU also provided for the

---

[28] ASF: Retail deposits are weighted at 95% (reciprocal weight used for the LCR), capital amounts are weighted at 100%. RSF: Government bonds eligible to the HQLA portfolio are weighted at 0%, as they are "as good as cash."

[29] The European System of Financial Supervision is responsible for supervising banks, insurance companies, and stock markets.

29.6 Control and Regulation of Banking Risks

creation of a "Resolution Regime" backed by the Single Resolution Board (SRB) which supersedes the diverse banking insolvency regimes applicable in each member State. The SRB drafts "Resolution Plans" to ensure readiness for potential bail-in actions to some targeted banks (usually G-SII/O-SII) in jeopardy with the use of the newly created "Resolution Fund". In the USA, the resolution powers and resolution plan ("living wills") are taken over by both the Fed and the FDIC which may dismantle ailing banks.

2. To reduce the moral hazard affecting rating agencies and improve their transparency. The amount of capital required when using the *standardized method for credit risk* depends strongly on the rating of assets. However, the *rating agencies* have proved to be deficient in many respects: Conflicts of interest can bias their judgment (they are clients of rated companies), their methodology is often opaque, and their errors are notorious, particularly on the rating of structured products. They are required to comply with a number of rules aimed at bringing transparency on their rating attribution, improving sovereign debt ratings, and avoiding conflicts of interest. In that respect, in the Euro-zone, the EBA regularly issues and updates the official list of recognized rating agencies, called *External Credit Assessment Institution* (ECAI) and provides the mapping with their rating scales with that to be used when applying the standardized method for credit risk.

3. To extend supervision and regulation to "shadow" financial institutions. Basel 2 did not apply (or only partially applied) to nonbank financial institutions such as collective investment vehicles (such as hedge funds, money market funds, and other investment funds), securitization vehicles, pension funds, certain credit distributors, and institutions located in offshore markets. These undertakings of major importance, commonly referred to as *shadow banking system* or *nonbank financial intermediation* (NBFI), are likely to constitute a systemic risk.[30] They are the counterparties of banks in operations designed to save regulatory capital (banks unloading their balance sheets on agents not subject to regulatory constraints on their capital). At the end of 2019, NBFI's assets represented nearly half of those of the banking system.[31] In brief, shadow banking plays a role similar to that of banks, with the important exception of deposit collection and money creation, but it is not regulated like banks. Some actors, such as pension funds or insurance company funds, have low financial leverage and stable resources. Others, such as most hedge funds, have high leverage and engage in maturity transformation activity, financing longer-term assets with short-term resources, which is dangerous when liquidity dries up. Basel 3 provisions set clear definitions for NBFI and impose stringent capital requirements on banks' exposures to shadow banking (e.g., full deduction from regulatory capital of

---

[30] On the "shadow banking system," reference can be made to the FSB's "Global Shadow Banking Monitoring Report 2013" or the *special report of* "The Economist" of May 10–16, 2014.

[31] https://www.fsb.org/2020/12/global-monitoring-report-on-non-bank-financial-intermediation-2020/

exposure to a hedge fund or a private equity fund) to alleviate systemic risk and prevent banks from taking excessive exposures.

4. To reduce moral hazard in securitization transactions. *In* addition to the problem previously mentioned, posed by the risks transferred to save regulatory capital, securitization is subject to a potentially very serious moral hazard: "originators" or credit distributors, who expect that the risk will be transferred to third parties, have no incentive to assess it accurately and become laxer. This moral hazard, inherent in securitization was one of the reasons for many setbacks during the *subprime* crisis. To limit its effects, "originators" must now keep on their balance sheet an amount that exposes them to bear some of the default attributable to the equity tranche that absorbs, for example, the first 3% of defaults (see Chap. 30). In parallel, in the Euro-zone, the *Simple Transparent and Standardized Securitization* (STS) regime fosters transparency on the securitized collateral and imposes additional Pillar 3 disclosures for banks originating and/or investing into securitizations. This more stringent regime is not yet applicable in the USA.

5. To prevent counterparty risks. A reorganization of some markets to reduce counterparty risk has been undertaken (see Chap. 30). In particular, clearinghouses have been set up in Europe and the USA to handle standard derivatives in order to prevent cascades of counterparty risk on OTC derivatives transactions.

29.6.5.2 Limits Inherent in the Modeling of Economic Phenomena

At a more fundamental level, the use of probabilities in economics is often contested. We have already introduced the distinction between uncertainty and risk (see Chap. 21, Sect. 21.4.4.2). In this distinction, originally made by Frank Knight (1921), "objective" probabilities are associated with risk situations, while uncertainty is characterized by the impossibility of attributing such probabilities. According to the opponents of "objective probabilism," the economic variables (prices, rates, etc.) that result from the actors' unstable and unpredictable behavior are themselves essentially unstable and must be considered as not amenable to probabilities in the following sense: it is futile to search, in the statistician's toolbox, for a law (Gauss, Student-t, Pareto, etc.) describing in a reliable and stable way the behavior of a market value. It is, among other things, all the quantitative finance developed since Markowitz and Merton that is being called into question.

Schematically and without entering into an in-depth epistemological debate, two situations can be considered:

- The "normal" situation that prevails most of the time, in which a very large number of more or less rational operators, with heterogeneous and more or less independent views, exchange financial products; this situation lends itself to statistics and the calculation of probabilities.
- An "abnormal" situation where herding and/or panicking operators, homogeneous in their views and behavior, copy each other and cause bubbles or crashes; the appearance of such a situation modifies all the dynamics of financial variables.

From this perspective, which we believe to be realistic, the models work well in normal periods and fail in exceptional ones, as has been empirically observed. Unfortunately, for risk management, models are most useful in preparing for crisis situations. Therefore, in the current state of financial science, the interest of the probabilistic approach seems indisputable but is limited. It must be complemented by other approaches, such as "stress testing," for example, which are at least partially free of probabilistic assumptions. In any case, no alternative solution is proposed by the opponents of probabilistic modeling, apart from the illusory and harmful recommendation to prohibit any risk taking by financial institutions and to concentrate all risks on equity. This viewpoint is illusory because even the traditional credit business is necessarily risky. It is harmful because, since risk cannot be avoided, the problem of its allocation between the different agents inevitably arises. Diversification of risks within large financial entities with adequate risk transfer tools to locate them optimally, therefore, appears to be the best solution. The view that risk must be borne entirely by shareholders is simplistic and misleading. A continuous and complex spectrum of financial products corresponds, as it should, to the "continuity" of risks in their degree and the diversity of their nature. Between risk-free Treasury bills and risky equities, located at the two extremes of the spectrum, there are risky bonds, hybrid products such as convertible bonds and subordinated products. Moreover, derivatives products (options, futures, CDS, etc.) are often held for hedging purposes and allow a convenient and welfare-improving transfer of risk to the highest bidder. The role of the regulator is therefore to adopt sound prudential rules, based on experience and reflection, and to ensure their implementation.

29.7 Summary

- The aim of Credit-VaR is to estimate the risk affecting the creditworthiness of a financial asset or a portfolio at a given horizon H. Two methods are proposed. In the MTM (*Marked to Market*) approach, assets are evaluated at their market value at time H, and the measured credit risk encompasses the loss due to the obligor's actual default and the depreciation due to a decrease in its creditworthiness. The DM (*Default Mode*) approach focuses only on losses due to the obligor's actual default. The MTM approach applies mainly to debt securities while the DM approach is applicable to loans (e.g., mortgages) and receivables as well as securities.
- The empirical approach is based on migration probabilities such as those provided by rating agencies and on valuation by discounting the contractual cash-flows using the appropriate yield curve for assets with the same credit rating. Useful, it suffers nevertheless from some drawbacks, such as the facts that rating transition and default probabilities are unstable over time, that there is no one-to-one relationship between rating migration or default probabilities and rating, and that the rating of assets from a given issuer is not reviewed continuously and thus may be stale.

- The theoretical approach aims at determining both migration probabilities and asset values using an analytical or a structural model. Different methodologies (MKMV, CreditMetrics™) are based on the dynamics of the value of the issuer's assets. The probabilities of default and migration are deduced from this value, typically by using a Merton-style structural model from which the parameters of the process governing the total assets' value can be inferred from those governing the dynamics of the sole equity value.
- Determining these migration probabilities constitutes the first step in calculating the VaR under an MTM model. The second step requires modeling the value of the assets on date H for each rating j. This evaluation is carried out by using a structural model or discounting the RN expectation of future cash flows. It uses the concepts of *standardized return* of the issuer's assets (distributed as a standard normal variate) and *distance to default DD* (the issuer is in default when the standardized return on its assets hits or goes below the threshold–DD).
- Credit-Metrics™ uses an external rating scale such as Moody's or S&P to calculate the distance to default while, conversely, MKMV uses the distances to default calculated using an endogenous equation (relation (29.3)) to derive a rating scale. The advantage of the first method is its simplicity as it does not require the estimation of the parameters of the process governing the value of the issuer's assets. Its main drawbacks lie in the heterogeneity of rating classes (in terms of default and transition probabilities) derived from the use of external ratings which are built on different criteria and in addition may not be updated frequently enough. The second method is more demanding computation-wise but MKMV distances to default and their associated *Expected Default Frequencies* are reliable tools to predict a default in the near future.
- The case of a *portfolio* of assets issued by *different* entities is more complicated as the diversification effect comes into play. First, in the Marked-to-Market (MTM) approach to VaR, the distribution of the portfolio value on date H, $B_p^d(H)$, can be simulated using a Monte Carlo approach. Many different values of the N-dimensional vector $(R_1, \ldots, R_N)$ are simulated, with R_i denoting the return on the total value of the ith issuer's assets between 0 and H, taking into account the dependence structure (typically the correlations) between its components.
- Second, semi-analytical Default Mode (DM) models aim at measuring the portfolio losses due to default on credit exposures. The two seminal models of Vasicek (2002) and Gordy (2003) belong to the asymptotic one-factor (*granular* or *fine-grained*) model family. Granular models are used to quantify the minimum capital requirements related to default risk. They lead to an analytical and tractable formula for the Credit-VaR of a portfolio by aggregating the systematic default risks of its individual exposures when the portfolio is deemed to be so well diversified that the largest individual exposure can be considered as an infinitely small share of the total portfolio exposure.
- These granular models can be extended to take into account the potential defaults occurring beyond time H and affecting the exposures still alive in H, or to accommodate non-Gaussian distributions, or to include any number of common

Appendix 1. Correlation of Defaults in a Portfolio of Debt Assets

factors, or to cope with the case of an *imperfectly diversified* portfolio. Also, copulas (Gaussian or not) can be used as an alternative approach to model the dependence structure for defaults. Finally, some counterparties establish *contractual netting* agreements to partially offset gains and losses upon one counterparty's default and thus to minimize accordingly the residual claim to be recovered and liquidation costs.

- The Credit-VaR of a portfolio composed of various exposures is used by the regulatory bodies and the managers of banks and other financial institutions to determine the equity capital needed to hedge the global credit risk. This requirement is grounded on the principle that potential losses should almost never exceed the amount of equity capital so as to avoid the bank's default. Therefore, to cover the losses due to the default of its credit exposures at an horizon H and a confidence level p, the bank's minimum capital requirement is set equal to VaR (H, p). More precisely, the capital required to cover the issuer's default risk concerns only the *Unexpected Loss*, the part of the *total loss* that exceeds the *Expected Loss* treated as a normal business cost compensated by the credit spread charged to the debt issuer. The VaR is calculated on this *Unexpected Loss* component only.

- As the banking industry is crucial to the functioning and performance of the whole economy which it is a part of, the control and regulation of banking risks, which fit into the prudential framework of banking activities, is a major concern for the government and regulatory bodies. From the unsatisfactory Basel 1 supervisory system to that of Basel 3 (and 4), new and more stringent rules have been adopted for determining the minimum required capital of banks, for better controlling shadow banking, for ensuring a better international integration in view of mounting systemic and counterparty risks, and for increasing transparency and efficiency of bank and non-bank risk models.

Appendix 1. Correlation of Defaults in a Portfolio of Debt Assets

As previously indicated in this chapter, the term *default correlation* is ambiguous since it refers indistinctly to the correlation of default *events*, obligors' asset *returns*, *losses* due to default, or even the correlation of *default dates*. It becomes unambiguous once we have a model in which the dependence structure within a set of well-defined random variables is captured through correlation parameters. Here, we characterize the correlation of default events and its connection with the correlation of asset returns, and how this connection is derived from a structural approach.

- Let us consider two exposures a and b (from two distinct obligors). A and B denote the default events and 1_A, 1_B their respective indicator functions: 1_A is equal to 1 if a is in default between dates 0 and T and 0 otherwise; and similarly for 1_B.

By definition, the correlation of default events A and B is equal to that of their indicators:

$$\mathrm{corr}(A, B) = \mathrm{corr}(\mathbf{1_A}, \mathbf{1_B}) = \mathrm{cov}(\mathbf{1_A}, \mathbf{1_B})/\sigma\,(\mathbf{1_A})\sigma\,(\mathbf{1_B}).$$

This coefficient, also called "binary correlation coefficient," is sometimes used to characterize the default correlation. Noting that the probability of an event is equal to the expectation of its indicator $E(1_A) = \mathrm{Proba}(A)$, and also: $1_A 1_B = 1_{A \cap B}$, $(1_A)^2 = 1_A$, we have:

$$\mathrm{cov}(\mathbf{1_A}, \mathbf{1_B}) = \mathrm{Proba}(A \cap B) - \mathrm{Proba}(A)\,\mathrm{Proba}(B);$$

$$\sigma^2(1_A) = \mathrm{Proba}(A) - (\mathrm{Proba}(A))^2;\ \sigma^2(1_B) = \mathrm{Proba}(B) - (\mathrm{Proba}(B))^2;$$

$$\mathrm{corr}(A, B) = \frac{\mathrm{Proba}(A \cap B) - \mathrm{Proba}(A)\mathrm{Proba}(B)}{\sqrt{\mathrm{Proba}(A)(1 - \mathrm{Proba}(A))\mathrm{Proba}(B)(1 - \mathrm{Proba}(B))}}. \tag{29.16}$$

- Let us now show, in a static structural model, how to derive the binary correlation of default events from the correlation of asset returns. Suppose that default events A and B occur if the standardized returns of assets Z_a and Z_b are, respectively, smaller than the distances to default $-DD_a$ and $-DD_b$. Let ρ be the correlation parameter between Z_a and Z_b and $N_2(x, y; \rho)$ the cumulative joint distribution function of two standard Gaussian variates with correlation ρ. We have:

$\mathrm{Proba}(A \cap B) = N_2(-DD_a, -DD_b;\, \rho);\ \mathrm{Proba}(A) = N(-DD_a);\ \mathrm{Proba}(B) = N(-DD_b)$.

The binary correlation coefficient of default events then writes as a function of ρ:

$$\mathrm{corr}(A, B) = \frac{N_2(-DD_a, -DD_b; \rho) - N(-DD_a)N(-DD_b)}{\sqrt{N(-DD_a)N(DD_a)N(-DD_b)N(DD_b)}}. \tag{29.17}$$

In absence of asset correlation: $\rho = 0$, then $N_2(-DD_a, -DD_b; \rho) = N(-DD_a)N(-DD_b)$, and we obtain the expected result: $Corr(A, B) = 0$. Generally, $\mathrm{corr}(A, B)$ is an increasing function of ρ and $\mathrm{corr}(A, B) < \rho$.

Example 6

To assess the order of magnitude of $\mathrm{corr}(A, B)$, let us consider two debt assets a and b with equal distances to default $DD_a = DD_b = 2$, and with correlation coefficient on asset returns $\rho = 0.3$. We have:

$N_2(-2, -2;\, 0.3) = 0.002041$ (this result is obtained, e.g., using a VBA, Matlab or equivalent software); $N(-2) = 0.02275$ (the probability of default for each asset is therefore 2.275%).

(continued)

$$\text{Hence}: \text{corr}(A, B) = \frac{N_2(-2, -2; 0, 3) - N(-2)N(-2)}{N(-2)N(2)} = 6.5\%.$$

The correlation coefficient of default events $\text{corr}(A, B)$ is therefore almost five times lower than ρ (asset return correlation).

As explained in Sect. 29.4.4, a one-factor copula model based on exogenous default probabilities $\text{Proba}(A)$ and $\text{Proba}(B)$ appropriately estimated at a prior stage would lead to a relationship similar to (29.17):

$$\text{corr}(A, B) = \frac{N_2\big(N^{-1}(\text{Proba}(A)), N^{-1}(\text{Proba}(B)); \rho\big) - \text{Proba}(A)\text{Proba}(B)}{\sqrt{\text{Proba}(A)(1 - \text{Proba}(A))\text{Proba}(B)(1 - \text{Proba}(B))}}.$$

$$(29.18)$$

The transition from (29.17) to (29.18) is made simply by interpreting ρ as the copula correlation (and not as the correlation of asset returns) and substituting Proba (A) and Proba(B) for $N(-DD_a)$ and $N(-DD_b)$, respectively.

Finally, recall that the static model that assigns a probability $N(-DD)$ to default at time T is not accurate as it disregards the defaults followed by a recovery before time T.

Appendix 2. Regulatory Capital, Market VaR, and Backtesting

On a daily basis, any bank running a trading book is required to report the 1-day market VaR computed at the 99% quantile. This market VaR is a theoretical and ex ante measure of potential losses and we denote by $\text{VaR}(d)$ the 1-day VaR for day d. The minimum regulatory capital requirement attributable to market risk on day d is based on the computation of the last 60 daily VaR and is equal to:

$$\sqrt{10}\,\text{Max}\left(\text{VaR}(d - 1), \kappa\frac{1}{60}\sum_{i=1}^{60}\text{VaR}(d - i)\right)$$

The factor $\sqrt{10}$ allows translating the 1-day VaR into a 10-day VaR and κ is a *conservativeness coefficient* set between 3 and 4. Thus, the minimum capital requirement is based on the average of the reported VaRs, except in periods of crisis when $\text{VaR}(d - 1)$ is larger than κ times the average VaRs over the last 60 days.

In addition, κ is increased if the back-testing process imposed by the regulators and monitored by the supervisory bodies reveals that the empirical VaR (observed ex post) differs significantly from the reported theoretical VaR.

More specifically, the regulatory process aimed at monitoring market risks operates as follows:

- The coefficient κ is equal to $3 + \theta$, where θ is an additional term set between 0 and 1.
- The conservativeness coefficient equal to 3 is justified in particular (if the market VaR is calculated from a Gaussian model) by the skewness and kurtosis (fat-tail distributions) that characterize the distribution of the loss L (which is not Gaussian empirically).
- The coefficient θ is determined from backtesting results which compare the discrepancies between the theoretical VaR (grounded on an in-house model) reported on a daily basis $VaR(d)$ and the actually observed losses. From a practical viewpoint, one counts the number of times over the last 250 days where the actual loss L_d on day d exceeds the reported $VaR(d)$. If the theoretical VaR computed with a 99% confidence level were accurate, the number of those occurrences should not exceed 2.5. The coefficient θ is based on this number of outliers, as shown in table below:

Number of outliers:	≤ 4	5	6	7	8	9	≥ 10
Value of parameter θ:	0	0.40	0.50	0.65	0.75	0.85	1 (in red zone)

Thus, this empirical back-testing penalizes banks with "overly optimistic" market VaRs by imposing an additional level of capital requirement, which reduces the overall, systemic risk.

Appendix 3. Calculation of Regulatory Capital under the IRB Approach: Adjustment to the Infinitely Grained One-Factor Model

The Basel Committee and the regulators do not apply Eq. (29.12a, 29.12b, 29.12c) of Sect. 29.4.2 of the previous chapter (granular one-factor model) as such to determine the Credit-VaR of a debt asset portfolio and its related capital requirement. They make some adjustments to address the main weaknesses and limitations of this model.

The correlation ρ to the systemic factor I is determined as a function of the probability of default p_{id} by the equation: $[\rho_i(p_{id})]^2 = \mu_1 \frac{1-e^{-kp_{id}}}{1-e^{-k}} + \mu_2 \left(1 - \frac{1-e^{-kp_{id}}}{1-e^{-k}}\right)$ where μ_1, μ_2, and k depend on the asset class to which i belongs. Since $\mu_1 < \mu_2$, $\rho_i(p_{id})$ is a decreasing function of p_{id}, i.e., the riskier is an obligor, the lower is its correlation with the systemic factor. In addition, the function $\rho_i(p_{id})$ depends on the asset class (e.g., banks and corporations) to which the obligor (retail or wholesale borrower) belongs (6 asset classes, hence 6 types of functions $\rho_i(p_{id})$, are distinguished).

- The Credit-VaR is expressed as the difference with the average loss (it is the *Unexpected Loss UL* defined in Sect. 29.5 of the previous chapter), over a one-year horizon. To obtain the *UL* of an exposure i, its Expected Loss $l_i(p_{id})$ is subtracted from its VaR. *UL* is calculated at the regulatory confidence level (quantile) of 99.9%, i.e., a minimum capital requirement set to cover the credit risk equal to *UL* (99.9%), which corresponds to a default probability of 0.1% and, approximately, a minimum rating of A-. This *UL* is then multiplied by an additional correction coefficient of 1.06 to obtain:

$$UL_i = 1.06 l_i \left[N \left(\frac{N^{-1}(p_{id}) - \rho_i(p_{id})N^{-1}(1-p)}{\sqrt{1 - \rho_i(p_{id})^2}} \right) - p_{id} \right].$$

- The granular one-factor model, being a Default Mode model, only addresses losses due to default. An adjustment is generally mandatory to account for the spread risk (deterioration of creditworthiness affecting the debt asset's market value and hence impacting the exposure value upon default). It simply consists in increasing the Credit-VaR of the asset by a coefficient that depends on its maturity and probability of default. One empirically observes that this spread risk increases with maturity T all the more because the probability of default p_d is *low*. To conform to these observations, the Basel Committee (whose proposal was endorsed by the European Commission) proposed to multiply the UL_i of each asset by a coefficient called *maturity adjustment*:

$$c(T_i, p_{id}) = \frac{1 + (T_i - 2.5)h(p_{id})}{1 - 1.5h(p_{id})} \text{ with } h(p_{id}) = \left[0.11852 - 0.05478 \times \ln{(p_{id})^2} \right].$$

Note that for some asset classes (e.g., retail borrowers) there is no maturity adjustment $(c(.) = 1)$.

By summing up the N adjusted UL_i of the portfolio's N components, we derive the formula used to determine the capital requirement for credit risk on a portfolio, under the IRB approach:

$$Capital\ requirement = adjusted\ UL$$

$$= 1.06 \sum_{i=1}^{N} c(T_i, p_{id}) l_i \left[N \left(\frac{N^{-1}(p_{id}) - \rho_i(p_{id})N^{-1}(1-p)}{\sqrt{1 - \rho_i(p_{id})^2}} \right) - p_{id} \right].$$

The reader can compare this formula with Eq. (29.12b) in Sect. 29.4.2), to assess the effect of the different adjustments.

In addition, the value of the *RWA* (Risk-Weighted Assets) is linked to the value of the capital requirement by the following equation $RWA = $ Capital/(Total Capital Ratio), e.g., $RWA = 12.5 \times Adjusted\ UL$, where Total Capital Ratio $= 8\%$ $(1/8\% = 12.5)$.

Suggestion for Further Reading

Books

Crouhy, M., Galai, D., & Mark, R. (2006). *Risk management.* McGraw-Hill.

Girling, P. X. (2013). *Operational risk management: A complete guide to a successful operational risk framework.* John Wiley & Sons.

Gregory, J. (2012). *Counterparty credit risk and credit value adjustment: A continuing challenge for global financial markets* (2nd ed.). Wiley.

Hull, J. (2018). *Risk management and financial institutions* (5th ed.). Wiley.

*Lando D., 2004, *Credit risk modeling*, Princeton University Press.

Matten, C. (2000). *Managing bank capital: Capital allocation and performance measurement.* Wiley.

*Schönbucher P., 2003, *Credit derivative pricing models*, Wiley .

Articles and Documentation

Bank of International Settlements. (2003). *Sound practices for the management and supervision of operational risk.*

Bank of International Settlements, Various technical documents on the Basel agreements

Basel Committee on banking supervision, Various technical documents on the Basel agreements

Brunel, V. (2014, July). Operational risk modeled analytically. *Risk, 27*(7), 55–59.

Cociuba, S. E. (2009). *Seeking stability: What's next for banking regulation? Federal Reserve Bank of Dallas.*

Credit Suisse Financial Products, Credit Management Framework. (1997)

Crouhy, M., Galai, D., & Mark, R. (2000). A comparative analysis of current credit risk models. *Journal of Banking and Finance, 24*, 57–117.

Duffie, D., & Singleton, K. (1999). Modeling term structures of defaultable bonds. *Review of Financial Studies, 12*, 687–720.

Duffie, D., & Zhu, H. (2011). Does a central counterparty reduce counterparty risk? *Review of Asset Pricing Studies, 1*, 74–95.

Gordy, M. (2003). A risk factor model foundation for rating-based bank capital rules. *Journal of Financial Intermediation, 12*, 199–232.

Gorton, G. (2009, January). The subprime panic. *European Financial Management, 15*(1), 10–46.

Hull, J. (July 2010). OTC derivatives and central clearing: Can all transactions be cleared? *Financial Stability Review, 14*, 71–89.

Hull, J. (2012, Fall). CCPs, their risks, and how they can be reduced. *Journal of Derivatives, 20*(1), 26–29.

Hull, J. (2014). The changing landscape for derivatives. *Journal of Financial Engineering, 1*(2), 1–8.

Hull, J., Predescu, M., & White, A. (2004, November). Relationship between credit default swap spreads, bond yields, and credit rating announcements. *Journal of Banking and Finance, 28*, 2789–2811.

Hull, J., Predescu, M., & White, A. (2005, Spring). Bond prices, default probabilities, and risk premiums. *Journal of Credit Risk, 1*(2), 53–60.

Kealhofer, S. (2003a). Quantifying credit risk I: Default prediction. *Financial Analysts Journal, 59*(1), 30–44.

Kealhofer, S. (2003b). Quantifying credit risk II: Debt valuation. *Financial Analysts Journal, 59*(3), 78–92.

Keys, B. J., Mukherjee, T., Seru, A., & Vig, V. (2010). Did securitization lead to lax screening? Evidence from subprime loans. *Quarterly Journal of Economics, 125*(1), 307–362.

Li, D. X. (2000, March). On default correlation: A copula approach. *Journal of Fixed Income, 9*, 43–54.

Litterman, R., & Iben, T. (1991). Corporate bond valuation and the term structure of credit spreads. *Journal of Portfolio Management, Spring*: 52–64.

Mian, A., & Sufi, A. (2010). The consequences of mortgage credit expansion: Evidence from the US mortgage default crisis. *Quarterly Journal of Economics, 124*(4), 1449–1496.

MKMV, Technical Documentation

Morgan, J. P. (1997). CreditMetrics^{TM}™ Technical Document.

Turner, A. (2009, January). The financial crisis and the future of financial regulation. *The Economist's Inaugural City Lecture, 21.*

Vasicek, O. (2002, December). *Loan portfolio value* (pp. 160–162). Risk.

Websites

www.bis.org/bcbs/
www.defaultrisk.com
www.moodys.com
www.moodysanalytics.com
www.riskmetrics.com
www.standardandpoors.com

Credit Derivatives, Securitization, and Introduction to xVA

30

This chapter presents an overview of the credit derivatives and securitization techniques used to transfer and manage the credit risk affecting most portfolios. At first glance, credit derivatives may appear simple from a mathematical viewpoint but they turn out to be perhaps the most complex asset class from a risk management perspective. First, they aim at transferring the credit risk borne by instruments such as loans, receivables, and debt securities. Over the past two decades, we observed the rise and near collapse of this market which has gradually recovered since 2010 with more transparency and more experimented market participants. Second, credit derivatives disrupted the securitization business. Third, they constitute an efficient tool to price the counterparty's default risk affecting a portfolio of derivatives.

This chapter is structured as follows. Section 30.1 first introduces the main financial instrument dedicated to transferring credit risk, namely the *Credit Default Swap* (CDS). In essence, this instrument displays similar features as an interest rate swap with the key difference that the parties exchange a "default" against a fixed coupon. CDSs started trading at the end of the 1990s and their volume soared until the subprime crisis in 2007–2008 and then started to decline as this crisis unveiled some complex features which were initially ignored.

Section 30.2 then provides a simple presentation of the securitization principle, which consists in *transferring the credit risk of non-tradable assets present in one's portfolio to external investors,* and some techniques grounded on this principle.

Finally, Sect. 30.3 introduces a concept which, in the aftermath of the subprime crisis, has attracted a huge academic and regulatory attention and has a major practical implication, namely the notion of *valuation adjustment.* We address in a sketchy manner the key current debates underlying the various proposed adjustments without presenting a comprehensive analysis. Those *adjustments* were first introduced to account for the *counterparty risk* which is equivalent to *credit risk* (see Chaps. 28 and 29) except that it captures the potential loss affecting the value of any derivative in case its counterparty defaults. This initial adjustment is called CVA (Credit Valuation Adjustment) which requires the measurement of the derivative's

© The Author(s), under exclusive license to Springer Nature Switzerland AG 2022
P. Poncet, R. Portait, *Capital Market Finance*, Springer Texts in Business and
Economics, https://doi.org/10.1007/978-3-030-84600-8_30

1279

exposure, and the default probability of the counterparty to be found in the CDS market.

Enduring disrupted market conditions following the subprime crisis questioned the notion of 'risk-free rate' and even jeopardized the role played by the presence/absence of collateral when it comes to valuing derivatives. This issue was fueled by a voluminous academic research and numerous market practitioners' discussions which concluded that collateralized and non-collateralized derivatives are not equivalent from a valuation perspective and required another layer of *valuation adjustments,* such as DVA, FCA, FBA, and other adjustments presented in this chapter. "xVA" embraces the initial CVA and subsequent adjustments thereof and today "xVA" pricing is a must to satisfy the key assumptions of arbitrage-free markets.

It is, however, worth noting that, with the advent of regulations fostering mandatory clearing (see Chap. 1) now applying to the vast majority of traded vanilla derivatives and counterparties, the xVA debate has become less prominent except for non-collateralized trades.

Although xVA is presented in Sect. 30.3, interested readers may directly start there, regardless of their prior knowledge of the CDS market, and will at some point understand the role played by CDSs in xVA measurement.

30.1 Credit Derivatives

The 2000–2010 decade saw the rise and the (partial) fall of a brand new asset class constituted by credit derivatives. It complements already mature asset classes such as Equity, Interest Rates, FX, and Commodities. In most banks, the "Fixed Income" department, also called sometimes "FICC" (Fixed Income, Currency and Commodities), welcomed the new growing family of Credit Derivatives as they *apparently* behaved as, in particular, interest rate swaps, and were risk-managed in a similar way. This turned out to be an uninspired choice as we will show later on.

In essence, Credit Derivatives make it possible to transfer to a third party the default risk (in the sense explained in Chap. 29) of an asset or a portfolio of assets subject to default risk without transferring the ownership of the asset itself and even in some instances without prior ownership of the asset. They opened the door to unrivalled financial innovation and, during the decade starting in 2000, an endless list of products was created (CDS, CLN, synthetic CDO, CDO^2, CPDO, First-to-Default, etc.). At the time of publication of this book, most of these complex products have disappeared but the financial mechanism supporting them remains.

In this section, we explain the features and valuation techniques of CDSs which remain at the heart of the credit derivatives business. We then briefly present some other credit derivatives still traded today.

30.1.1 General Principles and Description of Credit Default Swaps

We start by presenting the key features applying to single-name CDSs and then focus on the contractual terminology which constitutes a key element.

30.1.1.1 Single-Name CDS: Basic Pay-off and Risk Transfer Mechanism

Single-name Credit Default Swap (CDS or single-name CDS without distinction in the remainder of this chapter) is the most common credit derivative and involves a *protection buyer* and a *protection seller* who *commit* to the following over the life of the contract: the protection buyer usually pays periodically[1] a series of fixed payments (called *running coupon* or *running spread*) to the protection seller (its counterparty) in exchange for a contractual payment (*compensation*) following a *credit event* affecting a *reference entity*.

In contrast to securities/bonds market where we mostly speak indistinctly of credit risk or default risk affecting an obligor, the phrasing credit event constitutes a contractual provision to precisely accommodate all forms of credit risk; we dedicate below a specific section to the notion of credit event as it lies at the core of the CDS market.

When a *credit event* occurs before the CDS maturity, it is worth noting that:

– The CDS ceases to exist and the *settlement* phase starts; a specific section below is dedicated to this settlement.
– The buyer ceases to pay the premium and the last coupon is then calculated pro rata *temporis* (*accrual*).

In absence of a credit event before the CDS maturity, the protection seller is not committed to make any payment. The amount of the contractual payment is called the CDS *notional* and, as for interest rate swaps, coupons are proportional to the notional (coupon expressed in percentage or basis points times the notional expressed in the relevant currency). The CDS coupon is contractually determined at inception. In the remainder of this subsection, we will stick to the terminology "coupon" while we will use "spread" when it comes to CDS valuation in Sect. 30.1.2.1.

In a single-name CDS, the *reference entity* designates a single legal entity subject to credit risk. The reference entity is the terminology used for CDSs and is equivalent to the notion of *obligor* exposed in Chaps. 28 and 29 devoted to the securities/bond universe. Those reference entities are Corporates or Sovereign States who have issued bonds and hence are subject to credit risk. Those securities are either called *reference securities*, *reference bonds* or even *reference assets*.

We address at the end of this section the case of a CDS written on an index (*CDS index*), where the index is a basket of reference entities (usually corporate obligors).

---

[1] Sometimes, the series of payments is replaced by an *up-front payment* equal to the present value of the series (see below at the end of Sect. 30.1.2.2).

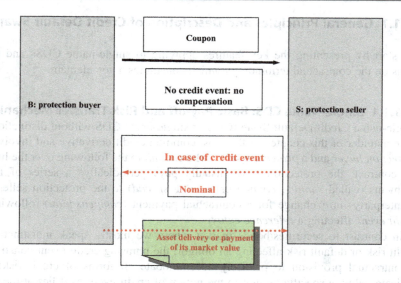

Fig. 30.1 Single-name CDS contractual cash-flows. In black: Fixed amount in block or periodical. In red: Credit event

It follows that a CDS is an off-balance sheet instrument similar to:

- An insurance contract against a certain type of default or credit risk (*credit event*) of an obligor.
- A put option on the reference security(ies) issued by a reference entity and is therefore sometimes referred to as a *Credit Default Put*.
- A *swap where* the fixed leg is the series of coupon payments and the variable leg consists of the possible compensation paid the protection seller upon the occurrence of a credit event (*see* Fig. 30.1); the payoff of the variable leg, therefore, is akin to that of an option.

In the usual terminology, buying the CDS means buying protection (buying the option and paying the premium).

Figure 30.1 illustrates the cash-flows sequence between the buyer and seller of a CDS. Thus the payoff of the CDS from the perspective of the protection buyer writes:

$$\begin{cases} -Coupon, \text{ as long as no credit event has affected the reference entity and hence the reference asset;} \\ \qquad 100 \text{ (notional)} - \textit{Value of the Reference Asset}, \text{ at the time of the credit event.} \\ \qquad \text{Or 0 if no credit event occurred.} \end{cases} \qquad (30.1)$$

Upon a credit event, if the contract is settled in cash, the protection buyer receives the difference between the notional value (100) and the market value of the reference asset. This difference is expected to be positive as, in case of a credit event, the

30.1 Credit Derivatives

reference asset of the defaulted obligor should reasonably trade below par. *Physical settlement* and *cash settlement* are theoretically equivalent provided the market value of the defaulting reference asset(s) can be determined accurately and indisputably.[2]

> **Example 1 Numerical Illustration of a Single-name CDS Mechanism**
> Let's assume that bank B has €300 million in outstanding securities issued by one of its major client x, *rated A^-* and operating in the food-processing business. To avoid such a concentration of risks, it can buy protection with, say, a €100 million CDS of 5 years maturity, implying the payment of an annual coupon of 90 bps, or €900,000 per year. Such a transaction will reduce its credit risk exposure to client x by one-third without affecting its business relationship with x and will enable it to alleviate its capital requirements. In the absence of credit event affecting x over the life of the CDS, the bank will have paid a coupon without receiving any compensation from its CDS counterparty (say, bank S). In the event of a credit event affecting x, the corresponding nominal amount of securities (regardless of their market price following the credit event) issued by x and held on bank B's balance sheet are delivered physically to bank S which in return pays €100 million corresponding the CDS notional. Assuming that the outstanding securities trade at 20% of par following the credit event, bank S would achieve a capital loss of €80 million $(100 - 20)$, only very partially compensated by the coupons that possibly accrued before the event.

30.1.1.2 Common Contractual Terminology Regarding the CDS Market

Amidst the financial turmoil triggered by the subprime crisis, many, if not all, large market participants trading CDS (and to a broader extent, credit derivatives) came to the conclusion that when it comes to addressing the credit event issue, contractual provisions are key. Under the threat of a looming credit event, many market players pinpointed that their lack of scrutiny regarding these provisions could be disastrous (see discussion in Sect. 30.1.2) and that the contractual framework should be streamlined.[3]

As such, the ISDA (*International Swap Dealers Association*) whose role is to set the contractual provisions so as to minimize potential disputes between CDS users

---

[2] In practice, both the value of the option to choose the delivered asset (to the benefit of the protection buyer in certain contracts settled by physical settlement; see below) and the risk of squeeze affecting the protection buyer (which is linked to the possibility of a lack of liquidity of the deliverable assets) must be considered. New procedures developed under the guidance of the ISDA (International Swap Dealers Association) and involving auctions (see below) aim at limiting this risk.

[3] https://www.isda.org/2011/11/08/the-first-rule-about-cds-dont-talk-about-cds-unless-youve-read-the-contract/

around the world, launched in 2009 the *Big Bang Protocol* aiming to achieve this degree of contractual standardization and also to encourage mandatory clearing (see Chap. 1) for CDS. All contractual provisions presented below apply in today's CDS markets and stem from the outcome of the Big Bang Protocol. Mandatory clearing and standardization triggered a sizeable volume and a reduction of complexity in the whole credit derivatives universe (see the statistics in 2.2.1). Barring exception, we will not address the contractual features prevailing prior to this Protocol.

(a) *The notion of credit event*

Recall that the protection seller of a CDS compensates the protection buyer if and only if a credit event occurs that directly or indirectly affects the credit worthiness of the reference entity.

By default, the ISDA provides a minimal set of contractual provisions prevailing in all CDS contracts, namely:

- *Failure to pay* on the principal or interests of the reference asset.
- *Bankruptcy* (or judicial liquidation) resulting in the dissolution of the reference entity or its inability to repay (partially or totally) its debt.
- *Restructuring* (e.g., reduction of interests or principal payment, moratorium, or rescheduling of the contractual cash flows attached to the reference entity's debt) that modifies the initial terms of the agreement between the reference entity and its creditors.[4]
- *Repudiation/moratorium* where the obligor fails to recognize its debt; this provision is mandatory on CDS whose reference entity is a Sovereign State.

We insist that these credit events mandatorily apply to all single name CDSs cleared through a CCP (Central Clearing Counterparty; see Chap. 1). Additional contractual credit events may also be freely added for bilateral non-centrally cleared (OTC) CDSs, namely:
- Non-compliance with a covenant of the debt contract.
- Down-grading of the reference entity's security(ies) below a given threshold by one or more recognized international *rating agencies*.

(b) *The notion of settlement*

Regarding the compensation settlement following a credit event, protection buyers can choose between *physical* settlement (delivery to the protection seller of one of

---

[4]This provision has been removed from CDS negotiated on US regulated markets but still applies by default on EU regulated markets since the enforcement of the Big Bang Protocol in 2009.

the eligible reference securities in exchange for the nominal value paid in cash to the protection buyer) and *cash* settlement.

Following the Big Bang Protocol, to avoid the risk of a *squeeze under the physical settlement mechanism*,[5] the ISDA has generalized the implementation of an auction process. The auction determines the undisputable price for the reference security(ies) and therefore the compensation paid in cash by the protection seller (equal to the difference between the reference debt par value and the auction price). For those market participants willing to opt for a physical settlement, the ISDA has imposed additional requirements in 2014 to further reduce the risk of a squeeze.[6]

Regarding the occurrence of a credit event, the agreement was generally the result of a discussion between the protection sellers. To clarify the situation, avoid potential disputes, ensure a seamless settlement mechanism among market players and facilitate the role of the CCP in this process, the ISDA has set up in accordance with the Big Bang Protocol in 2009 a *determination committee*. Its role consists in declaring the occurrence of a credit event and, if need be, determining the eligible reference debt securities whose price is to be determined through the auction.

30.1.2 Single-Name CDS Valuation Techniques

The valuation of a single-name CDS is part of the credit risk analysis framework developed in previous Chaps. 28 and 29. It makes use of static replication arguments and to the use of credit risk models under the reduced approach. Valuation parameters are derived in the risk-neutral (RN) world, which enables to derive implicit RN default probabilities from quoted CDS spreads. In this subsection, we devote a specific discussion to the RN framework and the replication arguments and explain how replicating a CDS differs from replicating an option.

We limit ourselves to the valuation of single-name CDSs and show that CDSs on indices obey to a large extent the same valuation rules. We provide a brief overview of CDS swaptions and Total Return Swaps at the end of Sect. 30.1.2.2.

30.1.2.1 The Valuation of a Single-Name CDS

The valuation of single-name Credit Default Swaps and that of vanilla interest rate swaps are based on fairly similar principles. We draw the reader's attention that despite their similarities, the methods exposed below do not imply that CDSs and vanilla interest swaps bear similar risks.

As vanilla interest rate swaps, single-name CDSs can be valued either by a stripping method or by static replication. We present these two approaches below

---

[5] A squeeze occurs if the number of open CDS contracts exceeds the number of existing underlying securities, which is frequently the case in the CDS market. For this reason, the Big Bang Protocol has encouraged cash settlement as the standard contractual provision.

[6] See "Single-name Credit Default Swaps: a review of the empirical and academic literature," ISDA, September 2016.

and also address the key following questions: what are the *quotation conventions*?[7] How is determined the *par spread/upfront lump sum payment* which leads to a zero value at the CDS inception? What is the market value of a CDS on any given date of its life?

We devote a specific paragraph at the end of this subsection (below) to insist on the hybrid nature of CDSs which behave as interest rate swaps when the probability of a credit event is remote (in that instance they are managed as any vanilla derivative subject to market risk only). However, when a credit event is looming, CDSs behave as an insurance contract offering protection against an ailing obligor. Due compensations may therefore be material and trigger losses which could lead the protection seller to the brink of bankruptcy.

Basic Principle: Breaking Down the CDS into Two Legs

We value separately each of the CDS legs (*see* Fig. 30.1). The first *leg (spread, or coupon, leg)* is denoted by CL. It consists of a series of fixed contractual coupons paid by the protection buyer over the life of the CDS, unless a credit event occurs prior to the CDS contractual maturity. The second leg (*default leg*), denoted by DL, corresponds to the payment due by the protection seller following a credit event.

At any time, the market value of the CDS, $V(CDS)$, is equal to the difference in the market values of the two legs, namely from the perspective of the protection buyer (long on DL and short on CL):

$$V(CDS) = V(DL) - V(CL). \tag{30.2}$$

Symmetrically, for the protection seller that is long on CL and short on DL, the value of the CDS (sold) is $V(CL) - V(DL)$.

In the following paragraphs, we will make extensive use of relationship (30.2) when using the stripping approach to derive the RN default probabilities.

CDS Pricing at Inception

Let 0 be the CDS inception date and s the spread paid by the protection buyer and contractually set at date 0. We present the key mechanisms used by market players to determine the spread s in accordance with the prevailing initial market conditions.

Recall that CDS has a value of 0 at inception to account for its non-optional feature (also applicable to any non-optional derivative). We present here how any protection seller and buyer willing to trade a CDS may use available market information to determine the coupon s required to come up with $V(CDS) = 0$ at inception under the no-arbitrage assumption. As for vanilla interest rate swaps, this process is called CDS *pricing*. This coupon (or spread) s *is* sometimes referred to as *spread at inception* or *market spread or par spread*. Once both parties agree on this spread, it contractually prevails over the life of the CDS.

---

[7] Recall that for a vanilla interest rate swap, an interest rate is quoted under the market convention that such a swap has zero value at inception.

30.1 Credit Derivatives

(a) *Pricing by static CDS replication, and summary of the different measures of credit spread*

- Let us first consider a CDS written on a given reference entity that has issued a single bond bearing a *floating interest rate*. A simple pricing approach consists in replicating the CDS spread s from the contractual spread s' of the reference asset. Consider the protection buyer's viewpoint. She may replicate the CDS cash flows by entering into a short position on the reference bond and invests the proceeds (i.e., lends on the money market) of this short position at the risk-free rate. Assume that the reference bond pays contractual interests equal to Libor + s' whilst the risk-free rate of lending is Libor.[8] The payoff of a short position on the former and a long position on the latter is equal to:

$$\begin{cases} \text{Libor} - (\text{ Libor} + s') = -s' \text{ as long as the reference asset is not in default;} \\ (100 - \text{reference bond market value}) \text{ in the event of default} \\ \qquad\qquad (\text{as the risk-free asset is at par}). \end{cases}$$

Recall that, as the reference is a bond, it is appropriate to speak of default and not of credit event. We will again discuss this difference below.

This payoff is that of a CDS buyer position, which, under the no-arbitrage assumption, implies that $s = s'$ (*see* payoff (1) above).

- Let us now consider the case of a reference asset bearing a *fixed interest rate and therefore subject to both interest rate risk and credit risk*. Pricing is performed in two steps. The first step consists in swapping the fixed interest rate against a floating interest rate + spread s'' through entering into an interest rate swap. This swap is called an *asset swap*. Recall that from a contractual perspective the reference asset and the interest rate swap are two separate contracts. The *asset swap* thus eliminates the interest rate risk (because it transforms the fixed-rate asset into a floating-rate asset), without altering the credit risk; the spread s'' over the (risk-free) floating rate reference Libor accounts for the credit risk borne by the reference asset and is called the *asset swap spread*.

The second step is similar to the one described above and assimilates the CDS spread s *to* the asset swap spread s''. Again, to prevent arbitrage, s *and* s'' should be equal. The asset swap mechanism is further detailed in Appendix 1.

---

[8] Actually, Libor is not exactly a risk-free rate, but rather corresponds to an interbank risk perceived since 2010 as around AA. However, for the ease of our discussion we assume that Libor is a good approximation of the risk-free rate.

Example 2

An investor buys a 3-year bond, issued by company (obligor) X, whose yield to maturity is 5.63%. Assume that the fixed rate of the vanilla *swap* (against 3-month Libor) is 5.1%. The *asset swap* contracted by the investor will consist in paying 5.63% to the *swap* dealer and receive Libor +0.53% ($s'' = 0.53\%$). The entire transaction (bond + asset swap) will yield Libor +0.53% provided that obligor X does not default. The spread of the CDS written on the same reference asset X protecting the investor against a credit event on company X should be equal to 0.53% (barring arbitrage) so that the risk-free rate is equal to Libor.

In this example, we assume that the spread (over Libor) of the *asset swap* is exactly equal to the difference between the yield of the bond and the fixed rate of the vanilla swap, both instruments having the same maturity. This assumption is only valid if the bond trades at par which is rarely the case. Nothing prevents to contractually set the value of the asset swap fixed rate equal to the fixed rate of the bond. However, determining the spread requires some additional computations to take into account the fact that the bond does not trade at par. Appendix 1 explains these computations, determines the theoretical value of the asset swap spread and makes the example more realistic.

If the aforementioned static replication method seems appealing from a theoretical perspective, we observe in practice that the difference between spreads s and s'' (or s') is rarely equal to zero. The difference $(s - s'')$ is called the *CDS-bond basis* by market practitioners. We insist that the CDS-bond basis does not violate the arbitrage-free assumption and may be explained by various factors such as:

- The counterparty risk, i.e., the risk of default of the protection seller (which reduces the value of the CDS spread s and does not affect s'');
- The liquidity prevailing on the CDS market, which makes it easier to sell protection rather than entering a short position on the reference asset (short-selling requires to buy back the reference asset at a shorter time horizon than the maturity of the CDS);
- The contractual and strictly delimited provisions defining a credit event for a CDS which do match the events of credit affecting a bond (e.g., if the credit events for a CDS solely embrace "bankruptcy" and "failure to pay" from a contractual perspective and, at some point, the obligor of the reference asset decides to forfeit on its debt, no credit event is triggered on the CDS and the protection buyer should continue paying the coupon whilst the bond would trade at a level close to zero);
- The option, sometimes granted to the protection buyer, to deliver the cheapest reference asset from a contractually predetermined basket (which increases the value of s *and does* not affect s''). Note, however, that since the enforcement of the Big Bang Protocol and mandatory clearing, the cheapest-to-deliver-option has almost disappeared from CDS customary contractual provisions.

30.1 Credit Derivatives

We have seen that an empirically observed CDS-bond basis and the approximate nature of the relationship $s = s''$ (or $s = s'$) do not contradict the no-arbitrage assumption but make more complex valuation techniques necessary, analogous to those implemented in credit risk management. *Moreover, the differences observed in practice between credit spreads justify the following alternative definitions.*

Definition 1: Ibor-Spread *When the risky security (bond) bears a floating vanilla-indexed rate, the credit spread is defined simply as the contractual spread over the vanilla rate (e.g. Libor, Euribor, OIS) used as the reference for the floating rate.*

Definition 2: Yield Spread *For a fixed-rate security (bond), the credit spread is often defined as the difference between its yield to maturity and the yield to maturity of a risk-free security with the same maturity (benchmark). The benchmark may be a government security or a fixed swap rate.*

This is the definition most frequently used in practice and implicitly used here, unless otherwise specified. As it involves a comparison of yields to maturity, the yield spread has the disadvantage of not taking into account the term structure of interest rates. For this reason, it is theoretically better to calculate the credit spread using the zero-coupon rate curve in accordance with definition 3 below.

Definition 3: Z-Spread or Option-Adjusted Spread (OAS)[9] *The z-spread on a risky security (bond) is the spread which, when added to the risk-free zero-coupon rates, recovers its quoted price (see Example 3).*

Example 3

Let us consider a 3-year bullet bond with a fixed rate of 8%, which has just paid its coupon and trades at 104.00 for a nominal value of 100 (in this instance, the dirty price equals the clean price). In addition, the risk-free zero-coupon yield curve is as follows: 1 year: 4.00%; 2 years: 4.50%; 3 years: 5.00%.

So we have: $104 = \frac{8}{(1+4.00\%+Z)} + \frac{8}{(1+4.50\%+Z)^2} + \frac{108}{(1+5\%+Z)^3} \Rightarrow Z = 2.49\%$

(Z can be determined by a solver).

Definition 4: Asset Swap Spread *This is the spread of the asset swap on this security (bond). (see detailed example 2 and Appendix 1). In practice, it is very close to the Z-spread (definition 3) and differs from it more as the security (bond) trades away from par.*

---

[9]OAS is a misnomer as no optional calculation is actually involved.

Definition 5: CDS Spread *This is just the premium (coupon) of the CDS that hedges the credit risk borne by the considered security (bond).*

Following the *ISDA Big Bang Protocol* of 2009, CDS quotation standards have shifted to standardized coupons, and enduring pressure was put on market participants to shift CDS trading to regulated markets (see Chap. 1) and mandatory clearing. Since then, the Dodd-Frank Act in the US and the EMIR regulation in the EU have been gradually enforced to increase price and volume transparency on the CDS market.

Before 2010, the prevailing format was to have CDS coupons determined as explained previously to satisfy, at inception, $V(DL) - V(CL) = 0$. Today, the market sets standardized fixed CDS coupons taking on the values: 20, 100, 300, or even 500 bps (paid on a running basis). Under these standards, it is unlikely that $V(DL)-V(CL)$ is nil at CDS inception. Therefore, an upfront lump sum payment (*up-front premium*), paid at inception, should be added (or subtracted) to the spread leg so that: $V(DL)$ +/− up-front premium − $V(CL) = 0$. By convention, the model developed by JPMorgan (see below)[10] is used to determine the up-front premium from the standardized CDS coupons and the prevailing market conditions. In other words, we see in the next section that there is a one-to-one relationship between these two quotation frameworks. The definition of a CDS coupon is thereby refined alternately as:

Definition 5.1: CDS coupon *is the contractual coupon determined at inception to come up with a CDS value of 0, paid on a running basis protecting against a credit event on the reference entity.*

Definition 5.2: CDS coupon *is the coupon set by the CDS quotation market standards paid on a running basis and protecting against a credit event on the reference entity. The upfront premium (algebraically) added to the CDS fixed leg nullifies the CDS value at inception.*

(b) *Stripping Method*

For the sake of simplicity, we retain Definition 5.1 above where the CDS spread (or coupon) $s(0)$ is computed at inception under the prevailing market conditions so that $V_0(CDS) = 0$ (absence of upfront premium). At the end of this subsection, we explain how to transform this spread (coupon) into the standardized coupon convention prevailing in today's market.

The value of the cash-flows constituting CL and DL may be calculated by using one of the discounting methods relevant in the context of counterparty risk (see Chaps. 28 and 29). For example, one can calculate the risk-neutral (RN) or forward-neutral (FN) probabilities of default implicit in the spread curve for zero-coupon

---

[10]The conversion is done with a flat spread curve and a standardized recovery rate of 40% for senior unsecured debt and 20% for subordinated debt (see the discussion below).

30.1 Credit Derivatives

securities issued by the obligor or by other obligors bearing similar risks (*see bond stripping* in Chap. 28) and then calculate the RN-FN expectations as in Chaps. 27 and 28.

Let $\gamma^*(t)$ be the survival RN-FN probability on date t. $p*_\theta = \gamma^*(\theta - 1) - \gamma^*(\theta)$ is, therefore, the probability of default[11] between dates θ–1 and θ. $\phi^*(\theta) = 1 - \gamma^*(\theta)$ denotes the cumulative probability of default between 0 and θ; α denotes the recovery rate so that in case of default the amount $(1 - \alpha)Nom$ is recovered, where *Nom* denotes the nominal value of the risky security (bond). Furthermore, s is expressed as an annual rate and therefore the coupon actually paid at each maturity is equal to $Nom \times s(0) \times D$, where D *is the time lag* between two payments (e.g. $D = 0.25$ for quarterly payments). Finally, r_θ denotes the zero-coupon risk-free rate for maturity θ.

We therefore come up with:

$$V(DL) = Nom \sum_\theta e^{-r_\theta \theta}(\gamma^*(\theta - 1) - \gamma^*(\theta))(1 - \alpha); V(CL)$$

$$= Nom \sum_\theta e^{-r_\theta \theta}\gamma^*(\theta)s(0)D. \tag{30.3a}$$

In a continuous-time model where the credit event occurs at the first jump of a Poisson process of RN-FN intensity equal to λ (see Chaps. 28 and 29), we have: $\gamma^*(\theta)=e^{-\lambda\theta}$. Therefore, assuming a flat zero-coupon risk-free rate curve at level r, *we obtain* for a CDS written on a reference asset with nominal *Nom*:

$$V(DL) = -Nom \int_0^T e^{-r\theta}(1 - \alpha)d\gamma^*(\theta)$$

$$= Nom(1 - \alpha)\lambda \int_0^T e^{-(r+\lambda)\theta}d\theta,$$

$$V(CL) = Nom\, s(0) \int_0^T e^{-(r+\lambda)\theta}d\theta,$$

which yields

$$V(DL) = Nom\,(1 - \alpha)\lambda\frac{1 - e^{-(r+\lambda)T}}{r + \lambda}; V(CL) = Nom\, s(0)\frac{1 - e^{-(r+\lambda)T}}{r + \lambda}. \tag{30.3b}$$

We can deduce the spread $s(0)$, called *par spread* or *market spread*, which makes the initial value of the CDS equal to 0 and remains contractually fixed for the entire life of the CDS. In discrete time, (30.3a) implies:

---

[11] We use here "default" instead of "credit event," as is customary in CDS terminology, to ease the exposition.

$$s(0) = \frac{(1 - \alpha)\sum_\theta e^{-r_\theta\theta}(\gamma * (\theta - 1) - \gamma * (\theta))}{D\sum_\theta e^{-r_\theta\theta}\gamma * (\theta)}. \tag{30.4a}$$

In continuous time (30.3b) leads directly to the credit triangle formula:

$$s(0) = (1 - \alpha)\,\lambda. \tag{30.4b}$$

In the current market quotation standards, when the CDS trades with an up-front premium plus a running standardized spread (coupon) s_s, we have:

$$Up-front\ premium + V(CL_s) = V(DL), \text{i.e.} :$$

$$Up-front\ premium = V(DL) - V(CL_s). \tag{30.4c}$$

Under the theoretical conditions set out in Eq. (30.3b) and in continuous time, we have:

$$Up-front\ premium = Nom((1 - \alpha)\lambda - s_s)\frac{1 - e^{-(r+\lambda)T}}{r + \lambda}. \tag{30.5a}$$

In discrete time, for the calculation of $V(DL)$ and $V(CL_s)$ we apply relation (30.3a) or a more accurate version of it (e.g., the JP Morgan-ISDA model exposed in Sect. 30.1.2.1.4).

Finally, in the rare case where the CDS contractually pays no spread (coupon), the value of the up-front premium is given by relation (30.5a) with s_s set equal to zero:

$$Up-front\ premium = Nom(1 - \alpha)\lambda\frac{1 - e^{-(r+\lambda)T}}{r + \lambda}. \tag{30.5b}$$

Par Spread and Value of a Single-Name CDS at any Time T

Let us consider, at time t, a CDS with maturity $t + T$ and contractual spread (coupon) $s(0)$ (determined at inception at date 0). Let $s(t)$ be the CDS spread that makes the market value of this CDS at time t equal to 0: $s(t)$ is the par spread or market spread determined on the basis of prevailing market conditions. In other words, a protection buyer who contracted at $s(0)$ and willing to exit the contract at t would sell the protection at $s(t)$. In general, the CDS value, equal to $V_t(CL) - V_t(DL)$ where the RN-FN are determined with the prevailing market conditions at time t, is not zero as the market spread $s(t)$ is different from $s(0)$. The difference $s(t) - s(0)$ may be due to various reasons: the time decay since inception, the potential evolution of the credit risk affecting the reference entity (e.g., a change in the obligor's rating), the variation of market prices of risk (due to a change in the market's average risk aversion). The evolution of the risk-free rate curve plays also a role but to a lesser extent.

The spread $s(t)$ can either be observed or calculated. When the CDS is quoted with a coupon format prevailing before the Big Bang Protocol (i.e., without upfront premium and standardized coupons), $s(t)$ *is directly observable*. The value

30.1 Credit Derivatives

1293

of the CDS at time t from the protection buyer's perspective (contractually committed at spread $s(0)$) is obtained by computing the sum of the discounted cash flows $D[s(t)-s(0)]$ over the remaining maturities $t + \theta$, $\theta = \theta_1, \ldots, \theta_T$, where D denotes the time lag between two coupon payments and *Nom* the CDS notional:

$$V(t, \text{CDS}) = Nom\, D\sum_{\theta} e^{-r_\theta\theta}\gamma^*(t + \theta)\,[s(t) - s(0)] \qquad (30.6a)$$

in discrete time. $\gamma^*(t + \theta)$ is determined in accordance with the prevailing market conditions at time t.

The CDS value from the protection seller's viewpoint is clearly given by the same expression changed in sign.

Alternatively, in a continuous-time model based on a Poisson process where the RN intensity is $\lambda = \frac{s(t)}{1-\alpha}$, the CDS value from the protection buyer's perspective position writes[12]:

$$V(t, \text{CDS}) = Nom\, [s(t) - s(0)] \int_0^T e^{-(r+\lambda)\theta}d\theta$$

$$= Nom\, [s(t) - s(0)]\, \frac{1 - e^{-(r+\lambda)T}}{r + \lambda}. \qquad (30.6b)$$

We draw the reader's attention that the protection buyer (the same remark applies to the seller) can exit the CDS contract in two ways: either by receiving on a running basis the coupon *s(t) until maturity, or paying/receiving (depending on the value of s(t)−s(0))* a lump sum amount equal to $V(t, CDS)$ determined in accordance with (30.6a) or (30.6b).

When $s(t)$ cannot be reliably observed, or when it comes to *pricing* the swap at t, we can derive its theoretical value from relations (4) (expressed at time t). Then, we obtain the market value of the CDS by using expressions (30.6a) or (30.6b). We can also use relations (3) (and extract from the reference asset values the RN-FN probabilities prevailing at time t) and calculate the CDS value as the difference between the values of its two legs:

$$V(DL) = Nom\, (1 - \alpha)\sum_{\theta} e^{-r_\theta\theta}(\gamma^*(\theta - 1) - \gamma^*(\theta));$$

$$V(CL) = Nom\, D\sum_{\theta} e^{-r_\theta\theta}\gamma^*(\theta)s(0);$$

$$V(\text{CDS}) = V(DL) - V(CL).$$

---

[12] The integral, which expresses the present value of cash flows received between t and $t + T$, uses the fact that: $\int_t^{t+T} e^{-a(x-t)}dx = \int_0^T e^{-a\theta}d\theta$ by the change of variable $(x-t) = \theta$.

Example 4

Let us consider a CDS at time t with a nominal value of €100, contracted at a previous date 0, whose coupon is 0.68%, with three years remaining maturity, and with an A-rated reference bond. The recovery rate on the reference asset is estimated at 43%. For the sake of simplicity, we assume that the coupon payments occur on an annual basis $(D = 1)$. The zero-coupon rate curve for A-rated securities and the zero-coupon risk-free rate curve (stripped from the interest *swap* fixed-rate curve) prevailing at time t are assumed to be known and provided in the following Table:

Maturity:	1 year	2 years	3 years
ZC rate (swap curve): r_θ	4.00%	4.20%	4.30%
ZC rate for A-rated bonds	4.60%	4.90%	5.10%
Spread s_θ	0.60%	0.70%	0.80%

– First, the cumulative RN probabilities of credit event $\phi^*(\theta)$ are derived using the credit triangle relationship $[s_\theta = \frac{1}{\theta} \phi^*(\theta)(1 - \alpha)]$ and the corresponding survival probabilities $\gamma^*(\theta) = 1 - \phi^*(\theta)$:

$$\phi^*(1) = 1.75 \times 0.6\% = 1.05\%, \text{i.e.} \gamma^*(1) = 0.9895;$$

$$\phi^*(2) = 2 \times 1.75 \times 0.70\% = 2.45\%, \text{i.e.} \gamma^*(2) = 0.9755;$$

$$\phi^*(3) = 3 \times 1.75 \times 0.80\% = 4.2\%, \text{i.e.} \gamma^*(3) = 0.958.$$

– The value of the two legs and then the value of the CDS are then deduced.
$$V(DL) = (1 - \alpha)Nom\sum_{\theta=1}^{3} \frac{1}{(1+r_\theta)^\theta}(\gamma^*(\theta - 1) - \gamma^*(\theta)), \text{i.e.:}$$

$$V(DL) = 0.57 \times 100 \times \left[\frac{1}{1.04} \times 0.0105 + \frac{1}{(1.042)^2} \times 0.0140 + \frac{1}{(1.043)^3} \times 0.0175\right]$$

$$= €2.190.$$

$$V(CL) = s(0)Nom\sum_{\theta} e^{-r_\theta\theta}\gamma^*(\theta), \text{i.e.:}$$

$$V(CL) = 0.0068 \times 100 \times \left[\frac{1}{1.04} \times 0.9895 + \frac{1}{(1.042)^2} \times 0.9755 + \frac{1}{(1.043)^3} \times 0.958\right]$$

$$= €1.832.$$

$$\text{Therefore}: V(CDS) = V(DL) - V(CL) = €0.358.$$

– On the market, as date t, the spread $s(t)$ of the CDS with a three-year remaining maturity and written on the A-rated reference entity is consistent

(continued)

30.1 Credit Derivatives

> **Example 4** (continued)
>
> with the theoretical relationship (30.4a) and writes: $s(t) = V(DL)/\sum_{\theta=1}^{3} \frac{1}{(1+r_\theta)^\theta}\gamma^*(\theta)Nom$.
>
> The denominator being equal to $V(CL)/s(0) = 1.832/0.0068 = 269.4$, we have:
>
> $$s(t) = 2.190/269.4 = 0.813\%.$$
>
> We observe that the value of the CDS can be also derived from this theoretical spread and relation (30.6a).
>
> Indeed, $V(CDS) = Nom\,[s(t) - s(0)]\sum_{\theta=1}^{3}\frac{1}{(1+r_\theta)^\theta}$ implies:
>
> $$V(CDS) = 100 \times 0.00133 \times \left[\frac{1}{1.04}\times 0.9895 + \frac{1}{(1.042)^2}\times 0.9755 + \frac{1}{(1.043)^3}\times 0.958\right]$$
>
> $$= €0.358.$$

JPMorgan Model: ISDA

The ISDA imposed a market standard of the discrete-time model (see Eqs. 30.3a, 30.4a and 30.6a). It frames the time lag D_θ between two CDS coupon payment dates (D_θ = (Number of days between $\theta - 1$ and θ)/360)). The cash flow due at time θ is therefore equal to $Nom \times spread \times D_\theta$. In addition, the default is assumed to occur at time $D_\theta/2$, which implies that a coupon equal to $i \times D_\theta/2$ is perceived by the protection seller.

We then have for the CDS valuation with a contractual spread of $s(0)$:

$$V(CDS) = V(DL) - V(CL),\text{ with :}$$

$$V(DL) = Nom\,(1 - \alpha)\sum_\theta e^{-r_\theta\theta}\,[\gamma^*(\theta - 1) - \gamma^*(\theta))],$$

$$V(CL) = Nom\,s(0)\sum_\theta e^{-r_\theta\theta}\left\{D_\theta\gamma*(\theta) + [\gamma*(\theta - 1) - \gamma*(\theta)]\frac{D_\theta}{2}\right\}.$$

In addition, the *par spread* $s(t)$ verifying $V(DL) - V(CL) = 0$ writes:

$$s(t) = \frac{(1 - \alpha)\sum_\theta e^{-r_\theta\theta}[\gamma*(\theta - 1) - \gamma*(\theta)]}{\sum_\theta D_\theta e^{-r_\theta\theta}\{\gamma*(\theta) + 0,5[\gamma*(\theta - 1) - \gamma*(\theta)]\}}.$$

The same model also applies to the valuation of a CDS bearing a standardized spread (20, 100, 300, or 500 bps). The initial value of the CDS then is not zero, since the contractual spread differs from the *par spread*: one of the parties, therefore, pays to the other an upfront lump sum equal to $V(CL) - V(DL)$. These formulas apply at time $t \le 0$, by interpreting the par spread $s(0)$ as the standardized contractual spread s_c.

Additional Provisions Regarding the CDS Recovery Rate

In all the previous formulas, α denotes the *recovery rate* (and $(1 - \alpha)$ the *loss given default*). Let us discuss the determination of this parameter when valuing a CDS. In absence of default, α remains unknown and both Relations (30.4a) and (30.4b) are equations with two unknowns (namely α and, respectively, $\gamma*$ or λ). Example 4 also introduces an assumption regarding the parameter α to determine the RN-FN default probabilities and finally the CDS value. Considering all these elements, it would be more appropriate to speak of an *expected recovery rate* when the default is remote.

When the occurrence of a default becomes more likely or is looming, upfront payments observed in the market reach values close to the recovery rate likely to be achieved at the auction.

Let us consider again the credit triangle formula, $s(0) = (1 - \alpha) \lambda$ and address the nature of the information embedded into the quoted coupon $s(0)$. Does the market convey through $s(0)$ a RN-FN default probability including a risk premium (assuming a constant recovery rate)? Or does it quote a price about the RN-FN expected loss $(1 - \alpha) \lambda$ (as λ is assimilated to a default probability) where those two parameters cannot be distinguished? There is no clear answer to this question and opinions from market participants vary. However, as a market convention, α is *assumed* to be equal to 40% on senior debt and 20% on junior debt, so as to derive RN-FN default probabilities in a "steady state" regime (i.e., when default is remote). We emphasize that this is only a market convention and not a contractual provision.

When a default is looming (noting however that the frontier with the "steady state" regime is fuzzy) there may be a shift in the informational content from an initial state where $s(0)$ represents a sheer RN-FN default probability to a state where $s(0)$ represents an RN-FN expected loss. Again, this proves the hybrid and complex nature of the CDS, which behaves as an interest swap when the credit event is remote and as a bond exposure (or an insurance payout) when it is close.

Specificities of CDS Hedging

We elaborate on CDS hedging strategies implemented on the market and present their respective risks and limits. In Sect. 30.1.2.1, we presented the static hedging strategy which consists in buying either a floating-rate (e.g. IBOR) bond with a spread s' or a package made of a fixed-rate bond and an asset swap with a spread s'', and simultaneously entering into a long CDS position with a coupon s. The cost of funding the bond is equal to the risk-free rate reference (e.g. IBOR). We proved that theoretically the relation $s = s' = s''$ should hold in an arbitrage-free market. The reverse strategy could also be carried out by shorting a floating-rate bond (or a fixed rate bond), investing its proceeds on the money market at the risk-free rate, and selling a CDS protection.

From the bond holder's viewpoint, the strategy offers a protection both against the default and market risks attributable to the credit worthiness fluctuations of the reference entity. However, the bondholder remains exposed to the market risk stemming from the CDS-bond basis. For instance, a bondholder being long in a CDS would make a profit in case the CDS-bond basis tightens (e.g., becomes less negative) and conversely make a loss if it widens.

30.1 Credit Derivatives

The static hedging strategy entails to buy the bond and therefore requires a funding capacity. It also sets market risk limits to avoid running a too large exposure to the gains or losses stemming from the CDS-bond basis fluctuations. This resembles bond trading, barring that the bond is protected against default.

Market participants often consider an alternative type of hedging strategy called *maturity mismatch* which implies CDSs only and therefore does not necessitate a funding capacity. It consists in selling the CDS protection on a given maturity (e.g., 5 years) and immediately purchasing the CDS protection with the same characteristics (notional, reference entity) but on a shorter maturity (e.g. 3 years). This strategy provides a hedge against the default of the reference entity *up to the CDS long position maturity* but exposes the investor to fluctuations in the *CDS forward spread* (e.g., the CDS forward spread beginning in 3 years for a 2-year period). This strategy may be scaled up to a portfolio of CDSs composed of various reference entities and maturities. However, as for a bond portfolio trading activity, the default risk is monitored by a set of limits avoiding too much concentration on any of the reference entities composing the portfolio. This strategy is in many aspects similar to managing the market risk of an interest rate swap portfolio composed of various maturities.

Some market participants implement imperfect hedging strategies by relaxing the condition bearing on notionals and are long a CDS protection with a notional smaller than the CDS short position. Indeed, the CDS coupon gap between the short and long positions is all the larger because the (upward sloping) CDS market spread curve is steep.

The previous discussion is summarized in Table 30.1 and illustrates the hybrid character of the CDS.

Table 30.1 Summary of usual hedging strategies

Hedging strategy	Static: Long the bond & long CDS protection	Static: Short the bond & short CDS protection	Maturity mismatch short the long-maturity CDS and long the short-maturity CDS
Default risk	Nil.	100% of the CDS notional.	No default risk up to the CDS long protection with the shortest maturity.
Market risk	Gain if CDS-bond basis tightens. Loss if CDS-bond basis widens.	Loss if CDS-bond basis tightens. Gain if CDS-bond basis widens.	Change in CDS forward spread.
Conditions	Sufficient funding capacity to purchase and hold the bond. No credit risk limits.	Sufficient funding capacity to buy back the bond. Market risk limits to frame CDS-bond basis fluctuations. Credit risk limits to frame the exposure on the reference entity.	No funding capacity to hold reference asset. Market risk limits to frame CDS forward spread fluctuations. Credit risk limits to monitor the exposure on the reference entity beyond the CDS long protection with the shortest maturity.

Let us remark that there is no constraint to enter into a maturity mismatch strategy as long as the market participant is ready to be exposed to both market and default risks (beyond a certain maturity). In contrast to bond trading, the absence of funding fostered the rapid growth of these strategies, which gradually became independent of static hedging strategies and bond trading.

Consequently, during the 2000–2010 decade, the outstanding notional stemming from long CDS positions exceeded the outstanding amount of *traded* debt securities issued by some reference entities, e.g. the European sovereign States of Greece, Ireland, Italy, Portugal, and Spain (collectively known as GIIPS). In case of a physical settlement following a default, market participants can only buy the securities actually available for trade and not those which willingly remain on their owners' balance sheets.

In this respect, the sovereign debt crisis that affected the GIIPS countries in 2011 led the CDS market to the brink of a systemic crisis. A bail-out operation, equivalent to a debt restructuring, had to be conducted on GIIPS's government bonds. However, the ISDA did not trigger any credit event to avoid compensations that would have certainly jeopardized the financial system stability. Indeed, the total net estimated outstanding notional[13] of sovereign State CDSs amounted to €0.15 trillion on the Euro zone and the GIIPS's bonds relative market share was sizeable. A credit event would have led to an unsustainable squeeze on this market as the physical settlement was customary at that time. Therefore, those protection buyers who did not previously hold the (defaulted) GIIPS bonds would have been forced to buy them *at any price* to obtain compensations from the protection sellers. This would have fueled an unexpected price increase on defaulted bonds as the market had already started to dry up due to increasing doubts as to the credit worthiness of GIIPS governments. It is worth noting that the supervisory bodies had at their disposal only *estimates* of the actual CDS net outstanding notional and not actual figures. Market participants entering into maturity mismatch strategies did not have accurate information either, and thus could not measure properly the risk of a squeeze.

After 2011, considering that the CDS market was not sufficiently regulated and permitted excessive risk taking, stringent regulations were enforced so as to:

- Prohibiting *short-selling* positions or equivalently forcing long CDS protection buyers to enter into static hedging strategies. This limits the volume of maturity mismatch strategies where CDS protections buyers must hold the same amount of underpinning debt securities, thereby avoiding the disastrous situation described above.

---

[13] Source: ICMA (International Capital Market Association)- Net notional is the sum of the net protection bought by net buyers (or equivalently net protection sold by net sellers), which represents the aggregate payments that would be made in the event of the default of a reference entity, assuming the market value of defaulting bonds is equal to zero.

30.1 Credit Derivatives

- Imposing CDS trade repositories[14] to provide the supervisory bodies with a centralized and accurate view on the CDS outstanding notional for each reference entity.
- Promoting cash settlement instead of physical settlement to limit the risk of squeeze when a default occurs.

Furthermore, the ISDA introduced in 2014, a new type of credit event, namely *Government Intervention* dedicated to EU financial entities which is triggered when a Government modifies (e.g., amends, postpones, cancels) the cash-flow payments related to its own reference obligations.

A similar phenomenon was observed with ailing Monolines,[15] which offered reinsurance contracts "structured" as CDSs to protect against the default of some complex securitization products (see Sect. 30.3 for more details). Monolines did not have the financial capacity to compensate their CDS counterparties when those products started to default. As Monolines were unregulated, supervisory bodies could not intervene and the ISDA did not acknowledge any credit event. Instead, it advocated for out-of-court settlements, on a bilateral basis, regarding the level of an acceptable compensation and the final amount of loss to be incurred.

30.1.2.2 Additional Elements on the Credit Derivatives Market

CDS Market: Some Key Contemporaneous Figures

At the outset of this chapter, we stated that the CDS traded volumes evolved as a roller-coaster over the past 15 years. We also provided some reasons for their decrease observed since the subprime crisis period. It may be attributable to the actual risks revealed by the crisis, the complexity of the contractual provisions which were to a large extent misunderstood, and price opacity. Despite the effort undertaken to promote price transparency through standardization and mandatory clearing, the CDS market did not recover the volumes observed before 2007.

We now illustrate our statement with some statistics regarding the volumes and the nature of traded CDS as well as other credit derivatives. Barring exception, all data displayed here stem from the Bank of International Settlements (BIS) located in Basel (Switzerland) which also houses the Basel Committee. Those data are collected from Central Banks and other national supervisory bodies. They present the advantage of being publicly available and having a long history,[16] although they may sometimes be less precise than those collected by the supervisory bodies in charge of supervising the trading venues where CCPs operate.

Figure 30.2 presents the outstanding notional of CDS traded (including single-name and indices), which corresponds to the sum of CDS absolute notional values

---

[14] For instance, the European Market Infrastructure Regulation (EMIR) has been gradually enforced since 2012.

[15] US unregulated reinsurance companies.

[16] https://www.bis.org/statistics/index.htm

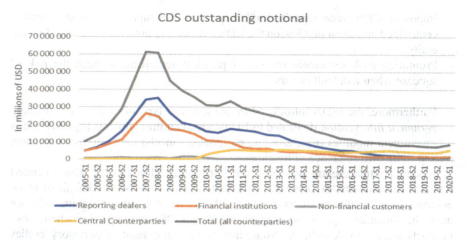

Fig. 30.2 CDS outstanding notional. *Source: BIS statistics*

on purchased and sold protections and is a measure of the size of the market at a given time t. The global notional peaked at $61 trillion in 2008 and underwent a constant decline to a steady volume of $8 trillion in 2019. We observe the gradually increasing share of CCP clearing (Central Counterparties on Fig. 30.2), which now represents 60% of the CDS outstanding notional (around $5.3 trillion), thereby reducing counterparty risk (see Sect. 30.3 below) and increasing price transparency. The remaining 40% are attributable to OTC transactions. Clearly, the role of dealers and financial institutions on OTC trades has strongly decreased since 2008.

Breaking CDSs by type, Fig. 30.3 shows that CDS Indices have gained popularity and now represent half of the market. According to ESMA[17] reports, the other "multi-name CDS" are Total Return Swaps on Indices (see below) for an outstanding notional of $0.3 trillion. The CDS options have always represented a small market share and accounts for around $0.1 trillion only.

As to Futures on CDS traded on listed markets (such as ICE Europe), the BIS does not provide any statistics. However, the ESMA[18] indicates that they represent in the Euro Zone $5 trillion in 2018. As this market is not negligible, we briefly introduce Futures on CDS thereafter.

Other Types of Credit Derivatives

As mentioned before, the credit derivatives market has undergone a massive wave of streamlining during the period 2010–2020 which left only a few instruments afloat, those with sufficient liquidity and transparency for all market participants. The

---

[17] The European Securities Market Authority, equivalent to US agencies such as the CFTC (Commodity Futures Trading Commission), is in charge of enforcing market abuse regulations. In particular, it receives mandatory reports on OTC and CCP traded derivatives volumes.

[18] https://www.esma.europa.eu/sites/default/files/library/esma50-165-1362_asr_derivatives_2020.pdf

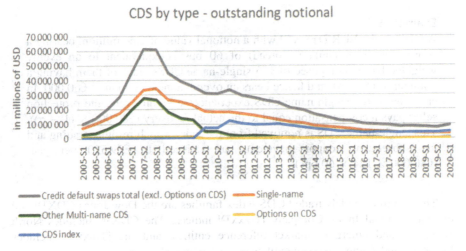

Fig. 30.3 CDS outstanding notional – by type. *Source: BIS statistics*

gradual imposition of mandatory clearing reinforced this trend, leaving only a small range of actually traded credit derivatives.

CDS Index

A CDS index enables market players to trade an instrument for which the reference entity is a basket of pre-defined single-name CDSs. Those indices are only traded on regulated markets (see Chap. 1) and are subject to contractual provisions in addition to those required on single-name CDSs (e.g. contractual standardized fixed coupons), namely: (i) a contractual maturity date beyond which the existing index ceases to trade and is replaced by a new one (a "Series" in market jargon), (ii) a limited range of contractual maturity dates (usually 3, 5, 7, and 10 years), and (iii) mandatory clearing on the most liquid maturities (mostly 5 and 10 years).

The value of the CDS index quoted by the market does not necessarily correspond to the (weighted average) sum of the quoted single-name CDS values composing it. The difference is called "*skew*" and enables arbitrageurs to enter into trades where they buy (sell) the protection on the CDS index and sell (buy) the weighted average of their components. CDS indices enable market participants to hedge the systematic component of the credit risk affecting a large and well-diversified bond portfolio.

In case a credit event is triggered on a single-name CDS composing the index, the compensation mechanism is described in the index contractual provisions (see Example 5). It is customary to determine the recovery rate through an auction and the compensation is then paid in cash. A credit event occurring on one (or several) single-name CDS component(s) does not entail the termination of the CDS index; it continues to trade until maturity, with a reduced number of components.

> **Example 5**
>
> A CDS on the EUR iTraxx™ with a notional value of €ten million, bearing a contractual fixed coupon (spread) of 60 bps is equivalent to an evenly weighted basket composed of 125 single-name CDSs, each of them having a nominal of €80,000, and hence bears an overall annual coupon of €60,000. In case one of the single-name CDS experiences a credit event, the protection seller pays a compensation equal to the actual recovery (€80,000 × (1 − α)). This single-name CDS is removed from the index which resumes trading and the subsequent coupon payments are reduced to €59,520 per year (€80,000 × 124 × 60 bps).

The two most widely traded CDS index families are the Dow Jones CDX™ and the International Index Company iTraxx™ indices. The CDX™ includes North American and emerging market reference entities, and the iTraxx™ includes European and Asian reference entities.

The two most popular benchmark baskets are the *CDX™ NAIG* (North American Investment Grade index) and the *iTraxx™ EUR* index. The CDX™ NAIG is composed of 125 North American Corporate reference entities whose individual rating are investment grade (>BBB). The iTraxx™ EUR has the same characteristics but is composed of 125 European Corporate reference entities. The components are evenly weighted. The most liquid maturities are 5 years and 10 years. Twice a year, a new "Series" of the index is issued (the basket of 125 may be revamped in a discretionary manner by the index administrator).

These 125 reference entities cover various industrial sectors. As for the EUR iTraxx™, they represent six sectors (automotive, consumer goods, energy, industrial products, technology media and telecom-TMT, financial sector), and are chosen among the most traded European Corporate reference entities on the CDS market.

Other *CDX™* or Itraxx™ indices referencing smaller and specifically dedicated baskets are also available, e.g. index on each of the six sectors just mentioned, reference entities chosen among the CDS displaying a high volatility (HiVol), or reference entities likely to shift from Investment Grade to Non-Investment Grade (Xover).

CDS Index Futures

The traction on the market regarding CDS Futures has become increasingly significant since the beginning of 2010 in the context of a more stringent regulatory environment. In this context, CDS Index Futures whose underlying is chosen among the CDX™ or Itraxx™ indices have reached a sizeable market share. It is also worth noting that there are no listed Futures written on a single-name CDS.

As any Futures, CDS Index Futures are listed on a regulated market, namely on the ICE and subject to clearing and daily margin calls. They enable market participants to purchase and sell protection on a given CDX™ or Itraxx™ existing series at a predefined maturity date.

30.1 Credit Derivatives

CDS Index Futures are available on CDX™ IG and Itraxx™ Xover and the Futures price is contractually set at a predefined level (100 bps for CDX™ IG and 500 bps for Itraxx™ Xover), which leads to an initial upfront payment when entering into the Futures contract.[19]

Options on CDS (Credit Default Swaptions)

CDS options (also called *credit default swaptions*) written on indices (such as CDX™ or Itraxx™) constitute the most popular options available on the CDS market. In contrast to CDS Index Futures, those options are not listed on a regulated market but are traded bilaterally on a trading venue (see Chap. 1) and subject to mandatory clearing.

Those options give the holder the right (but not the obligation) to enter into a CDS at time T and expiring at T' ($T' > T$) where an Index is the reference entity. The strike k corresponds to the CDS Index coupon to be paid/received at time T if the option expires in the money. If a credit event affects one of its components during the life of the swaption, the index is subject to the same mechanisms as those previously described for Indices and the swaption resumes trading (i.e. there is no *knock-out* effect). In contrast, a CDS option written on a single name reference entity, albeit less frequently traded on the market, has the knock-out feature whereby the options cease to trade if a credit event affects the reference entity.

Two types of CDS Index options exist:

- A *payer default swaption grants* the holder of the option at maturity T to enter into a CDS where he pays the coupon k determined at the option inception date and gets compensated if a *default* occurs between T and T' on any component of *the underlying index.*
- *Receiver default* where the holder has the option at T to enter into a CDS where she receives the coupon k determined at the option inception and pays the compensation if a default occurs between T and T' on any component of the underlying index.

Total Return Swaps

Total Return Swaps, albeit less popular than CDS indices, still have a significant market share.

The payer (protection buyer) of a *Total Return Swap* (TRS) owns the reference asset x and passes its coupons on to the *receiver* (protection seller) in exchange for a floating reference (Euribor/Libor) + TRS spread. At maturity, the change in the market value of the reference asset since the TRS inception is paid by the protection buyer to the protection seller if positive (capital gain) and is paid by the seller to the buyer if negative (capital loss).

---

[19] The interested reader may access the contractual provisions on such a contract at: https://www.theice.com/publicdocs/IFUS_Eris_iTraxx_Crossover_Credit_Futures_5_Year.pdf

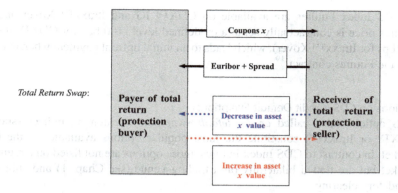

Fig. 30.4 Cash-flows of a Total Return Swap

A *TRS, therefore,* enables to exchange the *total return* on a reference asset (all received coupons plus a capital gain or loss) against a short-term forward-looking (pre-determined) rate (e.g., Libor/Euribor plus a spread). The protection buyer thus is fully hedged against credit risk as well as, if the reference asset bears a fixed rate, interest rate risk. The TRS spread is set so that the present value of the expected cash-flows to be exchanged over the life of the TRS are equal at inception (the TRS thus is said to be *unfunded*) as illustrated in Fig. 30.4.

The reference asset is generally a bond index[20] (e.g., Iboxx™) which is similar to the Itraxx™ except that the reference basket is composed of corporate bonds and not single-name CDSs. TRSs on an index are transacted on trading venues and subject to clearing obligations. TRSs written on single bonds are traded OTC.

TRS parties must also set contractual provisions in case a default occurs on the reference asset (or one of the assets composing the reference index), namely:

- Either to leave the *swap* alive, with the protection seller immediately compensating the buyer for the loss of market value subsequent to the default of the obligor of the underlying asset or one obligor of the reference basket,
- Or to cancel the *swap where the protection buyer delivers* the defaulted bond (of the underlying basket *for an index)* to the seller. This case would prevail when it is impossible to determine indisputably the fair market value of the defaulted bond.

TRS are often used as financing instruments by the receivers and as long positions in a bond or a basket immunized against credit and interest rate risks by the payers, as illustrated in the following example.

---

[20] https://ihsmarkit.com/research-analysis/iboxx-trs-comes-of-age.html

30.1 Credit Derivatives

> **Example 6 A TRS**
> Let us assume that R (the receiver) wants to invest €100 million in a fixed-rate bond x and borrows €100 million at Libor + 30 bps to finance this investment. The bank P (the payer) will acquire €100 million of bond x, accounted as an asset on its balance sheet, and will retain ownership although the transaction is in fact being carried out on behalf of R. R and P will enter into a TRS where: (i) R receives all the coupons from bond x and, if applicable, the *gain on* the market value of x when the TRS matures, and (ii) P receives Libor +30 bps and bears the potential loss on the market value of x. In terms of cash-flows, R has a long position on bond x financed by a loan at the floating rate Libor +30 bps. P holds a mirroring position where he grants a loan of €100 million to R, bearing Libor +30 bps and is short on bond x.

Unfunded and Funded Credit Derivatives

The credit derivatives presented so far are called *unfunded*, as they do not require any funding capacity. Recall that an outright position in a CDS, TRS, or even a maturity mismatch strategy solely requires from a purely financial viewpoint to pay or receive CDS coupons. In contrast, *funded* credit derivatives require a funding capacity exposed to the default of the reference entity or the index. We briefly introduce the Credit-linked-Note (CLN) which constitutes one of the most popular funded credit derivatives and analyze its differences with a debt security.

Investor INV purchases (for a price of 100 at inception in Fig. 30.5) the CLN from the issuer, also called the *originator* and denoted by AB. In turn, AB enters into a

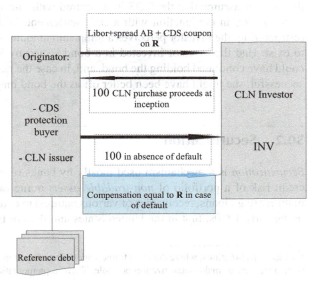

Fig. 30.5 CLN structure and cash flows

short CDS position on behalf of the investor, written on a reference entity R. The investor is exposed to the risk of default of both AB and R, and thereby perceives a total reward made of the *sum* of AB's cost of funding (e.g., IBOR + credit spread AB, assuming that it borrows at a floating rate) and the CDS coupon (spread). For the sake of clarity, if AB is a bank and R a corporate operating in a nonfinancial sector, the investor bears all at once the default risks in two distinct and presumably independent industries and is rewarded accordingly.

The originator may propose an alternative structure for the CLN, where it invests the proceeds received at inception (100) in a portfolio of highly rated bonds (e.g., European Government bonds), whereby the investor is exposed to the risk of default of those bonds instead of AB's. As in a CDS, in case R defaults before the CLN matures (assuming that AB's default risk is largely smaller than that of R), the investor either receives the defaulted bonds (in case of a physical settlement) or the compensation equal to the recovery rate α achieved on the debt securities R (in case of a cash settlement). In both cases, the investor bears a loss of $100 \times (1 - \alpha)$. The originator may also propose to use a CDS index as the reference.

From the originator's viewpoint, the CLN permits to offer an enhanced reward by combining two default risks in one single structure and to attract investors who are not allowed to invest in unfunded credit derivatives for various regulatory and/or contractual reasons. In case the originator is a bank bearing a sizeable loan portfolio on its client R, the CLN enables it to both refinance itself and alleviate the concentration on R (if a CDS market exists for R) by receiving the funding from the investor and transferring the default risk, while continuing to hold the loans. Recall that loans are not tradable securities and therefore cannot be easily sold.

From the investor's perspective, the larger reward offered by the CLN is balanced by the exposure to two default risks (AB and R) instead of one. Moreover, the investor should be wary as to the CDS provisions regarding the credit event definition. Let us assume that the CDS is structured with one single credit event, namely *restructuring* in conjunction with a cash settlement. When the ISDA declares the restructuring, the investor perceives a lump sum amount in cash and the CLN ceases to exist. Had the investor invested in a bond issued by R instead of the CLN, she could have continued holding the bond, and, in case the restructuring turned out to be successful, she might have been better off as the bond may have recovered in value.

30.2 Securitization

Securitization is a mechanism used mostly by banks or corporations to transfer the credit risk of a portfolio of *non-tradable assets* borne on their balance sheet until maturity (e.g., loans, receivables of various natures) to external investors. It appeared in the early 1980s, first in the United States and then in Europe.[21]

---

[21] In some jurisdictions, where the legal framework derives from the Roman law, some amendments were necessary to make securitization possible. Their implementation thus was initially delayed.

30.2 Securitization

Securitization hence enables banks or corporations to alleviate their balance sheet by receiving immediate cash from external investors against the transfer of ownership of non-tradable assets. The amount of cash received corresponds to the value of the portfolio less a discount to compensate the investors for the transfer of risk.

By definition, a non-tradable asset cannot be sold directly in the market due to its contractual nature (e.g., a mortgage loan). While the sale of non-tradable assets may occasionally take place, it is a nonpublic bilateral trade and often entails a lengthy and expensive process. In contrast, securitization can be executed in a timely manner and targets various external investors (instead of one single informed investor), which better suits the needs of banks and corporations.

The key feature of securitization consists in transforming the portfolio (also called *pool*) of non-tradable assets into tradable securities which can be sold on OTC primary and secondary markets among various investors. Those tradable securities were initially called *Asset Backed Securities* (ABS) where the word *Asset* refers to the non-tradables. As the ABS market became more popular during the 1990s, other types of assets were *securitized* and other terms were used such as *Collateralized Debt Obligations* (CDOs), *Collateralized Loan Obligations* (CLOs) but the basic mechanism is the same.

At the dawn of 2000, securitization almost disrupted the traditional commercial banking industry whose historical purpose was to grant loans to retail or corporate customers and hold those loans to maturity (and thus bear the default risk). Securitization shifted the historical banking business model (*"originate and hold"*) to an *"originate to distribute"* model where most of the loans were contemporaneously securitized into ABS sold to external investors.

This originate to distribute model was fueled by the appearance of *subprime* loans (i.e., residential mortgage floating rate bearing loans granted to US retail clients whose credit worthiness was below *prime quality*, thereby called subprime) securitized into so-called *subprime ABS* sold to various investors globally. Interest rates rose in 2007–2008 to a point where these clients could not repay their loan installments making them all insolvent at once and triggering the subprime crisis in 2008.

In parallel to this hectic growth of the ABS market, tradable assets such as bonds and even single-name CDSs started to be securitized on a large scale and gave birth respectively to *Collateralized Bond Obligations* and *synthetic Collateralized Debt Obligations* also called "correlation products". These new structures shifted even more the ABS mechanisms outside of their traditional patterns. In the aftermath of the financial turmoil, most of these complex financial instruments disappeared (hence will not be further discussed) and only some of the classic ABSs survived where the asset side is sufficiently transparent for external investors. Besides, regulatory requirements became more stringent and imposed on banks to bear a sizeable amount of loss potentially occurring on the pool of their securitized assets.[22]

---

[22] Synthetic CDOs still exist today but with limited and decreasing volumes, with the asset side represented by the Itraxx™ or CDX™ index. Those instruments remain OTC but are traded on trading venues and subject to mandatory clearing to increase their price transparency.

The first subsection is devoted to the generic securitization mechanism and the second to risk re-allocation to external investors through the specific *tranching mechanism*.

30.2.1 Introduction to Securitization and ABS

Securitization articulates around 2 types, *on-balance sheet* and *off-balance sheet*:

- *On-balance sheet securitization is a mechanism used exclusively by banks that refinance a dedicated pool of loans present on their balance sheet* (mostly retail mortgage loans granted to high-credit worthiness clients also named "above par" or "premium," or public sector loans) through a bond issuance. These securities are *backed* by the pool of loans which plays the role of *collateral*. They are called *covered bonds* as the collateral (pool of loans) is legally restricted.[23] Besides, the pool of loans eligible as collateral is legally framed (i.e., the type and quality of collateral is subject to legal and not contractual requirements) and in most jurisdictions the pool of collateral remains on the bank's balance sheet. Covered bonds, unlike the *off-balance sheet* ABS discussed below, therefore benefit from a double guarantee: The covered bonds' issuer (namely the bank) is committed to pay the bonds' installments and, in case of the bank's default, the covered bondholders own the collateral. Covered bonds thus are considered virtually free of default risk (and often rated AAA). Banks use these instruments as refinancing tools and not as credit risk transfer instruments. *Covered bonds exist in many countries and are known under other names such as Pfandbriefe* in Germany, *Cédulas in* Spain, and *obligations sécurisées in* France.

Over the decade 2010–2020, the European central bank (ECB) has accepted to refinance certain covered bonds (which may hence be brought as eligible refinancing collateral) thereby fostering the development of the Covered Bond market in the Euro zone.

- *Off-balance sheet securitization is a mechanism whereby* an entity (a bank but also a corporation, called the *originator*) sells a pool of designated non-tradable assets on its balance sheet to a dedicated entity called a *Special Purpose Vehicle (SPV)*, also variously called *Special Purpose Entity (SPE), Special Investment Vehicle* (SIV), or *Trust*. The SPV acquires the pool of asset and refinances it by the proceeds stemming from a simultaneous issuance of ABS purchased by external investors. Thus, the asset side of the SPV's balance sheet is composed of the acquired *pool of assets* and its liabilities are made up of the issued ABS.

---

[23] The pool of loans is contractually segregated on the bank balance sheet and, in case of the bank's default, it will be owned by the covered bond investors and not be part of the bank's generic liquidation process.

The SPV collects the installments[24] (interests and principal repayments) from the pool of assets and pays the installments related to the ABS. An investor holding ABS owns securities that contractually deliver a series of cash-flows subject to default risk. The ABS investor, therefore, is exposed to the default risk of the backing pool of assets and *not* to the default risk of the originator: these two default risks are *disconnected* by the SPV.[25] The SPV is said to be *bankruptcy remote*, i.e. independent of the credit worthiness of the originator.

> **Example 7 Simple Securitization (Without Tranche Structuring)**
> A bank wishes to issue a bond secured by a portfolio P of mortgage loans worth €100 million.
>
> – It sells to the SPV, created for that purpose, the mortgage loan portfolio P.
> – The SPV issues €100 million of ABS (bonds), the interests and principal repayments of which will stem from the cash flows generated by the mortgage loan portfolio P.
> – The €100 million proceeds raised by the SPV from the ABS issue are used to purchase the pool of mortgage loans P from the bank.
>
> As the SPV is an intermediation entity, the possible default of the obligor will have no impact on the investors. In principle, the credit quality of the ABS is that of the underlying mortgage loan portfolio. The scheme is displayed in Fig. 30.6.

In the simple securitization scheme shown in Fig. 30.6, called *pass through,* the SPV issues *identical* ABS displaying the same seniority: Each and every investor has a claim on the *same* cash-flow sequence (proportional to the number of ABS held) generated by the interests and capital repayments of pool P. These ABS are said to be *pari-passu* (i.e., of same seniority).

In more complex securitization schemes, called *pay through,* all ABS are *structured in tranches with* different characteristics, particularly in terms of seniority and therefore credit risk, as shown in Fig. 30.7. Investors holding the senior ABS tranche perceive a low interest rate and are subject to minimal credit risk, whilst investors holding the most junior tranche (called *equity tranche*) earn the highest interest rate but bear the first losses affecting the pool of assets and thus take the risk to lose the totality of their investment (more on this in Sect. 30.2.2).

---

[24] In fact, it is customary to contractually assign the originator, on behalf of the SPV, to perform the *servicing* of the pool of assets (e.g., collecting the cash flows and allocating the losses of ailing assets to the ABS holder of unpaid receivables). Provisions regarding servicing are contractually defined at inception of the SPV.

[25] The pool of assets sold to the SPV can no longer be used in the liquidation proceeds should the originator default. Therefore, the credit spread of the ABS only hedges investors against the credit risk attached to the pool of assets, not that of the originator.

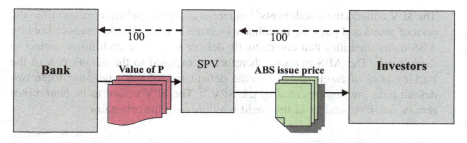

Fig. 30.6 Simple securitization scheme

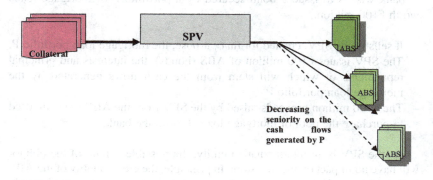

Fig. 30.7 Securitization structured in tranches (*pay through*)

Although any asset or pool of assets generating cash flows may, in principle, be securitized, securitization has mainly concerned certain types of loans or receivables (e.g., mortgage loans, consumer loans, car leases, trade receivables). For instance, ABS backed by retail or commercial mortgages loans are called, respectively, *Mortgage Backed Securities (MBS) and Commercial Mortgage Backed Securities (CMBS)*.

Originally, securitization was mostly used by credit institutions or insurance companies to manage the size of their balance sheet and diversify their source of funding. Then large corporations started to securitize some pools of receivables to also manage their risks.

From the originator's perspective, securitization allows:

- To transform non-tradable assets into tradable securities which can be sold to a broad range of investors in the market;
- To diversify sources of funding (e.g., not solely rely on the interbank market);

30.2 Securitization

- To maintain confidentiality (the client is not informed of the disposal of its loans or receivables to an SPV, which avoids deteriorating the business relationship);[26]
- To alleviate the size of the balance sheet and, when the originator is a bank, to improve its solvency ratio.[27]

From the investors' perspective, securitization broadens the traditional range of investment instruments (shares, bonds) and thus allows for better portfolio diversification.

In the past (at least until 2010), one of the criticisms raised towards securitization was that it may favor "moral hazard" from the originator. Indeed, credit institutions (or corporations) may be tempted to be less rigorous when defining or applying their lending criteria if they know that the loans/receivables are to be soon sold out to investors. In addition, the financial turmoil triggered by the subprime crisis also cast doubt on the methods applied by rating agencies and their capacity to check the quality of the pool of assets backing the ABS. More generally, lenders benefit from a competitive advantage as to the true value of the loans/receivables they grant as they entertain a customer relationship with their clients. This information asymmetry generates both moral hazard and adverse selection. Recent regulations, by imposing stringent rules on the ABS market and bringing more transparency as to asset pool eligibility, aim at reducing these two drawbacks.

30.2.2 ABS Tranching Structuration

Although the installments received by all ABS holders are *globally* equal to those paid by the pool of collateral assets, it is possible to structure them in tranches that differ by their maturity and seniority, hence their level of credit risk. This mechanism is called *tranching*. The tranches are obtained through a contractual subordination that allocates a range of priorities over the repayments made by the SPV to the ABS holders. As illustrated in Fig. 30.6, cash flows collected from the asset pool are comparable to a *waterfall* as they are used to repay the senior debt first (*senior tranche*), then the second tranche, and so on until the last tranche (*equity tranche*), which contractually collects the remaining cash-flows, which can possibly be nil (hence the somewhat misleading name of *equity*) if the pool is subject to a high credit risk.

In a structured securitization, the credit risk, and the offered compensation, of an ABS thus depend on the *tranche* to which it belongs. In reverse order, the *waterfall* comprises the *equity tranche*, which absorbs most of the risk (like the equity of a

---

[26]The bank does not renounce its client-facing role, continues on behalf of the SPV the servicing of the pool of transferred assets (see Footnote 32) and passes the collected cash-flows on to the SPV.

[27]Note, however, that Basel III rules impose on originators to hold a sizeable amount of ABS junior tranches issued by the SPV. These requirements limit the impact of securitization and narrow the spectrum of potential (non-regulated) investors in junior ABS tranches.

company), then one or more *mezzanine* and *junior tranches*, then a *senior tranche* bearing little credit risk and sometimes a *super senior tranche,* often considered as virtually free of default risk. Generally, the equity tranche is not given a rating, whereas the mezzanine, junior and senior tranches are rated, with the senior debt often being AAA.

Example 8 Securitization Structured in Tranches

The following structuring is fairly typical of a *pay through* securitization. The asset pool (collateral) is globally rated BBB+ and the ABS are structured in four tranches, with the following *waterfall*:

- The senior tranche represents 87% of the SPV issuance and is rated AAA.
- The subordinated junior tranche represents 5% of the SPV and is rated A+.
- The mezzanine tranche represents 5% of the SPV and is rated BBB.
- The equity tranche represents 3% of the SPV and has no rating.

It is worth noting that, due to the subordination mechanism, the asset pool must bear losses in excess of 13% of the total SPV value to trigger losses on the senior tranche. Such a situation is considered extremely unlikely in the case of a BBB+ pool of assets. Similarly, losses attributable to defaults on the asset pool must exceed 8% of the SPV value to trigger losses on the subordinated junior tranche (as respectively 3% and 5% are absorbed by the equity and mezzanine tranches), which is rather unlikely, at least in what are considered to be normal market conditions.[28]

The credit risk affecting *the whole* ABS issue is in principle that of the asset pool. To reduce this risk for the investors, insurance companies may offer a guarantee of up to a certain amount of loss affecting the asset pool and be compensated by a fee in return. This mechanism is called *insurance wrapping* and is equivalent to an improvement of the credit worthiness of the asset pool (known as *credit enhancement*).

Other credit risk mitigation techniques can be used to enhance the rating of the various tranches issued by the SPV, such as:

- *Over-collateralization*, by which the value of the SPV issue (i.e., the sum of the nominal tranches held by the investors) is less than that of the asset pool, the gap constituting a "safety cushion" dedicated to absorbing the first losses.
- Creation of a *cash reserve* financed by the originator;
- *Financial guarantee* granted by a third party such as a Monoline company (AAA non-regulated insurance company specialized in this type of guarantee).

---

[28] Unlikely but not impossible, as the crisis in the US subprime housing market has demonstrated; in a few days some AAA rated tranches were downgraded to CCC or even defaulted.

The available tranches are tailored to attract a broad range of investors in accordance with their respective risk/reward preferences. Subordinated tranches reduce the probability of loss for the most senior securities thereby reducing their cost of funding. The subtlety of tranching consists in striking the right balance between the overall cost of funding (which should be minimized) and the spectrum of investors willing to invest in the offered tranches. The ratings bestowed on the various tranches (except the equity one) are also key and depend on the criteria used by the rating agencies.

30.3 The "xVA" Framework

In Chap. 29, we introduced the notion of default risk (of the obligor) for various types of credit exposures (e.g., loans, receivables, or bonds) except for derivatives. In this respect, *counterparty risk* is the dedicated wording when it comes to measuring *default risk* on derivatives exposures.

We present the key concepts underpinning the counterparty risk measurement and how it disrupted the derivatives' valuation theory by introducing a series of *valuation adjustments* commonly known as "xVA." We do not address its whole realm and solely focus on the main mathematical and practical challenges that are still prevailing today.

In 2008, the collapse of Lehman Brothers or the out-of-court settlement of CDS transactions with some ailing Mono-lines (see Sect. 30.1.2.1) brought under the limelight the need to evaluate the counterparty risk on derivatives. Therefore, many market participants rapidly appreciated that the derivatives' values should be adjusted to avoid being arbitraged against by the market. The first adjustments to be introduced were the *Credit Value Adjustment* (CVA) and its mirroring concept the *Debt Value Adjustment* (DVA). They represent the *expected loss* in case one derivatives' counterparty defaults (CVA) while the other survives (and reciprocally for the DVA).

This collapse also questioned the anchored notion of the *risk-free rate* that has been extensively used throughout this book to discount the cash-flows of the derivatives under the risk-neutral probability. Recall that the risk-free rate adopted as a market standard was the *interbank* lending/borrowing rate (e.g., Euribor, Libor) where those banking counterparties were AA-rated and were therefore perceived as virtually riskless. How to consider the interbank risk-free rate when banks start to collapse or near-collapse (not only Lehman Brothers but also other banks in various locations)? Those events have jeopardized the notion of risk-free rate which was replaced by the Eonia/OIS and triggered additional valuation adjustments, called *funding valuation adjustments* (FVA).

The xVA thus not only embraces the CVA, DVA, and FVA but also subsequent adjustments we will allusively address. Numerous research articles as well as new accounting and regulatory requirements fueled the extension of the xVA on the decade 2010–2020. However, since 2015–2016 onwards, the role of the xVA has become less important as the introduction of mandatory clearing has gradually framed its usage to *non-collateralized* trades. Despite this trend, it is worth understanding how it has triggered significant and durable changes on derivatives valuation.

The remainder of this section is organized as follows. First, we introduce the concepts required to measure counterparty risk exposure as well as its available mitigation techniques (Sect. 30.3.1). Second, we analyze the modeling techniques to measure the exposures stemming from derivatives that underpin the xVA computation (Sect. 30.3.2). Third, we present statistics to illustrate the impact of the mandatory introduction of Central Counterparties (CCP) clearing in the OTC derivative market (Sect. 30.3.3). Fourth, we introduce the CVA, DVA, and FVA and the key issues they raise (Sects. 30.3.4–30.3.6).

30.3.1 Counterparty Risk Exposure Measurement and Risk Mitigation Techniques

Let us first introduce some useful notations we extensively refer to in the remainder of this section. A and B are two counterparties entering into one or more derivative transactions. V_t denotes the market value (also called mark-to-market or fair value) of a derivative or a portfolio of derivatives at time t. The sign of this value depends on the counterparty's viewpoint. More specifically, a derivative is accounted as an asset in A's balance sheet when V_t is positive from A's viewpoint and as a liability in B's balance sheet and in the opposite way when V_t is negative.

For instance, if A sells a vanilla call option to B, A is always liable to repay the payoff, while B will always receive a non-negative payment. In contrast, the value of an interest rate swap, a CDS or a forward contract fluctuates overtime and may be either positive or negative, depending on the prevailing market conditions.

For the sake of clarity, V_t^+ and V_t^- respectively denote the derivative's positive value and negative value. For instance, let us adopt the viewpoint of A engaged in a vanilla interest rate swap with B, where A receives the fixed rate. If at time t, the present value of the fixed rate leg is larger than that of the floating rate leg, then A accounts for the swap as an asset equal to V_t^+ and B a liability equal to V_t^-. In case B defaults and A survives, A has obviously an exposure to B and is entitled to claim the amount V_t^+ in the liquidation process. A may end up recovering $V_t^+ R_B$ (or equivalently end up losing $V_t^+ (1-R_B)$) where R_B denotes the recovery rate achieved on B. Using the wording of Chap. 29, we can state that A lends to B or B borrows from A. However, as the lending position is a constituted by a derivative (i.e., a *contingent* claim, not a claim as a loan or a debt security), A is said to bear *a counterparty risk* on B instead of *a credit or a default risk* although in essence these two risks are analogous.

Consider now a portfolio composed of two derivatives and assume that:

- $V_{t,1}^+ = 100$ and $V_{t,2}^- = -90$ where $V_{t,1}^+$ and $V_{t,2}^-$ respectively denote the derivatives value from A's viewpoint.
- $V_{t,1}^- = -100$ and $V_{t,2}^+ = 90$ from B's viewpoint.

30.3 The "xVA" Framework

Let us determine the amount that A is entitled to claim in case B defaults. In absence of additional assumptions, the legal insolvency provisions in force in many jurisdictions impose A to pay *immediately* an amount of 90 to B. Then, A can claim an amount of 100 on B and waits, sometimes for years, until the insolvency proceedings come to an end and recovers eventually 100 R_B or equivalently bears a loss of $100(1- R_B)$.

This outcome appears somehow unfair as, from an economic viewpoint, A has a *net* claim of 100–90 = 10 on B. However, in case A and B had previously entered into a *contractual agreement* that permits the *netting* between assets and liabilities when one of them defaults, the previous reasoning would be legitimate and reconcile the economic and contractual perspectives. A *contractual netting agreement* thus reduces the net claim and constitutes a *credit/counterparty risk mitigation* (CRM) technique. Therefore, the amount to be eventually lost is equal to $(V_{t,1}{}^{+} + V_{t,2}{}^{-})\,(1- R_B)$ and, in this example, A would claim an amount of 10.

We now assume that A and B have entered into a contractual netting agreement covering a portfolio composed of M derivatives. V_t^{Tot} denotes the total *net value* of the portfolio being either positive (an asset) or negative (i.e. a liability). Hence,

$$V_t^{Tot} = \sum_{i=1}^{M} V_{t,i}^{+/-}, \text{ where } V_{t,i}^{+/-} \text{ denotes the value (positive or negative) of each}$$

derivative composing the portfolio. Assume that V_t^{Tot} is positive from A's perspective and can be claimed in case B defaults. A may further reduce its potential exposure by an additional CRM technique called the *collateral mechanism or collateralization*, a collateral agreement supplementing the provisions of the netting agreement whereby B *posts* (or, equivalently, A *calls*) an amount of collateral called the *variation margin*. It is denoted by VM_t and leads to a *residual* or *net exposure* equal to $V_t^{Tot} - VM_t \geq 0$ for A at time t. When the residual exposure is zero, namely $V_t^{Tot} - VM_t = 0$, the portfolio has a *zero net exposure* it is said to be *fully or perfectly collateralized*.

Let us assume that at time t, A and B have not defaulted. Counterparty A recognizes an asset on its balance sheet equal to the received variation margin VM_t posted by B and in turn the corresponding debt to acknowledge that it still remains due to A. Symmetrically, B recognizes a claim to A equal the variation margin posted, VM_t. In case B defaults and A survives at time t, A is entitled to take immediate possession of the collateral amount which is not part of the general claim included in the insolvency proceedings for B. In practice, the variation margin is left on a third party's account (neither A nor B) to further facilitate its access if A or B defaults.

Conversely, we assume that this collateral agreement also provides that if V_t^{Tot} is negative at time t, A posts an amount VM_t to B, bearing a residual exposure of $V_t^{Tot} - VM_t \geq 0$ (or a zero exposure in case of a perfect collateralization). Such a collateralization mechanism is called *bilateral* or *two-way* to underline that symmetrical conditions apply to both parties. *Unilateral* or *one-way* mechanisms also exist in the market, especially when a counterparty is a sovereign State or a supranational organization that never posts any variation margin but calls it to reduce its exposure.

The composition of the variation margin (e.g., cash only, cash or debt securities, range of eligible debt security issuers, of ratings and maturities, etc.) and its frequency (intra-daily, daily, weekly, ...) are defined contractually. It is also possible to add two provisions to lower the number of cash/debt security transfers between A and B. These provisions, respectively called the *threshold* (TH) and the *minimum transfer amount* (MTA), are defined as follows:

- TH: the counterparty starts posting the collateral when $| V_t^{Tot} | \geq TH$ and the first margin call is equal to $V_t^{Tot} - TH \geq 0$. Namely, A accepts to bear an enduring exposure up to TH.
- MTA: counterparties exchange additional collaterals when the fluctuation of the derivative portfolio's value exceeds the MTA between t and t', i.e., $\left| V_{t'}^{Tot} - V_t^{Tot} \right| \geq MTA$, where t and t' are two contractual consecutive (variation) margin calls dates.

Example 9 Threshold and Minimum Transfer Amount

A and B have traded at $t = 0$ two interest rate swaps, whose respective values are zero at time 0. They have entered into a contractual netting agreement where: MTA= \$0.5 million and TH=\$1 million. We analyze the margin calls from A's perspective.

$V_0^{Tot} = 0$.

$V_1^{Tot} = \$0.9$ million; no margin call occurs as the exposure is inferior to TH.

$V_2^{Tot} = \$1.3$ million; a margin call of \$0.3 million is called by A (B posts the margin call to A) as $V_2^{Tot} \geq TH$.

$V_3^{Tot} = \$1.9$ million; an additional amount of \$0.6 million is called by A as $V_3^{Tot} - V_2^{Tot} \geq MTA$. We observe that A bears a residual exposure equal to TH.

$V_4^{Tot} = \$1.1$ million; A returns an amount of \$0.8 million to B as $\left| V_4^{Tot} - V_3^{Tot} \right| \geq MTA$.

$V_5^{Tot} = \$0.5$ million; A returns to B the amount of \$0.1 million as the exposure falls below TH.

The provisions addressing the variation margin are laid out in a *credit support annex* (CSA) supplementing the netting agreement. Derivative trades performed under a netting agreement but in absence of a CSA are known as *uncollateralized derivatives* and represent today a small fraction of the derivatives markets (see Sect. 30.3.3). This follows the enforcement of new regulations which have since 2015–2016 gradually imposed collateralization either through clearing obligations with a CCP or mandatory CSAs on the OTC derivatives market.

For the sake of clarity, we describe the variation margin dynamics between two contractual margin call dates denoted again t and t', assuming that TH = MTA = 0. We come up with: $\left(V_{t'}^{Tot} - VM_{t'} \right) - \left(V_t^{Tot} - VM_t \right) = 0$ or $V_{t'}^{Tot} - V_t^{Tot} =$

30.3 The "xVA" Framework

$VM_{t'} - VM_t$. The variation margin is rebalanced at time t' to offset the variation of the portfolio that has occurred between t and t' (*margin call*). Example 9 has shown how the margin call behaves when MTA and TH are positive.

Recall that in case of a perfect collateralization, if we assume without loss of generality that V_t^{Tot} is positive from A's viewpoint, it does not bear any loss if B defaults exactly at the very time t that $V_t^{Tot} - VM_t = 0$. This situation is unfortunately purely theoretical. In practice, B defaults at time $t + \varepsilon$, where $t \leq t + \varepsilon \leq t'$ implying that $V_{t+\varepsilon}^{Tot} \neq V_t^{Tot}$, thereby leaving A unprotected from an amount equal $V_{t+\varepsilon}^{Tot} - V_t^{Tot}$ if it is positive. If it is negative, A would be fully protected.

In brief, the variation margin mechanism is sufficient to mitigate (if not eliminate) the counterparty risk at each contractual variation margin call date but not between two margin call dates, as: (i) $V_{t+\varepsilon}^{Tot}$ (value of the portfolio on the default date) is very likely to differ from V_t^{Tot} due to market fluctuations, (ii) the larger the contractual time lag set in the CSA between t and t' is, the larger the unprotected value $V_{t+\varepsilon}^{Tot} - V_t^{Tot}$ may be, (iii) not to mention that, in case of imperfect collateralization, we have $V_t^{Tot} - VM_t > 0$ which makes A even less protected, and (iv) even if V_t^{Tot} remains unchanged, when the variation margin is constituted of debt securities, their market value may fluctuate independently from the derivative portfolio's value.

We here introduce an additional layer of protection, with a CRM technique called the *Initial Margin* (IM_t) or *deposit* (albeit this denomination is less frequent) which aims to protect a counterparty at time t from the unknown (and potentially large) fluctuations $V_{t+\varepsilon}^{Tot} - V_t^{Tot}$ in case one counterparty defaults between two consecutive variation margin calls (t and t'). We emphasize that the variation margin and the initial margin differ by nature: the former reduces (eliminates in case of perfect collateralization) the counterparty risk measured on the *known* value of the portfolio V_t^{Tot}, while the latter reduces from time t the counterparty risk attributable to the *unknown* fluctuations $V_{t+\varepsilon}^{Tot} - V_t^{Tot}$ (under a given level of confidence, as we deal with a distribution of portfolio values).

Let us now consider again how the default works in practice and assume that A has no residual exposure (perfect collateralization) to B at time t, i.e., $\left(V_t^{Tot} - VM_t\right) = 0$, and that B defaults between t and t'. Counterparty A needs to find in a timely manner a sound counterparty, say B', to take on the defaulted portfolio. It is also possible that if A runs a large portfolio of derivatives, B' may not accommodate the whole defaulted portfolio and A would either need to reallocate the remaining portion among additional counterparties or to offer a price discount to B'. It is also frequent that, upon the default of a counterparty, the market operates under stressed conditions and the time-lapse needed to perform the reallocation may end up resuming at time t'', where the derivative portfolio has a value $V_{t''}^{Tot}$ much larger than V_t^{Tot} in the worst case. In practice, A is bound to incur additional costs, called *close-out costs* (COC) equal to $\left(V_{t''}^{Tot} - V_t^{Tot} > 0\right)$ to transfer its derivatives portfolio under the prevailing market conditions. In absence of IM, the COC are claimed to B and A is subject to a loss equal to $\left(V_{t''}^{Tot} - V_t^{Tot}\right)(1 - R_B)$ at the end of the insolvency procedure, which in many instances lasts for years.

The time lapse between t and t'' (or t' if the counterparty does not default between two consecutive variation margin call dates) is called the *margin period of risk* (MPOR). The IM aims to further mitigate (if not eliminate) the potential incremental exposure occurring during the MPOR and therefore the quantity ε is conservatively set not at the potential default date but at the MPOR ($\varepsilon = MPOR$). We simply present in a sketchy manner the key ideas underpinning the determination of IM_t. Recall that IM_t measures the unexpected upcoming market fluctuations under stressed conditions, which supposedly rarely occur. Such a definition suggests evaluating the IM_t by using the Value-at-Risk (VaR) or the Expected-Shortfall (ES) approach presented in Chap. 27. In case a VaR approach is considered, we come up with IM_t being a quantile of the portfolio distribution: $\mathbb{P}^B\left(|V^{Tot}_{t+MPOR} - V^{Tot}_t| \geq IM_t\right) = \alpha$ (e.g., $\alpha = 1\%$). Regulators also provide simpler analytical formulas for those entities willing to determine the IM but who do not have sufficient modeling capabilities.

We do not investigate further IM_t modeling techniques but we add some practical details. Common practice and regulators consider that the MPOR should range between 5 and 20 days, thereby calibrating the VaR/ES models accordingly. IM_t is by design dependent on the derivative portfolio and its value evolves overtime in accordance with the portfolio composition and the prevailing market conditions. We draw the reader's attention that the terminology IM, albeit customary among market practitioners and regulators, is misleading as it wrongly suggests that the initial margin remains fixed over time, which is obviously not the case.

Let us illustrate how practically the initial margin at time t complements the variation margin to further reduce the counterparty risk. First, the IM is currently not subject to the TH or MTA and is called only by the counterparty bearing an exposure. Second, the composition (e.g., cash and or debt securities) of the IM is laid out in the CSA. Namely, if A has an exposure to B (i.e., $V^{Tot}_t \geq 0$) and IM_t is applicable pursuant to the CSA's provisions, the residual exposure of A at time writes $max\left(V^{Tot}_t - VM^A_t - IM^A_t; 0\right)$ assuming TH $=$ MTA $= 0$, where IM^A_t denotes the IM posted by B to A.

Upcoming regulations will state that IM_t is binding to both parties, whether they bear an exposure or not. More precisely, A and B will need to hold (on a third-party account) permanently the amounts IM^A_t and IM^B_t which cannot be deducted from the variation margins already received or posted. In this context, even if $V^{Tot}_t \geq 0$, A will be required to post an amount IM^B_t to account for unexpected changes of the sign of the exposures during the MPOR making B potentially exposed to A. IM^B_t will be larger as V^{Tot}_t is closer to zero. The residual exposure of A will then become: $max\left(V^{Tot}_t - VM_t - IM^A_t; 0\right) + IM^B_t$. This bilateral IM mechanism thus creates an ongoing exposure for *both* parties.

After adding all available CRM techniques, whether one party eliminates its exposure in all situations or not depends on the contractual provisions laid out in the CSA. In various jurisdictions, laws or regulations not only impose the mandatory signing of a CSA (except for certain small and/or non-financial counterparties, such as small and medium entities or sovereign States) but also frame its contractual provisions by setting minimal values for *MTA*, *TH*, the contractual time lag between two variation margin calls and what constitutes an eligible collateral.[29] These requirements intend to minimize residual exposures and limit counterparty risk. Some counterparties may nevertheless decide to impose even more stringent CSA provisions to bear a virtually negligible counterparty risk on OTC derivatives.

Finally, it is worth noting that CCPs impose equal contractual conditions to all their counterparties (*CCP members*) through a publicly available *rule book* in which:

– The variation margins are only in cash and exchanged on a daily basis (sometimes intra-daily).
– The IM is bilateral and only composed of debt securities eligible according to the terms of the rule book.
– TH = MTA = 0 for both VM and IM.
– The IM models are also calibrated to cover a 10-day MPOR.

Moreover, CCPs are built to ensure the timely reallocation of the derivative portfolio from an ailing CCP member to the other members and therefore to minimize their potential losses. In addition, a CCP disposes of its own regulatory capital to cover any extreme loss (as does the clearing house of a futures and/or option exchange traded market. See Sect. 1.4.3.1). CCPs then are virtually riskless and residual exposures may be considered as immaterial.

30.3.2 Counterparty Risk Exposure Modeling Techniques

We present here the notations and techniques widely used to measure counterparty risk over the life of a derivative portfolio. We start with the key notion of *Expected Positive Exposure* (EPE) lying at the core of this measurement. Again, we do not delve into the complex modeling techniques underpinning the EPE simulations but rather present the most relevant concepts and computational challenges.

Let us again consider the portfolio composed of M derivatives maturing at time T whose value at time 0 is $V_0^{Tot} = \sum\limits_{i=1}^{M} V_{0,i}^{+/-}$. We assume that the derivatives mature before T and will be not replaced by new ones. We assume that a netting agreement is in place but, barring exception, with no CSA (i.e. no VM and no IM) to ease the notation and facilitate the understanding. The introduction of a CSA does not alter the subsequent reasoning.

The first step consists in simulating the value of the derivative portfolio at some future dates with a Monte-Carlo approach as follows:

---

[29] In Europe, it is the regulation 648/2013 (EMIR). In the US, its equivalent is the sub-part I of Part 23 of the CFTC Regulations and The Margin Requirements for Uncleared Swaps for Swap Dealers and Major Swap Participants published in January 2016 ('final margin rule').

- Set a time grid $(t_0 = 0, t_1, t_2, ..., t_k, ..., t_K = T)$ where the value of each derivative is calculated at time t_i. In practice, the grid is granular on the short-term horizon (time step of 15 days or one month) and gradually coarser (quarterly of even yearly time steps) on the longer term.
- For each point t_k of the grid, simulate NxM derivatives values $V_{t_{k,i}}^{+/-}(j)$ where $1 \le i \le M$ (number of derivatives composing the portfolio) and $1 \le j \le N$ (number of simulations).
- Compute the portfolio value $V_{t_k}^{Tot}(j) = \sum_{i=1}^{M} V_{t_{k,i}}^{+/-}(j)$ by adding the positive and negatives derivative values (justified by the presence of a netting agreement).
- Repeat the procedure for each time t_k, where $1 \le k \le K$ which implies a total number of NxMxK derivative simulations.

From A's viewpoint, we define by $PE(t_k) = \max\left(V_{t_k}^{Tot}(j), 0\right)$ the *positive exposure* to B simulated at t_k and resulting from the simulation number j. Symmetrically, we define by $NE(t_k) = \min\left(V_{t_k}^{Tot}(j), 0\right)$ the *negative exposure* to B simulated at t_k and resulting from the simulation number j.

Recall that in Chaps. 28 and 29 the wording *"exposure"* explicitly refers to a loan, a receivable or a claim accounted as an *asset* in A's balance sheet, whereby counterparty A is *exposed* to a loss in case B defaults. The wording *"negative exposure"* may appear as a misnomer as it refers to the negative value of the derivative accounted as a *liability* in A's balance sheet but does not expose A to a loss when B defaults. Recall that from a legal perspective, in case B defaults, A mandatorily repays immediately the negative value of derivative, or leaves B's liquidator to take possession of the variation (and the initial margin if applicable).

Notwithstanding the previous discussion, PE and NE denote, respectively, an asset and a liability and are extensively used in the literature as they convey two different kinds of information; this will prove to be useful when we address the DVA and FVA puzzles (see Sects. 30.3.5 and 30.3.6).

For each point in time t_k, we then define (recall that we do not include VM and IM to ease the notation):

$$EPE(t_k) = \frac{1}{N} \sum_{j=1}^{N} \max\left(V_{t_k}^{Tot}(j), 0\right).$$

The $EPE(t_k)$ is hence the average over the simulated net positive exposures of A to B.[30]

Similarly, we define the Expected Negative Exposure (ENE) as:

---

[30] The terminologies EPE and Expected Exposure (EE) are used equivalently in the literature. Note also that the notion of average exposure also exists and is defined by the quantity $\frac{1}{N} \sum_{j=1}^{N} V_{t_k}^{Tot}(j)$.

30.3 The "xVA" Framework

$$ENE(t_k) = \frac{1}{N} \sum_{j=1}^{N} \min \left(V_{t_k}^{Tot}(j), 0 \right)$$

The value $ENE(t_k)$ is the average of the simulated net negative exposures (i.e. residual liabilities) stemming from the derivatives portfolio of A to B at time t_k.

We then obtain two arrays $(EPE(t_1), \ldots, EPE(t_K))$ and $(ENE(t_1), \ldots, ENE(t_K))$ that we will use to compute the CVA, DVA, and FVA (see Sect. 30.3.4, 30.3.5, and 30.3.6).

We go back to the computation steps to illustrate the difficulties met in practice. The computation of EPE and ENE necessitates to first simulate $Nx\ MxK$ values $V_{t_{k,i}}^{+/-}(j)$ for a derivative portfolio composed of M derivatives. For the sake of simplicity, we assume below that this portfolio is solely composed of M interest rate swaps. We need to simulate the corresponding number of discount rates curves to compute the swap values at each future date and it requires using an interest rate curve diffusion model. We have examined in several chapters various models to value derivatives and recall that they need to be calibrated on the market data prevailing at t_0.

Using a complex diffusion model such as a 3-factor model would rapidly become intractable. To ease the computational burden, market participants use models of a lesser accuracy, as the objective for the EPE/ENE computations is to be "roughly right" in a timely manner. If we consider a portfolio composed of non-optional and optional derivatives written on a large range of asset classes, a similar choice of the appropriate diffusion models must be considered. Besides, for those derivatives whose market value should be simulated on each date t_k, running in parallel two sets of Monte-Carlo simulations (one for EPE and ENE and one for the derivative value $V_{t_{k,i}}^{+/-}(j)$ itself) is highly time-consuming. It is therefore necessary to approximate the derivative value with a closed formula without too much compromising on the accuracy of the overall computation.

Over the decade 2010–2020, opinions varied whether the derivatives valuation models underpinning the EPE and ENE should be calibrated on implicit market data, thereby using a risk-neutral probability, or on historical data. Presently those models are calibrated on implicit market data. This is because the CVA, DVA, and FVA enter into the market price of the derivatives as we will see thereafter and therefore cannot be calibrated on historical data.

Let us now compute the EPE (and likewise ENE) in presence of a CSA with no IM. Assuming that TH and MTA are equal to zero, we need to compute at each time t_k the quantity $max \left(V_{t_k+MPOR}^{Tot}(j) - VM_{t_k}(j), 0 \right)$ and hence depart from $max \left(V_{t_k}^{Tot}(j) - VM_{t_k}(j), 0 \right)$ as this latter quantity is close to zero by construction. Accounting for the MPOR reflects what may happen in practice when the liable counterparty defaults between two consecutive margin calls, thus entailing a residual exposure for the non-defaulted counterparty.

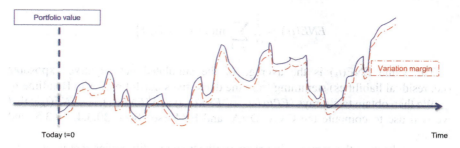

Fig. 30.8 Exposure in presence of a CSA with no IM

The simulations of quantities $V^{Tot}_{t_k+MPOR}(j)$ and $VM_{t_k}(j)$ may be performed as follows (see Fig. 30.8):

- We consider the initial time grid $(t_1, t_2, ..., t_k, ..., t_K = T)$ and make it finer it by imposing additional dates $(t_1, t_1+MPOR, t_2, t_2+MPOR, t_k, t_k+MPOR, ..., t_K = T)$ to account for the time decay between the last margin call and the fluctuation of the portfolio value until the end of the MPOR.
- Assume that at each time t_k, $VM_{t_k}(j) = V^{Tot}_{t_k}(j)$ to come up with no exposure at each (variation) margin call date.

This approach has the merit of simplicity and, from a computational viewpoint, is not too complex to implement. However, assuming that $VM_{t_k}(j) = V^{Tot}_{t_k}(j)$ does not take into account the existence of a non-zero TH and/or MTA and thus underestimates the actual exposure at time t_k. Actually, their presence implies a path dependency on $VM_{t_k}(j)$, this latter being posted or returned in accordance with the fluctuations of $V^{Tot}_{t_k}(j)$. In case the MTA is small, the previous simplifying assumption is acceptable. If not, more complex techniques are used to better capture the margining effect (see the references at the end of the chapter).

We can add to the CSA a one-way IM, where solely the counterparty bearing an exposure receives the amount IM_{t_k}. In case IM_{t_k} is calibrated accordingly to cover the MPOR, we have $V^{Tot}_{t_k+MPOR}(j) - IM_{t_k} = 0$. In practice, the MPOR is set to 5, 10, or 20 days and the quantity $max\left(V^{Tot}_{t_k+MPOR}(j) - VM_{t_k}(j) - IM_{t_k}(j), 0\right)$ is almost virtually equal or close to zero at each future date t_k and so are the $EPE(t_k)$ and $ENE(t_k)$.

Recall that VM and IM require some funding. Therefore, when counterparties discuss the CSA's provisions, they have to strike the right balance between the level of counterparty risk they are ready to bear and their capacity to mobilize the appropriate level of funding. When both counterparties have a limited funding capacity, they may set TH and/or MTA at a high level for the VM and a small MPOR, then $V^{Tot}_{t_k+MPOR} \geq V^{Tot}_{t_k} \geq VM_{t_k}$ and come up with material $EPE(t_k)$ and $ENE(t_k)$ which will in turn impact the CVA, DVA (and FVA) values. In contrast, when

30.3 The "xVA" Framework

both counterparties have strong funding capacities, they are keener to include low TH and /or MTA, bilateral IMs, and a large MPOR whereby the $EPE(t_k)$ and $ENE(t_k)$ are again virtually equal to zero.

Since the EPE and the ENE are computed on a portfolio basis, some issues arise as to measuring the marginal contribution of a new trade. The obvious procedure is as follows:

- Recall that $(EPE^M(t_1), \ldots, EPE^M(t_K))$ and $(ENE^M(t_1), \ldots, ENE^M(t_K))$ denote the exposures for a portfolio composed of the existing M derivatives.
- Let $(EPE^{M+1}(t_1), \ldots, EPE^{M+1}(t_K))$ and $(ENE^{M+1}(t_1), \ldots, ENE^{M+1}(t_K))$ denote the exposures for this portfolio to which we contemplate to add a new derivative.
- $(EPE^{M+1}(t_1) - EPE^M(t_1), \ldots, EPE^{M+1}(t_K) - EPE^M(t_K))$ and $(ENE^{M+1}(t_1) - ENE^M(t_1), \ldots, ENE^{M+1}(t_K) - ENE^M(t_K))$ then capture the marginal effect of the additional derivative.

In the absence of CSA, the new derivative may significantly distort the existing portfolio profile resulting in markedly different EPE/ENE. The CVA (DVA and FVA as well) is therefore highly dependent on the portfolio composition of each counterparty entering into a non-collateralized trade, thereby leading to "fragmented" markets, each of them being driven to a large extent by the counterparty's portfolio composition. The very concept of *arbitrage-free market* for non-collateralized trades becomes blurred when it comes to CVA (DVA and FVA) pricing. In contrast, the collateralization mechanisms reduce this dependency on the EPE/ENE and promote arbitrage-free valuation and price transparency. We illustrate thereafter that the regulatory pressure to impose CCP clearing and collateralization mechanisms for non-cleared OTC derivatives leaves little room for non-collateralized trades. They are mostly represented by derivatives concluded between non-financial counterparties (corporations, sovereign States) and banks or dealers.

To summarize, Table 30.2 provides a simple overview of what might be expected in terms of EPE/ENE magnitude, and consequently expected xVA materiality, according to the use of CRM techniques. We emphasize that introducing stringent collateralization methods entails a shift from counterparty risk to funding risk, as

Table 30.2 Materiality of EPE/ENE according to the CRM technique

CRM technique	No CSA	CSA with TH MTA and no IM	CSA with TH MTA and one-way IM	CSA with TH=MTA=0 and one-way IM	CSA with TH=MTA=0 and bilateral IM	CCP trades
EPE/ENE order of magnitude	High	Medium to low	Low	Low to negligible	Negligible	Negligible to nil
Funding requirements	Nil	Negligible to low	Medium to low	Medium	Medium to high	High

these techniques require mobilizing a funding capacity to finance the VMs and IMs (see Sect. 30.3.3).

30.3.3 Collateralized vs Non-collateralized Trades: Some Statistics

We briefly present a set of statistics related to the current size of the OTC and listed derivatives markets to better illustrate the issues associated with counterparty risk measurement in a market where collateralization has constantly extended over the recent years. The present derivatives market may be split into the following components presented in decreasing order of importance according to their *outstanding notional* (see Fig. 30.9):

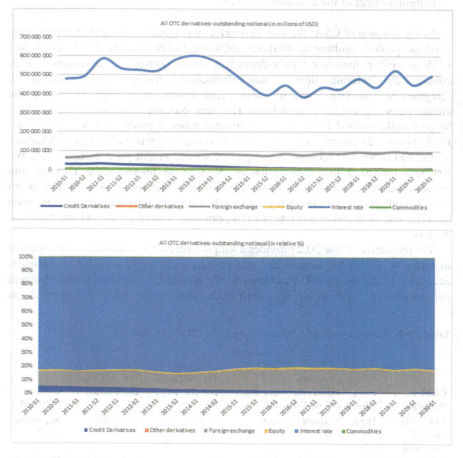

Fig. 30.9 OTC derivatives – evolution of the market size measured by the outstanding notional. Source: BIS

30.3 The "xVA" Framework

- The OTC interest rate derivatives (mostly swaps) are the most traded (80% of the market) and bear by far the longest maturity (up to 30–40 years). Recall that these products are used by the banks and corporates to hedge their interest rate exposures and bear the heaviest xVA costs among all derivatives' asset classes.
- FX products rank second (cross-currency swaps, FX swap) with 15% of the market. Recall that these products have shorter maturities (predominantly 5 years) than those of interest rate derivatives and may display a lighter amount of xVA.
- CDS and Equity derivatives represent less than 5% of the market, they have short maturities (5 years and less) and hence do not bear significant xVA amounts.

The statistics in Fig. 30.10 are key to grasp the evolution of the market, as mandatory clearing is still expanding and will lead to EPE/ENE amounts virtually close to zero. It hence shifts away the counterparty risk from the bilateral OTC derivatives to a systemic risk borne by the CCPs and to a lesser extent to bilateral OTC under a CSA.

This shift is not a "free lunch" as it solely transforms the counterparty risk into a funding risk. CCPs require strong funding capacities from their members to fund the VMs and IMs and that increases the liquidity risk and the costs to manage them. Besides, the CCPs are regulated entities and should be sufficiently capitalized and supervised to ensure that their systemic risk is adequately monitored. The counterparties which remain under CSA without CCPs clearing need also to rely upon strong funding capacity.

The magnitude of the EPE/ENE, especially for the non-collateralized derivatives, depends on their maturity and increases with it. We propose the following classification regarding the level of xVA with a specific focus on OTC and interest rate and FX derivatives which represent the largest part of the derivative market (Fig. 30.10):

- Eighty percent of the OTC interest rates derivatives are cleared through CCPs and represent 80% of the OTC derivative market (Fig. 30.9), which entails that 64% of the OTC derivative market is subject to perfect collateralization. The xVA amounts are virtually equal to zero and it considerably limits the scope of the previous discussions. OTC FX derivatives are rarely cleared through CCP.
- Bilateral (i.e., non-cleared) OTC derivatives traded under a CSA (with other financial institutions and reporting dealers) approximately represent 15% of the market of interest derivatives and practically 80% of FX derivatives. In both cases, the xVA amounts should remain limited and the computational challenge is driven by the CSA provisions (e.g., cash/debt securities eligible as collateral, margin call frequencies).
- OTC derivatives with non-financial counterparties (e.g., sovereign States, Corporates) represent less than 5% of the market for interest rate derivatives and 20% for FX derivatives and would concentrate the most complex xVA challenges. It is worth noting that in various jurisdictions (especially in the Euro zone) the introduction of a CSA has gradually become mandatory. By 2022–2023, solely the sovereign States and the entities trading small volumes

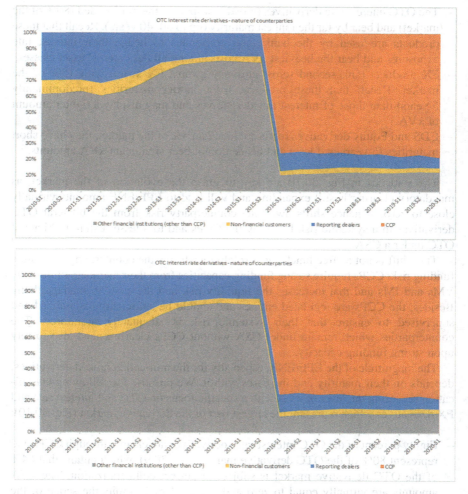

Fig. 30.10 OTC interest rate and FX derivatives—evolution of the nature of the counterparties. Source: BIS

of derivatives will remain exempted from CSAs to avoid putting pressure on their funding capacities. The other non-financial entities will be required to introduce collateral mechanisms or shift to CCP clearing.

Finally, Fig. 30.11 illustrates that the volumes of listed (exchange traded) derivatives, subject by construction to CCP clearing (see Chap. 1), remain limited in comparison with the other OTC cleared derivatives.

30.3 The "xVA" Framework

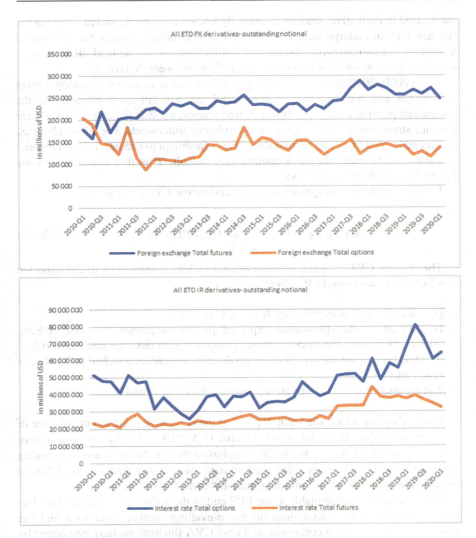

Fig. 30.11 Exchange traded derivatives—evolution of the outstanding notional for Interest Rate and FX derivatives. Source: BIS

30.3.4 Introduction to CVA

We start with the first value adjustment introduced into derivative pricing in the dawn of the 2000–2010 decade. At that time, the most advanced institutions[31]

---

[31] For instance, universal banks and investment firms dedicated to capital market activities, neither being entitled to take on client deposits or granting loans.

specialized in derivatives trading, commonly known as *dealers*, started to revisit the key assumption underpinning the valuation of derivatives, namely the absence of default of the counterparty. Recall that until Chap. 28 we derived all valuations formulas by assuming that all contractual cash flows were delivered.

In the 2000–2010 decade, the signing of a CSA was not customary among market participants and the rare early adopters did not pay sufficient attention to the contractual provisions (e.g., margin call frequencies, eligible collateral). They were also not streamlined by an appropriate regulatory framework. The most advanced dealers addressed this issue by introducing an adjustment to account for the potential loss they would incur should a counterparty default. This first adjustment is the *Credit Value Adjustment* (CVA).

Let us consider counterparties A and B and define $CVA_{A->B}$ as:

$$CVA_{A->B} = (1 - R_B) \int_0^T D(t) EPE_A(t) \mathrm{d}\mathbb{P}^B(t). \tag{30.7a}$$

The amount $CVA_{A->B}$ is a positive quantity which corresponds to the expected loss from A's viewpoint to B, where:

- $(1 - R_B)$ is the loss incurred by A when B defaults.
- $D(t) EPE_A(t)$ is the discounted expected positive exposure (as seen in Sect. 30.3.1), $D(t)$ being the discount factor computed with risk-free rate curve $r(t)$.
- $\mathrm{d}\mathbb{P}^B(t)$ is the *instantaneous RN (density) default probability* attached to B between time t and $t + dt$.
- T is the maturity of the derivatives portfolio between A and B.

From A's perspective, the value $V_A^{Tot}(0)$ of the derivative portfolio in presence of a counterparty risk, therefore, writes at time 0: $V_A^{Tot}(0) - CVA_{A->B}$, and in turn writes $V_B^{Tot}(0) - CVA_{B->A}$ for B. We emphasize that in the presence of a netting agreement, which constitutes the predominant case in the market, the CVA is computed on a portfolio and not on a trade-by-trade basis.[32]

CVA fluctuations attributable to the EPE and/or the default probabilities translate into a profit or a loss depending on the prevailing market conditions and the evolution of the portfolio composition. These CVA fluctuations may be hedged by dynamic strategies which imply the simultaneous hedge of both the EPEs and the default probabilities. The interested reader may consult the Suggested References to learn more about CVA hedging.

More generally, counterparty A faces multiple counterparties. Therefore, if we assume that A trades with a number L of counterparties, the total CVA writes:

$$CVA_{Total} = \sum_{l=1}^L CVA_{A->l}.$$

---

[32] It is possible but very burdensome to allocate the CVA to every derivative, as this requires computing the marginal contribution of each deal to the EPE (a convex function).

30.3 The "xVA" Framework

In practice, we use a time grid ($t_0 = 0, t_1, t_2, ..., t_k, ..., t_K = T$) to discretize Eq. (30.7a) and compute the EPE array ($EPE(t_1), ..., EPE(t_K)$) as defined in Sect. 30.3.1, which leads to:

$$CVA_{A->B} \approx (1 - R_B) \sum_{k=1}^{K} EPE_A(t_k)D(t_k)\left(\mathbb{P}^B(t_k) - \mathbb{P}^B(t_{k-1})\right). \tag{30.7b}$$

This relationship states that the CVA is (the sum of) the discounted expected loss $(1 - R_B)EPE_A(t_k)D(t_k)$, idem provided that the counterparty B has "survived" until t_{k-1} and defaults in the interval (t_{k-1}, t_k). $\mathbb{P}^B(t_k)$ denotes the cumulative RN probability of default of B until time t_k. As illustrated in Table 30.2, the EPE is close to zero if the CSA stipulates that TH = MTA = 0 with daily margin calls and if the portfolio is composed of liquid derivatives to minimize the MPOR and hence the close-out costs.

Equations (30.7a and 30.7b) embed specific and not innocuous assumptions, namely:

- The risk factors which govern the derivatives portfolio value are independent from the default of the counterparty, thereby enabling to compute separately $EPE_A(t_1), ..., EPE_A(t_K)$ and $\mathbb{P}^B(t_1), ..., \mathbb{P}^B(t_K)$.
- If the collateral is composed of securities, the default risk of their respective obligors is independent from that of the counterparty.

When the default risk of the counterparty and the risk factors are not independent, their dependence is called the *wrong-way risk* (WWR) and the formula to compute CVA is derived in a more generic setting that incorporates the joint density of the risk of B's default and the risk factors governing the portfolio of derivatives. It writes in a general setting with a CSA (i.e. with VM and IM):

$$CVA_{A->B} = (1 - R_B) \int_0^T \mathbb{E}^{(V_A^{Tot}(t) \times \tau_B)} \left(I_{(\tau_B \leq t)}D(t)\, max\left(V_A^{Tot}(t) - VM_t - IM_t, 0\right)\right)dt \tag{30.8}$$

where:

- $I_{(\tau_B \leq t)}$ denotes the indicator function and τ_B the time of B's default[33] (which must occur before the maturity of the portfolio).
- $D(t)\, max\,(V_A^{Tot}(t) - VM_t - IM_t, 0)$ is the residual exposure at the time of default.
- $\mathbb{E}^{(V_A^{Tot}(t) \times \tau_B)}$ is the joint (density) of B's default probability and the risk factors governing the portfolio of derivatives.

---

[33]Technically, τ is a *stopping time*.

We recover Eq. (30.7a), if we assume again that B's default probability is *independent* from the risk factors of the derivatives, where (30.8) writes (in case of independence, the joint expectation is the product of the individual expectations):

$$CVA_{A->B} = (1-R_B) \int_0^T \mathbb{E}^{(V_A^{Tot}(t))} (D(t)max(V_A^{Tot}(t) - VM_t - IM_t, 0)) \mathbb{E}^{(\tau_B)} \left(I_{(\tau_B \le t)} \right) dt$$

We do not address the WWR measurement any longer as those techniques are beyond the scope of this book. Actually, practitioners avoid bearing WWR as it is very complex to evaluate and consequently implement risk mitigation techniques. This is achieved either through CSA contractual provisions or by standard risk management aiming at pinpointing at the *pre-trade* stage any potential incoming trades tainted with WWR and potentially refusing them. Risk mitigation may involve:

- To ensure that no legal link exists between the risk factors driving the derivative value and the risk of the counterparty. For instance, when trading an equity derivative, the underlying stock should not be legally connected with the counterparty. Equity derivatives or single-name CDS where the underlying / reference entity is a subsidiary of the counterparty should for example be banned. This risk is called the *specific WWR* by the regulators and is easily identifiable.
- To ensure that the risk factors driving the underlying value do not materially increase the credit worthiness of the counterparty. This risk requires a more refined analysis of the counterparty's sensitivity to the derivatives' risk factors. Customary situations embrace interest rate swaps or FX derivatives with a one-way CSA or no CSA when the counterparty is a sovereign State or a corporation. Assume for example that a sovereign State has contracted a receiver fixed rate interest rate swap (payer floating rate) with a dealer to hedge its fixed rate debt. If interest rates soar, the value of the swap becomes increasingly negative for the sovereign State and positive for the dealer. Since the State does not post collateral due to one-way CSA provisions, the dealer runs a "winning" non-collateralized trade on paper but faces an increasing probability of default by the State. This situation is typical of what regulators call *generic WWR*.
- To ensure that the risk factors affecting the value of the received collateral are independent from the risk of default of the counterparty. This may be achieved by the prohibition of certain types of collateral (e.g., Government bonds issued by the very sovereign State which is the counterparty). Note that in case of dependence between B's default probability and the risk factors governing the value of the portfolio, the relevant Eq. (30.8) would be even more complex as the joint density between the risk of B's default and the risk factors governing the value of the collateral should be introduced.

Let us now address the computation of the default probabilities $\mathbb{P}^B(t_1)$, ..., $\mathbb{P}^B(t_K)$. As CVA is a component of the derivative valuation, it must be determined in accordance with the available market information to reflect the market price of the default. Therefore, according to our previous discussion in Sect. 30.3.2, the most appropriate information consists in using the CDS of the counterparty to determine its RN default probability. Let us assume that B has a traded CDS, denoted by s^B.

30.3 The "xVA" Framework

Equation (30.7b) rewrites:

$$CVA_{A->B} \approx (1 - R_B) \sum_{k=1}^{K} EPE_A(t_k)D(t_k)$$

$$\times \left(\exp\left(-\frac{s^B(t_{k-1})t_{k-1}}{(1-\alpha)} \right) - \exp\left(-\frac{s^B(t_k)t_k}{(1-\alpha)} \right) \right).$$

We used the triangle formula and adapted the notations of Sect. 30.1. Let us define $1 - \gamma^*(\theta) \equiv \mathbb{P}^B(\theta)$. This leads to $\mathbb{P}^B(\theta) = 1 - exp\left(-\lambda^B\theta\right) = 1 - exp\left(-\frac{s^B(\theta)\theta}{(1-\alpha)}\right)$, due to the triangle formula expressed at time θ.

We may remark that the CDS curve conveys information that does not necessarily correspond to a default probability, but rather to a credit event probability, making it perhaps not the best candidate to compute the CVA. However, we discussed and concluded previously (see Sects. 30.1 and 30.1.2.1) that the information embedded in other default risk measures (e.g., the asset-swap spread or the z-spread) is less accurate than that of the CDS, leaving us no other choice than to stick to the CDS spread.

Remark that we use α in conjunction with R_B, whereas those two quantities denote the recovery rate of the counterparty B and should presumably be equal. This statement, however, does not necessarily hold. R_B is the recovery rate expected to be achieved in case B defaults, while α is a market convention used to strip the default probability from the quoted CDS spreads and used here to evaluate the CVA. Recall that we must therefore stick to $\alpha = 40\%$ or $\alpha = 20\%$ to justify that the quantity $1 - exp\left(-\frac{s^B(\theta)\theta}{(1-\alpha)}\right)$ corresponds to a default probability (and not an expected loss), as long as the default is not looming (see Sect. 30.1.2.1).

Besides, the prevailing accounting standards addressing the xVA[34] measurement impose to extensively use market-implied parameters unless they do not exist or cannot be reliably approximated, thereby leading to $R_B = \alpha$ in absence of specific information. However, if counterparty A has some additional rights (such as a more senior position than some other debtors) or benefits from specific guarantees in B's insolvency proceedings (e.g., customary practice in project financing), it may be entitled to adjust R_B accordingly and come up with $R_B \neq \alpha$. We must distinguish those rights applicable only in case of default from the provisions of the CSA whose role is to reduce the EPE/ENE over the life of the portfolio.

If a counterparty is not part of a CDS, the default probability may or must be approximated. The use of a proxy is mandatory in various jurisdictions and especially in the Euro zone. A proxy consists in determining the spread of a counterparty by mapping it onto the average CDS of a portfolio composed of entities with similar attributes (rating, industry, and region). Even if this method has some drawbacks (e.g., the number of available CDSs does not always permit to determine an appropriate proxy CDS spread), it is often sufficient to determine a suitable default probability.

---

[34] IFRS 13 regarding the fair value measurement is in force in many jurisdictions.

We conclude this subsection by considering again CVA determination when the counterparty is a CCP. Earlier on, we stated that this CVA would be challenging to assess accurately as it is expected to be very small due to the stringent collateral mechanism put in place which leads to an EPE virtually equal to zero. In addition, the determination of the default probability $\mathbb{P}^{CCP}(\theta)$ and the recovery rate R_{CCP} is virtually (or almost) impossible. First, CDSs on CCPs do not exist since CCPs do not issue debt. The proxy approach is irrelevant for the same reason and there is no industry comparable to that of the CCP. Second, the recovery rate determination would be purely subjective and anyway close to 100% as a CCP has a relatively large equity by regulation. Market participants therefore generally assume that the CVA is zero in this case.

30.3.5 Introduction to DVA

Let us consider again the portfolio of derivatives traded between two counterparties A and B. We previously stated that its value from A's viewpoint wrote $V_A^{Tot}(0) - CVA_{A->B}$ and $V_B^{Tot}(0) - CVA_{B->A}$ from B's viewpoint. We may also reasonably assume, in accordance with the market statistics in Sect. 30.3.3, that if the portfolio is mostly composed of vanilla or semi-exotic derivatives, A and B use similar models and market parameters to determine $V_A^{Tot}(0)$ and $V_B^{Tot}(0)$. Therefore, we should have $V_B^{Tot}(0) = -V_A^{Tot}(0)$ or at least $V_B^{Tot}(0) \approx -V_A^{Tot}(0)$.

However, taking into account the CVA, which is always a positive quantity, makes A and B disagree on the valuation of their portfolios, since $\mid V_A^{Tot}(0) - CVA_{A->B} \mid$ obviously differs from $\mid V_B^{Tot}(0) - CVA_{B->A} \mid$ except if both CVAs were zero (along with the assumption $\mid V_B^{Tot}(0) \mid = \mid V_A^{Tot}(0) \mid$).

We introduce the *Debit Value Adjustment* (DVA), also called the *Debt Value Adjustment*, to recover the symmetry between counterparties and allow consistent portfolio valuation in presence of counterparty risk.

In the same way, CVA is defined from EPE, DVA is defined from ENE:

$$DVA_{A->B} = -(1 - R_A) \int_0^T D(t) ENE_A(t) d\mathbb{P}^A(t) \text{ and}$$

$$DVA_{B->A} = -(1 - R_B) \int_0^T D(t) ENE_B(t) d\mathbb{P}^B(t),$$

where:

- $DVA_{A->B}$ and $DVA_{B->A}$ are the (positive) DVA attached to A and B.
- R_A and R_B respectively denote the recovery rate on A and B.
- $ENE_A(t)$ and $ENE_B(t)$ are the ENEs determined from A and B's viewpoints, respectively.
- $\mathbb{P}^A(t)$ and $\mathbb{P}^B(t)$ are the RN (or FN) default probabilities.

30.3 The "xVA" Framework

We then have the following portfolio valuations in presence of both CVA and DVA:

- From A's perspective: $V_A^{Tot}(0) - CVA_{A->B} + DVA_{A->B}$.
- From B's perspective: $V_B^{Tot}(0) + DVA_{B->A} - CVA_{B->A}$.

We should therefore have: $|V_A^{Tot}(0) - CVA_{A->B} + DVA_{A->B}| = |V_B^{Tot}(0) + DVA_{B->A} - CVA_{B->A}|$ to reconcile the derivatives portfolio values in presence of counterparty risk so that A and B may trade on a single price assuming that $V_B^{Tot}(0) = -V_A^{Tot}(0)$. This equality implies that $|DVA_{B->A}| = |CVA_{A->B}|$ and similarly that $|DVA_{A->B}| = |CVA_{B->A}|$. The DVA for one counterparty thus is equivalent to the CVA of the other party and reciprocally.

Recall also from Eq. (30.7a) that $CVA_{A->B} = (1 - R_B)\int_0^T D(t)EPE_A(t)d\mathbb{P}^B(t)$ and likewise, we posit $CVA_{B->A} = (1 - R_A)\int_0^T D(t)EPE_B(t)d\mathbb{P}^A(t)$. We hence should have $|EPE_A(t)| = |ENE_B(t)|$ and $|EPE_B(t)| = |ENE_A(t)|$ barring the frictional computations that exist in practice.

The DVA *fluctuations* are accounted as profits or losses according to market conditions. However, the DVA seems less intuitive to understand than the CVA. Let us continue to assume that B is the borrowing counterparty, i.e. $V_B^{Tot}(0) < 0$, and that B is extremely risky while A's credit risk remains unchanged. If we rewrite as in Eq. (30.7b) the equation defining the DVA from B's viewpoint, we have:

$$DVA_{B->A} = -(1 - R_B)\sum_{k=1}^K ENE_B(t_k)D(t_k)$$
$$\times \left(\exp\left(-\frac{s^B(t_{k-1})t_{k-1}}{(1-\alpha)} \right) - \exp\left(-\frac{s^B(t_k)t_k}{(1-\alpha)} \right) \right),$$

where $s^B(t_k)$ denotes the CDS spread (coupon) of B at time t_k. Assuming, to ease the understanding, a constant s^B and $R_B = \alpha$ yields the following approximation:

$$DVA_{B->A} \approx -\sum_{k=1}^K ENE_B(t_k)D(t_k)s^B(t_k - t_{k-1}).$$

All other things being equal, $DVA_{B->A}$ (positive since ENE is negative) is, therefore, larger when B's credit spread s^B is higher. The value $V_B^{Tot}(0) + DVA_{B->A}$ then is less negative (since $V_B^{Tot}(0) < 0$), which *reduces* B's liability to A. Thus, somewhat counter-intuitively, B should record *a profit* when its own credit worthiness (measured by its CDS spread) *deteriorates* and conversely a loss when it improves. The DVA, therefore, represents B's *own default risk* (and by extension each counterparty's own default risk) as it uses its *own CDS spread* and its *own recovery rate*, in contrast to the CVA which uses both the counterparty's CDS spread and its recovery rate.

To regain intuition, let us observe that in case B defaults and $V_B^{Tot}(0) < 0$, B's shareholders will not be requested to repay the entire value of their liabilities,[35] $V_B^{Tot}(0)$, but solely a fraction equal to $V_B^{Tot}(0) + DVA_{B->A}$ which corresponds to making a profit from their perspective. In case B does not default, this (paper) profit will never translate into cash for B's shareholders.

Accounting-wise (both under IFRS and US GAAPs), this profit should be booked to respect the symmetry with CVA and valuation consistency. In contrast, and against financial theory, the regulators deem that this profit is fictitious as solely accruing to the shareholders in case of default and thereby should not be recognized as part of the banks' regulatory capital. This stand against sound financial theory may be explained by a precautionary motive on the part of regulators. We revisit the DVA in the following Sect. 30.3.6 when we discuss the FVA puzzle.

30.3.6 The FVA Puzzle

Throughout this book, we evaluated all the derivatives by discounting their contractual cash-flows at the risk-free rate. We asserted that the prevailing interbank money market references (IBOR), should be used as the appropriate risk-free rate. The ongoing benchmark reform which aims to replace the IBOR with new references (e.g., Euribor+, SOFR) by 2023 does not alter the following analysis.

We have purposely ignored hitherto the contractual collateral remuneration prevailing in any CSA. The counterparty posting the collateral (IM or VM) incurs a financing cost which is contractually compensated by the interest rate paid by the counterparty receiving the collateral. The prevailing interest rate for all CSAs is the *Overnight Indexed Swap* (OIS) which is the compounded fixed rate of the *overnight interbank funding rates*, for instance: the €STER (formerly the Eonia in the Euro zone) or the *Sonia* (for the GBP). This reference has been adopted in all CSAs, as it represents the fair collateral remuneration which constantly fluctuates overtime (see Chap. 7).

We now illustrate the utmost importance of the OIS for cash-flow discounting. Suppose that counterparty A lends \$1 at the risk-free rate r_T^{Ibor} to counterparty B until time T and receives \$1 as collateral being contractually remunerated at r_T^{OIS}. If we assume that A does not charge any commercial margin and that B is riskless, A may borrow \$1 at r_T^{Ibor} on the interbank market at inception and lend this amount to B, while lending the received collateral at r_T^{Ibor}.

The cash flows at maturity, from A's viewpoint, are as follows:

- A receives from B: $exp\left(r_T^{Ibor}T\right)$.
- A repays on the interbank market the amount initially borrowed: $-exp\left(r_T^{Ibor}T\right)$.

---

[35] Assuming that B is a limited liability company, which is always the case for banks and other counterparties issuing debt securities on the market.

30.3 The "xVA" Framework

- A receives from the interbank market the collateral remunerated at the risk-free rate: $exp\left(r_T^{Ibor}T\right)$.
- A returns to B the collateral contractually remunerated at OIS: $-exp\left(r_T^{OIS}T\right)$.

The net cash flow at time T, positive or negative depending on the sign of the *IBOR-OIS spread*, is equal to $exp\left(r_T^{Ibor}T\right) - exp\left(r_T^{OIS}T\right)$.

The sign of this quantity determines whether A is incurring an undue (or "unfair") interest gain or loss. We insist that the OIS is no more a risk-free rate than the IBOR from a financial viewpoint. Indeed, the OIS is the compounded *overnight* interbank market rate until time T, while the IBOR is the average interbank lending/borrowing rate prevailing over the same period. More precisely, until 2008 market participants believed that lending money on the interbank market for a T-period (up to 1 year, but in practice mostly ranging between 3 and 6 months) was equivalent as rolling an overnight lending position on the same period as the interbank money market was virtually risk-free (i.e. banks could not default). This belief was so unquestionable that the *IBOR-OIS spread* has traded steadily for a long time at around 2 basis points (bps) only.

In 2008, the turmoil of the subprime crisis questioned the banks' credit worthiness and the IBOR-OIS spread peaked at 200–300 bps and has remained between 20 and 50 bps onwards. Also, the liquidity on the interbank market practically disappeared and did not recover except on the overnight market. Market participants running large collateralized derivatives portfolios with an enduring *negative* value, i.e., $V^{Tot} < 0$ (corresponding to a borrowing position), observed that the IBOR reference used as the discounting rate led to incur significant losses on a running basis, as the value $exp\left(r_T^{OIS}T\right) - exp\left(r_T^{Ibor}T\right)$ became strongly negative. They rapidly switched to OIS discounting to accommodate this situation for the future. The change of the discounting curve required to reevaluate the derivative portfolios led almost all market participants to book profits or losses, sometimes significant, equal to the difference $V^{Tot,\ IBOR} - V^{Tot,\ OIS}$.

This preliminary discussion justifies the following statements:

- All collateralized derivatives cash flows should be discounted with the CSA prevailing reference rate, namely the OIS.
- All non-collateralized, or partially collateralized (e.g., one-way CSA, CSA with weekly variation margin calls, high values for the TH or the MTA) derivatives cash flows should be valued as a collateralized derivative and *adjusted* with the appropriate valuation adjustments set out below.
- Collateralization mechanisms create per se segregated markets (e.g., perfectly collateralized, imperfectly collateralized, not collateralized) which have their own specificities and are impossible to reconcile. In other words, the non-arbitrage rule is to be appreciated for each market, knowing that the fully and perfectly collateralized market represented by CCP-cleared trades accounts today for the vast majority of the derivatives trades.

Incidentally, let us go back briefly to the choice of the reference rate prevailing in the CSA to discuss one issue so far omitted. The CSA provisions, albeit referencing the OIS as the rate to be applied to the collateral, were initially tailored to leave the maximum contractual flexibility to each party with regards to its currency or its nature (cash and /or debt securities). Each party may change the currency or the nature of the collateral over the life of the portfolio to take advantage of the prevailing market conditions. This contractual leeway is an embedded *option* at the hand of both counterparties. It is more complex to evaluate as the range of the contractual provisions gets larger. It should be accounted for with an additional adjustment called the *Collateral Valuation Adjustment* (CollVA). However, it is too complex to evaluate in practice, especially in disrupted market conditions and may impede price transparency and hence liquidity. From 2013 onwards, the ISDA initiated discussions to streamline contractual provisions (e.g., align the derivative and variation margin currencies, permit only cash collateral for the variation margin, limit the eligible range of securities for the initial margin) into a *standardized CSA* (SCSA). The CollVA then became of a lesser importance from a mathematical viewpoint, as numerous counterparties adopted the SCSA for their non-CCP trades still running. The interested reader will find useful references as to CollVA valuation techniques at the end of the chapter.

Let us now consider again the derivative portfolio between counterparties A and B in absence of a CSA. Assume that $V_A^{Tot}(t) \geq 0$ until maturity T. A bears an exposure to B, and needs to finance its position until T. The financing cost departs from the fully collateralized situation and has to be accounted for through an additional adjustment called the *Funding Cost Adjustment* (FCA) over the risk-free rate (OIS). By contrast, if $V_A^{Tot}(t) \leq 0$, A is liable to B but must not post any collateral and saves financing costs. A thus benefits from a favorable funding situation, which has also to be accounted for with a *Funding Benefit Adjustment* (FBA). The *Funding Value Adjustment* (FVA) is the sum of those two components: FVA= FBA+FCA. This quantity is also determined at the portfolio level.

To determine the cost of financing the positive exposure for A, we need to simulate the exposure value. We use the same array $EPE(t_1)$, $\ldots$, $EPE(t_K)$ as previously estimated. We denote by $FCA_{A->B}$ the financing cost of the uncollateralized exposure incurred by A, which is approximately equal to:

$$FCA_{A->B} \approx \sum_{k=1}^{K} D_{OIS}(t_k) EPE_A(t_k) \left(\exp\left(-X \, t_{k-1}\right) - \exp\left(-X \, t_k\right) \right),$$

where $D_{OIS}(t) = exp\left(-r^{OIS} \, t\right)$. Define $D_X(t) = exp\left(-X \, t\right)$, where X is the funding spread for A *over the risk-free rate* (equal to OIS). We assume to ease the notation that r^{OIS} and X are constant over time. If we approximate[36] once again the previous relationship, we have:

---

[36] $\exp(-u) \approx 1 - u$.

30.3 The "xVA" Framework

$$FCA_{A->B} \approx \sum_{k=1}^{K} D_{OIS}(t_k)EPE_A(t_k)X(t_k - t_{k-1}).$$

Similarly, $FBA_{A->B}$ is approximated as:

$$FBA_{A->B} \approx -\sum_{k=1}^{K} D_{OIS}(t_k)ENE_A(t_k)X(t_k - t_{k-1}).$$

We finally come up with a total value (neglecting the CollVA) for A's derivative portfolio equal to:

$$V_A^{Tot}(0) - (CVA_{A->B} + FCA_{A->B}) + (DVA_{A->B} + FBA_{A->B}),$$

where the CVA and FCA are accounted as costs, therefore reducing the value of the portfolio, and the DVA and FBA alleviate these costs.

Assume A is a bank (or a dealer). Taking a closer look at the nature of the funding spread X, we observe that it corresponds to the net cost of financing for A if it borrows money on the market at a total cost of $r^{OIS} + X$. To put it differently, bank A sets X as its *target funding spread* over OIS when issuing debt securities. It represents the spread that the bank is ready to pay investors for bearing its default risk. This target funding spread is continuously revised by the bank to reflect the prevailing financing conditions.

The spread X necessarily embeds a *systemic interbank market cost,* which is to some extent captured by the IBOR-OIS spread, and an *idiosyncratic* cost of fund attributable specifically to A. X reflects the capacity to access the funding market and A's own default risk. There is then some redundancy in the expression $(DVA_{A->B} + FBA_{A->B})$ as X contains a fraction of A's default risk already captured by the DVA. One may think that this portion could be easily isolated, for instance by using the CDS written on A, but this is not necessarily the case for technical reasons beyond the scope of this book.

We may simply decompose X as: $X = s^{Market} + s^{Id}$, where s^{Market} represents the average interbank market funding spread over OIS and s^{Id} the idiosyncratic funding spread of the bank embedding a compensation for the investor taking the default risk. There is in fact no ideal solution to split X appropriately to avoid overlapping between DVA and FBA, and market participants use rather practical approaches.

The first approach consists in aligning X to the target funding spread determined by the bank and in assuming that $s^{Id} = s_{CDS}$, so that $s^{Market} = X - s_{CDS}$. In this case, there is no overlap between DVA and FBA and the DVA is reported in accordance with the prevailing accounting requirements, namely:

$$FBA_{A->B} \approx -\left(FBA^{Market}{}_{A->B} + DVA_{A->B}\right)$$

$$FBA_{A->B} \approx -\left(\sum_{k=1}^{K} D_{OIS}(t_k)s^{Market}ENE_A(t_k)(t_k - t_{k-1}) \right.$$

$$\left. + \sum_{k=1}^{K} D_{OIS}(t_k)s_{CDS}ENE_A(t_k)(t_k - t_{k-1}) \right).$$

To simplify this relationship, we use the same approximation as before and we assume that $R_A = \alpha$ as, in contrast to the CVA, we use the bank's own CDS spread in the DVA:

$$DVA_{A->B} \approx (1 - R_A) \sum_{k=1}^{K} D_{OIS}(t_k)ENE_A(t_k)$$

$$\times \left(\exp\left(-\frac{s_{CDS}t_{k-1}}{(1-\alpha)}\right) - \exp\left(-\frac{s_{CDS}t_k}{(1-\alpha)}\right) \right)$$

$$\approx \sum_{k=1}^{K} D_{OIS}(t_k)ENE_A(t_k)s_{CDS}(t_k - t_{k-1}).$$

Then, $FVA_{A->B} = FCA_{A->B} + (FBA^{Market}_{A->B} + DVA_{A->B})$ and again we have: $V_A^{Tot}(0) - (CVA_{A->B} + FCA_{A->B}) + (FBA^{Market}_{A->B} + DVA_{A->B})$. This approach is customary called *bilateral* CVA as it contains an explicit DVA component.

The second approach consists in disregarding completely the DVA and aligning X on the target funding spread, which yields:

$$FVA_{A->B} = FCA_{A->B} + FBA_{A->B}$$

and again from A perspective:

$$V_A^{Tot}(0) - (CVA_{A->B} + FCA_{A->B}) + FBA_{A->B}.$$

This approach fully embraces the costs and benefits from a non-collateralized portfolio but is not compliant with the accounting requirements as it completely disregards the DVA, although the latter adjustment does not translate into a profit unless the bank defaults.

We end with two remarks. First, the previous analysis relies upon the core assumption ignored so far that a bank borrows and lends money at the *same spread* X over the risk-free rate OIS (symmetric funding). To assess whether this assumption is realistic or not, we can go back to the statistics of the BIS. Vanilla interest rate swaps remain the largest market and its non-collateralized (or one-way collateralized) part is represented by non-financial counterparties (corporations, sovereign States). As interest rate swaps are the instruments that by far have the largest maturities (up to 30 years and sometimes more), the sign of a portfolio predominantly composed of interest rates swaps is likely to fluctuate over the years. If so, assuming on average a symmetric funding spread X for banks appears reasonable as a first approximation.

Second, the arbitrage-free market assumption is somewhat problematic when it comes to *uncollateralized derivatives*. This market is represented by trades between banks or dealers on the one hand and corporations and, to a lesser extent, sovereign States on the other. Since the xVA components (CVA, DVA, FBA, FCA, etc.) are highly dependent upon the portfolio of derivatives upon consideration to determine the EPE/ENE, the notion of "market" is blurred. In effect, arbitrage opportunities are difficult to define and consider since market participants are subject to different conditions as they manage very different portfolios whose composition affects the value of the xVA.

30.4 Summary

- Credit derivatives are an asset class whose popularity and outstanding notional peaked during the 2000–2010 decade before declining by 85% since the aftermath of the subprime crisis. Credit derivatives allow to transfer to a third party the credit/default risk of an asset or a portfolio of assets subject to default risk without transferring or even owing the ownership of the asset itself.
- **Credit Default Swaps** (CDS) are the most important credit derivative. A CDS is a swap with a notional *Nom* and a maturity T involving two counterparties, the *protection buyer* and the *protection seller*. The former pays periodically a series of fixed payments (*coupon* or *spread*, $s(0)$, determined at inception) to the latter in exchange for a compensation equal to *Nom* $(1-\alpha)$ where α denotes the recovery rate achieved on a *reference entity* following its *default*. Thus, the CDS has a fixed leg, the *coupon leg CL*, and a floating leg, the *default leg DL*. At any time t, the market value of the CDS, $V_t(CDS)$, is equal (from the protection buyer's viewpoint) to $V_t(CDS) = V_t(DL) - V_t(CL)$. At inception, $V_0(CDS)$ is equal to 0, as for any swap.
- For a *single-name* CDS, the *reference entity* is a *single* entity (also called obligor) issuing *debt security(ies)* (also called *reference asset(s)*) on the market (corporate, bank, sovereign State). The reference may also be an *index* composed of *various entities* (e.g. corporations belonging to different industrial sectors); the CDS then is called *multi-name* CDS or CDS *on index*.
- A single-name CDS generates the following sequence of cash-flows:

$$
\begin{cases}
-s(0)(coupon), \text{ as long as the reference asset is not in default;} \\
100 \text{ (notional)} - Value\ of\ the\ Reference\ Asset, \text{at the time of default;} \\
\quad\quad\quad \text{or 0 if no default occurred before } T.
\end{cases}
$$

- Single-name and multi-name CDS each represent 45% of the outstanding notional on the CDS market. Other types of CDS exist but are of much lesser importance (e.g., Total Return Swaps, Credit-Linked Notes, options on CDS).

- The CDS coupon $s(0)$ may be determined by *static replication*. The CDS protection buyer takes a *short position* on the reference entity's debt security (bearing a spread $s'(0)$ over Libor) and invests the proceeds on the money market at Libor. The no-arbitrage rule implies in theory that $s(0) = s'(0)$. In practice, $s(0)$ *differs from* $s'(0)$, and the difference, called the *CDS-bond basis*, is explained by several factors such as the respective liquidity of the bond and the CDS markets, or the default risk affecting the CDS protection seller.
- Single-name CDS may be priced with the *stripping method*. With $\gamma^*(t)$ denoting the survival Risk Neutral or Forward Neutral probability on date t, $p*_\theta = \gamma^*(\theta - 1) - \gamma^*(\theta)$ *is the probability of default* of the *reference entity* between dates $\theta-1$ and θ. With $\phi^*(\theta) = 1 - \gamma^*(\theta)$ denoting the cumulative probability of default between 0 and θ, D the time lag between two coupon payments and r the flat risk-free rate curve, *in* a continuous-time model where the default occurs at the first jump of a Poisson process of intensity λ and $\gamma^*(\theta)$ is set equal to $e^{-\lambda\theta}$, we have:

$$V(DL) = -Nom \int_0^T e^{-r\theta}(1 - \alpha)d\gamma * (\theta) = Nom(1 - \alpha)\lambda \int_0^T e^{-(r+\lambda)\theta}d\theta = Nom (1 - \alpha)$$
$$\lambda \frac{1-e^{-(r+\lambda)T}}{r+\lambda} \text{ and } V(CL) = Nom\, s(0) \int_0^T e^{-(r+\lambda)\theta}d\theta = Nom\, s(0) \frac{1-e^{-(r+\lambda)T}}{r+\lambda}.$$ At inception, since $V_0(CDS) = V_0(DL) - V_0(CL) = 0$, we recover the *credit triangle formula:* $s(0) = (1 - \alpha)\,\lambda$. The value α is set by market convention at 40%.
- *Securitization* is a mechanism used mostly by banks or corporations to transfer the credit/default risk of a portfolio of *non-tradable assets*, also called *pool*, to external investors. Securitization consists in transforming the portfolio of *non-tradable assets* into *tradable securities* that can be sold on a financial market. These tradable securities were initially called *Asset Backed Securities* (ABS), where the underlying assets are for instance *residential mortgages (RMBS) or commercial mortgages (CMBS)*.
- Securitization articulates around 2 types, on-balance sheet and off-balance sheet. *On-balance sheet* securitization is used exclusively by banks that refinance a dedicated pool of loans (*collateral*) present on their balance sheet (mostly retail mortgage loans) through a *covered bond* issuance. In case the bank defaults, the bondholders take possession of the collateral without taking part to the liquidation process. Covered bonds thus are considered virtually default-free (and thus often rated AAA).
- *Off-balance sheet* securitization is a mechanism whereby an entity (a bank but also a corporation, called the *originator*) sells the *pool* to a dedicated entity called a *Special Purpose Vehicle* (SPV), which in turn refinances it by simultaneously issuing ABS purchased by external investors. Although the payments received by the ABS holders (from the SPV) are *globally* equal to those paid by the pool of assets, one can structure them in *tranches* (*tranching mechanism*) that differ by their maturity and seniority, hence their level of credit risk. The cash flows collected from the asset pool are allocated as a *waterfall* whereby the most *senior* ABS holders are repaid first (*senior tranche*), then the second tranche (*mezzanine*), and so on until the last, most risky, tranche (*equity tranche*).
- The **xVA framework** has been elaborated to cope with the *counterparty risk* which an investor is exposed to on its positions in derivatives. Let A and B be the

30.4 Summary

counterparties of a portfolio composed of M derivatives, $V_{t,i}^{+/-}$ the value of each derivative at time t, $V_t^{Tot} = \sum_{i=1}^{M} V_{t,i}^{+/-}$ the value of the portfolio and T the longest maturity of the derivatives composing the portfolio. If $V_t^{Tot} \geq 0$ from A's viewpoint, A is exposed to B's *counterparty risk*: if B defaults at t, A incurs a loss equal to $V_t^{Tot}(1 - R_B)$ where R_B is the recovery rate achieved on B. The roles of A and B are symmetric if we consider that $V_t^{Tot} \geq 0$ from B's viewpoint.

- Counterparty risk can be mitigated with *contractual provisions* known as *Credit Risk Mitigation* (CRM) techniques which take the form of margin calls. The *variation margin* (VM) is an amount of cash or securities posted by the indebted counterparty (B here) to reduce the exposure of A to $\left(V_t^{Tot} - VM_t\right) \geq 0$. The *initial margin* (IM) complements the variation margin at time t to protect against the *unknown* fluctuations of V_t^{Tot} until the next (variation) margin call date t', in case one counterparty would default between these two dates.

- Current regulations in force in Europe and the US impose either the *clearing* with Central Counterparties (CCP) where perfect collateralization, daily margin calls, and initial margin are mandatory, or *bilateral implementation* of CRM techniques that mimic those of a CCP. Currently, only a small fraction of the derivatives traded (embracing all asset classes) is free from these mechanisms, especially the derivatives traded by corporations and sovereign States.

- For these counterparties still exempt from CRM and hence subject to counterparty risk, market practices and accounting standards have gradually imposed *valuation adjustments*, known collectively as "xVA", to prevent arbitrage. Expected positive exposure and expected negative exposure are defined, respectively, by:

$$EPE(t_k) = \frac{1}{N} \sum_{j=1}^{N} max\left(V_{t_k}^{Tot}(j), 0\right) \quad \text{and} \quad ENE(t_k) = \frac{1}{N} \sum_{j=1}^{N} min\left(V_{t_k}^{Tot}(j), 0\right)$$

where j denotes the jth simulated value of the portfolio at time t_k, with ($t_0 = 0$, $t_1, t_2, ..., t_k, ..., t_K = T$).

- *Credit Value Adjustment* (CVA) is the *expected loss* incurred by A (in case B defaults and A survives) on the derivative portfolio: $CVA_{A->B} \approx (1 - R_B) \times$

$$\sum_{k=1}^{K} EPE_A(t_k) D_{OIS}(t_k)\left(exp\left(-\frac{s^B(t_{k-1})t_{k-1}}{(1-\alpha)}\right) - exp\left(-\frac{s^B(t_k)t_k}{(1-\alpha)}\right)\right), \text{ where } D_{OIS}(t_k) \text{ is}$$

the discount factor computed with the risk-free OIS rate curve, $s^B(t_k)$ is the CDS spread of B at time t_k derived with the previous credit triangle formula.

- Similarly, the *Debit Value Adjustment* (DVA) borne by A to B is the *expected profit* incurred by A (in case B defaults and A survives): $DVA_{A->B} =$

$$(1 - R_A) \sum_{k=1}^{K} ENE_A(t_k) D_{OIS}(t_k)\left(exp\left(-\frac{s^A(t_{k-1})t_{k-1}}{(1-\alpha)}\right) - exp\left(-\frac{s^A(t_k)t_k}{(1-\alpha)}\right)\right).$$

- The absence of variation margin entails either a cost to finance the exposure $V_A^{Tot}(t) \geq 0$ or a financial benefit when $V_A^{Tot}(t) \leq 0$. Like for the CVA and DVA, the respective expected financing costs and benefits should be accounted for by *a Funding Value Adjustment* (FVA), decomposed into a *Funding Cost Adjustment* (FCA) and a *Funding Benefit Adjustment* (FBA), with FVA = FBA + FCA. From A's point of view, we thus have: $FCA_{A->B} \approx \sum_{k=1}^{K} D_{OIS}(t_k)EPE_A(t_k)X(t_k - t_{k-1})$

and $FBA_{A->B} \approx - \sum_{k=1}^{K} D_{OIS}(t_k)ENE_A(t_k)X(t_k - t_{k-1})$ where X denotes the (flat) cost of lending or borrowing funds on the market for A.

Appendix 1

Asset Swap Analysis

The *asset swap* mechanism is more complex than what has been presented in a sketchy manner at the beginning of this chapter. We introduce below in a more comprehensive manner the customary features prevailing for so-called *par asset-swap* trades.

Par asset-swap trades embrace two distinct operations from a contractual perspective:

1. Operation #1: The investor buys a fixed rate bond B whose market value is V_B and nominal value is *Nom*, which contractually pays a series of fixed cash-flows denoted by $\underline{FC}$ (composed of fixed-coupons payments and nominal repaid at bond maturity). The actual flows paid by the bond are random since subject to default risk and denoted by $\underline{B}$. This entails $\underline{B} \le \underline{FC}$.

In fact, the investor aims to transform the fixed coupons into floating coupons, which explains the second operation.

2. Operation #2: The investor enters into an interest rate *swap* agreement whose fixed leg cash-flows match exactly those of the bond thereof (same coupons, nominal, and coupon and maturity dates). This contract involves three series of cash-flows: the sequence $\underline{FC}$, the cash-flow paid or received at inception $V_B - Nom$, and the sequence $\underline{VC}$:

$\underline{FC}$ (fixed leg): The investor pays to the *swap counterparty* the sequence of fixed coupons equal to those of the bond; these cash-flows are mandatory payments due even if the bond defaults (recall that the bond and the swap are two distinct instruments from a contractual perspective);

$V_B - Nom$: If the bond trades above par, the investor receives, at swap *inception, the* difference between the price of V_B and par (i.e., *Nom*), and if the bond trades below par the investor pays this difference (*Nom* $- V_B$) to his *swap* counterparty;

$\underline{VC}$ (variable, or floating, leg): The investor receives a sequence of cash-flows composed of $\underline{Libor}$ + spread corresponding to the floating leg of the *swap*;

s_a denotes the asset swap spread computed at inception so that the sum of the net present value of the three series of cash-flows generated by the swap *is* equal to zero.

Appendix 1

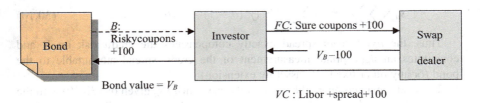

Fig. 30.12 Cash flows implied by a Par Asset Swap

As usual, underlined notations (e.g., $\underline{FC}$ or $\underline{VC}$) denote a series of cash-flows whilst a non-underlined notation denotes a single cash flow. For the sake of clarity, we consider a nominal (*Nom*) of 100 and this amount is added to both the final cash flows of the fixed and variable legs (these two cash-flows then exactly offset).

To summarize:

$\underline{FC}$ = contractual coupons, 100 (*at maturity*), certain cash-flows if the investor is free of risk.

$\underline{VC}$ = Libors + s_a, 100 (*at maturity*)

$\underline{B}$ = coupons and terminal capital, actually received from bond B (these are random)

$V_B - 100$ = the lump sum received or paid by the investor at inception

These cash flows are shown in Fig. 30.12.

We assume that the investor is not subject to counterparty risk whilst the bond is subject to default risk, implying that the investor is bound to pay the sequence $\underline{FC}$ whereas she may not receive all the cash-flows $\underline{B}$ from the bond (dotted line in Fig. 30.12). V_{FC} denotes the present value of $\underline{FC}$, i.e., the value of a risk-free bond paying the same contractual cash flows as B: hence, $V_{FC} > V_B$.

V_{VC} denotes the value of the swap variable leg and $V_{VC} = 100 +$ Present Value of spreads s_a (since the present value of $\underline{Libors}$ + 100 (nominal repayment at maturity) = 100).

- Let us first consider all the cash flows exchanged with the *swap* dealer (right side of Fig. 30.12): $\underline{VC} + (V_B - 100) - \underline{FC}$.

At the *swap inception*, the spread s_a is calculated so that the net present value of the cash-flows is equal to zero, i.e.: $V_{VC} = V_{FC} + 100 - V_B$;

Therefore: $100 +$ Present value of spreads $= V_{FC} + 100 - V_B$, or:

$$100 \times s_a \times \sum_\theta e^{-r_\theta \theta} = V_{FC} - V_B. \qquad (30.9)$$

If the bond is subject to credit (default) risk, the present value of the asset-swap spreads must therefore be equal to the difference between the value of the risk-free bond and that of the risky bond paying the same contractual coupons. In addition, $V_{FC} - V_B =$ present value of losses attributable to potential defaults may also write: present value of spreads = present value of losses due to defaults, i.e., for a \$1 nominal *Nom* and with the usual notations:

$$s_a \times \sum_{\theta} e^{-r_\theta \theta} = \sum_{\theta} e^{-r_\theta \theta} (\gamma^*(\theta - 1) - \gamma^*(\theta))(1 - \alpha). \tag{30.10}$$

Thus, the *asset swap* spread exactly compensates the credit risk on B and constitutes an appropriate measurement of the credit spread attributable to the bond (or any other fixed-rate asset by extension).

The reader may compare the value of the *asset swap* s_a given by (30.10) with the value of the spread of a CDS written on the same reference asset B (*see* Eq. (30.4a) with $D = 1$):

$$s_{CDS} \times \sum_{\theta} e^{-r_\theta \theta} \gamma^*(\theta) = \sum_{\theta} e^{-r_\theta \theta} (\gamma^*(\theta - 1) - \gamma^*(\theta))(1 - \alpha).$$

The difference between the l.h.s. of the formulas is attributable to the survival probability $\gamma^*(\theta)$ multiplying the discount factors of the CDS spreads which we do not find in Eq. (30.10). This difference is due to the fact that the protection seller ceases to pay the CDS spread when a credit event is triggered on the reference entity B (i.e. the CDS ceases to exist) while the *asset swap* spread is paid until the end of the transaction, even if B defaults. The *asset swap* spread is therefore theoretically lower than the CDS spread. However, for transactions displaying a short-term maturity and a low risk of default, the survival probabilities are close to 1 and the difference between s_a and s_{CDS} is not material.

- Let us now consider *all* the cash-flows induced by the *par asset swap* (see the right and left parts of Fig. 30.12 above): $\underbrace{-100}_{\text{Initial investment}} + B - FC + VC.$

In the absence of default on the bond, we have $B = FC$, and in case of default, $FC - B = Losses$ due to the default. Therefore, an investment of 100 yields for the investor:

$VC = Libors + spreads + 100$ (at maturity date), in the absence of default of the bond,

$VC - (FC - B) = Libor + spreads + 100$ (at maturity date) $- losses$ in the event of default.

The value of these cash-flows must be equal to the initial investment, i.e., 100. All of these conclusions are recovered, in particular, the present value of the spreads must be equal to that of the losses due to defaults (since the present value of *Libor* + 100 (at the terminal date) $= 100$).

Example 10

Let us consider the previous Example 2 and tweak it to match the actual *par asset swap* mechanism. An investor buys a 3-year bond B, bearing a contractual fixed rate of 6%, which has just paid its coupon, trading at 101.00 for a nominal value of 100. The fixed rate of a vanilla *swap* is 5.1% for all maturities

(continued)

(and we assume a flat discounting curve). We note that the rate of return of this bond (IRR of the sequence $-101, 6, 6, 106$) is equal to 5.63%.

The investor who enters the *asset swap* with a nominal value of \$100 pays: $\underline{FC} = (6, 6, 106)$; she receives $101 - 100 = 1$ at inception, as well as $\underline{VC} = 100$ (Libor(1) + s_a, Libor(2) + s_a, Libor(3) + s_a + 1).

$$V_{FC} = \frac{6}{(1 + 5.10\%)} + \frac{6}{(1 + 5.10\%)^2} + \frac{106}{(1 + 5.10\%)^3}$$

$$= 102.45 \text{ and } \sum_{i=1}^{3} \frac{1}{(1.051)^i} = 2.72.$$

According to (30.9), the *fair value of* the asset swap spread is $s_a = \frac{V_{FC} - V_B}{100 \sum_\theta e^{-r_\theta \theta}} = \frac{1.45}{272} = 0.53\%$.

We note that the same result is obtained in the simplified example (11) and that the spread of the *asset swap* is equal to the yield spread (5.63–5.10%). In fact, the equality between these two spreads is an approximation, both theoretically and empirically. We would have come up with a slight difference if we had taken the example of a longer and riskier security (therefore with a higher spread), whose price significantly differs from par.

Suggestion for Further Reading

Books

Anson, M., Fabozzi, F., Choudhry, M., & Chen, R. R. (2004). In F. J. Fabozzi (Ed.), *Credit derivatives: Instruments, applications, and pricing* (Series). John Wiley & Sons.

**Bielecki, T., Jeanblanc, M., & Rutkowski, M. (2009). *Credit risk modelling*. Lectures Notes, Osaka University.

**Brigo, D., & Mercurio, F. (2007). *Interest models – Theory and practice, with smile, inflation and credit* (2nd ed.). Springer.

Dongsheng, L. (2016). *The XVA of financial derivatives: CVA, DVA and FVA explained*. Palgrave Macmillan.

Duffie, D., & Singleton, K. (2003). *Credit risk: Pricing, measurement and management*. Princeton Series in Finance.

Gregory, J. (2020). *The xVA challenge: Counterparty risk, funding, collateral, capital and initial margin* (4th ed.). Wiley.

Hull, J. (2018a). *Options, futures and other derivatives* (10th ed.). Prentice Hall-Pearson Education.

Hull, J. (2018b). *Risk management and financial institutions* (5th ed.). Prentice Hall - Pearson.

**Schönbucher, P. (2003). *Credit derivative pricing models: Models, Pricing and implementation*. Wiley Finance.

Articles

Andersen, L., & Sidenius, J. (2004). Extensions to the Gaussian copula: Random recovery and random factor loadings. *Journal of Credit Risk, 1*, 29–70.

Blanchard, O. (2009, April). The crisis: Basic mechanisms and appropriate policies. International Monetary Fund *working paper*.

Brennan, M., & Schwartz, E. (1980). Analyzing convertible bonds. *Journal of Financial and Quantitative Analysis, 15*, 907–929.

Briys, E., & de Varenne, F. (1997). Valuing risky fixed rate debt: An extension. *Journal of Financial and Quantitative Analysis, 32*, 239–248.

Das, S. (1995). Credit risk derivatives. *Journal of Derivatives, 3*, 7–23.

Duffie, D. (2008, July). Innovations in credit risk transfer: Implications for financial stability. Bank for International Settlements *working paper*.

Duffie, D. (2009). Policy issues facing the market for credit derivatives. In J. D. Ciorciari & J. B. Taylor (Eds.), *The road ahead for the fed*. Hoover Institution Press.

Duffie, D., & Zhu, H. (2009, May). When does a central clearing counterparty reduce counterparty risk? *Working paper*, Graduate School of Business, Stanford University.

Gregory, J., & Laurent, J. P. (2003). I will survive. *Risk, 16*, 103–107.

Hellwig, M. (2008, December). Systemic risk in the financial sector: an analysis of the subprime mortgage financial crisis. *Working paper*, Max Plank Institute.

Hull, J., & White, A. (2004). Valuation of a CDO and nth-to-default *Swap* without monte carlo simulation. *Journal of Derivatives, 12*(2), 8–23.

Hull, J., & White, A. (2006). Valuing credit derivatives with an implied copula approach. *Journal of Derivatives, 14*, 8–28.

Hull, J., & White, A. (2013). Libor vs. OIS: The derivatives discounting dilemma. *Journal of Investment Management, 11*(3), 14–27.

Jarrow, R., & Turnbull, S. (1995). Pricing derivatives on financial securities subject to credit risk. *Journal of Finance, 50*(1), 53–86.

Laurent, J. P., & Gregory, J. (2005). Basket default Swaps, CDOs and factor copulas. *Journal of Risk, 7*, 1–20.

Li, D. (2000, March). On default correlation: A copula function approach. *Journal of Fixed Income, 9*, 43–55.

Longstaff, F., & Schwartz, E. (1995, June). Valuing credit derivatives. *Journal of Fixed Income, 5*, 6–12.

Morini, M., & Prampolini, A. (2011, April). Risky funding with counterparty and liquidity charges. *Asia Risk, 2011*, 58–63.

Website: defaultrisk.com.

Bank of International Settlements – Statistics: https://stats.bis.org/

Credit Derivatives – Correlation Trading, *Merrill Lynch*, 2003.

From OTC to the organized market: the new regulation of CDS, *Natixis*, special report, March 2009.

Guide To Exotic Credit Derivatives, *Lehman Brothers*, 2004.

The *J.P. Morgan* guide to credit derivatives, 1999.

Index

A

Absence of arbitrage opportunity (AAO), 8–19, 32, 225, 319, 320, 322, 362, 364, 366–368, 443, 651–653, 655, 659, 660, 694, 698, 797, 798, 800, 802–805, 807–809, 817, 822, 824, 825, 827–829, 836, 838, 842–844, 863, 897, 963, 966, 967, 969, 970, 972–974, 985, 986

Absolute risk aversion, 879, 915, 1014

Active management, 482, 952, 991, 1031, 1033, 1035, 1043–1045, 1047, 1055, 1056

Actual/actual convention, 43, 44, 612

Actual/360 convention, 43

Adapted processes, 767, 779, 790

Additive spread, 224, 227

Adjusted shares, 137, 263–267, 269, 709, 712

Adjustment coefficient (for a share), 264

Admissible portfolios, 369

AEX, 300

Agency markets, 29

Allais (Maurice), 913

Alpha (of a portfolio), 980, 986, 1038, 1040, 1043, 1047, 1055

Alternative investment, 980, 1048–1049, 1051–1054

Alternative management, 1031, 1048, 1052

AMC, *see* Asset Management Company (AMC)

American auction, 112

American bond market, 143

American calls, 365–367, 518–520

American options, 361, 364–365, 367, 391, 501–545, 559, 708, 710, 1000, 1064, 1072, 1089, 1090, 1101, 1206, 1208, 1216

  early exercise, 502–516

  and Monte Carlo, 1216

  stopping rule, 525

American puts, 366, 511–514

  one-dividend paying asset, 513–514

AMEX (American Stock Exchange), 294

Amortization (of a loan), 83

Amortization (step-down) swaps, 250

Amortization schedule (for a loan), 79–84

Amortization schemes, 123, 130

Amortization swap, 683

Annual percentage rate (APR), 125

Annual percentage yield (APY), 55, 125

Annual proportional rate (APR), 55, 79

Annual proportional yield (APY), 79

Antithetic variables, 1086, 1100, 1140

APT, *see* Arbitrage Pricing Theory (APT)

Arbitrage portfolio, 968, 970, 971, 973, 985

Arbitrage Pricing Theory (APT), 798, 806, 835, 843, 847, 850–853, 855, 856, 860, 866, 963, 966–976, 978, 982, 985, 986, 988–989, 1233

Arbitrages, 1, 9, 10, 12, 18, 32, 104, 127, 141, 231, 232, 319–324, 364–368, 618–624, 648, 650–654, 750, 797, 801–809, 824, 827–829, 843, 856, 970, 971, 975, 985, 986, 1000, 1019, 1049, 1287, 1288, 1339, 1341

  conversion, 362

  inverse conversion, 362

ARCA (Archipelago Exchange), 294

ARCH, 283, 305–307

Arithmetic Brownian motion, 278

Arithmetic returns, 274, 275, 302, 1112, 1113

Asian options, 567, 585, 589

Asset allocation puzzle, 1010

Asset Backed Commercial paper, 101

Asset Backed Securities (ABS), 123, 144, 1307–1313, 1340

Asset Management Company (AMC), 288

Asset-or-nothing option, 552

Asset Redundancy, 896–897

© The Author(s), under exclusive license to Springer Nature Switzerland AG 2022
P. Poncet, R. Portait, *Capital Market Finance*, Springer Texts in Business and
Economics, https://doi.org/10.1007/978-3-030-84600-8

1347

1348   Index

Asset swap, 1287–1289, 1296, 1331, 1342–1345
At a discount, 95, 321, 338, 340
At a premium, 321, 338, 340
At any price (market order), 303
At par, 122, 123, 132, 133, 147, 207, 216, 218, 223, 224, 227, 233–235
Attainable claims, 803, 805
Average options, 589

B

Backward-looking, 211, 222–224, 251
Backward-looking rate, 115, 208, 211, 222, 235, 247, 646, 647, 649, 661
Ban on short positions, 908–909
Ban on short selling, 916
Bank discount rate, 50, 51, 98, 99, 644, 645
Banker's acceptance, 101
Barone-Adesi and Whaley (valuation of US options), 525, 528, 544, 559
Barrier option with strike reset, 572
Barrier options, 567–573, 588, 589, 594–595, 704, 1202
Basel Committee, 1104, 1116, 1119, 1143, 1258–1260, 1274, 1275, 1299
Basel II, 1104
Basel III, 1104
Bases, 49, 53, 128, 130, 136, 145, 238, 348, 689, 693–695, 809, 919, 920, 1052, 1091, 1105, 1172–1176, 1181, 1197–1200, 1202, 1235, 1273, 1274, 1281, 1286, 1288–1290, 1292–1294, 1296, 1297, 1299, 1319, 1323, 1328, 1335, 1340
Basis, 43, 49–54, 95, 99, 130, 313, 321, 324, 325, 330–332, 336, 337, 342, 621, 659
Basis points (bips, bps), 99, 104, 112, 126, 150, 151, 191, 198, 211, 234, 247, 622, 642, 643, 655, 1235, 1281, 1335
Basis risk, 331, 332, 348, 634, 637, 658
Basis swaps, 205, 247, 250
Behavioral finance, 874, 906, 911–916, 922
Benchmarks, 13, 19, 113, 115, 135, 183, 206, 209–211, 213, 223, 232, 248, 290, 295, 303, 609, 948, 950, 996, 1031–1043, 1045–1047, 1049, 1052–1059, 1289, 1302, 1334
Bergomi (stochastic volatility model), 749
Bermuda swaptions, 749
Bernoulli, 393, 874, 875
Best of option, 555–556, 588

Beta, 280–282, 325, 330, 347, 466, 473, 474, 850, 855, 891–893, 934, 935, 937–940, 944, 945, 950, 952, 955, 957–961, 966, 974, 979, 980, 983, 989, 1035, 1038–1042
Beta (stability of), *see* Beta
Bid-ask spread, 28, 234, 291, 482
Binary option, 551
Binomial model, 367, 377–392, 529–533, 541, 544, 578, 1211
   calibration, 389–390
   Cox, Ross and Rubinstein (model), 367
Binomial trees, 379, 531, 533, 537, 541, 543–545, 1091, 1101
Black (formula), 426, 444
Black (model of), 1043
Black-Litterman (model of), 906–907
Black–Scholes formula, 386–391, 393, 395, 402–406, 420, 449–451, 459
   BS pricing function, 412
Black–Scholes formula (extensions), 412–442
Black–Scholes formula (proof), 409–412
Black–Scholes–Merton (model of), 429, 430, 443, 517, 520, 550, 667, 668, 867, 1237
   See also BSM (model)
Black–Scholes model, 399–412, 1197
   continuous dividends, 413–418
Black's two-fund separation theorem, 902, 910, 916
Black's zero-beta CAPM, 935
   See also CAPMs
Block trading, 31
Blue chips, 301
Bond, 119
   clean price, 131, 132
   dirty price, 131
   full price, 131
   indexed bonds, 135–138
   quotation of bonds, 131–135
Bond indices, 135, 1054, 1304
Bond options, 667
Bond subscription warrants, 713
Bonds with warrants, 136, 143
Book value, 266
Book-to-market ratio, 980, 983, 986
Bootstrap, 1130
Box-Muller, 1065, 1066, 1099, 1139
Brace, Gatarek, Musiela (BGM) model, 738
Break-even point, 481
Brennan and Schwartz (interest rate model), 729, 859, 860, 1013
Brennan and Schwartz (model), 729

Index

Brokers, 17, 19, 26–29, 31, 33, 105, 139, 144, 284, 286, 1048

Brownian (motion), 277, 278, 769–774, 794, 799, 848, 1008, 1066, 1139, 1189, 1218, 1229

BSM (model), 429, 430, 441, 443, 533, 550, 668, 670, 674, 691, 711, 714, 745, 749, 753, 1208, 1237

BSM-price (model), 434, 667, 668, 670–672, 686, 694, 734, 750

BSM-rate (model), 434, 667, 670, 672, 686, 687, 693, 714, 716, 738, 739, 744, 747, 751

Bull spread, 456

Bullet, 81, 82, 123, 132, 145, 147, 177, 179–180, 182

Bullet (security or profile), 145, 146, 151, 164, 608

Bullet sequence, 65

Butterflies, 486

Buy and hold (strategy), 296, 906, 910

C

CAC 40, 300

Calculus rules (for Brownian motions), 775–777

Calendar spreads (options), 484
See also Horizontal spreads (options)

Calibration of the binomial model, 393–394

Calibrations, 386–391, 393, 441, 442, 531, 541, 544, 545, 547, 745

Callable bonds, 713, 715

Call down-and-out (CDO), 567–572, 598, 1280

Call on call, 560

Call on put, 560

Call-put parity, 358–362, 364, 366, 388, 392, 406, 412, 415, 416, 419, 424–426, 464, 467, 469, 470, 489, 490, 495, 502–516, 553, 556, 580, 582, 583, 588, 604, 734

Call ratio backspread, 456

Calls, 23, 72, 98, 120, 152, 354–362, 366, 391, 436, 485, 493, 496, 503–509, 512, 517, 523, 526, 528, 544, 561, 572, 589–590, 611, 642, 643, 654, 691, 703, 733, 734, 802, 806, 815, 832, 848, 856, 998, 999, 1003–1005, 1007, 1027, 1050, 1053, 1072, 1087–1089, 1092, 1095, 1096, 1135, 1136, 1186, 1197–1199, 1202, 1209, 1211, 1213, 1218, 1314, 1315, 1330
premium, 356, 357

Cap, 212, 688–705, 714, 740–742, 748, 1012

Capital, 119, 150, 255, 800, 992, 1149, 1173, 1198, 1202, 1222, 1242, 1257, 1271, 1319, 1343

Capital adequacy, 1104

Capital gain or loss, 401

Capital issues, 266

Capital market line (CML), 931–933, 939, 948, 960, 961

Capital Requirement Directives, 1259

Capital Requirement Regulations, 1254, 1259

Caplets, 667, 673–677, 689, 691, 692, 700, 703–705, 714, 740–743, 745, 748, 751

CAPM (applications of), 943–953

CAPM (extensions), *see* CAPMs

CAPM (limits of), *see* CAPMs

CAPM (tests of), *see* CAPMs

CAPM-APT Comparison, 983–984

CAPMs, 280, 473, 901, 907, 929–945, 948, 950, 951, 955–961, 963, 966, 967, 974–976, 979, 981–983, 985, 988–989, 1038, 1233, 1238

Caps and floors (non-standard), 693, 702–705

Caps and floors with barriers, 703–704

Caps and floors with steps, 703

Cap spread, 702–703

Carhart (model of), 982

Carries, 104, 618–621, 623, 801, 1014, 1076, 1078, 1093, 1182, 1248

Carry trade, 102

Cash-and-carry, 319–325, 336, 337, 565, 608, 618–621, 623, 645, 650–653, 659, 660, 1013

Cash-and-carry arbitrage, 339

Cash-and-carry (reverse), 320, 323, 337, 339, 608, 621–624, 652, 653, 660

Cash flow sequences, 39–40

Cash settlement, 314, 315, 607, 640, 647, 1285, 1299, 1306

Cass and Stiglitz (separation theorems), 910

Cauchy–Dirichlet PDE, 864–865

CDS (basket), 1250

CDS n to default, 1078

CDS on indices, 1301

Central banks, 21, 94, 105–114, 117, 287, 1258, 1299, 1308

Central clearing counterparties (CCP), 315

Central Counterparties (CCP), 26, 27, 245, 1173, 1284, 1285, 1300, 1314, 1316, 1319, 1323, 1325, 1326, 1332, 1341

Centralized and Decentralized Markets, 29–30

Central Security Depositary (CSD), 106, 117

Certainty-equivalent, 220, 252, 878, 941, 946

Certificate of deposit (CD), 100, 101, 117

1350 Index

Change of numeraire, 583, 826
Changes in probability, 444–445
Cheapest to deliver (CTD), 608, 611, 613–618, 620–622, 624–628, 632, 633, 659, 660, 662, 664
Chicago Mercantile Exchange, 338
Choleski, 1073–1085
Choleski decomposition, 1073–1075, 1100, 1241
Chooser options, 567, 586–587, 589
CIS, *see* Collective Investment Schemes (CIS)
Clean bond price, 613
Clearing, 284, 286–288
Clearing houses, 26, 27, 33, 245, 311, 607, 642, 1173, 1319
Closed-end funds, 289
CME Group, 34, 311, 335, 340, 341, 607, 638, 641, 647
Coherent risk measures, 1141–1155
Collars, 699–701, 714
Collateral Valuation Adjustment (CollVA), 1336, 1337
Collaterals, 102–104, 109, 114, 115, 117, 119, 129, 146, 171, 225, 246, 315, 1173, 1181, 1265, 1268, 1280, 1308, 1311, 1312, 1315, 1316, 1319, 1325, 1326, 1328–1330, 1332, 1334–1336, 1340
Collective Investment Schemes (CIS), 289, 290
Combo option, 549, 561, 588
Commercial paper, 100, 101, 105, 106, 117, 129
Commodity futures, 335, 349, 1300
Commodity Trading Advisors (CTA), 1051
Common sense rules (strategic allocation), 992, 993, 997
Common stock, 255, 260
Compatibility of CAPM and APT, 988–989
Complete markets, 797, 798, 802–805, 807, 809, 816–818, 828–832, 836, 840, 843, 1022–1024
Compo call, 565–566
Compo option, 588
Composition of an index, 295
Compound interest, 42, 44, 45, 47, 48, 53, 125, 648, 654
Compound options, 559–560, 588
Compound rate, 54, 55
Conditional VaR, 1141, 1243
Conditional VaR (expected shortfall), 913
Condors, 486
Consol (or perpetuity), 66
Constant maturity swaps (CMS), 248, 685–686
Contango, 325

Contingent barrier options, 589
Contingent claims, 3, 354, 443, 799, 800, 802–805, 813–816, 823, 842–844, 1314
Contingent premium, 705
Continuous dividends, 416, 492, 848
Continuous processes, 768
Continuous rates, 53, 54, 61, 149, 157, 158, 168, 185, 186, 194, 274, 443, 720, 1071, 1179
Continuous time APT, 851–853
Contract Euro-bund, 609, 610, 613, 617
Contract T-bond, 622
Control variate, 533, 585, 1086–1088, 1101
Convenience yields, 335–338, 421, 422, 443, 531
Convergence of the basis, 313
Conversion factors, 608, 611–616, 621, 638, 639, 659
Conversion premium, 710
Conversion rate (for convertible bond), 708
Conversion rights, 709, 1214
Convertible arbitrage (strategy), 1050
Convertible bond (CB), 23, 708–711, 713–715, 1050, 1207, 1208, 1210–1216, 1218, 1219, 1264
Convex strategies (portfolios), 996, 1027
Convexity, 167–169
Convexity adjustments, 317, 623, 645, 652, 653, 664, 665, 667, 676–679, 685, 686
Cooke ratio, 1221
Copula (Gaussian), 1076, 1078, 1080, 1081, 1100, 1250–1252
Copula (of Student), 1080–1081, 1100, 1159
Copulas, 1073, 1076–1078, 1080–1081, 1100, 1159, 1239, 1250–1252, 1271, 1273
Corner portfolios, 909
Cornish-Fisher (approximation), 1113, 1138, 1156–1158, 1166
Correlated processes, 542
Correlation coefficients, 328, 543, 557, 558, 885, 923, 1074, 1076–1078, 1082, 1158, 1247, 1255, 1272
Correlation of defaults, 1253, 1274
Correlation products, 556
Correlation risks, 330–332, 634, 637, 658
Correlations, 18, 29, 545, 556, 563, 564, 566, 778, 885, 891, 954, 964, 968, 979, 984, 985, 1051, 1053, 1054, 1056, 1074, 1077, 1078, 1080, 1081, 1083, 1087, 1100, 1155, 1159, 1221, 1224, 1239–1242, 1247, 1250–1253, 1255, 1256, 1270–1274, 1307
Cost of capital, 946

Index

Cost of carry, 321–323, 336, 337, 342, 348, 621
Counterparty markets, 29
Counterparty risks, 25–27, 102, 163, 228, 231, 234, 241, 245–246, 311, 312, 315, 316, 994, 1000, 1010, 1122, 1172–1174, 1185–1189, 1197, 1217, 1265, 1268, 1271, 1279, 1288, 1290, 1300, 1313–1325, 1332, 1333, 1340, 1341
Coupon bond, 724
Coupons, 6, 7, 20, 93, 119–122, 124–128, 130, 132–134, 136, 140, 141, 145–147, 172, 179, 181, 182, 205, 207, 240, 253, 362, 419, 443, 503, 517, 529, 544, 608, 609, 613, 614, 625, 693, 709, 800, 805, 826, 848, 1005, 1012, 1174, 1190, 1193, 1194, 1197, 1202–1204, 1206–1208, 1213–1216, 1226, 1262, 1279, 1281–1283, 1286, 1288–1297, 1301–1306, 1333, 1339, 1340, 1342–1344
Covariances, 327, 662–664, 826, 891, 923–925, 934, 935, 938, 939, 941, 942, 965, 966, 968, 977, 984, 1023, 1075, 1082, 1115, 1116, 1240
Covenants, 119–121, 125, 129, 135–138, 145, 147, 225, 226, 1284
Covenants (bonds with optional), 713
Covered warrants, 712
Cox, Ingersoll and Ross (model), 725, 749, 858, 859, 942
Cox, Ross and Rubinstein (model), 377
Cox-Huang (model), 1022
CPPI (portfolio insurance), 1009, 1027
Crank–Nicolson, 534, 544
Credit/counterparty risk mitigation (CRM), 1315
Credit Default Swap (CDS), 1078, 1250, 1269, 1279–1307, 1313, 1314, 1325, 1330–1333, 1337–1341, 1344
Credit derivatives, 705, 1174, 1186, 1228, 1238, 1240, 1250, 1252, 1253, 1279–1344
Credit events, 1174, 1190, 1191, 1193, 1195, 1281–1288, 1290, 1291, 1296, 1298, 1299, 1301–1303, 1306, 1331, 1344
Credit-linked-Note (CLN), 1280, 1305, 1306
Credit Metrics, 1236, 1270
Credit risks, 126, 127, 129–131, 149–175, 187, 191, 192, 204, 208, 212, 213, 216, 224–227, 241–247, 646, 651, 653, 1104, 1105, 1171–1175, 1179, 1183, 1185, 1186, 1189, 1198, 1216–1218, 1221–1271, 1275, 1279, 1281–1283, 1287, 1289, 1290, 1292, 1297, 1301, 1304,

1306, 1308, 1309, 1311, 1312, 1333, 1340, 1341
Credit spread risk, 241, 246–247
Credit spreads, 130, 131, 147, 149, 170–172, 174–176, 178, 187, 191–192, 225, 651, 1173, 1185, 1187, 1199, 1207, 1213, 1214, 1217, 1219, 1222, 1224, 1236, 1237, 1255, 1256, 1271, 1289, 1306, 1309, 1333
Credit support annex (CSA), 1316–1319, 1321–1323, 1325, 1328–1331, 1334–1336
Credit triangle, 172–174, 1174
Credit triangle (formula), 175, 1195, 1292, 1296, 1341
Credit Value Adjustment (CVA), 1279, 1280, 1313, 1314, 1321–1323, 1327–1334, 1337–1339, 1341
Credit-VaR, 1104, 1105, 1221–1275
Cross-currency swap, 1325
Cumulative options, 573, 589
Currency basis swaps, 250
Currency contracts, 338–340
Currency forward contract, 338
Currency futures, 338, 349
Currency-interest swaps, 248, 249, 252
Currency risk, 954
Currency swaps, 248–250, 252
C-VaR, 1145–1148

D

DAX index, 300, 948
Dealers, 27–29, 33, 105, 106, 143, 245, 284, 1283, 1288, 1300, 1323, 1325, 1328, 1330, 1337, 1339, 1342, 1343
Debit Value Adjustment (DVA), 1262, 1280, 1313, 1314, 1320–1323, 1332–1334, 1337–1339, 1341
Decomposition method, 235, 237, 240, 242
Default intensity, 1212
Default probability, 174–176
Default risks, 100, 103, 147, 149, 170, 172, 177–179, 247, 651, 653, 1171, 1179, 1185, 1187, 1188, 1193, 1218, 1221, 1222, 1237, 1239, 1242, 1255, 1270, 1271, 1279–1281, 1297, 1298, 1306–1309, 1312–1314, 1329, 1331, 1333, 1337, 1339, 1340, 1342, 1343
Deflators, 836, 840–842, 1019
Deliverable securities (DS), 608–609, 611–617, 622, 624, 633, 638, 639, 659, 661, 662, 664

1352

Index

Deliveries, 103, 104, 106, 285, 286, 313, 315, 316, 318, 320, 335, 338, 339, 607–614, 616, 618–620, 622, 626, 638–641, 1284
 physical delivery, 314
Delivery option, 613, 622, 659–662
Delivery price, 659
Delta, 464–466, 472, 476–479, 482, 485, 489, 536, 538, 1121, 1135, 1136, 1141
Delta-gamma method, 1123, 1134–1138
Delta-neutral, 465, 476–485, 488
Delta-normal method, 1130–1134, 1141
Delta valuation, 1122, 1131, 1142
Deposits, 119, 315, 316, 341, 1173, 1258–1260, 1265–1267, 1317
Diagonal model, 891–893
Diagonal spreads (options), 457
Differential swaps, 251
Diffusion matrix, 778
Diffusion parameters, 279, 772, 779
Diffusion processes, 730, 750, 776, 779–786, 788, 791, 795, 847, 851, 857, 866, 1020, 1023, 1028, 1066, 1073, 1083, 1100, 1213
Digital barrier option, 573–575
Digital calls, 694, 704
Digital options, 551–555, 573, 588, 676, 704
Digital puts, 694, 704
Dilution, 261–263
Dirac gamma, 468
Dirty price, 131, 135, 140, 608, 611, 613, 614, 618, 624, 625, 628, 659, 671, 1289
Discount factors, 58, 180, 181, 183, 668, 669, 713, 734, 840, 844, 1193, 1328, 1341, 1344
Discounted dividends, 270
Discounted value, 46
Discounting, 39–86, 96, 99, 213, 267, 832, 1003, 1070, 1179–1189, 1193, 1197, 1217, 1224, 1225, 1228, 1269, 1270, 1290, 1334, 1335, 1345
Discounting under uncertainty, 864–866
Discount rates, 63–67, 72–74, 84, 272, 280, 941
Discounts, 50, 54, 56, 58, 59, 62, 63, 67, 72–74, 85, 95–99, 101, 125, 127, 180, 183, 241, 261, 1003, 1004, 1019, 1071, 1184, 1187, 1193, 1195, 1217, 1237, 1307, 1313, 1317, 1321, 1328, 1341, 1344
Discount window, 113
Discrete dividend, 418–421
Disintermediation, 19, 94
Displaced diffusion LMM, 749
Disposition effect (behavioral finance), 914

Distribution of default dates, 1252
Diversifiable (non-systematic) risk, 893
Diversifiable risks, 893, 937, 956, 959, 960, 964, 971, 972, 986, 1242
Diversification, 288, 335, 873, 882, 885, 886, 890–893, 915, 937, 952, 954, 957, 960, 986, 991, 994, 995, 1051, 1052, 1056, 1106, 1116, 1120, 1130, 1142, 1143, 1148, 1158, 1159, 1228, 1239, 1246, 1247, 1252, 1253, 1256, 1269, 1270, 1311
Dividend (continuous), 508–512, 531
Dividend (discrete), 510
Dividend detachment, 507, 508
Dividend swap, 472
Dividend yield, 413, 414
Dividends, 6, 104, 257, 267, 268, 271, 272, 280, 362–363, 443, 472, 502–508, 517–520, 523, 529, 531, 543, 544, 709, 800, 801, 813, 815, 816, 826, 979, 981, 1215, 1216, 1218, 1264
Double barrier options, 589
Double digital option (DDO), 553, 554, 588
Dow Jones indices, 299, 300, 1156
Down-and-in option, 567
Down-and-out option, 1202, 1218
Downside risk measure, 912
Drift adjustment, 867
Drifts, 279, 302, 400, 407, 528, 730–732, 750, 772, 773, 779, 781, 811, 818, 822, 823, 831, 832, 843, 1066, 1067, 1082, 1199, 1204, 1215, 1237
Duffie-Singleton (model), 1195, 1217
Duplication, 235, 496–498, 693–707, 1285, 1287, 1288
Durations, 42, 44, 98, 101, 103, 112, 130, 149–177, 217, 616–618, 624–626, 628–633, 635, 636, 656–660, 662, 694, 992, 994, 1005, 1090, 1108, 1179, 1206, 1210
Durbin-Watson test, 959
Dutch auction, 112
Dynamic hedging, 634
Dynamic portfolio strategies, 1014
Dynkin operator, 521, 538, 791–792, 796, 849, 851, 863, 865–867

E

Early exercise of options, 517, 543, 1101
EBITDA, 69, 256, 257, 979
ECB (European Central Bank), 110, 111
Economic capital, 1247, 1254–1257, 1260
EDGE, 294

Index 1353

Efficiency, 1, 5, 8–19, 31–33, 538, 930, 932, 933, 955–956, 961, 991, 1036, 1064, 1071, 1086, 1087, 1089, 1100, 1271

Efficient frontiers, 873, 887–889, 901, 903, 904, 908, 909, 916, 930–932, 960, 1096

Efficient portfolios, 882–884, 888, 889, 894–906, 908, 909, 916, 931–933, 948, 960, 961, 991, 994, 1014, 1019

EFFR, 210

EGARCH, 307

Elasticity, 17

Elasticity (of an option), 466, 473, 474
   See also Omega (of an option)

ELS (equity-linked swap), 251

Endowment effect (behavioral finance), 914

EONIA, 210

Equilibrium, 1, 8–19, 25, 30, 32, 108, 109, 111, 127, 128, 188, 189, 280, 284, 324, 338, 607, 851, 930, 937–942, 948, 960, 961, 963, 964, 966, 967, 975, 983

Equilibrium (partial *vs.* general), 942

Equilibrium model, 725, 726

Equilibrium price of assets, 940–943

Equity, 302

Equity-linked swaps (or equity swaps), 251

Equity Market Neutral (strategy), 1049

Equity premium puzzle, 281

Equivalence (of interest rates), 53–55

Equivalent (interest rates), 48

ESCB European System of Central Banks, 110

€ster, 223, 247, 656

ETF (exchange-traded fund), 291, 292

EUREX, 34, 140, 311, 341, 609, 611, 638–641, 647, 656, 660

Euribor contracts, 644, 647, 660

Euribors, 98, 102, 113–115, 118, 136, 171, 182, 198, 207, 209, 210, 214, 222, 223, 225, 229, 242, 248, 641, 675, 688, 1173, 1289, 1303, 1304, 1313, 1334

Euro-bund futures contract, 638

Eurobond market, 122, 140, 143, 147

Euro Commercial Paper, 94, 106

Eurodollars, 641–644, 647, 649, 651, 660

Euronext, 21, 25, 29, 33, 293, 303

European bond markets, 144–146

European call, 405

European options, 23, 358, 362–364, 391, 401, 416, 443, 501, 517, 528, 531, 533, 544, 550, 576, 589, 671, 802, 803, 998, 1000, 1070, 1100

European puts, 366, 405

Euro Overnight Interest Average (EONIA), 34, 113, 114, 116, 118, 136, 210, 247, 656, 1313, 1334

Exchangeable bond, 713

Exchange options, 436, 555–557, 588, 662, 710, 1208
   See also Margrabe (formula)

Exchange Traded Funds (ETF), 1034

Exotic options, 457, 549–589, 1012, 1100

Expectations hypothesis, 188–191

Expected Default Frequency (EDF), 1233–1238, 1252

Expected loss, 173

Expected negative exposure (ENE), 1320, 1321, 1323, 1325, 1331–1333, 1339, 1341

Expected positive exposure (EPE), 1319, 1321, 1323, 1325, 1328, 1329, 1331, 1332, 1339, 1341

Expected Shorfall (ES), 913, 1070, 1100, 1103–1168, 1221, 1222, 1226, 1263, 1318

Expected spot prices, 324–325

Expected utility, 873–882, 906, 910, 911, 913, 915–922, 1014, 1016, 1017, 1020, 1021, 1023–1025, 1027, 1028

Exponential martingales, 809–814, 820, 830

Exposition to default, 1175, 1181

Extreme values, 276, 304, 922, 1081, 1138, 1146, 1147, 1155, 1156, 1159–1164, 1166

F

Fair value, 167, 266, 1314, 1331, 1345

Fama and French (model of), 958, 976–982, 986, 1049

Federal fund rate, 113, 247

Federal fund target rate, 113
   See also Key driving rates

Feynman–Kac (formula of), 847, 854, 866–868

Feynman–Kac theorem, 864–866

Filtration, 525, 766, 798, 799, 803, 804, 806, 810, 816, 817, 820, 842

Financial tables, 59, 60, 87–92

Finite differences method, 534–539, 541
   standard implicit method, 535–538

Fitting the yield curve, 726–727

Fixed income arbitrage (strategy), 1050

Fixed-rate bonds, 121–135, 143, 147

Fixed-rate payer, 228, 229, 232, 252

Fixed-rate receiver, 229, 233, 238, 252

Fixings, 30–31, 33, 284, 316, 648, 692, 1243

Flat volatilities, 692, 740, 741

Flexi-caps, 705

Flexi-floors, 705

Float, 17

Floaters, 205, 206, 208, 211, 213, 214, 218

Floating market capitalizations, 258, 296

Floating-rate bonds, 147, 206, 1296

1354 Index

Floating rates, 21, 120, 121, 135–138, 140, 143, 145–147, 205–252, 667, 673, 680, 681, 689, 690, 692, 693, 696–698, 1287, 1296, 1305–1307, 1314, 1330

Floorlets, 674–676, 689, 693, 700, 703–705, 714, 751

Floors, 211, 212, 688–706, 709, 714, 998, 1000, 1005–1012, 1027, 1208, 1214, 1263

Floor spread, 702–703

Foreign exchanges, 338–340

Forex, 349

Forward, 4, 22, 102, 163, 285, 312, 315–317, 319, 341, 343, 364, 534, 549, 607, 668, 829, 1001, 1072, 1183, 1225

Forward and futures prices (relationship), 349–352, 662–665

Forward contracts, 311, 314–317, 340–347, 363

Forward contracts (on a Market Index), 346–347

Forward contracts (on fixed-income securities), 343–344

Forward interest rates, 188

Forward-looking, 251

Forward-looking rate, 208, 235, 237

Forward-neutral probability, 325, 432, 667, 738, 829, 1183, 1340

Forward price, 319–325

Forward rate agreement (FRA), 229, 341, 344–346, 644–647, 660

Forward rate method, 236, 239

Forward rates, 177, 184–188, 190, 191, 204, 641–643, 645, 651, 653, 661, 673–674, 677–679, 682, 686, 729–733, 1225–1227

Forward rates dynamics, 729–749

Forward start options, 549–551, 588

Forward swap measure, 746

Forward swaps, 682–683

Forward yield curve, 184

Fractal, 773, 774

Frame (behavioral finance), 914

Free boundary, 538–539

Free shares, 263, 264

FTSE index (Footsie), 300, 341, 347

Full valuation, 1122–1123, 1131, 1135–1137, 1141

Fundamental Reform of the Trading Book (FRTB), 1263

Funding Benefit Adjustment (FBA), 1280, 1336, 1337, 1339, 1342

Funding Cost Adjustment (FCA), 259, 1280, 1336, 1337, 1339, 1342

Funding Value Adjustment (FVA), 1313, 1314, 1320–1323, 1334–1339, 1342

Future value, 46

Futures, 22, 285, 315–317, 319, 424, 534, 607, 622, 802, 994, 1072, 1184, 1225, 1300

Futures contracts, 140, 311, 313–317, 340–347, 364, 424, 514, 544, 607, 609, 611, 639, 645, 647, 1034, 1092, 1095, 1303

Futures contracts on an average overnight rate, 647–656

FX, 349
    See also Forex

G

Gamma, 467–468, 477, 479, 482, 485, 489, 536, 1135, 1136

Gamma monitoring, 480

Gamma/theta tradeoff, 480, 482, 488

Gap, 161–167

Gap (interest rate), 637

Gap option, 553, 588

GARCH, 283, 305–307, 1131

Garman-Kohlhagen (formula), 422, 423, 460

Gaussian models, 437, 444, 494, 670–671, 675, 724, 1239, 1250, 1274
    See also Black–Scholes–Merton (model of)

General forward swap measure, 747, 759

Generalized Pareto distributions, 1159–1160

Generalized scenarios, 1150

Geometric average strike, 583–584

Geometric Brownian Motions, 277–280, 282, 283, 302, 400, 407, 438, 443, 535, 579, 585, 589, 591, 781, 787–788, 795, 812, 1008, 1009, 1067, 1068, 1218, 1229

Geometric progressions, 85, 86

Girsanov (theorem), 444, 445, 451, 590, 596, 603, 755, 759, 813, 818–825, 836, 843, 862

Global macro (strategy), 1051

Gordon-Shapiro (formula), 269–271, 302, 945

Gordy (model of), 1242, 1270

Granular asymptotic model, 1246

Greek parameters, 453, 463–486, 488–493, 571, 1073, 1100, 1134, 1136

Grey market, 32

Grossman and Stiglitz (paradox), 14

Growth stocks, 980

Guaranteed capital funds, 550

Gumbel (law), 1168

Index

H

HARA Utility, 874, 880, 910, 1015

Hazard rate, 1176, 1179, 1181, 1192, 1193, 1195, 1197, 1217, 1219

Heath-Jarrow-Morton (model of), HJM, 623, 664, 667, 729–733, 735, 737, 750

Hedged position, 327, 330, 635

Hedge effectiveness, 328

Hedge efficiency, 332

Hedge funds, 104, 292, 980, 1031, 1047–1054, 1056, 1143, 1267, 1268

Hedge portfolio, 370

Hedge ratios, 328, 329, 370, 374, 376, 487, 630, 632, 634–636, 657–658, 660

Hedging, 100, 103, 104, 325–334, 348, 369–372, 404, 465, 549, 559, 565, 566, 571, 629–631, 633, 634, 636, 637, 640, 641, 656–658, 660, 661, 692, 1012, 1255, 1269, 1296–1299, 1328

Hedging an expected position, 634–637

Hedging with futures, 608

Heston (model of), 438–442, 444, 496

See also Stochastic volatilities

High water mark, 1048

Historical (or physical) volatility, 458

Historical probabilities, 376, 377, 562–564, 670, 674, 813, 818, 820, 822, 823, 825, 826, 829, 832, 836, 857, 859, 1019, 1024, 1028, 1176–1177, 1181, 1185, 1217, 1229, 1237, 1239

Historical simulation, 1141

Historical volatility, 453, 487

HML factor, 981

Horizontal spreads (options), 457

See also Calendar spreads (options)

Hot money, 108, 110

See also Monetary base

Hull and White (models of), 438, 533, 539, 541–543, 667, 675, 726–728, 749, 750, 755, 756

Hull and White-Vasicek (models), 727, 732

Hybrid instruments, 23

Hybrid products, 707–713

Hyperbola (efficient frontier), 885, 887, 901, 904, 909, 916

I

IAS, 167

IBEX, 300

ICE (InterContinental Exchange), 293, 294, 335, 341

ICE futures, 656

ICE Futures Europe, 311, 341

Idiosyncratic risks, 1244

Implicit volatilities, 480, 483, 487

Implied volatilities, 437, 442, 453, 457–463, 469, 483, 487

Importance sampling, 1071, 1088–1089, 1101

In advance, 84, 96, 97

In arrears, 49–54, 70, 78, 82, 84, 95–97, 101, 117, 251, 644, 645, 674, 684, 694

In arrears swaps, 684–686

Index futures, 1302–1303

Index management, 295

Index options, 1303

Indices, 94, 117, 209–211, 950, 979, 1012, 1013, 1054, 1240, 1285, 1299–1303

Indirect quotation (of exchange rates), 340

Inflation, 8, 58, 74–76, 85, 108, 109, 120, 137, 144, 146, 188, 861, 936, 965, 979, 986, 994

Information ratios, 950, 1038–1043, 1045, 1055

Initial public offering (IPO), 126, 260

Instantaneous forward rates, 721, 729, 730, 732, 750

Intensity models, 1189–1197, 1206, 1208, 1210–1212, 1252

Interest rate derivatives, 738, 747, 750, 1325

Interest rate options, 25, 137, 749

Interest rate risks, 99, 127, 131, 136, 147, 149–169, 172, 174, 177, 192–203, 206, 207, 212, 213, 222, 229–230, 242, 245, 623, 629, 630, 633, 635, 641, 656–659, 721, 994, 1225, 1227, 1287, 1304

Interest rate sensitivity, 152–156

Interest rate sensitivity (of futures prices), 624–629

Interest rate swaps, 22, 25, 115, 183, 205, 222, 223, 228, 241–247, 1050, 1253, 1279–1281, 1285–1287, 1297, 1314, 1316, 1321, 1330, 1338, 1342

Internal rate of return (IRR), 46, 56, 59, 63–78, 85, 90–92, 123–125, 147, 265, 946

International CAPM, 954–955

Intervention rates, 117

See also Key driving rates

Intrinsic value (IV), 267, 354–358, 366, 502, 503, 509, 511, 532, 537, 543, 544, 1098, 1216

IRB (credit risk approach), 1242, 1246–1248, 1256, 1263, 1274, 1275

ISDA, 234

Issue premium, 261

Issue price, 261

Index

Issuer call, 709, 711
Iterated expectations, 380
Itô calculus, 400
Itô integral, 783
Itô processes, 279, 779–786, 795, 799, 810–811, 813, 843, 860, 1015
Itô's lemma, 278, 351, 402, 408, 422, 439, 538, 563, 729, 731, 785–789, 791–792, 795, 796, 811, 831, 835, 849, 857, 866
iTraxx EUR, 1302

J

Jäckel and Rebonato (interest rate model), 755
Jarque-Bera statistic, 305
Jarrow-Turnbull (model of), 1217
Jensen (index of), 833
Jensen's alpha, 950–953, 961, 980, 1038, 1055
Jump processes, 768, 793–794
Junk bonds, 129, 1051

K

Kahneman and Tversky (behavioral finance), 913
Kawai (interest rate model), 755
Kelly's criterion, 1019, 1020
Key driving rates, 113
  See also Intervention rates
Knight (Franck), 913
Kuhn and Tucker (conditions of), 927
Kurtosis, 276, 304–305, 1053, 1156–1159, 1274

L

Lagrange multipliers, 898–900, 925, 1024, 1036, 1037
Lagrangian (operator), 898–900, 925, 926, 1037
Langetieg (model of), 729, 861
Legs of a swap, 232, 233
Leptokurtic, 276, 283, 748, 1142
Leverage, 85, 272
Leverage (financial), 71, 72
Leverage effects, 56, 257, 341, 1007, 1049
Libor, 102, 114, 115, 118, 136, 171, 207, 209, 218, 222, 225, 229, 248, 249, 543, 641, 667, 688, 1173, 1287–1289, 1305, 1342–1345
Libor in arrears swaps, 251
Libor Market Model (LMM), 667, 675, 688, 691, 692, 737–751, 754–761
LIBOR-OIS Spread, 246–247
LIFFE, 294

Limit order, 285, 303
Liquidation value, 267
Liquidity, 1, 8–19, 25, 27–29, 32, 33, 93, 105, 108–110, 112, 148, 225, 255, 258, 259, 263, 285, 312, 979, 1122, 1217, 1223, 1257–1261, 1265–1267, 1288, 1300, 1325, 1335, 1336, 1340
Liquidity premium, 171–173
Liquidity traders, 458
Logarithmic portfolio, 798, 832–836, 840, 1018, 1021, 1022, 1028
Logarithmic returns, 273, 1112, 1117
Logarithmic utility, 833, 876, 880, 1017–1020, 1025
Log-returns, 274–276, 279, 302, 304–305
London Metal Exchange, 335, 337
London Stock Exchange (LSE), 21, 33, 293, 303
Long positions, 104, 162, 194, 312, 883, 999, 1002, 1003, 1013, 1027, 1049, 1186, 1198, 1253, 1287, 1297, 1304, 1305
Long/short equity (strategy), 1049
Longer-Term Refinancing Operations (LTRO), 112
Long-short strategies, 1049
Longstaff and Schwartz (interest rate model), 729
Longstaff and Schwartz (regression), 1090, 1091, 1094–1098
Long-term loans, 76–84, 120
Lookback call, 598–601
Lookback options, 576–578, 589
Loss aversion, 874, 911–915
Loss given default (LGD), 173, 175, 1175, 1263, 1296

M

Macaulay duration, 149, 153, 157, 174, 183, 195
Main Refinancing Operations (by a Central Bank), 111
Mandatory convertible bond, 713
Mappings, 799, 804, 823, 824, 827, 828, 837, 843, 845, 1080, 1121, 1122, 1132, 1133, 1151, 1172, 1230, 1235, 1257, 1267, 1331
Margin calls, 245, 312, 315, 316, 351, 424, 607, 619, 623, 633, 635, 641–645, 652, 655, 656, 659–664, 1166, 1173, 1174, 1302, 1316–1319, 1321, 1322, 1325, 1328, 1329, 1335, 1341
Margin period of risk (MPOR), 1318, 1319, 1321–1323, 1329
Marginal risk, 1153

Index

Margrabe (formula), 434, 436, 556, 662
   See also Exchange options
Margrabe (model of), 434, 436, 555, 556, 662,
   711, 715
Marked-to-market, 161, 737, 1120, 1171, 1222,
   1223, 1239–1241, 1269, 1270
Marked to model, 1222
Market capitalizations, 139, 256–263, 268, 293,
   295–300, 303, 495, 907, 933, 955, 979,
   980, 986, 1033
Market completeness, 855–856
Market efficiency, 768
Market makers, 18, 28, 29, 33, 139, 143, 146,
   234
Market model, 891–893, 957
Market orders, 284, 285, 288, 303
Market price of risk, 374–377, 444, 721, 723,
   797, 805–809, 853, 858, 933, 941, 948,
   960, 974, 985
Market risk premium, 280, 281
Market risks, 933, 944, 954, 961, 1104, 1105,
   1171, 1260, 1262, 1273, 1286, 1296,
   1297
Market timing, 952, 1035, 1038–1042, 1055,
   1057–1059
Market volatility, 983
Markov processes, 749, 751, 767–769, 771
Markowitz (model of), 275, 906, 991, 1014
Martingale measure, 446–448, 450
Martingales, 351, 379–381, 383, 392, 407, 408,
   424, 425, 435, 443, 444, 446, 447, 450,
   563, 583, 589, 590, 662, 665, 668–670,
   673, 687, 713, 715–716, 735, 739, 740,
   746, 747, 751, 758, 771, 783, 795, 797,
   798, 809–818, 820, 821, 825–832, 836,
   840, 842–844, 868, 1019, 1024, 1028,
   1179, 1191, 1197, 1217
Maximum hedging, 326–333
McDonough Ratio, 1261
Mean-reverting process, 438, 722
Mean-variance criterion, 874–882, 910, 916,
   966, 1036
Mean-variance efficient, 930
Memoryless process, 768
Mental accounting (behavioral finance), 914
Merton model, 416
Merton's Intertemporal CAPM, 953–954
   See also CAPMs
Mezzanine (debt), 1312, 1340
MIB, 300
Migration matrix, 1224–1228
Minimum bid rate (set by a Central Bank), 111
Minimum transfer amount (MTA), 1316–1319,
   1321–1323, 1329, 1335
Minimum variance (portfolio), 901, 1036, 1037

Min-Max parity, 588
Min-Max parity (for multi-underlying options),
   557
Mismatch, 165
MKMV (approach, model), 1200, 1203, 1207,
   1221, 1222, 1228, 1232, 1233, 1235,
   1236, 1240, 1270
MMI (Major Market Index), 300
Modified durations, 149, 153, 154, 157, 174,
   195, 217, 218, 241–245
Momentum, 285
Monetary base, 117
   See also Hot money
Money market rate, 95
Money market yield, 98, 99, 102
Moneyness (of an option), 559, 748
Monte Carlo, 1064, 1086–1098, 1141, 1213
Monte Carlo (simulation), 533, 549, 558, 585,
   589, 1063, 1138–1140, 1208, 1212,
   1213, 1216, 1239, 1252, 1321
MSCI (Morgan Stanley Composite Index)
   indices, 301
MTF (multilateral trading facilities), 294
Multi-dimensional Brownian motions, 777–778
Multi-dimensional diffusion processes,
   789–792
Multi-dimensional Itô processes, 789–792, 796
Multifactor duration, 204
Multi-factor models, 745, 859–861, 867,
   963–987
Multifactor sensitivity, 194–203
Multifactor variation, 194–204
Multiplicative factor, margin, 227
Multiplicative margin, 207, 213, 222, 224, 227
Mutual funds, 20, 100, 105, 289, 290, 292, 303,
   904, 945, 947, 1033, 1048, 1054

N

NASDAQ, 29, 34, 293, 294, 303
NAV, *see* Net Asset Value (NAV)
Negotiable Debt Instrument, 1253, 1262
Net Asset Value (NAV), 290, 291
Net present value (NPV), 59, 61, 63, 64, 66–69,
   74, 85, 87, 90–92, 261, 873, 946, 1342,
   1343
Net Settlement system, 288
Netting, 245, 1239, 1253–1254, 1315, 1316,
   1319, 1320, 1328
Nikkei Index, 299–301, 341
Nominal rates, 56, 74–76, 121, 161, 163, 1214
Non-parallel shift of yield curves, 244
Nonstandard swaps, 250–251, 680–688
No regret option (lookback option), 576
Normal backwardation, 324, 325

1358 Index

Notional bonds, 611, 638
Notional security, 608–609
Numeraires, 6, 7, 431, 432, 435, 436, 446–448,
    450, 555, 578, 583, 596, 598, 602, 662,
    669, 713, 715, 716, 739, 746, 747, 755,
    797, 798, 809, 813–815, 825–833, 836,
    839, 840, 842–844, 1015, 1028
NYMEX, 335
NYSE, 21, 143, 293, 294, 303

O

OAT (French bonds), 137, 139, 145, 1005
OBPI (portfolio insurance), 997–999, 1011,
    1027
Offsetting swap method, 236, 238, 242
OIS, 210–212, 229, 240, 1289
OIS indices, 116, 117
Omega (of an option), 466, 473
    See also Elasticity (of an option)
Open-end funds, 289, 290
Open Market operations (OMO), 106, 108, 110,
    111, 118
Operational risks, 1260–1262
Optimal growth portfolio, 833, 834,
    1017–1021, 1024, 1028
Optimal hedging, 326
Option, 20, 22, 137, 231, 354, 496, 589, 800,
    848, 1000, 1067, 1068, 1071–1073,
    1083, 1086–1088, 1091, 1095–1099,
    1121, 1171, 1174, 1186–1189,
    1197–1199, 1202, 1204, 1208, 1214,
    1237, 1288, 1289, 1314
    American, 354
    at the money, 358
    at-the-money-forward, 496
    bond option, 671–672
    convexity, 367
    dividend risk, 472
    European, 354
    exercise price (*see* Strikes)
    expiry (date), 354
    implicit correlation, 556
    in the money, 358
    intrinsic value (IV), 356–358, 361, 392, 459
    moneyness, 460
    multi-underlying, 555–558
    option value, 369–372
    out of the money, 358
    path independent, 550–566
    payoff, 354
    premium, 354, 355, 357
    rainbow, 555–558

    sensitivity to dividend rate, 472
    speculative value, 358, 392
    strike, 358, 359
    time value, 357, 358, 361, 392, 459
    underlying asset, 354
    valuation, 381–386
Option at-the-money, 465, 496, 1002–1004
Option at-the-money-forward, 465, 485
Option Combo, 589–590
Option down and out, 1202
Option forward start, 550
Option in-the-money, 366, 438, 455, 459, 465,
    469, 528, 533, 544, 1088, 1091, 1092,
    1097, 1098
Option lookback, 577
Option on average price, 601–602
Option on futures or forward, 424–427
Option out-of-the-money, 438, 455, 459, 465,
    473, 494, 496, 498, 1088, 1091, 1095,
    1097, 1098
Option valuation, pricing, 1090, 1199
Options on arithmetic means (averages), 584
Options on averages (Asians), 578–585
Options on bond futures contracts, 735–737
Options on CDS, 1300, 1303
Options on forward bond contracts, 733
Options on maximum, 556, 588
Options on minimum, 556, 588
Options on spot bonds, 734
Options with an average strike, 602–604
Order books, 17, 25, 30, 31, 284, 285
Ornstein-Uhlenbeck (process), 722, 729, 749,
    751, 752, 781, 787–789, 795, 858, 859,
    861, 1083
Overnight indexed swap (OIS), 115, 116, 183,
    184, 209, 210, 223, 225, 229, 234, 240,
    241, 246–248, 543, 554, 648, 651, 652,
    654, 661, 1313, 1334–1338, 1341
Over-the-counter (OTC), 10, 24–27, 32, 33,
    100, 139, 143, 144, 146, 245, 259, 266,
    312, 314, 315, 341, 688, 1174, 1233,
    1253, 1268, 1284, 1300, 1304, 1307,
    1314, 1316, 1319, 1323–1326

P

P&L (profit and loss) matrix, 454, 486, 488
    See also Risk matrix
Parabola (efficient frontier), 901
Parisian options, 572, 589
Partial barrier options, 589
Partial differential equation (PDE), 402–408,
    440, 443, 491, 520–524, 526, 527, 534,

535, 538, 539, 544, 847–868, 1203, 1204, 1210, 1219
Partial valuation, 1122–1123, 1130–1138
Par yield, 134, 147, 233, 234, 687
Path (or trajectory), 766, 771
Path dependent, 501, 549, 573, 576, 587, 588, 998, 1071–1072, 1087, 1202, 1218
Path dependent option, 567–587
Path-independent, 549, 587, 588
Payer swap with profit-sharing, 707
Payoff replication, 496–498
Payoffs, 355, 356, 391, 457, 496, 498, 525, 537, 538, 549, 552, 554, 555, 584, 589, 650, 673, 691, 693, 694, 802, 803, 805, 816, 823, 832, 833, 1023, 1025, 1070–1072, 1087, 1088, 1091, 1094, 1097–1100, 1179, 1183, 1186–1188, 1195, 1197, 1198, 1204, 1217, 1218, 1238, 1254, 1282, 1287, 1314
Peaks over threshold (POT), 1159
Perfect hedge, 326
Perfect markets, 18–19, 827, 930
Performance (measures of), 947–953, 955, 1039, 1052, 1053, 1055, 1056
Periodic premiums, 694, 700
Perpetual American put, 527
Perpetual annuity, 150, 154–156
Perpetual annuity, perpetuity, 87, 155, 169
Poisson (law, process), 1078, 1189–1195, 1206, 1213, 1217, 1291, 1293, 1340
Poisson processes, 793, 796, 1078, 1189–1194, 1206, 1210, 1217
Portfolio insurance, 992, 996–1005, 1010, 1011, 1013, 1027
Portfolio of assets and liabilities, 164
Post-determined, 115
    See also Backward-looking; Backward-looking rate
Pre-determined interest, 115
Pre-determined rate, 208
Preferred stock, 260
Premiums, 23, 79, 83, 171–173, 176, 189, 191, 401, 689–695, 807, 933, 944, 959, 1000, 1005, 1083, 1174, 1180–1183, 1185, 1217, 1233, 1262, 1281, 1282, 1290, 1292, 1308
Present values, 46, 51, 56, 58, 59, 61, 62, 73, 75, 76, 81–85, 87, 91, 125, 127, 156, 181, 182, 194, 198, 203, 268, 271, 272, 280, 810, 814, 815, 1087, 1097, 1098, 1227, 1304, 1314, 1343, 1344
Price-Earnings Ratio (PER), 266–272, 302, 1215

Pricing a swap, 233, 234
Pricing kernels, 836, 840, 842
Primary markets, 6, 32, 93, 97, 101, 119, 138, 142, 255
Primitive assets, 802, 809, 816, 817, 822, 834, 835, 839, 843, 855
Principal, 173
Probability of default, 170, 175, 257, 1176, 1177, 1179, 1183, 1192, 1199, 1200, 1207, 1217, 1227, 1229, 1230, 1232, 1233, 1235, 1237, 1238, 1242, 1246–1250, 1255, 1257, 1272, 1274, 1275, 1291, 1329, 1330, 1340
Probability of loss, 1106, 1313
Probability of survival, 1219
Profitability, 172
Profit-sharing, 705–707
Profit-sharing contracts, 705–707
Promised rate, 170, 1172, 1175
Property rights, 255, 256, 302
    See also Shares
Proportional rates, 53–55, 61, 149, 157, 178, 185, 187, 720
Prospect theory, 912, 914
Pseudo-arbitrages, 482–484, 488
Public auction, 139, 147
Put, 12, 23, 111, 120, 354–360, 362, 386, 391, 457, 511, 512, 523, 526, 527, 533–535, 537, 544, 580, 691, 810, 837, 848, 999–1003, 1013, 1027, 1042, 1090, 1096–1098, 1188, 1198, 1264, 1266, 1282, 1290, 1332, 1337
    premium, 357
Putable bonds, 713, 715
Put on call, 560
Put on put, 560
Put ratio spread, 456

Q

Quadratic utility, 880, 882, 916
Quality option, 613
Quality spread differential (QSD), 231, 232, 252
Quality spreads, 225, 230–232
Quantiles, 1077, 1086, 1089, 1105, 1107–1108, 1110–1112, 1123, 1124, 1128, 1129, 1132, 1135–1137, 1143, 1146, 1149, 1152, 1156–1158, 1161, 1225, 1230–1232, 1244, 1245, 1248, 1250, 1273, 1275, 1318
Quantitative easing, 110–113
Quanto, 561, 564–566, 588

1360

Index

Quanto call, 562–565
Quanto option, 549, 560, 564–566, 588
Quanto product, 561
Quanto swaps, 251, 561

R
Radon-Nikodym derivative, 444–448,
    818–825, 827, 829, 830, 836, 843
Rainbow options, 588
Range notes, 706
Ratchet caps, 705
Ratchet floors, 705
Ratchet option, 572
Rate risk, 164
Rate sensitivity, 157, 162
Rate variation, 157
Rating agency, 129, 170, 1173, 1176, 1224,
    1267, 1269, 1284, 1311, 1313
Ratings, 101, 103, 113, 124, 126, 127,
    129–131, 136, 138, 147, 170, 174, 192,
    204, 225, 230, 1171–1179, 1182, 1184,
    1188–1190, 1193, 1216, 1217,
    1223–1228, 1230–1238, 1240–1242,
    1255–1258, 1263, 1264, 1267, 1269,
    1270, 1275, 1284, 1292, 1294, 1302,
    1311–1313, 1316, 1331
Raw materials, 323, 335
Realized volatility, 480, 481
Real rates, 56, 74–76, 137, 861
Real-Time Gross Settlement (RTGS), 287
Rebate, 588
Rebonato (interest rate model), 755
Receiver swap with profit-sharing, 706
Recombining tree, 392
Recovery rates, 172, 174–176, 1174,
    1180–1182, 1186–1189, 1193–1197,
    1200, 1202, 1217, 1233, 1237, 1238,
    1242, 1247, 1255, 1291, 1294, 1296,
    1301, 1306, 1314, 1331–1333, 1339,
    1341
Reduced approach, 1189
Reduction of equity capital, 260
Redundant assets, 9, 10, 32, 359, 897
Regressions, 200, 201, 281, 923, 934, 940, 944,
    951, 952, 956, 958, 959, 974, 978, 987,
    1038, 1039, 1054, 1055, 1088,
    1090–1097, 1101, 1214
Regret (behavioral finance), 914
Relationship between forward and futures
    prices, 735
Relative risk aversion, 879, 880, 910, 915, 916,
    1028

Reopening, 138–141, 144–146, 148
Replicable, 222–224
Replicable claim, 220, 805
Replicable swap, 233–237
Replication, 10, 213, 214, 228, 232–241, 369,
    382, 405, 457, 496, 498, 501, 696–698,
    1033–1035, 1055, 1121, 1285, 1288
Replication of a swap, 232–233
Repo, 101–104, 110, 115, 117, 171, 225, 622
Resilient, 17
Return, rate of return, 59, 67, 69, 70, 72, 73, 85,
    273, 800, 998, 1016, 1018, 1019, 1172,
    1175, 1200, 1216, 1344
Reverse cash-and-carry, 324, 339
Reverse collars, 699, 701
Reverse convertible bond, 713
Reverse floaters, 706
Rho (of an option), 1072
Risk-adjusted discount rate, 946
Risk-expected return tradeoff, 473–474
Risk matrix, 485
RiskMetrics, 1104, 1116, 1119, 1121, 1122,
    1132–1134, 1141
Risk-neutral measure, 759
Risk-neutral probability, 325, 351, 372–374,
    376, 377, 379, 382, 385, 387, 392, 407,
    443, 445, 450, 530, 535, 551, 562, 563,
    576, 602, 662, 665, 731, 822, 844, 1024,
    1313, 1321
Risk-neutral universe, 407–409, 862–864
Risk premiums, 73, 97, 126, 130, 171–174,
    213, 245, 246, 271, 280, 281, 302, 303,
    324–325, 348, 374–377, 407, 444, 474,
    797, 807, 857, 862, 878, 879, 915, 933,
    936–939, 941, 944, 946, 948, 960, 961,
    963, 971, 974, 979–981, 983, 985, 986,
    1041, 1174, 1185, 1217, 1233, 1296
Risk profile (strategic allocation), 992–994,
    1027
Risk-return trade-off, 890
Risk tolerance, 334
Roll (criticism of CAPM), 950, 955–956
Roll option, 572
Roll over, 101, 116, 117, 210, 213–215, 217,
    222, 223, 225, 240, 246, 251, 332, 648,
    650–653, 655, 739
Roll's critique, 1047

S
SABR, 444
Saint-Petersburg paradox, 875
SARON, 656

Index

Savage, 922

Secondary markets, 6, 21, 32, 93, 97, 100, 101, 119, 138–139, 142, 144, 148, 255, 261, 1307

Securitizations, 93, 106, 1267, 1268, 1279–1342

Security Market Line (SML), 938, 939, 946, 951, 960, 961

Security picking, 950, 952, 1035, 1038–1042, 1055, 1057–1059

Self-financing, 264, 266, 299, 382, 392, 401–405, 414, 415, 417, 419, 432, 436, 443, 446, 447, 715, 801, 802, 805, 806, 811, 813–816, 823–826, 832, 843–845, 1016, 1019, 1022

Self-financing portfolios, 402–405, 409, 414, 416, 422–424, 445, 446, 476, 801, 806, 811, 825, 826, 843, 1019

Semi-annual coupons, 125, 129, 132, 146

Semi-strong efficiency, 13, 15

Seniority (rules), 1309, 1311, 1340

Sensitivity, 152, 156–167, 169, 174, 202, 656

Sensitivity to interest rate risk, 149–151, 193

Separating funds, 1021, 1022

Separation into two funds (Black), 1014

Separation into two funds (Tobin), 931

Separation theorems, 889, 902, 910, 931, 953, 994

Serial autocorrelation, 1116–1120

Serial independence of log-returns, 275

Settlement price, 660

Settlements, 105–108, 121, 139, 140, 237, 245, 284, 286–288, 303, 607, 608, 610–613, 640, 643, 649, 650, 659, 661, 680, 681, 692, 1259, 1281, 1283–1285, 1298, 1299, 1306, 1313

SFOR, 113

Shadow price (of a constraint), 926

Shanghai Stock Exchange (SSE), 301

Share book value, 266

Share issuing, 256–260

Share market price, 262

Share repurchase, 260

Shares, 6, 10, 18, 21, 136–138, 145, 147, 255, 256, 258–273, 302, 369, 708–712, 715, 1013, 1027, 1050, 1067, 1202, 1208, 1210–1216, 1218, 1233, 1234, 1242, 1262, 1264, 1270, 1298, 1300, 1302, 1303, 1311

Shares (free), 259

Share subscription warrants, 711, 712

Share value, 261–262

Sharpe (market model), 916, 940

Sharpe (model of), 891, 966, 984, 985

Sharpe ratio, 901, 948–950, 961, 1038, 1043, 1046, 1053, 1055

Short positions, 18, 104, 161, 162, 194, 312, 883, 904–906, 908, 909, 1001, 1007, 1013, 1049, 1121, 1187, 1198, 1253, 1287, 1288, 1297, 1340

Short selling, 104, 366, 1049, 1288, 1298

Short-Term Interest Rate contracts (STIR), 641–658, 660, 661, 665

Simple interests, 42, 44, 47–49, 53, 84, 96

Skewness, 275, 302, 304–305, 881, 1156–1159, 1274

Skews, 437, 441, 442, 453, 457–463, 487, 494, 495, 692, 738, 748, 751, 1301

SMB factor, 981

SMI indices, 300

Smile, 437, 438, 441, 442, 453, 457–463, 483, 484, 487, 494, 495, 692, 738, 748, 749, 751

Smile curve, 483

Smooth pasting, 508, 510, 523, 524, 526, 527, 543, 545–546

Snowballs, 706

callable snowballs, 706

SOFR, 210, 223, 229, 247

Soft barrier, 572

SONIA, 113, 114, 116, 118, 210, 223, 247, 655, 656

Spanning, 496

Special Investment Vehicles (SIV), 1308

Special Purpose Vehicle (SPV), 93, 1308, 1309, 1311, 1312, 1340

Specific (diversifiable) risk, 892

Specific risks, 892, 952, 953, 966–968, 970, 971, 985, 1263

Speculation, 23, 333–334

Spontaneous conversion, 1209, 1218

Spot, 19–21, 185, 189, 190, 362, 365–367, 443, 544, 607, 740–743, 815, 1183

Spot asset, 503–508, 517–520

Spot-forward parity, 319–324, 335, 336, 339, 342, 344, 347, 348, 608, 670, 671

Spot measure, 739

Spot price, 313, 319–325

Spot rate curve, 184–187

Spot rates, 177–188, 191, 642, 644, 660, 673–674, 678, 861

Spot rates dynamics, 720–729

Spot volatilities, 692, 740, 741, 743, 751

Spot yield curve, 184, 203

Spread risk, 226

1362 Index

Spreads, 17, 28, 29, 99, 130, 136, 147,
170–173, 191, 192, 195–197, 199–201,
204, 208, 209, 211–227, 486, 558, 622,
640, 651, 658, 690, 981, 1122,
1172–1176, 1179–1189, 1194, 1195,
1197, 1199, 1200, 1202, 1206–1208,
1211, 1213, 1214, 1216, 1217, 1236,
1275, 1281, 1285–1297, 1302–1304,
1306, 1331, 1333, 1335–1345
Spreads (option), 457
Spreadsheets, 90, 91, 156, 1139
Standard CAPM, 954
    See also CAPMs
Standardized returns, 1228–1232, 1242, 1246,
1250, 1251, 1255, 1270, 1272
Standing Facilities (at a Central Bank),
112–113
State variables, 721, 722, 728, 729, 731,
847–848, 850–853, 855–857, 860, 861,
866, 954, 981, 1014, 1020–1022, 1028,
1210
Static hedging, 634
Static options strategies, 457, 496
Static portfolio strategies, 910
Static strategies, 454–457
Stationary process, 770, 771
Step down swap, 683, 684
Step up or down swaps, 251
Sticky caps, 705
Sticky floors, 705
STIRS, 346
Stochastic calculus, 277
$STER, ESTER, 113, 210
Stochastic differential equation (SDE), 278,
401, 402, 407, 421, 423, 425, 476, 579,
601, 772, 773, 778–781, 785, 787–789,
794–796, 810–812, 814, 822, 829, 847,
848, 861, 1068
Stochastic discount factors, 669, 823, 840, 842,
844
Stochastic integrals, 773, 782–785, 795, 808,
810, 811, 816
Stochastic processes, 255, 273, 276–283, 407,
748, 766–769, 779–782, 794, 797–799,
808, 825, 840, 843, 1189, 1217
Stochastic volatilities, 429, 437–440, 442, 444,
496, 749, 751, 1213
Stock (or share) issues, 261–263
Stock index, 295
Stock market indices, 294–301
Stock market prices, 273–283
Stock markets, 260, 293–294
Stock picking, 950, 1040, 1047

Stock prices, 273–276, 769
Stock returns, 273–276
Stock risk premium, 281
Stocks, 11, 18, 21, 29, 255–304, 387, 448, 768,
774, 824, 980, 981, 991–997,
1013–1015, 1026, 1027, 1057, 1068,
1070, 1072, 1083–1085, 1266, 1330
Stock split, 263, 297
Stop-loss, 303, 997–999, 1013, 1027
Stopping time, 570
Stoxx, 301
Straddles, 455, 486, 559, 586
Strangles, 486, 559
Strategic asset allocation, 991, 992
Stratified sampling, 1089–1090, 1101
Stress tests, 1155, 1164–1166
Strikes, 23, 354, 359, 360, 362, 363, 387, 391,
437, 455, 456, 461, 533, 565, 570, 572,
576–578, 583, 588, 589, 672, 676, 691,
693, 695, 701, 704, 748, 802, 815, 1072,
1088, 1092, 1097, 1198, 1208, 1303,
1322
Stripping, 122, 140–141, 144, 148, 681, 691,
1182, 1183, 1185, 1189, 1217, 1285,
1286
STRIPS, 139–141, 144
Strong efficiency, 13, 15
Structural models, 1189, 1218, 1222, 1224,
1228–1239, 1251, 1252, 1270, 1272
Structured products, 23, 554, 575, 667–715,
1267
Structured products on interest rates, 705–707
Structuring ABS, 1309, 1311, 1312
Style analysis (investment), 1049–1051, 1056
Sub-additive risk measure, 1146
Sub-additivity (of a risk measure), 1143, 1148
Subordination, subordinated debt, 1227, 1262,
1290
Subscription rights, 261–264, 302, 708
Subscription warrants, 711–712, 715
Swap forward, 680
Swap Market Model (SMM), 737–749, 751,
754–761
Swap notes, 608, 640
Swap option, swaption, 680, 686, 716
Swap rate, 183
Swap rate curve, 234
Swap zero-curve, 234
Swaps, 22, 346
Swap spreads, 234
Swaptions, 680–688, 716, 738, 743–749, 751,
755–756, 1285, 1303
SWIFT, 288

Index

Syndication, 138, 147
Synthetic option, 370
Systematic (non-diversifiable) risk, 891
Systematic risks, 893, 937, 945, 950, 952, 955, 956, 960, 964, 966, 970, 971, 985, 986, 1224, 1242, 1244, 1246
Systemic risk, 26, 245, 971, 1245, 1264, 1266–1268, 1274, 1325

T

Tactical asset allocation, 991, 992, 1010, 1031–1059
Tail correlation, 1158
Tail-VaR, 1145–1148
Tangent portfolio, 1055
T-bill futures, 641
T-bond futures contract, 638
TEC, 1214
Tenor, 229, 680, 686, 746
Terminal forward measure, 747, 756
Term premiums, 188–191, 204
Term structure of interest rates, 62, 177–204, 751, 857–861
Term structure of volatilities, 437, 457–463, 487, 692, 693
Thaler, 913, 922
Theta, 470–471, 477, 485, 490, 535
Theta (of an option), 490, 491, 535, 559, 1135, 1136
Three-dimensional trees, 534, 541–543, 545–547
Three-fund separation theorem, 953
Threshold (TH), 509, 523, 997, 998, 1006, 1150, 1162, 1163, 1202–1204, 1206, 1207, 1218, 1229, 1231–1235, 1245, 1250, 1255–1257, 1263, 1270, 1284, 1316–1319, 1321–1323, 1329, 1335
Threshold ARCH, 307
Thunderballs, 706
Time value, 354–358, 366, 502–505, 508, 509, 511, 513, 523, 543, 578, 1072, 1215
TIPS, 137, 144
Tobin's two-fund separation, 960
Tobin's two-fund separation theorem, 904–906, 910, 916
Tokyo Stock Exchange, 294
Tolerance, 1036
Total return swaps, 1285, 1300, 1303–1305, 1339
Trackers, 1033–1035
Tracking errors, 1032–1034, 1036–1039, 1041, 1055–1059

Trajectory (path), 766, 771, 774
Brownian trajectory, 774
Tranche equality, 1312
Tranche equity, 1268, 1309, 1311, 1312, 1340
Tranche Mezzanine, 1312
Tranche Senior, 1311, 1312, 1340
Transaction costs, 17–19, 31, 33, 288, 292, 800, 998, 999, 1010, 1011, 1015
Transition matrix, 1177, 1178, 1230
Treasury bill, 100, 101
Treasury bill, bond, 105, 107, 117, 936, 1269
Treynor (index), 935–936, 1043
Trinomial trees, 534, 539–541, 545
Tunnel, 699
Two-fund separation, 902–904
Two-fund separation theorem, 1036

U

UCITS, 1033
Uncovered warrants, 712
Underlying asset price, 354
Unexpected losses, 1222, 1254–1257, 1262, 1263, 1271, 1275
Unwinding a position, 317–318, 348
Up-front premium, 690, 691
Utilities, 2, 8, 144, 146, 834, 875, 876, 880, 911, 920, 922, 941, 996, 1007, 1014, 1015, 1017, 1020, 1021, 1024, 1025, 1028
Utility function, 876–880

V

Valuation of fixed-income securities, 859–861
Valuation partial differential equation (PDE), 853–856, 862–864, 867
Valuations, 94–100, 206, 212–227, 232–241, 266–272, 339–343, 501, 511, 529, 544, 549, 667–671, 673, 674, 680, 693, 738, 797, 809, 1001, 1019, 1078, 1089, 1090, 1172, 1179, 1189–1216, 1218, 1219, 1222, 1224, 1225, 1236–1238, 1240, 1252, 1253, 1269, 1279–1281, 1285–1299, 1313, 1314, 1321, 1323, 1328, 1330, 1332–1336, 1341
Value at risk, 194, 197, 913, 1068, 1104–1120, 1318
Value Line index, 300
Value matching, 526, 543
Value stocks, 980, 983
Valuing a swap, 234
Vanilla caps, 689–692, 714

1364 Index

Vanilla floors, 714
Vanilla option, 436
Vanillas, 25, 27, 115, 183, 206, 213, 214, 216, 217, 221, 237, 239, 251, 354, 441, 494, 496, 498, 501, 520, 556, 559, 565, 567, 568, 571–573, 577, 578, 588, 596, 667, 673–674, 687, 690, 692–693, 714, 724, 740, 1173, 1174, 1280, 1285, 1286, 1288, 1289, 1314, 1332, 1338, 1344
Vanilla swaps, 228–251, 346
VaR, 197, 1068, 1070, 1086, 1104–1158, 1162–1164, 1242
Variable rate, 697
Variance-covariance matrix, 778, 790, 808, 834, 839, 844, 848, 896, 906, 930, 966, 1015, 1023, 1028, 1073–1075, 1082, 1131, 1140, 1151
Variances, 279, 283, 307, 326–328, 333, 334, 770–772, 777, 811, 891, 907, 923–925, 956, 966, 968, 977, 978, 991, 995, 1055, 1067, 1075, 1082, 1085–1089, 1094, 1095, 1106–1107
Variation, 152, 156–158, 177
Variation (to interest rate), 624
Vasicek (model), 722, 724, 727–729, 750, 858, 861
Vasicek-Gordy one-factor model, 1242
Vega, 468–470, 482, 485, 490
Vertical spreads (options), 457
Volatilities, 279, 280, 282–283, 302, 305–307, 389, 390, 393, 400, 457, 482, 571, 671, 672, 693, 728, 730, 732, 759, 797, 806, 813, 818, 823, 829–832, 834–836, 840, 843, 998, 1002, 1004, 1005, 1009–1011, 1013, 1050, 1066, 1072, 1084, 1085, 1122, 1136, 1199, 1230, 1233, 1234, 1302
 determinstic volatility, 427–429
Volatilities implicit in Libor caps, 740
Volatility of the volatility, 438
Volatility risk, 439
Volatility skews, 442
Volatility surface, 453, 457, 461–463, 483, 487, 495, 748, 751
 feedback effect, 495
Volatility term structure, 740
Von Neuman and Morgenstern (Axiomatic), 874, 876, 916–922

W

Warrants, 120, 135–138, 146, 262, 294, 418, 711–713
Weak efficiency, 13, 15, 18, 275
Weighted average cost of capital, 73, 85, 267, 272

Wiener processes, 278, 546, 665, 771, 772, 774, 775, 778, 785, 794
Wild card, 622
Worst-case scenario, 1166
Worst of options, 555–556, 588

Y

Yield curve (forward), 661
Yield curve models, 194–203
Yield curves, 62, 158, 174, 177–179, 182, 183, 187, 188, 190, 192–194, 202, 203, 244, 421, 624, 656, 660, 720–724, 726, 727, 730, 741, 750, 751, 981, 986, 1050, 1122, 1173, 1175, 1176, 1224, 1269, 1289
Yield curve swaps, 250
 See also Basis swaps
Yield to maturity (YTM), 56, 59, 63, 65, 77, 78, 85, 121–125, 127–130, 132–134, 147, 152, 155, 156, 158–160, 170, 171, 174, 177–184, 203, 242, 612, 624, 632, 635, 1173, 1288
Yields, 46, 54, 55, 69–71, 73, 95, 100–102, 114, 127, 130, 135, 155, 158, 172, 178–180, 192–194, 199, 200, 202, 204, 234, 241, 321, 544, 616, 627, 660, 662, 665, 672, 676, 686, 692, 693, 727, 728, 802, 805, 811, 823, 826–832, 843, 899, 1004, 1009, 1035, 1096, 1100, 1150, 1173, 1180, 1182, 1184, 1190, 1207, 1214, 1224, 1231, 1232, 1241, 1251, 1269, 1289, 1291, 1333, 1338, 1344, 1345

Z

Zero-beta (portfolio), 935
Zero-coupon, 154, 158, 359, 550, 626, 663, 676, 1003, 1173, 1175, 1185, 1194, 1198, 1200, 1202, 1205
 See also Zero-coupon rate
Zero-coupon bond, 154–157, 168, 728
Zero-coupon curve, 194–197
Zero-coupon rate, 62, 85, 158, 177, 179–184, 192, 196, 203, 359, 515, 631, 633, 657, 668, 723, 724, 861, 1175, 1193, 1216
Zero-coupon, rate or bond, 664, 665, 668, 669, 673, 723, 724, 748, 826, 844, 1121, 1175, 1179, 1181, 1183, 1184, 1193, 1194, 1289
Zero-coupon rate curve, 180, 181, 245, 626, 631, 633, 657, 1175, 1289, 1294
Zero-coupon swap curve, 245